T0320641

TRANSACTIONS OF THE INTERNATIONAL ASTRONOMICAL UNION VOLUME XXVIIB

PROCEEDINGS OF THE TWENTY SEVENTH GENERAL ASSEMBLY RIO DE JANEIRO 2009

COVER ILLUSTRATION

The origins of astronomy in Brazil: in 1827, emperor Don Pedro I (1798-1834) decreed the start of the Imperial Observatory which was finally installed by Don Pedro II (1831-1889) in 1845, at Morro do Castelo, where it stood till 1922, with a 92 mm telescope. Don Pedro II was an astronomy enthusiast and he observed the total solar eclipse of April 25th, 1865 from his palace at São Cristóvão.
Source: Ângelo Agostini, Revista Ilustrada, 30 September 1882

INTERNATIONAL ASTRONOMICAL UNION

UNION ASTRONOMIQUE INTERNATIONALE

TRANSACTIONS OF THE INTERNATIONAL ASTRONOMICAL UNION VOLUME XXVIIB

PROCEEDINGS OF THE TWENTY SEVENTH GENERAL ASSEMBLY RIO DE JANEIRO, BRAZIL, 2009

Edited by

IAN F. CORBETT
General Secretary

CAMBRIDGE UNIVERSITY PRESS
Cambridge, New York, Melbourne, Madrid, Cape Town,
Singapore, São Paulo, Delhi, Tokyo, Mexico City

Cambridge University Press
The Edinburgh Building, Cambridge CB2 8RU, UK

Published in the United States of America by Cambridge University Press, New York

www.cambridge.org
Information on this title: www.cambridge.org/9780521768313

First published 2010

A catalogue record for this publication is available from the British Library

ISBN 978-0-521-76831-3 Hardback

PRESIDENTS OF THE INTERNATIONAL ASTRONOMICAL UNION

Présidents de l'Union Astronomique Internationale

Catherine J. Cesarsky

2006–2009

Robert Williams

2009–2012

Table of Contents

Preface xvii

CHAPTER I INAUGURAL CEREMONY ....

Opening Address by President Catherine J Cesarsky 1

Address by the President of the Brazilian Academy of Science 4

Address by the Mayor of the City of Rio de Janeiro 4

Address by the Minister of Science and Technology of Brazil 5

Address by the Governor of the State of Rio de Janeiro 6

Presentation of Sponsors 6

Peter & Patricia Gruber Foundation 7

Presentation of Gruber Fellowship 2009 7

Presentation of Gruber Cosmology Prize 2009 7

Citation and Award 10

Response of Awardees 10

Presentation: "Astronomy and World Heritage" *Clive Ruggles* 12

Presentation: "Astronomy in Brazil" *S. O. Kepler* 18

CHAPTER II BUSINESS SESSIONS of the GENERAL ASSEMBLY 27

First Session ....

1. Opening and Welcome 27
2. Representatives of national Members 27
3. Appointment of Official Tellers 31
4. Admission of new National Members 31
5. Revision of Statutes and Bye-Laws 31
6. Appointment of Finance Committee 31
7. Appointment of Nominating Committee 33
8. Report of Executive Committee 2006-09 33
9. Report of Special Nominating Committee 34
10. Proposals to host XXIX General Assembly 2015 34
11. Closure of Session 34

Second Session ....

1. Opening and Welcome 35

2. Individual Members 35
3. Deceased Members 35
4. Appointment of Official Tellers 36
5. Resolutions 36
Report of Resolutions Committee 36
Presentation of Resolutions and Voting 36
6. Appointment of Resolutions Committee 2009-12 37
7. Place and date of IAU XXIX General Assembly 2015 37
8. Divisional and Commission Matters 37
IAU Divisions 37
Changes to Divisions, Commissions and Working Groups 38
Commission Presidents and Vice-Presidents 2009-12 39
9. Financial Matters 40
Accounts 2006-08 40
Membership Dues 41
Budgets 2010-2012 42
10. Appointment of Finance Sub-Committee 2009-12 44
11. Appointment of Special Nominating Committee 2009-12 44
12. Election of Executive Committee 2009-2012 45
13. Closure of Session 45

CHAPTER III CLOSING CEREMONY ....

1. In honour of Francisco Xavier de Arajo 47
2. The IAU Strategic Development Plan 48
3. Invitation to XXVIII General Assembly, Beijing, 2012 48
4. Address by the retiring General Secretary 49
5. Address by the incoming General Secretary 50
6. Address by the retiring President 51
7. Address by the incoming President 52
8. Emblematic songs and rhythms of Brazil 53

CHAPTER IV - RESOLUTIONS ....

1. Resolutions Committee 2006-09 55
2. Approved Resolutions 55

Resolution B1 on the IAU Strategic Plan 55

Resolution A1 on implementing the IAU Strategic Plan 56

Resolution B2 on 2009 Astronomical Constants 57

Resolution B3 on the Second Realization of the International Celestial Reference Frame 58

Resolution B4 on Supporting Women in Astronomy 60

Resolution B5 in Defence of the Night Sky 61

3. Résolutions Approuvées 63

Résolution B1 63

Résolution A1 64

Résolution B2 64

Résolution B3 65

Résolution B4 67

Résolution B5 68

CHAPTER V REPORT of EXECUTIVE COMMITTEE 2006-09 ...

1. Executive Committee 71

2. Membership of the Union 75

3. Divisions, Commissions, Working/Programme Groups 76

4. IAU Scientific Meetings 77

5. IAU Publications 77

6. Educational Activities 82

7. Relations with International Scientific Organisations 83

8. Financial Matters 84

9. Administrative Matters 85

10. Relationship with the Peter & Patricia Gruber Foundation 86

11. Relationship with the Norwegian Academy of Science and Letters 87

12. UNESCO and the IAU 88

13. Appendix I UN Resolution on IYA2009 90

14. Appendix II- Statement of Income 2006-2008 91

Appendix III- Statement of Expenditure 2006-2008 92

15. Appendix III- Agreement with Gruber Foundation 97

16. Appendix IV- Agreement between the Norwegian Academy of Science and Letters and the IAU 99

17. Memorandum of Understanding between the IAU and UNESCO on Astronomy and World Heritage 101

CHAPTER VI - REPORTS on DIVISION, COMMISSION and WORKING GROUP MEETINGS ....

Division I ....

Division I plus Working groups 107

Commission 4 116

Commission 7 120

Commission 8 123

Commission 19 130

Commission 31 140

Commission 52 142

Division II ....

Division II including Commissions and WG 146

Division III ....

Division III 158

Commission 15 168

Commission 16 173

Commission 20 175

Commission 22 177

Commission 51 180

Commission 53 182

WG Small Bodies Nomenclature 184

WG Planetary Systems Nomenclature 186

Division IV ....

Division IV 188

Commission 26 191

Commission 29 193

Commission 35 195

Commission 36 197

Commission 45 199

Div IV-V WG Active B Stars 201

Div IV-V WG Ap stars 205

Division V . . . .

Division V 207

Commission 27 209

Commission 42 211

Division VI . . . .

Division VI 213

Div VI WG Planetary Nebulae 215

Division VII . . . .

Division VII 216

Commission 33 218

Commission 37 219

Division VIII . . . .

Division VIII 223

Division IX . . . .

Division IX 225

Commission 25 227

Commission 25 WG IR astronomy 229

Commission 30 233

Commission 54 236

Div IX-X WG Antarctic Astronomy 239

Division X . . . .

Division X 240

Div X WG RF Interference Mitigation 243

Div X-XII WG Historic Radio astronomy 246

Division XI . . . .

Division XI 248

Division XII . . . .

Division XII 249

Commission 5 250

Commission 6 259

Commission 14 261

Commission 41 263

Commission 41 WG World Heritage 267

Commission 46 270

Commission 50 273

Commission 55 274

CHAPTER VII - STATUTES, BYE-LAWS and WORKING RULES

Statutes 279

Statutes - Version franaise 284

Bye-Laws 290

Working Rules 296

CHAPTER VIII RULES and GUIDELINES for SCIENTIFIC MEETINGS ....

Introduction 309

IAU Symposia and General Assembly Symposia 309

Regional IAU Meetings 320

Educational Aspects of IAU Scientific Meetings 320

Co-sponsoring of Meetings 321

CHAPTER IX COMPOSITION of DIVISIONS, COMMISSIONS, and WORKING GROUPS 323

Division I 324

Division II 326

Division III 328

Division IV 330

Division V 332

Division VI 333

Division VII 334

Division VIII 335

Division IX 336

Division X 338

Division XI 340

Division XII 340

Executive Committee Working Groups 344

CHAPTER X NATIONAL MEMBERSHIP 345

CHAPTER XI INDIVIDUAL MEMBERSHIP by NATIONAL MEMBERS ....

Argentina 355
Armenia 356
Australia 356
Austria 357
Belgium 358
Brazil 358
Bulgaria 359
Canada 360
Chile 361
China Nanjing 362
China Taipei 364
Croatia 365
Cuba 365
Czech Republic 365
Denmark 365
Egypt 366
Estonia 366
Finland 366
France 367
Germany 371
Greece 374
Honduras 375
Hungary 375
Iceland 375
India 375
Indonesia 376
Iran 377
Ireland 377
Israel 377
Italy 378
Japan 381

Korea, Republic of 384
Latvia 385
Lithuania 385
Lebanon 385
Lebanon 385
Malaysia 385
Mexico 385
Morocco 386
Netherlands 386
New Zealand 388
Nigeria 388
Norway 388
Peru 388
Philippines 388
Poland 388
Portugal 389
Romania 390
Russian Federation 390
Saudi Arabia 392
Serbia 392
Slovakia 392
South Africa 393
Spain 393
Sweden 395
Switzerland 395
Tajikistan 396
Thailand 396
Turkey 396
Ukraine 396
United Kingdom 398
United States 401
Uruguay 415
Vatican City 415

Venezuela 415

Vietnam 415

Other Individual Members 416

CHAPTER XII INDIVIDUAL MEMBERSHIP by COMMISSIONS

Division I Commission 4 420

Division XII Commission 5 421

Division XII Commission 6 422

Division I Commission 7 423

Division I Commission 8 425

Division II Commission 10 427

Division II Commission 12 431

Division XII Commission 14 434

Division III Commission 15 436

Division III Commission 16 439

Division I Commission 19 441

Division III Commission 20 443

Division IX Commission 21 443

Division III Commission 22 446

Division IX Commission 25 447

Division IV Commission 26 449

Division V Commission 27 451

Division VIII Commission 28 455

Division IV Commission 29 462

Division IX Commission 30 465

Division I Commission 31 466

Division VII Commission 33 467

Division VI Commission 34 470

Division IV Commission 35 476

Division IV Commission 36 479

Division VII Commission 37 482

Division X Commission 40 484

Division XII Commission 41 491

Division V Commission 42 493
Division XI Commission 44 496
Division IV Commission 45 502
Division XII Commission 46 503
Division VIII Commission 47 506
Division II Commission 49 511
Division XII Commission 50 513
Division III Commission 51 514
Division I Commission 52 516
Division III Commission 53 517
Division IX Commission 54 518
Division XII Commission 55 519

Preface

It is a great pleasure to introduce these Transactions which record the proceedings of the XXVII IAU General Assembly which took place in the Centro de Convenções SulAmérica, in the magnificent city of Rio de Janeiro, Brazil, hosted by the Brazilian Astronomical Society (Sociedad Astromomica Brasiliera, SAB). There were 2434 registered attendees, of whom 796 were IAU members and 526 were students. They came from 76 nations, with 414 from Brazil, and 667 were female (27.4%).

The National Organizing Committee, co-chaired by Professors Daniela Lazarro and Beatriz Barbuy, had 14 members and was supported by 8 sub-committees. Together they organized a highly successful assembly with a full program throughout the 10 days, complemented by an interesting and diverse social program. The dynamic and dramatic Inaugural Ceremony was honoured by addresses from national dignitaries and saw the award of the Peter and Patricia Gruber Foundation Cosmology Prize 2009 and the Foundation Fellowship 2009. Sadly, a dynamic member of the NOC and of the Executive Sub-Committee, Francisco Xavier de Araújo, passed away a month before the General Assembly. For six years he had been at the heart of the organisation of the Assembly: he was sorely missed.

The General Assembly took place as usual in two sessions. The first session admitted four new National Members with Interim Status - Costa Rica, Honduras, Panama and Vietnam. We extend a warm welcome to them and note that they bring the total number of National Members to 68. At its second session General Assembly approved the membership of the nominees, bringing the total individual membership of the Union to 10,144. Finally, it was agreed that the 2015 General Assembly would be in Hawai'i under the auspices of the American Astronomical Society, and the new Executive Committee was elected.

The XXVII Assembly offered a rich scientific programme organized by the General Secretary Karel. A. van der Hucht. There were six Symposia (IAUS 262 - 267), 16 Joint Discussions, 10 Special Sessions, and 7 scientific sessions in the course of Divisional Business meetings. A major, and very successful, innovation was the inclusion of Plenary Reviews at the start of the day, the speakers beng drawn from the 6 symposia. This is a feature likely to be repeated at future General Assemblies. Divisions, Commissions and Working groups held their usual Business meetings spread throughout the GA. National Members' Representatives held 2 meetings, and there were 2 meetings of the Finance Committee and of the Nominating Committee. During the scientific programme over 700 papers were presented and more than 1500 posters displayed.

Financial support for a limited number of the participants was provided by the IAU, and the invaluable support of all the sponsors, listed later in this volume, is gratefully acknowledged.

The proceedings of the six IAU GA Symposia IAUS 262 - 267 will be published in 2010 in the regular *IAU Symposium Proceedings Series* by Cambridge University Press. The Plenary Review corresponding to each symposium will be included in the proceedings of that symposium.

The proceedings of the Invited Discourses, Joint Discussions and Special Sessions will be published by CUP during 2010 in the *Highlights of Astronomy*, Volume 15.

In addition to its scientific programme, the IAU XXVII General Assembly hosted the regular Business Meetings of the EC, the Divisions, Commissions and Working Groups. The present volume, the *Transactions of the International Astronomical Union, Volume XXVIIB*, records the organizational and administrative business of the IAU XXVI General Assembly and the status of the Membership of the IAU. The information in these Transactions was correct and complete at the time of production, but up-to-date information on all matters concerning the IAU can be found on the IAU website <http://www.iau.org>.

It is my pleasure to thank those IAU Division and Commission presidents and Working & Program Group chairpersons who have provided reports of their Business Meetings, and the staff of the IAU Secretariat, Mme Vivien Reuter and Mme Jana Zilova, for their invaluable assistance in assembling and checking these Transactions. Finally, and most importantly, we are all most grateful to Daniela Lazarro, Beatriz Barbuy and every member of the National Organizing Committee, and its sub-committees, for a most memorable XXVII General Assembly.

Ian F. Corbett
IAU General Secretary
Paris, November 2009

Preface

Transactions IAU, Volume XXVIIB
Proc. XXVII IAU General Assembly, August 2009
Ian F. Corbett, ed.

doi:10.1017/S1743921310004771

CHAPTER I

TWENTY SEVENTH GENERAL ASSEMBLY

INAUGURAL CEREMONY

Tuesday 4 August 2009, 14.00 - 16.00

Centro de Convenções SulAmérica, Rio de Janeiro

1. Opening Address by DR CATHERINE CESARSKY, President, International Astronomical Union

Minister of Science and Technology
Governor of the State of Rio de Janeiro
Mayor of the City of Rio de Janeiro
President of the Academy of Sciences of Brazil
Mrs. Patricia Gruber
Ladies and gentlemen, Friends and colleagues

It is wonderful to be here, in Rio de Janeiro, in the company of our distinguished visitors in a country with a very active astronomical community, and it is a great pleasure for the IAU to hold its 27th General Assembly in this spectacular city. We are delighted that you have been able to join us for this Opening Ceremony.

The rise in scientific activity in Brazil is related to the access to independence, in 1822. In particular, the first Emperor, Pedro I, authorized in 1827 the construction of a national observatory. The construction lingered, but finally took place on the hill of Morro del Castello, on the grounds of the Eschole Militar. The place became really active under the second Emperor, Pedro II, who had a keen interest in astronomy and knew how to operate telescopes; - he was an amateur astronomer and had an observatory at the So Cristvo palace. In 1882, he ordered the participation of Brazilians in international scientific expeditions to three different points on the Earth, the aim being to take advantage of a rare transit of Venus in front of the sun to accurately determine the distance between the Earth and the Sun.

As it happens, the Parliament did not authorise the trips. A member of Parliament said: "the people are not concerned about finding out what is going on in the stars that is a luxury." But promptly private money was found, two rich farmers funded the trip, and Brazil indeed took part in the expedition on three outposts.

Despite these early beginnings, modern astronomy really started in Brazil in the sixties, again through international connections, as gifted Brazilian students went to obtain degrees abroad. There are now 300 PhD and 500 astronomers, involved among others in prestigious international projects such as the GEMINI and SOAR telescopes, building frontline instrumentation and involved in the most up to date research topics. Future steps towards the enhancement of our discipline in Brazil, such as the potential construction of a VLBI addition to the millimetre and submillimeter world project ALMA, or a participation to the E-ELT project, are being studied in the framework of the Commission on projects for the future set up by the Ministry of Science and Technology. It is thus highly appropriate, in addition to being extremely pleasant, that we, astronomers from all corners of the world, find ourselves in beautiful Rio de Janeiro for our 2009 General Assembly.

But let us come back to the debate about the use of basic science in general, and of astronomy in particular. Today, in the era of earth observing satellites, interplanetary missions and GPS systems, we know the practical value of accurate measurements of quantities such as the distance between the Earth and the Sun. In the twentieth century, many discoveries about our universe ended up having a strong impact on our daily life. One of the most illustrious examples is the understanding by Hans Bethe of the source of energy which allows stars to shine over long periods: nuclear energy, promptly domesticated for our daily usage. In the 21st century, the incredible progress of astronomy and of our understanding of the universe derives directly from the excellent use astronomy makes of the latest developments in high tech technology. But astronomy not only profits from these advances, it fosters and even enhances them, with two effects. One is that empowers industry by helping it to succeed in realizing ultimate developments, raising its capabilities and skills. The other is that almost invariably the high tech developments fostered by astronomy in ground or in space, or even by theory and numerical modelling, end up having unexpected applications in our daily lives. Here the most obvious examples are the very numerous applications to medicine of imaging techniques, in both hardware and software.

But of course the indirect bonuses of astronomical research are not its main driving force. Astronomy is a natural human response to deep yearnings, the wish to know about our collective origin and our fate, to be part of the universe. As anybody who attempts to popularize our science and make it accessible to the public intuitively knows, all men and women are born astronomers. Some of us have the immense luck to have transformed this inner impulse to search into a profession, that eats our lives but also makes them exciting and fulfilling. We owe it to the public in general, from the taxpayers around us who pay for our research to the underprivileged all over the world to share our knowledge and our wonder.

This is the motivation of the International Year of Astronomy 2009, a global celebration of astronomy and its contribution to scientific development and cultural enrichment. This follows an original idea by former IAU President Franco Pacini, to coincide with the 400th anniversary of Galileo Galilei's first observations through the telescope he had just built, In the IAU GA of 2003 the General Assembly emitted a resolution in favour of IYA. In December 2007 the United Nations proclaimed 2009 as the International Year of Astronomy,

"The General Assembly
Convinced that the Year could play a crucial role, inter alia, in raising public awareness of the importance of astronomy and basic sciences for sustainable development, promoting access to the universal knowledge of fundamental science through the excitement generated by the subject of astronomy, supporting formal and informal science education in schools as well as through science centres and museums and other relevant means, stimulating a long-term increase in student enrolment in the fields of science and technology, and supporting scientific literacy
Decides to declare 2009 the International Year of Astronomy"

The IYA2009 activities take place at the global and regional levels, and especially at the national and local levels. National Nodes in each country have been formed to prepare activities for 2009, with global coordination by the IAU through its IYA Secretariat set at ESO Headquarters.

Now, we have passed the middle of IYA, and the results so far surpass all our expectations: 145 countries have set up IYA nodes, compared to the 63 national members of IAU. At least 2 million people have participated in the activities offered. Quoting Lars Lindberg Christensen and Pedro Russo, IYA coordinators: "Never before has such a network of scientists, amateur astronomers, educators, journalists and scientific institutions come together. IYA2009 is truly the largest network in astronomy", and, I shall add, in science. At this GA, a Special session is devoted to IYA, posters will be visible all along, and dedicated movies will be projected. I recall here the goals we set ourselves, years ago, for this very particular year:

- Increase scientific awareness.
- Promote widespread access to new knowledge and observing experiences.
- Empower astronomical communities in developing countries.
- Support and improve formal and informal science education.
- Provide a modern image of science and scientists.

Many of the IYA goals and objectives are related to education at all ages, and to the global development of astronomy. As we are discussing the IYA 2009 legacy, we see clearly that in it a prominent role should be played by educational programmes. Thus, this is an opportune time to review the long-term strategy of the IAU in development and education. The IAU Executive Committee regards has always given importance to stimulating astronomy education and development throughout the world , and Commission 46 has achieved impressive results, despite the scarcity of resources. Now, under the leadership of one of the vice-presidents, George Miley, a strategic plan for IAU involvement in development and education during the next decade has been emitted. This plan addresses the rationale for astronomy development, education at the primary, secondary and tertiary levels, public outreach and the development of an infrastructure for research. The plan will be presented in several occasions during the GA, and everybody is invited to discuss it at a Town meeting Friday at lunch time. This discussion is important since the Executive Committee has submitted two resolutions about this plan which will be voted at the end of the meeting.

400 Years after Galileo's first glimpse of the sky through a telescope, astronomy is in its golden age. To put things in perspective: 400 years after the discovery by Galileo of Jupiter satellites, space agencies all over the world, NASA, ESA, JAXA, and the Russian agency, are joining forces to send a probe which will visit two of these satellites, Europa and Ganymede.

The pace of discovery is so fast that even astronomers have difficulties in following; IAU General Assemblies give us all a chance to catch up on the news in the various fields every three years, especially with the new feature we have introduced here of having an up to date plenary review for each of the Symposia. An additional opportunity to hear first hand of new discoveries is through the IAU regional meetings. The ones from Latin America, LARIM, and from Asia Pacific, APRIM, already are fully established, and keep growing in quality, number of attendees, and in interest. This triennium we have inaugurated with success, in April last year in Cairo, a third type of regional meetings, Middle East and Africa: MEARIM.

This month, the IAU is celebrating its 90th birthday. The IAU was in fact born at the same time as the International Research Council, which later became ICSU. Born in difficult historic times, IAU at first comprised only countries that had been allies of the winning side of the First World War. With time, IAU has succeeded in federating a large number of countries, united by a common interest in astronomy, Astronomy - and all of Science - has changed enormously over those dramatic 90 years, and we can see in this General Assembly that the IAU has followed and reflects these changes. The International Year gives us the opportunity to celebrate and share in this progress.

So once again, welcome to the IAU General Assembly 2009. The Universe is yours to discover!

2. Performance of the Brazilian National Anthem

The Brazilian National Anthem was performed by *Atràs da Nota Choir (Rio de Janeiro City Hall)* and *Choir Association CPRM, (Geological Survey of Brazil)*, conducted by Mario Assef and assisted by Silvia Sobreira.

3. Address by JACOB PALIS JUNIOR, Presidente da Academia Brasileira de Cincias

President of the Brazilian Academy of Science

Authorities of the Federal Government, of my state and of my city, colleagues, ladies and gentlemen:

It is very gratifying to welcome all of you to this beautiful city and beautiful state at the moment when Brazilian science is at its highest level of performance. This has to do with the persons present here today, and the President of the Republic who has been supporting science in a very special way which I would say is unique in our history.

Soon, we hope to have the same support from the city of Rio de Janeiro and I must say that the Mayor and its Secretary of Science and Technology are looking forward to joining forces with the state agency that is now spending 2% of its budget on science and technology in the state, which is unique in Brazil.

On behalf of the Brazilian Academy of Science and the Brazilian Astronomical Society I extend to you a very warm welcome and I wish you success in your work in what is probably the most charming science of all, and in celebrating Galileo, whom I think is the father of modern science.

Thank you!

4. Address by EDUARDO PAES, Prefeito da Cidade do Rio de Janeiro

Mayor of the City of Rio de Janeiro

Dear friend Governor Sergio Cabral,
Madame President of the International Astronomical Union, Dr. Catherine Cesarsky,
Mr. Jacob Palis, President of the Brazilian Academy of Science,
Mr. Sergio Resende, Minister of Science and Technology, Mr. Alexandre Cardoso, Secretary of Science and Technology of the State of Rio de Janeiro,
Mr. Rubens Andrade, Secretary of Science and Technology of the City of Rio de Janeiro,
Mrs. Jandira Feghali, Secretary of Culture of the City of Rio de Janeiro,
Mr. Paulo Jobim, Secretary of Administration of the City of Rio de Janeiro,
Mr. Kepler Oliveira, researcher on astronomy and member of the National Organizing Committee,
Mr. Clive Ruggles, chair of the commission of the IAU and the UNESCO World Heritage working group on astronomy and world heritage,
ladies and gentleman, dear friends who have come to our beautiful city of Rio de Janeiro:

First, I would like to say that with Mr. Palis, who spoke just before me, I have already lost a lot of money in this first introduction. He wants me, as he said, to join efforts with the federal government of Mr. Lula, President of the Republic, which since it came into power in 2003, has much invested in the area. A great effort has also been made by the State in the past two and half years, since Mr. Cabral came into power. So after 6 -7 months of government, he is still expecting us to put a lot of money into science and technology. And that will happen anyway, so, first thing, it is a great pleasure to be here and I do not mind losing the money, but I need to get my budget into a better situation.

I was listening to Dr. Cesarsky's speech and I think there is a complete connectivity between the City of Rio and this event. First, because the city of Rio is the city in Brazil where most of the research institutes, as well as federal, state and private universities, are located. Rio is a huge research center, that is one of our main goals for the city, that is one of our natural capabilities, a natural way for the city to be developed. So that to host events like this one, concerning science and technology, Rio is a great place. And the other way that this event has connection with the city is in research, or trying to understand things that are not simple to regular people, like us. Rio is a city that inspires higher dreams, is a city that instigates the desire of the people to know more about where they came from and what is happening all around the Universe. The only thing Rio is missing in these days is what we call "astro rei", I do not know how to say it in English,"astro-king", which is our Sun. For sure your presence here it will make the Sun come out and illuminate this great event.

Thanks for coming to Rio. Enjoy the city! Welcome to our marvelous city. Thank you very much.

5. Address by SERGIO MACHADO RESENDE, Ministro de Ciência e Tecnologia do Brasil

Minister of Science and Technology of Brazil

Good afternoon to everyone. I would like to welcome Dr. Catherine Cesarsky and on your behalf all foreign participants. Let me greet the Governor of the State of Rio de Janeiro, Sergio Cabral, the Mayor, Eduardo Paes, the President of the Brazilian Academy of Science, Jacob Palis, the Secretaries Alexandre Cardoso, Rubens Andrade and Jandira Feghali. Let me also greet Beatriz Barbuy, and on your behalf greet all Brazilian participants.

I was very happy to hear the Mayor of Rio de Janeiro speaking about engaging the city in the effort of science and technology. This comes together with a plan that has been under development for the last few years, to put Brazilian science and technology on a higher level. In fact, Professor Jacob Palis had very favourable words about President Lulas government on this respect. But one thing that we are trying to do is to get the federal government, the state governors and the municipalities engaged in this effort because it is very clear, not to many people, but to most of the governors and some of the mayors, that this is the way to change the course of Brazilian development.

As you all know, and as the foreign participants also know, Brazil was a late comer in science. Although we had the National Observatory created almost 200 years ago, we had a very, very slow development in professional astronomy. In fact, perhaps you, our visitors, do not know that Brazil started giving doctors and masters degrees only 40 years ago. Before that we had an amateur system of science and technology. However, in the last forty years big progress has been made in many areas, for instance in astronomy. In 1975 we had something like 15 astronomers. Today we have about 250 astronomers with PhDs. Moreover, we now have 12 institutions educating people to Master and PhD degrees and the activity in all areas, including astronomy, is increasing very fast.

Brazil has 5 institutes under federal governorship that have astronomy and astrophysics as one of their research areas. In Rio de Janeiro we have three of them: the Brazilian Center for Physics Research (CBPF), the National Observatory (ON) and National Museum of Astronomy (MAST). We also have one institute in So Jos dos Campos, the National Space Center (INPE) and one in Minas Gerais, the National Astrophysics Laboratory (LNA), which provides most of the connection to some of the international projects like the Gemini, in which Brazil is an active participant, and also the SOAR.

Since we are concerned with the current development in this area, we have just created a few weeks ago a national committee to propose a plan for the next steps in the development of Brazilian astronomy. The committee is formed by people from our institutes but also from researchers from universities and other centres. In a few months we will have a plan to propose to the government and we know that among the proposals is joining the new big Astronomical Observatory ALMA, in Chile. I am sure that the Government will support the proposals, so that we can change, not only keep up the developing pace of the work in astronomy, but we are actually at a point where we have to change the level of the development of astronomy in Brazil.

On behalf of the Brazilian Government and specifically President Lula, who could not be here today, I would like to welcome all the foreign participants and wish you all a very good week of work in this XXVIIth General Assembly of the IAU. Thank you very much.

6. Address by SRGIO CABRAL, Governador do Estado do Rio de Janeiro

Governor of the State of Rio de Janeiro

Madame Catherine Cesarsky, President of the International Astronomical Union, Mr. Paes, Mr. Resende, Mr. Palis, ladies and gentlemen:

First, I would like to welcome all the participants of the General Assembly of the International Astronomical Union and to give my compliments to the organization for taking the right decision in choosing Rio de Janeiro to host such wonderful event. As I recall, for the first time in Brazil.

It is indeed an honour to receive the worlds most important event in the field of astronomy. In fact, the connections between astronomy and Rio de Janeiro can be traced back to 1827, as Catherine Cesarsky said, when the Imperial Observatory of Rio de Janeiro, today the National Observatory, was officially inaugurated introducing the scientific study of the stars and other celestial bodies in Brazil. Going back in time even further, we must recognize the vital importance of the study of astronomy for the very discovery of Brazil. The Portuguese were only able to reach our territory, in the year of 1500, thanks to the development of the technology inherited from the Arabians, for a long time very influential in the Iberian Peninsula, and which permitted the localization of warriors in the desert using the position of the stars. It was, I might say, the grand-grandfather of our modern hand-held GPS. Later on, the advent of the compass and of the astrolabe guaranteed the success of the great voyage.

Talking about history, I am convinced that this General Assembly will be a very important part of it, and I believe that the Union will have a difficult mission trying to find a city as beautiful as Rio to host the event in the future. The "cariocas", natives of Rio de Janeiro, have the unique ability to combine hard work and healthy leisure. This is one of the features that make our people so well known all over the world and our city such a joyful place to live, to work and, of course, to watch the stars. I am certain that the beauties of the marvellous city will conquer the hearts and minds of those dedicated to conquering space.

Finally, I would like to ask you to ask the stars during the meeting about a very important challenge for us: in Copenhagen, on October 2nd, we are in the running to host the 2016 Olympic Games, along with Madrid, Tokyo and Chicago. If we win, it will be the first time in South America. Please, ask the stars to support us!

Thank you very much.

7. Presentation of Sponsors

The National Organiziang Committee is pleased to acknowledge the support of the following institutions and sponsors:

- Astronomy and Astrophysics - Sponsor of Poster
- CAPES-MEC - Participant Support
- CNPq-MCT - Participant Support
- CNRS - Participant Support
- EMBRATUR - Sponsor for Promotion
- European Southern Observatory - Participant Support
- European Space Agency - Participant Support

- FEPEMIG - Participant Support
- FAPERJ - Organization Support
- FAPESP - Participant Support
- Gruber Foundation - Inaugural Ceremony Support
- IUPAP - Participant Support
- Laboratrio Nacional de Astrofsica - Organization Support
- LOral Corporation Foundation - Sponsor of Lounge
- Ministry of Science and Technology - Organization Support
- Norwegian Academy of Science and Letters - Sponsor of Education Luncheon
- Observatrio Nacional - Organization Support
- Prefeitura do Rio de Janeiro - Organization Support
- Rede Nacional de Ensino e Pesquisa - Sponsor of Media Streaming
- US National Science Foundation - Sponsor of YA and WiA Luncheon

8. Brazilian Celebration, Yes!

Performed by *Companhia Folclórica do Rio-UFRJ (The Rio-UFRJ Folkloric Company)* of the School of Physical Education at the Federal University of Rio de Janeiro (UFRJ) under the General Director, Eleonora Gabriel, Musical Director, Ronaldo Alves and Production Director, Katia Iunes

9. Peter and Patricia Gruber Foundation - Presentation of Cosmology Prize 2009 and Foundation Fellowship

Catherine Cesarsky
As President of the IAU, I would like to welcome The Peter and Patricia Gruber Foundation to the Inaugural Ceremony of our 27th General Assembly. We are very pleased that they again form part of our program, and we continue to appreciate the fruitful collaboration between our organizations, which dates back a decade. It is now my pleasure to present Patricia Murphy Gruber, president of the Gruber Foundation.

Patricia Gruber
Welcome to the presentation of the 10th annual Cosmology Prize, honoring a leading cosmologist, astronomer, astrophysicist or scientific philosopher for theoretical, analytical, or conceptual discoveries leading to fundamental advances in our understanding of the universe. On behalf of my husband, Peter Gruber, myself, and all of us at the Foundation, we are pleased to be here in Rio de Janeiro to present this Prize at the 27th General Assembly of the International Astronomical Union. Thank you, Catherine Cesarsky, for your warm welcome.

The Cosmology Prize was established in 2000 as the first Gruber international prize, and I would like to gratefully acknowledge the vision and leadership of Peter Gruber in establishing this and the other prizes. The Cosmology Prize is presented in conjunction with the International Astronomical Union. It is my pleasure to introduce Dr. Karel van der Hucht, Secretary General of the IAU, who will say a few words about this fruitful collaboration.

Karel van der Hucht
The IAU is pleased to collaborate with the Gruber Foundation on the Cosmology Prize. The primary goal of the IAU is the development of astronomy world-wide.

The collaboration between the PPGF and the IAU consists not only of the Cosmology Prize, but also an annual Fellowship established in the Gruber Foundation name. The annual fellowship is selected by the IAU and awarded competitively to a postdoctoral researcher. The stipend is to be used to further his or her research.

The Gruber Foundation contributes 50,000 US dollars annually to this fellowship program, which is administered by the IAU. Awards are presented to promising young scientists of any nationality to pursue education and research at a center of excellence in their field; the IAU selects recipients from applications received from around the world. The fellowship has been awarded to scientists from Poland, India, Spain, Greece, the Russian Federation, Mexico, and the United States.

The 2009 Gruber Fellow is Thijs Kouwenhoven, from the Department of Physics and Astronomy at the University of Sheffield in the United Kingdom. His work focuses on the stability and dynamical evolution of newly formed planetary systems in young, gas-rich star clusters. I am happy to introduce him on this occasion and invite him to say a few words.

Thijs Kouwenhoven
First of all my thanks to the Peter and Patricia Gruber Foundation for the award of this Fellowship, which I will use to go to the newly established Kavli Institute for Astronomy and Astrophysics in Beijing, China. I am aware that it is still uncommon for western astronomers to go to China, but the world is changing and China is becoming more important in all fields, including astronomy, asatrophysics, and space science. The majority of the the staff of this new Kavli Institute are from outside China, mainly from Western Europe and North America, and I am proud to be joining this community. Thank you.

Patricia Gruber
Thank you Thijs.

We are here to honor the achievements of Wendy Freedman, Robert Kennicutt, and Jeremy Mould, leaders of the Hubble Space Telescope Key Project on the Extragalactic Distance Scale. But first let me tell you a little about the company they are keeping.

The Foundation's prize program, established in 2000, now presents five annual 500,000 US dollar prizes in the fields of Cosmology; Genetics; Neuroscience; Justice; and Women's Rights. Each prize recognizes achievements and discoveries that produce fundamental shifts in human knowledge and culture.
On September 24th, the Justice Prize will be presented to the European Roma Rights Centre and Bryan Stevenson, at the Cumberland School of Law in Birmingham, Alabama.
On October 18th, at the annual meeting of the Society for Neuroscience, the Neuroscience Prize will be presented to Jeffrey Hall, Michael Rosbash, and Michael Young.
The Genetics Prize will be presented at the annual meeting of the American Society of Human Genetics on October 23rd, to Janet Rowley.
And on the 29th of October, in St. Thomas, United States Virgin Islands - where the Foundation is headquartered - Leymah Gbowee and the Women's Legal Centre will receive the Women's Rights Prize.

Returning to Cosmology, the 2009 Prize recipients were selected by a distinguished Cosmology Prize advisory board:

- Jacqueline Bergeron
- Ronald Ekers
- Peter Galison
- Andrei Linde
- Julio Navarro

- James Peebles
- Roger Penrose

Owen Gingerich and Virginia Trimble also serve as special cosmology advisors to the Foundation. Peter and I deeply appreciate the knowledge, commitment, and enthusiasm that the advisors bring to the judging process. Let me now invite the advisory board Chair, James Peebles, to say a few words about the history of the Gruber Cosmology Prize on its tenth anniversary.

James Peebles

The first Gruber Cosmology Prizes were presented in the year 2000. In my estimation this is when cosmology was completing its transition to a predictive theory.

Cosmology was a real physical science well before that, and it was predictive, in part, but largely adaptive. We saw the need for dark matter, so we added it. We saw the need for a cosmological constant, so we added it too. But in about the year 2000 the growing evidence had greatly restricted our freedom to adapt and shown that the relativistic hot big bang cosmology had become a predictive theory. The Gruber Cosmology Prizes have recognized many of the steps that made this so.

The first two Prizes were awarded to Alan Sandage and me. Here's a measure of Alan's insight. A half century ago, in his his paper *The Ability of the 200-Inch Telescope to Discriminate Between Selected World Models*, he placed particular emphasis on the redshift-magnitude test and the expansion time test. Both matured four decades later, at the time of the first Gruber Cosmology Prizes.

There were many challenges to putting together the pieces of a theory of the universe. The third Prize went to Martin Rees, who had the deep physical insight to do this, time and again.

I remember being delighted by Vera Rubin's vivid demonstrations of the flat rotation curves of isolated spiral galaxies, an indication of dark matter. The citation for the fourth Prize mentions one aspect of a full and productive life in astronomy, beginning in 1951 with a pioneering discussion of what she termed the local ellipsoidal agglomeration of galaxies, and Gerard de Vaucouleurs renamed the Local Supercluster. She's still publishing.

Another edifying life in astronomy is recognized by the 2003 award for a particular advance, Rashid Sunyaev's analysis with Yakov Zel'dovich of the effect of hot plasma in clusters of galaxies on the sea of thermal radiation that fills space. The effect has become a valuable tool in the exploration of the large-scale structure of the universe.

When I was a postdoc there were dark corners of cosmology we usually discussed only in private. What was the universe doing before the big bang, assuming there was one? Why do well-separated regions that in theory have not been in causal contact since the big bang look so similar? Alan Guth and Andrei Linde were honored in 2004 for their parts in creating the promising and influential answer, inflation.

When I was a postdoc the Palomar Sky Survey photographs of the sky were a valuable tool for astronomy. In 2005 Jim Gunn was honored for his vision, and his technical skills and leadership, that gave us a vastly improved tool, the Sloan Digital Sky Survey. It is fueling research on topics from the smallest stars to the most distant quasars.

The 2006 Prize recognized the demonstration that the sea of thermal radiation that fills space has the critical signature of a fossil from the early universe. The Prize named the COBE satellite Science Working Group, with John Mather as Project Scientist. Organizers of major prizes must learn how to recognize fundamental advances that increasingly depend on essential contributions by many. Maybe this is a step in the learning curve.

The redshift-magnitude relation is discussed in Richard Tolman's book on cosmology published in 1934. The test matured seven decades later, again at the time of the first Gruber Cosmology Prizes. The 2007 Prize recognizes the teams in the final push, and their leaders, Saul Perlmutter and Brian Schmidt.

Precision measurements show that the distribution of the sea of thermal radiation left from the hot big bang has properties that require a cosmological constant, consistent with the redshift-magnitude and expansion time tests. They also require a baryon mass density consistent with what is inferred from light element production in a hot big bang, and dark matter, consistent with what we had supposed when cosmology was a looser art. Learning how to make these measurements was part of the growth of cosmology into a mature big science, and it was the work of many. Dick Bond, who was awarded the Gruber Prize in 2008, stands out for his many years of leadership in emphasizing the importance of these measurements and how to interpret them.

Patricia Gruber
I'd now like to introduce Ronald Ekers, who will present the official citation and introduce the scientific accomplishments of our Cosmology Prize recipients. Dr. Ekers is a Federation Fellow of the Australia Telescope National Facility and also an advisor to the Cosmology Prize.

Ronald Ekers
The official citation reads:

The Peter and Patricia Gruber Foundation proudly presents the 2009 Cosmology Prize to Wendy Freedman, Robert Kennicutt and Jeremy Mould for the definitive measurement of the rate of expansion of the universe, Hubble's Constant. This parameter effectively determines the age of the universe at the current time and underpins every other basic cosmological measurement.

An accurate measurement of the expansion rate was one of three major goals of the Hubble Space Telescope when it was launched in 1990. From meticulous measurements of a particular kind of variable star, the Cepheids, Freedman, Kennicutt and Mould met this goal, resolving one of the longest-standing debates in the history of modern cosmology. Let me add an additional comment. It is so very appropriate in this, the International Year of Astronomy, when we are also celebrating the achievement of Gilileo, that we are awarding this prize to Freedman, Kennicut, and Mould. Galileo used the first telescope to establish the place of the Earth in the Universe, todays Gruber Prize recipients used the first telescope in space to answer humankind's quest to know how big and how old is our Universe. I invite the recipients to come forward.

Wendy Freedman
Thank you very much, and deep thanks to the Peter and Patricia Gruber Foundation, and to Peter and Pat personnaly, for this award. All three of us were surprised to be selected for this profound honour. I would like to recognise the contribution and support of the Carnegie Institute of Science which, through its observatories, has supported the work of people such as Hubble, Sandage, Rubin, and many other distinguished astronomers for more than a century. I would also like to pay tribute to Henrietta Swan Leavitt, who discovered the relationship between the period and luminosity of Cepheid variables which radically changed the theory of modern astronomy and underpins all work on the distance scale of the Universe since the time of Hubble. She could not obtain a Ph.D. in those days, and I am grateful that we now live at a time when women can rise to the levels of Directors or Presidents. I would also like to recognise the work of the engineers and technicians whose work made the project possible. We all work in large teams nowadays, and we benfited from a talented and dedicated team working together for over a decade. We should also pay tribute to Marc Aaronson who lead this project at the start but who died tragically in an accident in 1987, and of course to my two co-recipienst. Finally, I would like to give special thanks to a colleague, Barry Madore, with whom I have worked for more than 25 years, and who was also a member of our team.

Robert Kennicutt
After you spend more than 10 years and over 100 telecons working together, you start thinking alike, so I will not repeat much of what Wendy has just said. We are very flattered, honoured

and yet humbled by this award. We appreciate that we are only three out of many people who have worked on the problem of the distance scale and the Hubble constant, and we recognise their contribution. I would like to thank the Peter and Patricia Gruber Foundation for their generous support of astronomy and for sharing our passion for cosmology - thank you - *obrigado.*

Jeremy Mould

It is a particular pleasure to thank the Peter and Patricia Gruber Foundation for the award of this prize. Cosmology is a fascinating subject, dealing as it does with the largest questions of the Universe, and we are lucky that the Gruber Foundation has chosen it as one of its prize topics. I would like to thank all my colleagues on the H0 team - one couldn't ask for a better team of people, on whose shoulders we stood. I would, of course, like to thank my family who allowed me the time to work of this challenging project. Research can seem a thankless task, but when a prize like this comes along it makes it all seem worthwhile. Peter and Patricia, thank you again.

Patricia Gruber

Please note that these three recipients will give a public lecture entitled "Measuring the Hubble Constant with the Hubble Space Telescope" at 12:45 pm tomorrow in this room. Thank you for attending the 2009 Cosmology Prize ceremony. This concludes our presentation.

Transactions IAU, Volume XXVIIB
Proc. XXVII IAU General Assembly, August 2009
Ian F. Corbett, ed.

doi:10.1017/S1743921310004783

Astronomy and World Heritage

Clive Ruggles[1]

[1]Emeritus Professor of Archaeoastronomy, University of Leicester, UK,
Chair, IAU Working Group on Astronomy and World Heritage
email: cliveruggles@btinternet.com

1. Introduction

UNESCO's World Heritage List http://whc.unesco.org/en/list exists to help identify, protect and preserve sites and landscapes that are considered to be of outstanding universal value to humankind. This means that their significance reaches beyond national and cultural boundaries, and (if our attempts at preservation are successful) will remain as a source of inspiration for many generations into the future.

World Heritage Sites fall into two categories: cultural and natural. Following the 2009 meeting of the World Heritage Committee, there are currently 890 properties on the List, belonging to 148 'State Parties' (countries that adhere to the World Heritage Convention). Of these, 689 are cultural and 176 are natural, the remaining 25 being mixed (both cultural and natural). The cultural properties provide numerous examples of creative masterpieces from the distant as well as the more recent past, including places as diverse as the Acropolis in Athens, the temple complex at Angkor in Cambodia, the historic centre of Saint Petersburg, and the stone statues of tiny Easter Island, home for many centuries to the most isolated human community on the planet.

Science heritage in general, and astronomical heritage in particular, are poorly represented on the World Heritage List. While it is true that the List includes several ancient sites that have an established or postulated connection with astronomy-for example the Neolithic passage tomb at Newgrange in Ireland Ruggles99, Stonehenge in the UK Ruggles97 (Fig. 1), the Great Pyramids of Giza in Egypt Krupp97; Belmonte01 (Fig. 2), and several Mesoamerican ceremonial centres in Mexico and Guatemala Aveni01 (Fig. 3)-each of these is included only because of its broader archaeological and cultural significance. The same is true for sites important in the history of modern astronomy Hoskin97: thus the fourteenth-century meridian arc from Ulugh Beg's observatory in Uzbekistan is only included because it forms part of the historic city of Samarkand, the Observatory of St Petersburg because it forms part of the Historic Centre of St Petersburg, and the Old Royal Observatory at Greenwich because it forms part of Maritime Greenwich.

In fact, only one property has been explicitly inscribed because of its astronomical significance (along with its importance for the history of the earth sciences and topographic mapping). This is the Struve Geodetic Arc, a 265-point triangulation network set up in the nineteenth century by the astronomer Friedrich Georg Wilhelm Struve to determine the precise size and shape of the Earth. Thirty-four nodes of this network still survive, and these comprise a so-called 'serial nomination'-a World Heritage Site consisting of set of places rather than a single one-spanning ten modern countries from Norway down to the Black Sea.

Concerned at the situation with regard to astronomical heritage, the UNESCO Regional Bureau for Science in Europe (ROSTE) took steps back in 2005 that resulted

Figure 1. Stonehenge, showing the famous axial alignment upon winter solstice sunset. The monument is included in the World Heritage List as part of the property "Stonehenge, Avebury and Associated Sites". Photo by Clive Ruggles.

Figure 2. The pyramids of Khufu, Khafre and Menkaure at Giza. The World Heritage property 'Memphis and its Necropolis - the Pyramid Fields from Giza to Dahshur' was inscribed in 1979. Photo by Clive Ruggles.

in the creation of a thematic initiative within UNESCO on Astronomy and World Heritage http://whc.unesco.org/en/astronomy. Its explicit aim was to identify, safeguard and promote properties connected with astronomy 'as a means to promote in particular nominations which recognize and celebrate achievements in science'.

In October 2008, the UNESCO World Heritage Centre and the IAU signed a Memorandum of Understanding-a formal agreement to jointly progress the Astronomy and World Heritage Initiative. This established the IAU as a full partner in developing the

Figure 3. The Caracol at Chichen Itza, which forms part of the World Heritage property 'Pre-Hispanic City of Chichen-Itza'. It is one of several buildings and complexes within Mayan ceremonial centres that have a proven or postulated astronomical connection and are already on the World Heritage List for other reasons. Photo by Clive Ruggles.

Initiative and shortly afterwards, in order to fulfil its commitment in this regard, the IAU set up a new Working Group on Astronomy and World Heritage.

This report will describe the work of the Working Group and what has been achieved to date.

2. Right and Wrong Directions

So, what has the Working Group achieved in its first nine months of existence? I would prefer to start by turning the question around, asking instead what the Working Group has most conspicuously not achieved. The reason for this is that the answer will come as a surprise to many people: we have not made any effort to construct a list of astronomical heritage sites that the IAU considers of greatest significance.

There is a very good reason for this. The process of nominating properties for inscription on to the World Heritage List has been carefully established over many years and is universally respected by all the State Parties to the Convention. It involves several stages, with State Parties first having to register sites onto their national 'Tentative Lists' and then putting forward at most one property for inscription in any one year. The nominating country not only has to justify why the site should be considered of outstanding universal value but also has to draw up a credible plan for its management and protection. When a property is nominated for inscription, it is assessed not by UNESCO but by an independent Advisory Body: in the case of cultural applications this is the International Council on Monuments and Sites (ICOMOS). In the light of the report from ICOMOS, the World Heritage Committee makes its decision.

If the IAU were to publish its own 'Tentative List' this would undoubtedly be seen as an attempt to undermine the whole process, and would immediately result in the IAU's input being ignored. In fact, Commission 41 had produced such a tentative list back in January 2008, but it clearly reflected the specialist knowledge and, to a significant extent, the countries of origin or domicile of the authors who contributed to it. The list has not been made public for this reason, although it has served to inform the work that has followed.

A "right way forward" did, however, emerge soon after the formation of the Working Group. The basic problem is that ICOMOS is called upon to assess World Heritage List applications from a scientifically objective and politically neutral standpoint, but there are no clear criteria that can be used to assess astronomical heritage sites. If the IAU could work with ICOMOS to help develop such criteria, then not only would this help ICOMOS itself; it would also give State Parties a better idea of the criteria by which any astronomical heritage application would be judged, and thereby encourage them to put forward such applications.

Over the past decade and a half, when faced with similar situation in other subject areas, ICOMOS has worked (either on its own or in collaboration with a suitable partner organization) to produce a 'Thematic Study' http://www.icomos.org/studies/. This is a book-length document aiming to' establish what qualities best demonstrate outstanding universal value in the field concerned. Following a series of meetings in January 2009, the Working Group began a co-operation with ICOMOS to produce a Thematic Study on the Heritage Sites of Astronomy and Archaeoastronomy in the context of the UNESCO World Heritage Convention. In doing so we will produce a document that will present our overall vision on astronomical heritage, using examples of properties worldwide as examples and illustrations of the broad themes. This is a fitting activity for the International Year of Astronomy and we hope to complete the task by the end of 2009. This Thematic Study will then be formally submitted to UNESCO for approval and adoption.

3. Differing Views on Heritage

The project to produce a Thematic Study on astronomical heritage has much broader implications. No Thematic Study yet exists that addresses any field of science heritage, despite the fact that the UNESCO World Heritage Centre is committed to improve the recognition and celebration of the achievements of science, engineering and technology on the World Heritage List. Being the first Thematic Study to explicitly address any field of science heritage, ours must tackle a number of broad issues of wider relevance to the whole of science heritage. Fortunately, the study's co-editor appointed by ICOMOS, Professor Michel Cotte, is a specialist in technology heritage, a related field in which ICOMOS has published three Thematic Studies (on the subjects of Canals, Bridges and Railways).

One of the trickiest issues is, in fact, encapsulated in the first of the Working Group's Terms of Reference:

"To work on behalf of the IAU to help ensure that cultural properties and artefacts significant in the development of astronomy, together with the intangible heritage of astronomy, are duly studied, protected and maintained, both for the greater benefit of humankind and to the potential benefit of future historical research." This statement rightly makes reference not only to the 'tangible immoveable' heritage of astronomy-monuments, buildings and fixed instruments-that are eligible for inscription onto the World Heritage List, but to the range of associated types of heritage that, taken together, are crucial in assessing the scientific significance of a place: moveable instruments, documents, archives, and of course the discoveries that were made at the place in question, the ideas that were generated, and the personalities involved. These other artefacts and associations are categorised in heritage terms as 'tangible moveable' and 'intangible' heritage.

Of course, astronomical knowledge-whatever the time period or social context-is inherently intangible. And herein lies the heart of the matter. In order to assess the significance of a heritage site relating to astronomy we need firm criteria for assessing how the intangible heritage of astronomy in all its forms -varying from indigenous calendars and cosmologies to modern scientific facts and theories-together with the various types of moveable heritage-portable instruments and artefacts, archives, etc-contribute to our assessment of the universal value of the immovable architecture. These are the sorts of issues with which the Thematic Study is trying to come to grips.

4. Increasing Awareness

Another of the Working Group's Terms of Reference is
"To work, in conjunction with IAU C41 (History of Astronomy), IAU C46 (Education) and other Commissions and Working Groups within the IAU as appropriate, to enhance public interest, understanding, and support in the field of astronomical heritage."
Outreach will be of vital importance in the longer term, but the Working Group has also been working to increase awareness throughout its first year of existence. Astronomy and World Heritage is one of twelve Global Cornerstone Projects for the International Year of Astronomy 2009. The September issue (no. 54) of UNESCO's quarterly magazine World Heritage Review is devoted to astronomical heritage and science heritage, with contributions by Working Group members. The year has also seen various conferences and meetings in which astronomical heritage and the work of the Working Group has been discussed and advanced, including a conference dedicated to the topic in Kazan, Russian Federation, in August.

What will perhaps prove to be the Working Group's most important activity is advancing thanks to a collaboration developed through the IYA Global Cornerstone Project. The objective is to create a public database of properties with a relationship to astronomy, not just those that are already on the World Heritage List (for which see UNESCO's Astronomy and World Heritage 'Timeframe'
http://whc.unesco.org/pg.cfm?cid=281&id_group=21&s=home). The first stage in this process is to transfer and modify the contents of the existing 'timeframe' and to transfer it on to the Working Group's own website where it will act as the basis for the new database.

But if the Working Group must avoid producing its own list of astronomical heritage sites that it considers most significant, lest it be seen to be undermining the established nomination process, how can it consider developing such a database? The answer is that it can still provide a forum for others to exchange information-i.e., to view and discuss sites from the widest possible range of places and times. The database will be accompanied by a prominent disclaimer making it clear that the inclusion or otherwise of any particular property has no bearing whatsoever upon the outcome should it be nominated for inscription onto the List. However, we do hope that in the longer term this forum will be of considerable assistance to State Parties considering the nomination of astronomical heritage sites of their own.

5. Dark Skies

A considerable proportion of the world's population today live in places where light pollution denies them the opportunity of ever experiencing a truly dark night sky. To such people, it may be all the harder to communicate the importance of cultural sites relating to the sky. This in itself makes the protection of dark sky places a matter of some relevance to anyone concerned with the protection of cultural heritage relating to astronomy.

However, one can go further. The presence of a dark night sky enhances the natural significance of a place. Why, then, should we not recognise the presence of a pristine night sky, free of light pollution, as a natural characteristic that contributes to universal value? As a result of the Starlight Initiative and the Starlight Declaration of 2007 (http://www.starlight2007.net/), UNESCO and the IAU are working to protect the dark night sky, for example through the creation of "Starlight Reserves". If one regards dark sky sites as places having a natural heritage relating to astronomy, then strong parallels emerge between the cultural and natural aspects of astronomical heritage and efforts to encourage their recognition. In particular, this suggests the need for strong links with the IUCN (the International Union for the Conservation of Nature), the Advisory Body to UNESCO that assesses nominations relating to natural rather than cultural heritage.

During 2009, the Working Group has developed strong links with the Starlight Initiative and been represented at two of its meetings. A significant outcome of the first, in Fuerteventura in March, was the establishment of a "Dark Skies Advisory Group" whose remit is to support IUCN activities relating to dark skies. This Group includes representation both from our Working Group and from ICOMOS.

It is conceivable that, just as some World Heritage sites derive their outstanding universal value from a combination of their cultural and natural heritage, so some places may achieve this by combining cultural heritage relating to astronomy with the natural heritage of the dark night sky.

6. Conclusion

Astronomical heritage in not just important because every human culture has a sky. Astronomy is nothing less than a fundamental reflection of how all people, past and present, understand themselves in relation to the universe. It is for this reason that we must take urgent steps to identify, protect and preserve the most outstanding manifestations of both our cultural and natural heritage relating to the sky. We can hope that the IAU, through its Working Group on Astronomy and World Heritage, working alongside UNESCO and its Advisory Bodies, will be able to take some significant steps towards ensuring that vital aspects of our astronomical heritage are not lost forever.

References

Aveni, A. F. 2001, *Skywatchers* (Austin: University of Texas Press)

Belmonte 2001, *Archaeoastronomy no. 26* (supplement to Journal for the History of Astronomy 32), S1

Hoskin, M. A. 1997, *The Cambridge Illustrated History of Astronomy* (Cambridge: CUP)

Krupp, E. C. 1997, *Skywatchers, Shamans and Kings* (New York: Wiley)

Ruggles, C. L. N. 1997, in: B. W. Cunliffe & A. C. Renfrew (eds.), *Science and Stonehenge*, (Oxford: OUP), p. 203

Ruggles, C. L. N. 1999, *Astronomy in Prehistoric Britain and Ireland* (New Haven and London: Yale University Press)

Transactions IAU, Volume XXVIIB
Proc. XXVII IAU General Assembly, August 2009
Ian F. Corbett, ed.

doi:10.1017/S1743921310004795

Astronomy in Brazil

S. O. Kepler[1]

[1]Departamento de Astronomia, Instituto de Física, UFRGS, Porto Alegre, RS, Brasil
email: kepler@if.ufrgs.br

Abstract. Astronomy in Brazil grew to around 500 astronomers in the last 30 years and is producing around 200 papers per year in refereed journals. Brazilian astronomers are participating in several international collaborations and the development of instrumentation is on the rise.

---

1. Introduction

Brazil is the fifth most populous nation in the world with nearly 200 million inhabitants. The history of Astronomy in Brazil starts around 1639, when the first observatory in Brazil was built under the Dutch government of Johann Moritz von Nassau-Siegen in occupied Recife, at the Northeast of Brazil. The observatory was located at the Fribourg palace in Antonio Vaz island. The astronomer in charge, Georg Markgraf (1616–1644), observed the solar eclipse of November 13th, 1640, published in his *Tractatus topographicus et meteorologicus Brasiliae cum eclipsi solaris*. The observatory was in fact the first observatory built in the Southern Hemisphere and in the Americas, except for the Maia's and Astek's monuments. When the Portuguese retook the region in 1654, the whole palace was razed to the ground.

A few later observations are know, like the observations of a comet in 1668 by the Jesuit priest Valentim Estanciel (1621–1705), explicitly mentioned by Isaac Newton in the Principia (1687), and the observations obtained at Morro do Castelo, in Rio de Janeiro, to measure the latitude and longitude of the city, from 1781 to 1788 by the Portuguese astronomer Bento Sanches Dorta (1739–1795), published in 1797. He observed the eclipses of Jupiter satellites with John Dollond (1706–1761) achromatic telescopes, one of 1 1/2 foot focal distance and another of 17 inches. With the 17 inch he also observed the lunar eclipse of November 10th, 1783.

2. Observatório Nacional

In 1827, emperor Don Pedro I (1798–1834) decreed the start of the Imperial Observatory which was finally installed by Don Pedro II (1831–1889) in 1845, at Morro do Castelo, where it stood till 1922, with a 92 mm telescope. Don Pedro II was as astronomy enthusiast and he observed the total solar of April 25th, 1865 from his palace at São Cristóvão, while Camilo Maria Ferreira Armond (1815–1882), Baron of Prados, observed it from the observatory. Antônio Luís von Hoonholtz (1837–1931), Baron of Teffé, observed the transit of Venus over the solar disk in December 6th, 1882 with a 16 cm equatorial telescope at San Thomas island, in the Antilles, leading one of the three groups sent by the observatory. In 1906, the observatory receives the name Observatório Nacional. By 1922 the observatory was moved to its present location, in São Cristóvão, Rio de Janeiro city. Henri Charles Morize (1860–1930), naturalized Brazilian as Henrique Morize, was the director from 1908 to 1930, and he organized the English

Figure 1. Observatory at Fribourg palace, in Recife

Figure 2. Observatory at Morro do Castelo, Rio de Janeiro.

expedition lead by Andrew Claude de la Cherois Crommelin (1865–1939) and Charles Rundle Davidson (1875–1970) to observe the total solar eclipse of May 29th, 1919, in Sobral, Ceará, to test the bending of light by mass predicted in Albert Einstein's general relativity. The eclipse occurred in the Hyades, with 13 bright stars in the field. The

Figure 3. Charge of Don Pedro II observing the eclipse.

expedition obtained 7 good pictures (Davidson 1922). The expedition to Brazil resulted in a measured bending of $1,98 \pm 0,12$" (internal) $\pm$0,30" (systematic). The metal primary telescope used in Sobral lost its focus due to the large temperature change during eclipse and the best observations where obtained with the secondary telescope. The expedition lead by Arthur Stanley Eddington (1882–1944) (Eddington 1919) to the isle of Principe obtained only two plates, measuring a bending of $1,61 \pm 0,30$" (internal). As Einstein's prediction was 1.73", both measurements confirmed the theory (Dyson, Eddington & Davidson 1920; Crelinsten 2006). When Einstein visited Brazil in 1925, he declared to the local newspapers: "The idea that my mind conceived was proven in the sunny sky of Brazil."

Figure 4. Observatório Nacional in Rio de Janeiro.

Figure 5. Telescopes used in Sobral on May 29th, 1919.

Figure 6. Observatório da Universidade Federal do Rio Grande do Sul.

3. The Observatory in Porto Alegre

In 1908, the construction of the observatory of the Universidade Federal do Rio Grande do Sul was completed in Porto Alegre, where it still stands as the oldest observatory in Brazil in its original location, still open to the public for visits every week, with its 190 mm equatorial telescope from Paul Ferdinand Gautier (1842–1909), *Constructeur d'instruments de précision, Paris.*

4. São Paulo

The main observatory in São Paulo was built between 1932 and 1941, part of Instituto Astronômico e Geofísico da Universidade de São Paulo.

Two important Brazilian researchers in astronomy were Mário Schenberg (1914–1990), who proposed the Urca process of energy loss by neutrino emission in stellar interiors with George Antonovich Gamow (1904–1968) (Gamow & Schoenberg 1940), and the Schenberg-Chandrasekhar limit for the end of the hydrodynamical equilibrium at the end of the main sequence life of a star, with Subrahmanyan Chandrasekhar (1910–1995) (Schönberg & Chandrasekhar 1942), and Cesare Mansueto Giulio Lattes (1924–2005), a Brazilian experimental physicist, co-discoverer of the pion, pi meson, studying cosmic

Figure 7. Observatório do Instituto Astronômico e Geofísico da Universidade de São Paulo.

rays (Lattes, Occhialini & Powell 1947). Both researchers spent most of their lives in Brazil.

5. Astrophysics

By the end of the 1960s, the first astronomy doctoral degree programs in Brazil started at São José dos Campos, São Paulo and Porto Alegre, and research in astrophysics really starts in Brazil. By 1971 a group of radio-astronomy from Universidade Presbiteriana Mackenzie builds a 13,7 m millimiter dish comprising the Rádio Observatório de Itapetinga. During the 1970's, a group lead by Observatório Nacional starts to build the Observatório do Pico dos Dias, with a 1.6 m optical telescope, later to become the Laboratório Nacional de Astrofísica. Other astronomy groups developed at Belo Horizonte, Natal and Instituto Nacional de Pesquisas Espaciais, in São José dos Campos. By 1981 Brazil had 41 astronomers with doctor degrees. Nowadays there are 284 doctors in astronomy hired by 41 institutions, plus 208 graduate students and around 60 post-docs. The groups in São Paulo and Rio de Janeiro are still the largest in Brazil, but there 11 institutions with graduate degree granting programs.

The Brazilian Astronomical Society, founded in 1974, currently has over 500 members.

Brazilian astronomers participate in a number of important international collaborations, including the construction and operation of the Gemini and SOAR optical and infrared telescopes in Chile, the Pierre Auger cosmic ray observatory and the Solar Submillimetric Telescope, in Argentina, the CoRoT space mission, and the Brazilian Decimetric Array. Brazil's participation as a 34% partner in the SOAR consortium has enabled the development of a Brazilian instrumentation program for this telescope with three world-class instruments currently under construction: a 1300 channel IFU fiber-fed

Figure 8. Rádio Observatório de Itapetinga.

Figure 9. Observatório do Pico dos Dias.

dual-beam spectrograph, a tunable filter imager (both to be operated with ground layer adaptive optics) and an echelle spectrograph. Although Brazil is one of the smallest Gemini partners, with only a 2.5% share, Brazilians author about 10% of the total number of papers in the Gemini partnership. The group at INPE has participated over the years in several international collaborations for the construction and operation of ballom and satellite observations, including the construction of a small gravitational wave detector. A Brazilian group is also associated with the Dark Energy Survey.

Figure 10. Photo at the annual meeting of the Brazilian Astronomical Society in 2008.

Table 1. Number of papers in refereed journals in Astronomy

1965	0
1970	8
1975	15
1980	25
1985	47
1990	74
1995	111
2000	205
2005	214
2008	219

Table 2. Areas of research of the 208 papers published in 2008

Optical and infrared stellar astronomy	63	28.8%
Theoretical cosmology	38	17.4%
Optical and infrared extragalactic astronomy	26	11.9%
Asteroid physics	12	5.8%
Theoretical stellar astrophysics	9	4.3%
Chemical evolution of stellar systems	9	4.3%
Dynamical astronomy	9	4.3%
Solar radio astronomy	7	3.2%
Instrumentation	7	3.2%
Exoplanets	6	2.7%
Others	29	13.2%

In terms of astronomy education and outreach, the most successful is the Brazilian Astronomical Olympiad, running since 1998, with 860 000 students participating in 2009 in more than 10 000 schools across the country.

Acknowledgements

The statistics presented here were collected by professor João Evangelhista Steiner, to whom we thank.

References

Crelinsten, J. 2006, *Einstein's Jury: the race to test relativity,* Princeton University Press
Davidson, C. R. 1922, *The Observatory*, 45, 224
Dyson, F. W., Eddington, A. S., & Davidson, C. 1920, *Philosophical Transactions of the Royal Society of London, Series A*, 220, 291
Eddington, A. S. 1919, *The Observatory*, 42, 119
Gamow, G. & Schoenberg, M. 1940, *PhRv*, 58, 1117
Lattes, C. M. G., Occhialini, G. P. S., & Powell, C. F. 1947, *Nature*, 160, 453
Newton, I. 1687, *Principia*, Tomo III, Prop. XLI, p. 507
Sanches Dorta, B. 1797, *Memórias da Academia Real das Sciencias de Lisboa*, 1, 345
Schönberg, M. & Chandrasekhar, S. 1942, *ApJ*, 96, 161

Transactions IAU, Volume XXVIIB
Proc. XXVII IAU General Assembly, August 2009
Ian F. Corbett, ed.

doi:10.1017/S1743921310004801

CHAPTER II

TWENTY SEVENTH GENERAL ASSEMBLY

BUSINESS SESSIONS

FIRST SESSION, 4 August 2009, 16.15-17.15

Centro de Convenções SulAmérica, Rio de Janeiro

1. Opening and Welcome

The President of the IAU, Dr. Catherine Cesarsky, welcomed the delegates and members to this the XXVII General Assembly. The delegates confirmed the agenda of the meeting as in the Program Book. The President invited the General Secretary, Dr. Karel A. van der Hucht, to present the first item.

2. Representatives of IAU National Members

The General Secretary listed the representatives of the IAU National Members: the first line applies to the first meeting, the second to the second meeting.

National Members	Category of Dues	Votes (a), (b)	National Representatives	Nominating Committee Members
Argentina	II	1, 3	Roberto Aquilano Andrea V. Ahumada	Roberto Aquilano Andrea V. Ahumada
Armenia	I	0, 0	Areg Mickaelian Areg Mickaelian	Areg Mickaelian Areg Mickaelian
Australia	IV	1, 5	Matthew Colless Matthew Colless	Warrick Couch Elaine M. Sadler
Austria	I	1, 2	Gerhard Hensler Gerhard Hensler	Gerhard Hensler Gerhard Hensler
Belgium	IV	1, 5	Anne Lemaître Anne Lemaître	Anne Lemaître Anne Lemaître
Bolivia	I	0, 0	-	-
Brazil	II	1, 3	Daniela Lazzaro Daniela Lazzaro	Eduardo Janot-Pacheco Daniela Lazzaro
Bulgaria	I	1, 2	-	-
Canada	V	1, 6	Gregory Fahlman Gregory Fahlman	Gregory Fahlman Gregory Fahlman
Chile	II	0, 0	René A. Méndez Bussard Mario Hamuy	René A. Méndez Bussard Mario Hamuy
China Nanjing	VI	1, 7	Gang Zhao Gang Zhao	Xiang-Dong Li Xiang-Dong Li
China Taipei	II	1, 3	Hsiang-Kuang Chang Lin-Wen Chen	Hsiang-Kuang Chang Lin-Wen Chen
Croatia	I	1, 2	-	-
Czech Republic	III	1, 4	Cyril Ron Jan Palouš	Jirí Borovicka Jirí Borovicka
Denmark	III	1, 4	Johannes Andersen Johannes Andersen	Johannes Andersen Johannes Andersen
Egypt	III	0, 0	-	-
Estonia	I	1, 2	Laurits Leedjärv Laurits Leedjärv	Laurits Leedjärv Laurits Leedjärv
Finland	II	1, 3	Ilkka V. Tuominen Ilkka V. Tuominen	Seppo Mikkola Seppo Mikkola
France	VII	1, 8	Daniel Rouan Corinne Charbonnel	Jean-Eudes Arlot William Thuillot
Germany	VII	1, 8	Andreas Quirrenbach Andreas Quirrenbach	Hans Zinnecker Christian Henkel
Greece	III	1, 4	-	-
Hungary	II	1, 3	Kristof Petrovay Lidia van Driel-Gesztelyi	Kristof Petrovay Lidia van Driel-Gesztelyi

National Members	Category of Dues	Votes (a), (b)	National Representatives	Nominating Committee Members
Iceland	I	1, 2	-	-
India	V	1, 6	S. Sirajul Hasan Ravi Subrahmanyam	S. Sirajul Hasan Ravi Subrahmanyam
Indonesia	I	1, 2	Chatief Kunjaya -	Chatief Kunjaya -
Iran	I	0, 0	-	-
Ireland	I	1, 2	Thomas P. Ray Thomas P. Ray	Thomas P. Ray Thomas P. Ray
Israel	III	1, 4	Morris Podolak Dina Prialnik-Kovetz	Morris Podolak Dina Prialnik-Kovetz
Italy	VII	1, 8	Ginevra Trinchieri Ginevra Trinchieri	Ginevra Trinchieri Ginevra Trinchieri
Japan	VII	1, 8	Norio Kaifu Norio Kaifu	Sadanori Okamura Sadanori Okamura
Korea RP	II	1, 3	Myung Gyoon Lee Myung Gyoon	Myung Gyoon Lee Lee Myung Gyoon Lee
Latvia	I	1, 2	Dimitijs Docenko Dimitijs Docenko	Dimitijs Docenko Dimitijs Docenko
Lebanon	Int.	0, 0	-	-
Lithuania	I	0, 0	Gražina Tautvaišiene Gražina Tautvaišiene	Gražina Tautvaišiene Gražina Tautvaišiene
Malaysia	I	1, 2	-	-
Mexico	III	1, 4	Alberto Carramiñana Alberto Carramiñana	Irene Cruz González Irene Cruz González
Mongolia	Int.	1, 0	-	-
Morocco	Int.	1, 0	-	-
Netherlands	V	1, 6	Thijs van der Hulst Thijs van der Hulst	Jan Lub Jan Lub
New Zealand	II	1, 3	Alan C. Gilmore -	Alan C. Gilmore -
Nigeria	I	1, 2	-	-
Norway	II	1, 3	Øyvind Sorensen Mats Carlsson	Øyvind Sorensen Mats Carlsson
Peru	Int.	1, 0	-	-
Philippines	Int.	1, 0	-	-

National Members	Category of Dues	Votes (a),(b)	National Representatives	Nominating Committee Members
Poland	III	1, 4	Edwin Wnuk Edwin Wnuk	Edwin Wnuk Edwin Wnuk
Portugal	II	1, 3	José P. Osório José P. Osório	José P. Osório José P. Osório
Romania	I	1, 2	Petre Popescu Petre Popescu	Petre Popescu Petre Popescu
Russian Federation	V	1, 6	Dimitrij Bisikalo Dimitrij Bisikalo	Nikolaj Samus Nikolaj Samus
Saudi Arabia	I	1, 2	-	-
Serbia, Rep. of	I	1, 2	Zoran Kneževic Zoran Kneževic	Zoran Kneževic Zoran Kneževic
Slovakia	I	1, 2	Jozef Ziznovsky Jozef Ziznovsky	Jozef Ziznovsky Jozef Ziznovsky
South Africa	III	1, 4	Patricia Whitelock Patricia Whitelock	Peter Martinez Peter Martinez
Spain	IV	1, 5	Valentin Martínez Pillet Valentin Martínez Pillet	Rafael Bachiller Rafael Bachiller
Sweden	III	1, 4	Nikolai Piskunov Dainis Dravins	Nikolai Piskunov Dainis Dravins
Switzerland	III	1, 4	Manuel Güdel Manuel Güdel	Manuel Güdel Manuel Güdel
Tajikistan	I	0, 0	Ibadinov Kursand -	Ibadinov Kursand -
Thailand	I	1, 2	Boonrucksar Soonthornthum Boonrucksar Soonthornthum	Saran Poshyachinda Saran Poshyachinda
Turkey	I	1, 2	Kutluay Yüce Kutluay Yüce	Solen Balman Solen Balman
UK	VII	1, 8	Andrew Fabian Andrew Fabian	Paul Murdin Paul Murdin
Ukraine	III	1, 4	Klim Churyumov Klim I. Churyumov	Klim Churyumov Klim I. Churyumov
USA	X	1, 11	Edward Guinan Edward Guinan	Sara Heap Sara Heap
Vatican City State	I	1, 2	Christopher Corbally Christopher Corbally	Guy Consolmagno Guy Consolmagno
Venezuela	I	1, 2	- Gustavo Bruzual	- Gustavo Bruzual

The National Representatives met on Monday 3 August and Wednesday 12 August. The Nominating Committee also met on met the same days.

3. Appointment of Offical Tellers

The Official Tellers were duly appointed:

Martha P. Haynes (chair)	USA
Patricia Cruz	Brazil
Christina Magoulas	Australia
Rogério Riffel	Brazil
José Vasquez Mata	Mexico
Joyce Yun	China

4. Admission of new National Members to the Union

The following new National Members were unanimously admitted to the Union:

	Category	Adhering Organization
Costa Rica	Interim	The University of Costa Rica
Honduras	Interim	Observatorio Astronómico Centroamericano de Suyapa
Panama	Interim	Universidad de Panamá, Facultad de Ciencias Naturales
Vietnam	Interim	Vietnam Astronomy Society, Hanoi National University of Education

5. Revisions of Statutes and Bye-Laws

The revisions to the Statutes and Bye-Laws were presented by the Assistant General Secretary, Dr. Ian F. Corbett.

One revision was proposed from the floor and was accepted: that "Republic of France" be changed to "France". With this revision the changes were approved. The revised Statutes and Bye-Laws are given in Chapter 7 of these *Transactions*.

6. Appointment of the Finance Committee

The Finance Committee was appointed as follows:

National Member	Category of Adherence	Votes (a), (b)	Finance Committee members
Argentina	II	1, 3	Roberto Aquilano/Andrea Ahumada
Armenia	I	0, 0	-
Australia	IV	1, 5	John O'Byrne/John O'Byrne
Austria	I	1, 2	Gerhard Hensler/Gerhard Hensler
Belgium	IV	1, 5	Anne Lemaître/Anne Lemaître
Bolivia	I	0, 0	-
Brazil	II	1, 3	Horacio Dottori/Horacio Dottori
Bulgaria	I	1, 2	-
Canada	V	1, 6	Gregory Fahlman/Gregory Fahlman
Chile	II	0, 0	René Méndez Bussard/Mario Hamuy
China Nanjing	VI	1, 7	Shuang-Nan Zhang/Shuang-Nan Zhang
China Taipei	II	1, 3	Hsiang-Kuang Chang/Lin-Wen Chen
Croatia	I	1, 2	-
Czech Republic	III	1, 4	Cyril Ron/Cyril Ron
Denmark	III	1, 4	Johannes Andersen/Johannes Andersen
Egypt	III	0, 0	-
Estonia	I	1, 2	Laurits Leedjärv/Laurits Leedjärv
Finland	II	1, 3	Seppo Mikkola/Seppo Mikkola
France	VII	1, 8	Monique Spite/Monique Spite
Germany	VII	1, 8	Hans Zinnecker/Svetlana Hubrig
Greece	III	1, 4	-
Hungary	II	1, 3	Kristof Petrovay/Lidia van Driel-Gesztelyi
Iceland	I	1, 2	-
India	V	1, 6	S. Sirajul Hasan/Ravi Subrahmanyam
Indonesia	I	1, 2	Chatief Kunjaya/-
Iran	I	0, 0	-
Ireland	I	1, 2	Thomas Ray/Thomas Ray
Israel	III	1, 4	Morris Podolak/Dina Prialnik-Kovetz
Italy	VII	1, 8	Ginevra Trinchieri/Ginevra Trinchieri
Japan	VII	1, 8	Toshio Fukushima/Toshio Fukushima
Korea RP	II	1, 3	Myung Gyoon Lee/Myung Gyoon Lee
Latvia	I	1, 2	Dimitijs Docenko/Dimitijs Docenko
Lebanon	Int.	0, 0	-
Lithuania	I	0, 0	Gražina Tautvaišiene/Gražina Tautvaišiene
Malaysia	I	1, 2	-
Mexico	III	1, 4	Alfonso Serrano/Alfonso Serrano
Mongolia	Int.	1, 0	-
Morocco	Int.	1, 0	-
Netherlands	V	1, 6	Jan Lub/Jan Lub
New Zealand	II	1, 3	Alan Gilmore/Alan Gilmore
Nigeria	I	1, 2	-
Norway	II	1, 3	Øyvind Sørensen/Mats Carlsson
Peru	I	0, 0	-
Philippines	Int.	1, 0	-
Poland	III	1, 4	Edwin Wnuk/Edwin Wnuk
Portugal	II	1, 3	José P. Osório/-
Romania	I	1, 2	Petre Popescu/Petre Popescu
Russian Federation	V	1, 6	Dmitrij Bisikalo/Dmitrij Bisikalo

National Member	Adherence Category	Votes (a),(b)	Finance Committee members
Saudi Arabia	I	1, 2	-
Serbia, Rep. of	I	1, 2	Zoran Kneževic/Zoran Kneževic
Slovakia	I	1, 2	Jozef Ziznovsky/Jozef Ziznovsky
South Africa	III	1, 4	Catherine Cress/Catherine Cress
Spain	IV	1, 5	Valentin Martínez Pillet/ Rafael Bachiller
Sweden	III	1, 4	Nikolai Piskunov/Dainis Dravins
Switzerland	III	1, 4	Manuel Güdel/Manuel Güdel
Tajikistan	I	0, 0	Ibadinov Kursand/-
Thailand	I	1, 2	Boonrucksar Soonthornthum/ Saran Poshyachinda
Turkey	I	1, 2	Kutluay Yüce/Solen Balman
UK	VII	1, 8	Paul Murdin/Paul Murdin
Ukraine	III	1, 4	Klim Churyumov/Klim Churyumov
USA	X	1, 11	Kevin Marvel/Kevin B. Marvel
Vatican City State	I	1, 2	Richard Boyle/Richard Boyle
Venezuela	I	1, 2	-/Gustavo Bruzual

The Finance Committe met on Monday 3 August and Wednesday 12 August.

7. Appointment of Nominating Committee

The Nominating Committee members are listed under Section 2 above.

8. Report of Executive Committee 2006 – 2009

The report of the Executive Committee was presented by the General Secretary, Dr. Karel A. van der Hucht, and is given in Chapter IV of these *Transactions*.

9. Report of the Special Nominating Committee

Nominations of the IAU Executive Committee for Triennium 2009 – 2012

President-Elect	Norio Kaifu	Japan
Assistant General Secretary	Thierry Montmerle	France
Vice-President	Matthew Colless	Australia
Vice-President	Jan Palouš	Czech Republic
Vice-President	Marta G. Rovira	Argentina

The SPECIAL NOMINATING COMMITTEE proposes the following slate of IAU members for the Officers and Members of the IAU Executive Committee for the triennium 2009 – 2012:

President	Robert Williams	USA
President-Elect	Norio Kaifu	Japan
General Secretary	Ian F. Corbett	UK
Assistant General Secretary	Thierry Montmerle	France
Vice-President	Matthew Colless	Australia
Vice-President	Martha P. Haynes	USA
Vice-President	George K. Miley	Netherlands
Vice-President	Jan Palouš	Czech Republic
Vice-President	Marta G. Rovira	Argentina
Vice-President	Giancarlo Setti	Italy
Adviser	Catherine J. Cesarsky	France
Adviser	Karel A. van der Hucht	Netherlands

10. Proposals to host the XXIX GA in 2015

Proposals were received from:

National Member	city	proposed by
Canada	Calgary	Canadian NCA, NRC, CAS
France	Paris	French NCA, SF2A
United States	Hawai'i	US NCA, NAS, AAS

These proposals were examined by the Executive Committee at its 86th meeting and the outcome presented to the 2nd Session of the General Assembly.

11. Closure of Session

There being no other business the President declared the session closed.

Transactions IAU, Volume XXVIIB
Proc. XXVII IAU General Assembly, August 2009
Ian F. Corbett, ed.

doi:10.1017/S1743921310004813

CHAPTER II

TWENTY SEVENTH GENERAL ASSEMBLY

BUSINESS SESSIONS

SECOND SESSION Thursday, 13 August 2009, 14:00–15:30 hr

Centro de Convenções SulAmérica, Rio de Janeiro

1. Opening and Welcome

The President of the IAU, Dr. Catherine Cesarsky, welcomed the delegates and members to the 2nd session of the XXVII General Assembly. The delegates confirmed the agenda of the meeting as in the Program Book. The President invited the General Secretary, Dr. Karel A. van der Hucht, to present the first item.

2. Individual Membership

The General Secretary reported that the IAU XXVI General Assembly, August 2006, had admitted 925 new Individual Members.

As of 13 August 2009 the number of IAU Individual Members was 9259 and IAU National Members nominated 882 new Individual Members. In addition IAU Division Presidents have nominated 5 new Individual Members. These were scrutinised by the Nominating Committee and were recommended for admission.

The General Assembly approved the membership of the nominees.

3. Deceased members

IAU Executive Committee is saddened to report the death of 159 Individual Members of the Union since August 2006. The General Secretary projected their names and the General Assembly held one minute silence as a mark of respect.

4. Appointment of Official Tellers

The Official Tellers appointed were:

Martha P. Haynes (chair)	USA
Patricia Cruz	Brazil
Christina Magoulas	Australia
Rogério Riffel	Brazil
José Vasquez Mata	Mexico
Joyce Yun	China

The Tellers confirmed that the quorum requirements were satisfied.

5. Resolutions

5.1. *Report of the Resolutions Committee by Chair Jocelyn S. Bell Burnell*

The Resolutions Committee had received five Type B resolutions and one Type A resolution. All had been correctly proposed according to the Working Rules. The Committee had examined the proposed resolutions and, in conjunction with the Executive Committe and the proposers, had refined the wording of several of them. The resultant resolutions had been published in the GA newspaper. The resolutions were:

B1 IAU Strategic Plan: Astronomy for the Developing World

A1 Implementing the IAU Strategic Plan

B2 IAU 2009 Astronomical Constants

B3 Second Realization of the International Celestial Reference Frame

B4 Supporting Women in Astronomy

B5 Defence of the night sky and the right to starlight

5.2. *Presentation and vote on Resolutions*

The resolutions were displayed to the General Assembly in both English (the language in which they had been proposed) and French, and voted separately. The results of the voting were as follows:

B1 Unanimous
A1 two against, no abstentions
B2 Unanimous
B3 Unanimous
B4 seven abstentions
B5 unanimous

6. Appointment of Resolutions Committee 2009–2012

The Resolutions Committee 2009–2012 was appointed:

Michel Dennefeld	France
Martha P. Haynes	USA
Zoran Kneževic	Serbia
Busaba H. Kramer	Thailand
Irina I. Kumkova	Russian Federation
Daniela Lazzaro, chair	Brazil
Silvia Torres-Peimbert	Mexico
Na Wang	China

7. Place and date of IAU XXIX GA in 2015

The Executive Committee had decided that the XXIV General Assembly 2015 would take place in Hawai'i, USA.

8. Division and Commission matters, adjustments to the Divisional structure

8.1. *The IAU DIVISIONS 2009–2012 are:*

Division I - Fundamental Astronomy
president Dennis D. McCarthy (USA)
vice-president Sergei A. Klioner (Germany)

Division II - Sun and Heliosphere
president Valentin Martnez Pillet (Spain)
vice-president James A. Klimchuk (USA)

Division III - Planetary Systems Sciences
president Karen J. Meech (USA)
vice-president Giovanni B. Valsecchi (Italy)

Division IV - Stars
president Christopher J. Corbally (Vatican City State)
vice-president Francesca d'Antona (Italy)

Division V - Variable Stars
president Steven D. Kawaler (USA)
vice-president Ignasi Ribas (Spain)

Division VI - Interstellar Matter
president You-Hua Chu (USA)
vice-president Sun Kwok (China - Hong Kong)

Division VII - Galactic System
president Despina Hatzidimitriou (Greece)
vice-president Rosemary F. Wyse (USA)

Division VIII - Galaxies and the Universe
president Elaine M. Sadler (Australia)
vice-president Franoise Combes (France)

Division IX - Optical and Infrared Techniques
president Andreas Quirrenbach (Germany)
vice-president David R. Silva (USA)

Division X - Radio Astronomy
president Russell A. Taylor (Canada)
vice-president Jessica M. Chapman (Australia)

Division XI - Space and High Energy Astrophysics
president Christine Jones (USA)
vice-president Noah Brosch (Israel)

Division XII - Union-Wide Activities
president Franoise Genova (France)
vice-president Raymond P. Norris (Australia)

8.2. *Changes to the Division structure*

Div. I / Comm. 8 / WG - Densification of the Optical Reference Frame
Proposed to be discontinued

Div. II / WG - Solar and Interplanetary Nomenclature
Proposed to be discontinued

Div. III / Comm.20 / WG - Distant Objects
Dormant since 2003. Proposed to turn this WG into a Task Group

Div.III / Commission 21 - Light of the Night Sky
Proposal: dissolve present C21 and form a new C21 on Galactic and Extragalactic Background Radiation locate the new C21 within Division IX

Div. IV-V-IX / WG - Standard Stars
Proposed to be discontinued

Div.VII / WG - Galactic Center
Proposed to be discontinued

Div. IX / WG - Detectors
Proposed to be discontinued: redundant in view of SPIE activities in the field.

Div. XII / Comm. 41 / WG - Astronomy and World Heritage
New Working Group

Div. XII / Comm. 46 / PG - World-Wide Development of Astronomy
Div. XII / Comm. 46 / PG - Teaching for Astronomy Development

Div. XII / Comm. 46 / PG - Exchange of Astronomers
These PGs will merge into one coordinated educational body

Div. XII / Comm.55 / WG - Best Practices
Proposed to be discontinued

Div. XII / Comm.55 / WG - New Ways of Communicating Astronomy with the Public
Proposed to be discontinued

Div. XII / Comm.55 / WG - New Media
New WG

Div. XII / Comm.55 / WG - Citizen Science
New WG

8.3. *IAU Commission Presidents and Vice-Presidents 2009–2012*

Comm. &Div.	President	Vice-President
4 I	George H. Kaplan (USA)	Catherine Y. Hohenkerk (UK)
5 XII	Masatoshi Ohishi(Japan)	Robert J. Hanisch (USA)
6 XII	Nikolay Samus (Russia)	Hitoshi Yamaoka (Japan)
7 I	Zoran Kneževic (Serbia)	Alessandro Morbidelli (France)
8 I	Dafydd W. Evans (UK)	N. Zacharias (USA)
10 II	Lidia van Driel-G. (France)	Carolus J. Schrijver (USA)
12 II	Alexander Kosovichev (USA)	Gianna Cauzzi (Italy)
14 XII	Glenn M. Wahlgren (USA)	Ewine van Dishoeck (Netherlands)
15 III	Alberto Cellino (Italy)	Dominique Bockelée-Morvan (France)
16 III	Melissa A. McGrath (USA)	Mark T. Lemmon (USA)
19 I	Haral Schuh (Austria)	Cheng-Li Huang (China)
20 III	Makoto Yoshikawa (Japan)	Steven Chesley (USA)
21 III	Jayant Murthy (India)	
22 III	Jun-ichi Watanabe (Japan)	Petrus Jenniskens (USA)
25 IX	Eugene F. Milone (Canada)	Alistair R. Walker (Chile)
26 IV	José A.D. Docobo (Spain)	Brian D. Mason (USA)
27 V	Gerald Handler (Austria)	Karen Pollard (New Zealand)
28 VII	Roger L. Davies (UK)	John S. Gallagher III (USA)
29 IV	Nikolai E. Piskunov (Sweden)	Katia Cunha (Brazil)
30 IX	Guillermo Torres (USA)	
31 I	Richard Manchester (Australia)	Mizuhiko Hosokawa (Japan)
33 VII	Rosemary F. Wyse (USA)	Birgitta Nordström (Denmark)
34 VI	You-Hua Chu (USA)	Sun Kwok (China Hong Kong)
35 IV	Corinne Charbonnel (Switzerland)	Marco Limongi (Italy)
36 V	Martin Asplund (Australia)	Joachim Puls (Germany)
37 VII	Bruce G. Elmegreen (USA)	Giovanni Carraro (Italy)

Comm. & Div.	President	Vice-President
40 X	Russell A.Taylor (Canada)	Jessica M. Chapman (Australia)
41 XII	Clive L.N. Ruggles (UK)	
42 V	Ignasi Ribas (Spain)	Mercedes T. Richards (USA)
44 XI	Christine Jones (USA)	Noah Brosch (Israel)
45 IV	Richard O. Gray (USA)	Birgitta Nordström (Denmark)
46 XII	Rosa M. Ros (Spain)	John Hearnshaw (New Zealand)
47 VII	Thanu Padmanabhan (India)	Brian P. Schmidt (Australia)
49 II	Nat Gopalswamy (USA)	Ingrid Mann (Japan)
50 XII	Wim van Driel (France)	Richard F. Green (USA)
51 III	William M. Irvine (USA)	Pascale Ehrenfreund (Netherlands)
52 I	Grard Petit (France)	Michael H. Soffel (Germany)
53 III	Alan P. Boss (USA)	Alain Lecavelier des Etangs (France)
54 IX	Stephen T. Ridgway (USA)	Gerard van Belle (Germany)
55 XII	Dennis Crabtree (Canada)	Lars Christensen (Germany)

The General Assembly endorsed these proposals.

9. Financial matters

9.1. *Report of the Finance Committee on IAU 2006–2009 accounts and 2010–2012 budget*

This was presented by the Chair of the Finance Sub-Committee, Dr. Paul G. Murdin

1. Accounts 2006-2008

The accounts show the usual triennial cycle, with a loss of 220k CHF in 2006 and surpluses of 84k and 1k in 2007 and 2008 respectively, roughly in balance, especially when considered against the 210k CHF of late payments from National Members that were outstanding in total over the financial years 2006-2008. The assets of the Union increased from 1041k CHF to 1396k in the period. It is prudent to maintain the assets at the level of one year's turnover, in order to provide a reserve for cash flow during the triennial cycle, to cover late payments by some National Members, and to provide some investment income.

Between 2009 and 2010 the IAU is changing its unit of account from Swiss Francs (CHF) to euros (EUR). Membership dues will in future be invoiced in euros. Accounts and budgets will make the transition during the year 2009. The FSC welcomes this change, which reflects the fact that most of the Union's expenditure is in euros and the largest number of the Union's National Members have their currency in euros. The change reduces exchange losses overall (to the Union and to many National Members) and spreads the risk of currency fluctuations, while costing the National Members outside the euro zone little (since the CHF and EUR are well correlated).

The FSC recommended that the reserve accounts should be brought primarily to euros as appropriate and timely; it will also prove to be more effective and convenient if the accounts are held in Paris in a fully international bank. The GS has started this process.

While the accounts have consistently passed the test of professional audit, investigations by the new administration discovered in 2008 that the Union has been subject to systematic fraud, cleverly perpetuated by an office staff member, using false or inflated invoices and unmandated, falsely attributed expenditure, principally on office expenditure and IT. The sums involved have amounted to 56k CHF in 2007 and similar, previously undiscovered sums in earlier years. The Union's administration has taken steps, as yet incomplete, to recover from its bank those unmandated payments that lie within the limit of the time period allowed for such recovery. It has changed its auditor to improve the standard of its auditing procedures. It has implemented changes to its accounting systems that should make recurrence of such a fraud more difficult to conceal in the future.

In all this, the administration has kept the Finance Sub Committee fully informed and consulted on the measures to be taken. The FSC believes that the fraud was cleverly enough perpetuated that it would not have been possible to discover the fraud before the sudden death in service that revealed it. The FC believes that the measures taken to prevent recurrence are appropriate to the scale of the Union's activities, and will in fact bring increased effectiveness to the Union's operation. The FC commends the GS and the new Administrative Assistant for the work done to identify and scope the fraud and supports the administration in its attempts to recover the sums stolen. The FC expects the Union's bank, Credit Lyonnais, to face up to its degree of responsibility in the matter.

The particular methodology of the fraud means that all the lost sums have been put through the accounts and that the assets of the Union are what is expected.

Noting the large sum of outstanding late payments from National Members, regretting the fraud, and noting the efforts of the Executive to recover what it can, the FC recommends the acceptance of the statement of accounts 2006-07-08.

2 Members' dues

The number of units of contribution from National Members was increased from 262 to 301.5 in 2009 when a number of Members accepted changes to their category of contribution that more reflected the size of their country's individual membership of the Union. Other National Members accepted compromise changes, others declined any changes at all on grounds of affordability.

Clearly the activities and scientific output of the Union correlate with the size of the membership, with associated cost. Given the historic and expected growth both in astronomy and in the size of the Union, the FC recommends that the National Members' category of contributions be kept under review in this way. Changes would be most readily accepted if they were not large, i.e. if the reviews were frequent so the changes were smaller. The FC recommends that the National Members' category of contributions should be reviewed as a regular matter each triennium. This procedure will also have the advantage that National Members will feel an incentive regularly to keep their lists of active members controlled and up to date.

3 Budgets 2010-2012

In the budgets, the proposed increase in the unit of contribution has been kept within inflation at a modest and realistic forecast of 3%, but the rise in the total number of units of contribution has meant that the income for the Union increased sharply from 2009 by approximately 180k CHF, 14%. In the proposed budgets this money has been shown as expenditure on matters within the proposed educational programme, developed as the Union's Strategic Plan. Some existing programmes of the Union (those not supported from earmarked funds from outside sources) have been included within these headings of expenditure. The apportionment of expenditure to individual programmes will be a matter for debate and discussion as the Strategic Plan is implemented and will depend on the degree of success of the Union in raising external funds to execute substantial parts of the Plan. This way to handle the development of the budgets, an effective balance in a developing situation between flexibility and control, has been implemented in the budgets in consultation with the FC.

Other features of the budget include control of office administration costs as a result of the elimination of the fraud referred to above and other measures to improve efficiency, even given the additional cost of measures to give increasing visibility to the Union's accounts in the future.

The FC recommends that the GA adopts the proposed unit of contribution as set out in the budgets, i.e. 2540 EUR in 2010, 2616 EUR in 2011 and 2695 EUR in 2012.

Noting the improved financial position of the Union and the ambition of the Strategic Plan to raise additional revenue for additional activities from outside sources, the FC recommends that the budgetary provisions for 2010-2011-2012 should be accepted.

This report was drafted by the Finance Sub Committee, which remains in session between General Assemblies to deal with urgent matters on behalf of the FC. The FSC members during 2006-09 were as follows:

Paul Murdin (UK), chair; Birgitta Nordstrom (DK); Xiangqun Cui (CN Nanjing); Cyril Ron (CZ); Kevin Marvel (US); John O'Byrne (AU)

9.2. *Budget 2009-2012*

The Union is changing over to an all-euro budget from 2010. The proposed budget for 2010-2012 is given in both Swiss Francs (CHF) and euro (EUR) for comparison.

	CHF	EUR
General Assemblies	610 000	385 000
Scientific Activities	990 500	627 000
Educational Activities	675 000	433 200
Coop with other unions	112 500	70 500
Executive Committee	521 950	325 650
Publications	69 000	45 000
Secretariat & Administration	1 100 525	694 500
Total	4 079 475	2 580 850

Comparison of the budget 2007–2009 and 2010–012 (CHF)

	2007-09	2010-12
General Assemblies	447 000	610 000
Scientific Activities	966 000	990 500
Educational Activities	315 000	675 000
Coop with other unions	111 000	112 500
Executive Committee	349 000	521 950
Publications	30 000	69 000
Secretariat & Administration	1 151 870	1 100 525
Total	3 072 870	4 079 475

More detailed budget figures are given in the EC Report, Chapter V of these *Transactions.*

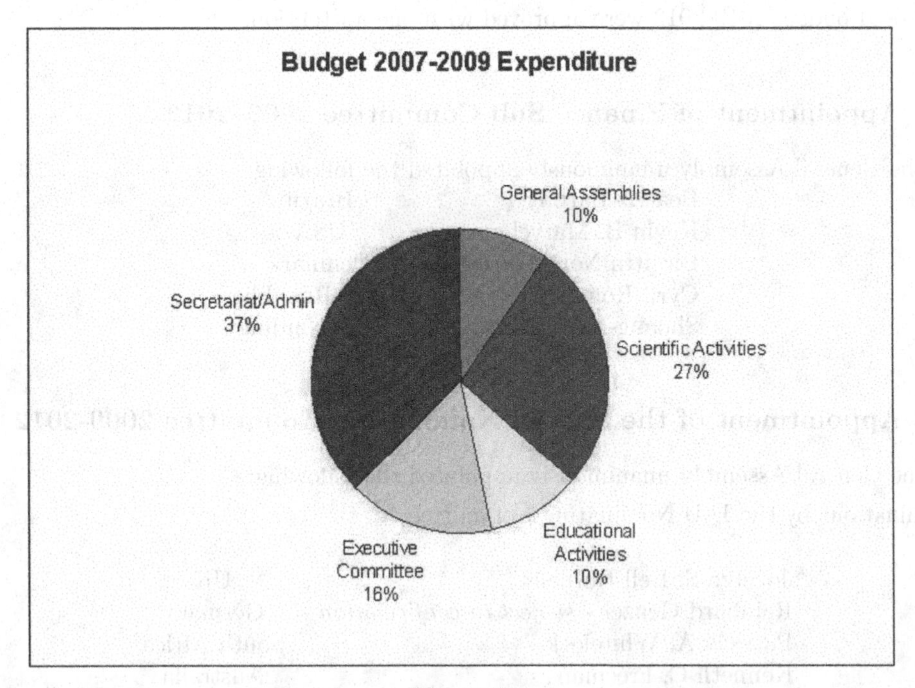

Figure 1.

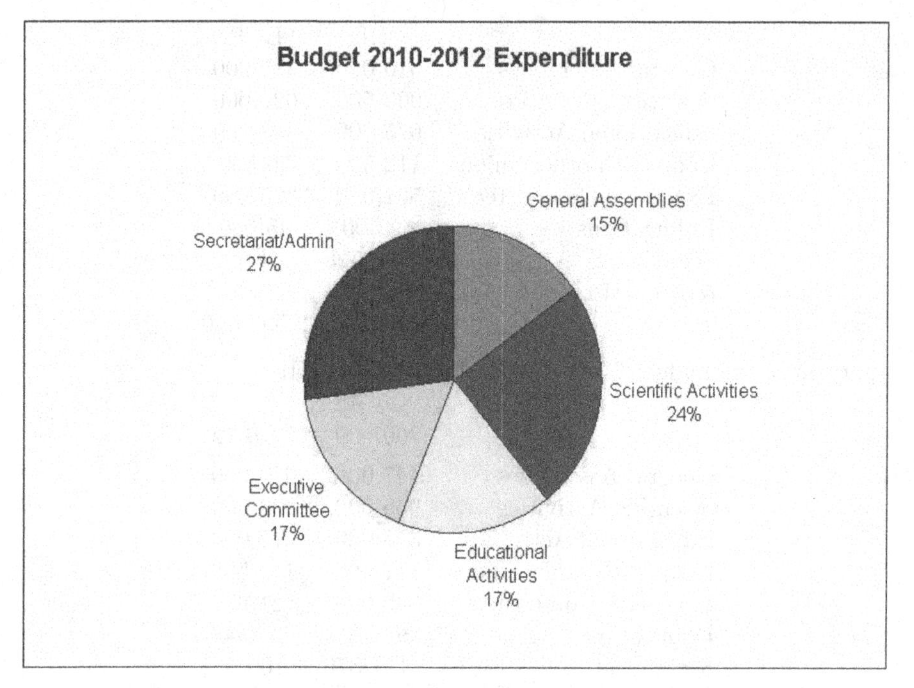

Figure 2.

9.3. *Vote on budget 2009-2012*

The Finance Sub-Committee's report, the statement of accounts 2006-2009, and the proposed budget 2010-2012 were approved with one abstention.

10. Appointment of Finance Sub-Committee 2009–2012

The General Assembly unanimously appointed the following:

Beatriz Barbuy	Brazil
Kevin B. Marvel	USA
Birgitta Nordstrm, chair	Denmark
Cyril Ron	Czech Republic
Shuang-Nan Zhang	China Nanjing

11. Appointment of the Special Nominating Committee 2009-2012

The General Assembly unanimously appointed the following:
Nominations by the IAU Nominating Committee

Jocelyn S. Bell Burnell	UK
Reinhard Genzel - *subject to confirmation* -	Germany
Patricia A. Whitelock	South Africa
Kenneth C. Freeman	Australia

Nomination by the IAU Executive Committee

Jian Sheng Chen China Nanjing

IAU ex officio members

Robert Williams (President) USA
Catherine J. Cesarsky (past-President) France

IAU Advisors

Ian F. Corbett (GS) UK
Thierry Montmerle (AGS) France

12. Election of the Executive Committee 2009-2012

The General Assembly unanimously elected the following:

President	Robert Williams	USA
President-Elect	Norio Kaifu	Japan
General Secretary	Ian F. Corbett	UK
Assistant General Secretary	Thierry Montmerle	France
Vice-President	Matthew Colless	Australia
Vice-President	Martha P. Haynes	USA
Vice-President	George K. Miley	Netherlands
Vice-President	Jan Palouš	Czech Republic
Vice-President	Marta G. Rovira	Argentina
Vice-President	Giancarlo Setti	Italy
Adviser	Catherine J. Cesarsky	France
Adviser	Karel A. van der Hucht	Netherlands

13. Closure of Session

There being no other business the President declared the Business Sessions of the XXVII GA closed.

Transactions IAU, Volume XXVIIB
Proc. XXVII IAU General Assembly, August 2009
Ian F. Corbett, ed.

doi:10.1017/S1743921310004825

CHAPTER III

TWENTY SEVENTH GENERAL ASSEMBLY

CLOSING CEREMONY

Thursday 13 August 2009, 15.45 - 17.00

Centro de Convenções SulAmérica, Rio de Janeiro

1. In honour of Francisco Xavier de Araújo

Address by *Professor Daniella Lazzaro*

"Chico"

On behalf of the National Organizing Committee, the IAU Executive Committee and all the Brazilian astronomical community I am here now to honor the memory of our dearest friend and member of the NOC, Dr. Francisco Xavier de Arajo, who passed away just one month ago.

Dr. Francisco, or Chico as was known by his friends, is the person who has been working for this meeting from the very first moment.

Since the end of 2002, when we started preparing the bid to host the GA here in Rio, up to few weeks ago, Chico has been the heart of the organization. His kindness, his natural skill in negotiation, his high commitment to the success of the meeting made him fundamental to the organization.

Although very difficult for us, we are here because this is what he would have wanted: for us to continue and turn into reality the dream of the Brazilian astronomical community to host an IAU GA.

Thank you, Chico, we miss you so much!

2. IAU Strategic Development Plan

Address by Vice-President *Professor George K. Miley*

On behalf of the Executive Committee and particularly of Bob Williams and Catherine Cesarsky, I thank you for passing the two resolutions endorsing our new decadal strategic plan "Astronomy for the Developing World".

I have described the plan at length on several occasions during this General Assembly and I shall not repeat this description here. The plan is an ambitious flexible and credible blueprint for using astronomy to stimulate sustainable global development. It contains a long term vision, achievable goals and a comprehensive new strategy for attaining these goals. An important element of the strategy is a regional bottom-up approach. The plan will be implemented with an effective and lean organisational structure subject to professional oversight.

Passing the resolutions is only the beginning. Now comes the hard work and I ask you to help with this. Please go home and discuss the plan with the directors of your institutes and the governing bodies of your universities. Is your institute willing to participate in a long-term twinning scheme with a physics department in a developing country or to fund your staff to give inspirational lectures in a poorer region of the world? Would your department be prepared to function as a regional node or maybe even provide matching funds and host the new Global Astronomy for Development Office? In seeking funding for such activities, please talk to your governments. The rationale for development activities is relevant to government departments, that you may not ever have approached before, such as Ministries of International Development or Foreign Affairs. Maybe a bilateral treaty involving your country could be used to fund astronomy development activities. If we in the IAU can help you in these fund raising activities, please let us know.

In seeking resources for the plan we must be creative. Among the organisations and people to approach are international and national development foundations, multinational companies operating in developing countries, international and regional development agencies and credible philanthropists.

I now address those of you who are involved in proposing new astronomical facilities, on the ground and in space. When writing your proposals, please consider earmarking a few percent of the budget for astronomy development programmes. Devoting a tiny fraction of astronomical resources to global development and education enhances the image of astronomy as a whole and make politicians more receptive to astronomy research proposals.

As we embark on raising funds in support of the plan, it is important to have strong arguments ready to use. In her beautiful address at the opening of this General Assembly Catherine Cesarsky touched on a few of the reasons why astronomy is of benefit to society. The closing ceremony is a fitting time to summarise more generally the rationale for astronomy presented in our plan and illustrated on its front and back covers.

Astronomy provides an inspirational and unique gateway to technology science and culture, three fundamental characteristics of developed nations.

3. Invitation to XXVIII General Assembly, Beijing, China August 2012

Address by President of Chinese Astronomical Society *Professor Gang Zhao*

It is my great pleasure to be here with you on this most important gathering in the world of astronomy, the IAU 27th GA in Brazil, and to say a few words on behalf of the Chinese Astronomical Society.

First, I would like to thank the International Astronomical Union, as well as the organizing committee of this assembly, for hosting such a grand meeting, and giving us this honorable opportunity to present our warm invitation for attending next IAU GA in Beijing, China. Being the capital of China, Beijing is well-known as a fast-growing, dynamic international metropolis with 3,000 years' history. It is one of the safest and the most peaceful cities in the world. The long history leaves Beijing precious cultural treasure, among which four sites have been designated as world cultural heritage by UNESCO: the Great Wall, the Summer Palace, the Forbidden City, and the Temple of Heaven. As Beijing will be the host city of the next IAU GA in 2012, the spirit of IAU will definitely promote the development of astronomy in China as well as in the world, and strengthen the friendship of our IAU family.

Since the adoption of 'open door policy' in the past 30 years, especially the last decade, great advance has been achieved in astronomical studies in China. We are now getting ready to present a memorable General Assembly for astronomers all over the world. With natural beauty, rich history and diverse culture, Beijing opens her arms to welcome your arrival in August of 2012!

Welcome to Beijing! Welcome to next IAU GA! Thank you!

Professor Gang Zhao then showed a short video illustrating the attractions of Beijing as a venue for an IAU General Assembly.

4. Address by retiring General Secretary, International Astronomical Union

Dr. Karel A. van der Hucht

The International Astronomical Union has been founded in Brussels 90 years ago, with as its mission to foster collaboration among scientists in the world. Young in comparison to the world's oldest science, the IAU is still sufficiently long-lived to have a rich tradition. I am pleased to be part of this venerable heritage. I have strived these past three years to serve you as General Secretary.

I give you three good reasons why I have enjoyed being your General Secretary.

- Trained as a research astronomer and specializing in the astrophysics of Wolf-Rayet stars, I used to focus my research mostly to that field, observing at all wavelengths from the ground and from space. Between 1990 and 2002 I helped organizing, among other things, four IAU Symposia in that field, and editing the proceedings. That may have been why someone suggested me for IAU General Secretary. Working for the IAU as General Secretary brought me in touch with colleagues working in all fields of astronomy. That enriched my life.
- I had the privilege to work with very motivated and dedicated fellow Officers and fellow EC members. That made my job as General Secretary very gratifying.
- Working for the IAU Secretariat as General Secretary means working for three years part-time in Paris: a rare opportunity to familiarize with a beautiful and fascinating city. However, due to circumstances beyond my control, I saw mostly the walls of my office at the IAU Secretariat. My wife got to see Paris. That's fine too.

Stating that by now I know all about the IAU may be too far fetched. In a Union with 63 National Members, over 10000 Individual Members organized in 12 Divisions, 40 Commissions and 75 Working Groups and Program Groups, there is always more to learn. The closest in getting to know the IAU is doing the editing of the General Secretary's trilogy: the Highlights of Astronomy, the Proceedings of the General Assembly (Transactions B), and the Reports of Astronomy (Transactions A). These volumes may be referred to by some of you as the books which no body reads, but editing them was for me an extremely useful and revealing exercise to get informed about the activities of the IAU membership.

Three years seem a long period in the beginning, but really feel like a short stretch at the end. The period from August 2006 to August 2009 was continuously dominated by expected and unexpected serious activities and developments. There is much to be grateful for and there are many to be grateful to. My sincere thanks go to: - my home institute SRON in Utrecht, the Netherlands, for giving me leave of absence for effectively 4.5 years, to work as AGS and as GS for the

IAU; - my fellow Officers Catherine Cesarsky, Bob Williams and Ian Corbett. In the past three years we have shared much of our work for the IAU, leading to a fruitful and effective cooperation; - all preceding General Secretaries, who were always there to give advice in a most constructive way; - the staff of the IAU Secretariat in Paris, in particular my Executive Assistant Mme Vivien Reuter. Mme Reuter joined the Secretariat at a very difficult time. Together we had to re-invent the Secretariat, often hanging on by our fingernails, but surviving. Vivien, without you the IAU Secretariat would not be in the good shape as it is today. You are for me formedable. - and our very motivated National Organizing Committee for this IAU XXVII General Assembly, co-chaired by Daniela Lazzaro and Beatriz Barbuy, who have invited us to this beautiful city, Rio de Janeiro, and who made the General Assembly happen. We would not be here without their intense efforts. But most of all I want to thank my wife Ritte.

5. Address by the incoming General Secretary

Dr. Ian F. Corbett

It is a great honour and a great challenge to be elected as General Secretary of the IAU, and I thank you most sincerely for your confidence. I greatly enjoyed my three years as Assistant General Secretary and I must thank publicly all the organisers, authors and especially editors of the 27 Symposia and 4 Regional Meetings held during my term of office. I also extend a warm welcome to the new AGS, Thierry Montemerle, and assure him that he will find the next three years interesting and rewarding.

Needless to say, I am very much looking forward to the next three years. I must, however, start by saying a few words of thanks to Karel, whose exemplary commitment and tireless dedication as General Secretary for the past three years has ensured that the Union survived major upheaval in the Paris Secretariat, emerging stronger and fitter as a result. We owe him a great debt of gratitude. I am doubly grateful, of course, because not only do I benefit as an ordinary IAU member but as incoming General Secretary I inherit a very healthy state of affairs and an extremely competent assistant in Vivien Reuter, whom you have all seen actively helping to run this GA with her assistant Jana. Thank you, Karel, for making my task so much easier, and for being such a diligent mentor over the past three years.

I should also say a word or two of appreciation for our outgoing President, Catherine Cesarsky, who has been the driving force - and the clich is truly appropriate in this case - behind the International Year of Astronomy. This is proving an outstanding world-wide success, perhaps more than we could ever have hoped, and is due to the hard work and dedication of many people all over the world, and particularly the IYA Secretariat hosted by ESO. However, Catherine, you led from the front in shaping the objectives, the strategy and the structure of IYA, and I think we can all agree that it could never have been the same with without you.

We should now look to the future. I think that three - or possibly four - known challenges face us in my term as General Secretary. Fortunately for me, we have an outstanding new President and Executive Committee, and I thank them all for agreeing to serve.

First of all, we must strive to implement the Strategic Plan so eloquently presented a few minutes ago by George Miley. Secondly, we must ensure that the International Year leaves a deep and lasting legacy. IYA has been, and continues to be, an unprecedented success - we have seen original and creative activities all over the world and witnessed the profound effect on public awareness. We are taking 'astronomy' into the lives of many people to "engage a personal sense of wonder and discovery", we are promoting astronomy with amazing success in developing nations, and we have brought the wonder of the telescope into the lives of hundreds of thousands of children - and adults, too. We must follow this up through the continuing work of the IYA Secretariat and, most importantly, nationally through the initiatives and structures established during the IYA. I know that work to build this legacy has already started, and I am confident that we will succeed.

The third challenge is more open ended - the IAU must continue to develop, to evolve, to adapt, but - above all - it must continue to reflect the needs and aspirations of its community. As Ron Ekers said in his opening address in Prague three years ago, "the IAU really works in the background to provide the lubrication for the wheels of the international machinery and international science". This is very much as it should be, but at the same time the IAU must achieve what I would call 'responsive leadership'. We need the active support and input of our National Members, who are far more than just our paymasters. We exist because we add value to what is being done in our National Members. We also depend very much on our Divisions and Commissions, and on their Presidents, Organising Committees, and members. These are the arteries and the nervous system of the Union. Through them, the Union moves forward and keeps in touch with the working astronomer, world-wide. I am confident that our new Division and Commission Presidents will rise to the challenge.

The fourth challenge may not be such a challenge after all. We have our next General Assembly in Beijing in 2012. We will have to match the very high standard set by our Brazilian hosts and Karel in organising this General Assembly. Yet such is my confidence in Professor Gang Zhao and his colleagues that I am sure we can all look forward to meeting again in Beijing in 2012, and in the course of another memorable General Assembly look back on a further three years of progress.

This has been a tremendous General Assembly and 2009 will be a momentous year in the history of the IAU. I thank you all, and I look forward to seeing you in Beijing.

6. Address by the retiring President

Dr. Catherine J. Cesarsky

For the 90th anniversary of IAU, the Secretary General has collected the memories of six past Presidents of the IAU in the Information Bulletin 104. Like them, I can state with confidence that being an IAU President is a light job: the entire burden of running IAU falls upon the General Secretary. I was fortunate in that "my" General Secretary was Karel van der Hucht, whose dedication to the union has been total. We suffered a bad blow when IAU Executive Assistant Monique Orine died in January 2008, leaving the office in a difficult situation. Luckily, we hired almost immediately Vivien Reuter to replace Monique, and Karel and Vivien together have orchestrated a fantastic recovery.

Like Karel and Ian, I can rejoice retrospectively at the excellent working atmosphere prevailing in our little group of Officers, as well as in the Executive Committee and also with the experienced Chair of the Finance sub-Committee, Paul Murdin, who has really been there for us in the hour of need. The Vice Presidents have each agreed to carry out a specific task for the Union. George Miley spearheaded the IAU Decadal Plan for Astronomy for the developing world; it remains for Bob and Ian, and then Norio and Thierry, to bring it to fruition: in particular to undertake the difficult task of raising the necessary funds, and, if possible, generate a generous host for the Global development office. Note that this plan covers only one subset of IAU activities, so there is more strategic thinking ahead; in fact, strategic plans always need to be reconsidered and adapted, but they remain an excellent method to trigger meaningful changes.

Like Karel, I have learnt a lot about the IAU these three years, and the more I know, the more I find the Union interesting and useful. I had a number of new experiences: I have taken part in a teaching venture in Kathmandu, and found it extremely gratifying. I advise all of you to try, as we have a great need for volunteers. I have also helped setting up the first MEARIM, in Egypt, and for this I am very pleased, as it was successful and will soon become a tradition at IAU.

But of course the highpoint of my presidency was the preparations and the launch of the International Year of Astronomy 2009 (IYA 2009). One of my early tasks was to lead the IAU delegation that went to New York, at the end of 2007, to lobby in favour of IYA at the United Nations. It was a momentous time. We were delighted to see that our proposal was extremely well received, and indeed the UN endorsement soon followed. In 2006 the Executive Committee established an IYA Working Group. I felt that this task was so important, in the years of my presidency, that I held the Chair of the Working Group, with the IAU Press Officer, Lars Christensen, as super efficient secretary. We soon agreed on the necessity of establishing an IYA Secretariat in charge of the global coordination of the activities. I was at ESO at the time, and we decided to put it at ESO, managed by Lars. We hired Pedro Russo, and later Mariana Barrosa, and, I am tempted to say, the rest is history. I have enjoyed working on a regular basis with Lars, Pedro and Mariana. I am proud of the work accomplished and of the results obtained. I will continue chairing the Working Group till the end of next year, which will be devoted to an evaluation of IYA activities and impact. The exchanges and feedback with Single Points of Contact, chairs of Cornerstones and of Special projects, astronomers and amateurs, have been an incredibly rewarding personal experience which I will never forget.

One of the goals of IYA is: "Provide a modern image of science and scientists", and I'll end this farewell as IAU President with a quote from the British newspaper *The Guardian*, on its editorial page on IYA2009, Saturday 25 July 2009: "Astronomers around the world compete, co-operate and confer; they are a global community, in the richest sense of the term, and we owe to them our understanding of space and time, and light, and mass, and gravity: in a word, everything."

7. Address by the incoming President

Professor Robert Williams

We are now more than halfway through the International Year of Astronomy and near the end of the General Assembly. This is a good time to take stock of where we are as a community and an organization. Let me begin by saying that I have been impressed with the different ways that many of you have employed to bring astronomy and discovery to the public. The efforts at all levels have been impressive, and there are many contributions that merit acknowledgement. From my standpoint on the Executive Committee I must begin by thanking Catherine Cesarsky for her efforts, which repeatedly took her around the globe. Her energy on behalf of the IYA and the Union has been remarkable.

Also deserving of comment is the unusual dedication without fanfare of the job that Karel van der Hucht has done as General Secretary. The Paris office went through serious turmoil with the death of our longtime executive assistant, and the way that Karel kept his focus on the job was exemplary. As one of the people who stood to be most affected by a meltdown in the Paris office, I spent some days fearing the worst. It did not happen and Karel's and our new assistant Vivien Reuter's steadfast, knowledgeable attention to details was a key part of the current healthy state of the IAU.

Then, there are you, the members: the core of the Union. American filmmaker Woody Allen once remarked: 85percent of success is showing up. You showed up—both for the IAU and IYA activities. We hope you continue to show up because this Union has many worthy programs that need your ingenuity and work, and we hope you participate in those for which you have interest and expertise.

This is an appropriate time to ask ourselves: where should the Union head in the future? Which programs might be expanded, and what new initiatives might we think of undertaking? The Executive Committee has been discussing Union priorities for the past two years, and several relevant events took place at this General Assembly. The first was the revision to the Bye-Laws that in the future allows for electronic voting on non-budgetary matters. We are developing tools that will enable more electronic communication and interaction between the divisions and commissions and the Paris office to take place. There is a divergence of opinions on whether e-voting should be used for scientific issues. The Executive Committee is still debating this issue. The basic question is whether the virtue of inclusiveness of members who are not at the GA outweighs the disadvantage of not being present for a live debate of issues. Some of us feel that both features can be accommodated by supplementing the GA debate with electronic debates on our website for a period after the GA, which would allow all members to vote on scientific issues. The option of allowing electronic votes on certain issues such as division and commission elections will definitely be pursued.

Another significant event of this General Assembly has been the adoption of a long-range strategy for the IAU. At this GA we conducted a town hall meeting to discuss the Strategic Plan that a group of members appointed by the Executive Committee and led by Vice President George Miley has developed. The Plan is titled 'World Wide Development of Astronomy', and it has just been approved by you by means of a resolution. Astronomy development is not a new activity for the Union but we will be expanding our efforts in support of the development of astronomy, especially in the area of science education. Specifically, we are trying to arrange for the funding of a Global Development Office that will coordinate a range of activities around the world to be carried out by IAU members.

The International Year of Astronomy has demonstrated the value of engaging the public. The IYA has brought the IAU further into public outreach than ever before. The constant struggle between science and superstition has important consequences for society, therefore the IAU should be deeply involved in educating the public. For some years the IAU has carried out development activities through Commission 46 initiatives. The Executive Committee wants to strengthen our programs of International School for Young Astronomers (ISYA), Teaching for Astronomy Development (TAD), and World Wide Development of Astronomy (WWDA), which have made a large contribution to science education in many countries. In order to expand Union efforts in this area we plan to seek external funding.

The IAU will continue to support the professional astronomy community through our long-standing sponsorship of symposia and issuance of travel grants. A large fraction of Union resources continues to flow back to the members through these grants. This is because the success of the Union depends not so much on the officers and the Executive Committee as on you, the members. We need your participation, and especially the ideas and energy of early career astronomers. Please continue to put forth your ideas and push your initiatives to keep the IAU active. Our mission is to understand the universe and pass this along to society. In doing so we expose them to a way of thinking that pushes back superstition and confronts preconceived ideas when they are not supported by facts, teaching the importance of being open to new facts and concepts.

This, I believe, is our core mission and I look forward to working with you to accomplish it.

8. Emblematic songs and rhythms of Brazil

Performed by *Atrás da Nota Choir (Rio de Janeiro City Hall)* and *Choir Association CPRM (Geological Survey of Brazil)*, conducted by Mario Assef, as performed in the Inaugural Program:

• Pantanal (Marcus Viana) - Arranged by Silvia Sobreira - *Sinfonia.* Evokes the myths and traditions of Pantanal, the Brazilian large central swampland ecosystem.

• Samba do Avião (Tom Jobim) - Arranged by Ju Cassu - *Bossa Nova.* This song celebrated the feeling of landing in Rio de Janeiro city after a long time away.

• A Voz do Morro (Zé Kéti) - Arranged by Marcos Leite - *Samba.* Tells about the importance of samba music as source of Brazilian happiness.

• Aquarela do Brasil (Ary Barroso) - Arranged by Jos Assunção Jr. - *Samba Exaltação.* Exaltation song about Brazilian beautiful landscape and people.

• Lata dagua na Cabeça (Lus Antonio e J. Júnior) - Arranged by Marcos Leite. *Samba.* Talks about typical slum hard women and her dreams

- Sabiá (Luiz Gonzaga and Zé Dantas) - Arranged by Silvia Sobreira - *Baião Medley.* Northeast Brazilian rhythmic that spread throughout Brazil with the forró dance, especially during June and July holidays.
- Cajuna (Caetano Veloso) - Arranged by Malu Cooper
- Patuscada de Gandi (Gilberto Gil) - Arranged by Marcos Leite - *Afroxé.* Afro-Brazilian rhythm from Bahia.
- Cidade Maravilhosa (André Filho) - Arranged by Silvia Sobreira - *Marcha Rancho.* Anthem of Rio de Janeiro city.

Transactions IAU, Volume XXVIIB
Proc. XXVII IAU General Assembly, August 2009
Ian F. Corbett, ed.

doi:10.1017/S1743921310004837

CHAPTER IV

RESOLUTIONS OF THE GENERAL ASSEMBLY

1. Resolutions Committee 2006 - 2009

During the triennium 2006 - 2009, the Resolutions Committee comprised:

Jocelyn	Bell Burnell	UK (Chair)
Michel	Dennefeld	France
Brian	Warner	South Africa
Rachel L.	Webster	Australia

The Report of the Resolutions Committee is given in Chapter II of these *Transactions.*

2. Approved Resolutions

RESOLUTION B1

The IAU Strategic Plan: Astronomy for the Developing World

The XXVII General Assembly of the International Astronomical Union,

recognizing

1. the goal of the IAU to encourage the development of astronomy and facilitate better understanding of the universe,

2. that the current activities of the International Year of Astronomy 2009 have made great strides in advancing knowledge of astronomy among citizens of all nations and awareness of its value to society,

3. that science education and research is an essential component of modern technological and economic development

resolves that the IAU should

1. place increasing emphasis on programs that advance astronomy education in developing countries,

2. approve the goals specified in the Strategic Plan "Astronomy for the Developing World" as objectives for the IAU in the coming decade.
3. assess programs undertaken during the IYA to determine which activities are most effective in advancing astronomy.

RESOLUTION A1

on Implementing the IAU Strategic Plan

The XXVII General Assembly of the International Astronomical Union,

recognizing

1. the goal of the IAU to encourage the development of astronomy and facilitate better understanding of the universe,

2. that the current activities of the International Year of Astronomy 2009 have made great strides in advancing knowledge of astronomy among citizens of all nations and awareness of its value to society,

3. that science education and research is an essential component of modern technological and economic development,

4. Resolution B1 adopting the IAU Strategic Plan Astronomy for the Developing World passed by the XXVII General Assembly,

resolves that the IAU should

1. give high priority to supporting the development of astronomy infrastructure in emerging nations,

2. proceed with the implementation of the IAU Strategic Plan Astronomy for the Developing World through the creation of a Global Development Office and seek appropriate additional resources for implementing the plan.

RESOLUTION B2

on IAU 2009 astronomical constants

The XXVII General Assembly of the International Astronomical Union,

Considering

1. the need for a self-consistent set of accurate numerical standards for use in astronomy,

2. that improved values of astronomical constants have been derived from recent observations and published in refereed journals, and that conventional values have been adopted by IAU GA 2000 and IAU GA 2006 resolutions for a number of astronomical quantities,

recognizing

1. the continuing need for a set of Current Best Estimates (CBEs) of astronomical numerical constants, and

2. the need for an operational service to the astronomical community to maintain the CBEs

Recommends

1. that the list of previously published constants compiled in the report of the Working Group on Numerical Standards of Fundamental Astronomy (see http://maia.usno.navy.mil/NSFA/CBE.html) be adopted as the IAU (2009) System of Astronomical Constants.

2. that Current Best Estimates of Astronomical Constants be permanently maintained as an electronic document,

3. that, in order to ensure the integrity of the CBEs, IAU Division I develop a formal procedure to adopt new values and archive older versions of the CBEs, and

4. that the IAU establish within IAU Division I a permanent body to maintain the CBEs for fundamental astronomy.

RESOLUTION B3

on the Second Realization of the International Celestial Reference Frame

The XXVII General Assembly of the International Astronomical Union,

Noting

1. that Resolution B2 of the XXIII General Assembly (1997) resolved *'that, as from 1 January 1998, the IAU celestial reference system shall be the International Celestial Reference System (ICRS)'*,

2. that Resolution B2 of the XXIII General Assembly (1997) resolved *'that the fundamental reference frame shall be the International Celestial Reference Frame (ICRF) constructed by the IAU Working Group on Reference Frames'*,

3. that Resolution B2 of the XXIII General Assembly (1997) resolved *'that IERS should take appropriate measures, in conjunction with the IAU Working Group on reference frames, to maintain the ICRF and its ties to the reference frames at other wavelengths'*,

4. that Resolution B7 of the XXIII General Assembly (1997) recommended *'that high-precision astronomical observing programs be organized in such a way that astronomical reference systems can be maintained at the highest possible accuracy for both northern and southern hemispheres'*,

5. that Resolution B1.1 of the XXIV General Assembly (2000) recognized *'the importance of continuing operational observations made with Very Long Baseline Interferometry (VLBI) to maintain the ICRF'*,

Recognizing

1. that since the establishment of the ICRF, continued VLBI observations of ICRF sources have more than tripled the number of source observations,

2. that since the establishment of the ICRF, continued VLBI observations of extragalactic sources have significantly increased the number of sources whose positions are known with a high degree of accuracy,

3. that since the establishment of the ICRF, improved instrumentation, observation strategies, and application of state-of-the-art astrophysical and geophysical models have significantly improved both the data quality and analysis of the entire relevant astrometric and geodetic VLBI data set,

4. that a working group on the ICRF formed by the International Earth Rotation and Reference Systems Service (IERS) and the International VLBI Service for Geodesy and Astrometry (IVS), in conjunction with the IAU Division I Working Group on the Second Realization of the International Celestial Reference Frame has finalized a prospective second realization of the ICRF in a coordinate frame aligned to that of the ICRF to within the tolerance of the errors in the latter (see note 1),

5. that the prospective second realization of the ICRF as presented by the IAU Working Group on the Second Realization of the International Celestial Reference Frame represents a significant improvement in terms of source selection, coordinate accuracy, and total number of sources, and thus represents a significant improvement in the fundamental reference frame realization of the ICRS beyond the ICRF adopted by the XXIII General Assembly (1997),

Resolves

1. that from 01 January 2010 the fundamental astrometric realization of the International Celestial Reference System (ICRS) shall be the Second Realization of the International Celestial Reference Frame (ICRF2) as constructed by the IERS/IVS working group on the ICRF in conjunction with the IAU Division I Working Group on the Second Realization of the International Celestial Reference Frame (see note 1),

2. that the organizations responsible for astrometric and geodetic VLBI observing programs (e.g. IERS, IVS) take appropriate measures to continue existing and develop improved VLBI observing and analysis programs to both maintain and improve ICRF2,

3. that the IERS, together with other relevant organizations continue efforts to improve and densify high accuracy reference frames defined at other wavelengths and continue to improve ties between these reference frames and ICRF2.

Note 1: The Second Realization of the International Celestial Reference Frame by Very Long Baseline Interferometry, Presented on behalf of the IERS / IVS Working Group, Alan Fey and David Gordon (eds.). (IERS Technical Note ; 35) Frankfurt am Main: Verlag des Bundesamts für Kartographie und Geodäsie, 2009. See http://www.iers.org/MainDisp.csl?pid=46-25772 or http://hpiers.obspm.fr/icrs-pc/ .

RESOLUTION B4

on Supporting Women in Astronomy

The XXVII General Assembly of the International Astronomical Union,

Recalling

1. the UN Millennium Development Goal 3: promote gender equality and empower women,

2. the IAU/UNESCO International Year of Astronomy 2009 goal 7: improve the gender-balanced representation of scientists at all levels and promote greater involvement by underrepresented minorities in scientific and engineering careers,

Recognizing

1. that individual excellence in science and astronomy is independent of gender,

2. that gender equality is a fundamental principle of human rights,

considering

1. the role of the IAU Working Group for Women in Astronomy,

2. the role of the IYA2009 Cornerstone Project 'She is an Astronomer',

Resolves

1. that IAU members should encourage and support the female astronomers in their communities,

2. that IAU members and National Representatives should encourage national organisations to break down barriers and ensure that men and women are given equal opportunities to pursue a successful career in astronomy at all levels and career steps.

RESOLUTION B5

in Defence of the night sky and the right to starlight

The XXVII General Assembly of the International Astronomical Union,

Recalling

1. the IAU/UNESCO International Year of Astronomy 2009 goal 8: facilitate the preservation and protection of the world's cultural and natural heritage of dark skies in places such as urban oases, national parks and astronomical sites,

2. the Declaration approved during the International Conference in Defence of the Quality of the Night Sky and the Right to Observe Stars (La Palma, Canary Islands, 2007),

Recognising that

1. the night sky has been and continues to be an inspiration of humankind, and that its contemplation represents an essential element in the development of scientific thought in all civilisations,

2. the dissemination of astronomy and associated scientific and cultural values should be considered as basic content to be included in educational activities,

3. the view of the night sky over most of the populated areas of the Earth is already compromised by light pollution, and is under further threat in this respect,

4. the intelligent use of unobtrusive artificial lighting that minimises sky glow involves a more efficient use of energy, thus meeting the wider commitments made on climate change, and for the protection of the environment,

5. tourism, among other players, can become a major instrument for a new alliance in defence of the quality of the nocturnal skyscape.

considering

1. the role of the IAU Division XII Commission 50 and its WG Controlling Light Pollution,

2. the role of the IYA2009 Cornerstone Project Dark Skies Awareness,

Resolves that

1. An unpolluted night sky that allows the enjoyment and contemplation of the firmament should be considered a fundamental socio-cultural and environmental right, and that the progressive degradation of the night sky should be regarded as a fundamental loss.

2. Control of obtrusive and sky glow-enhancing lighting should be a basic element of nature conservation policies since it has adverse impacts on humans and wildlife, habitats, ecosystems, and landscapes.

3. Responsible tourism, in its many forms, should be encouraged to take on board the night sky as a resource to protect and value in all destinations.

4. IAU members be encouraged to take all necessary measures to involve the parties related to skyscape protection in raising public awareness - be it at local, regional, national, or international level - about the contents and objectives of the International Conference in Defence of the Quality of the Night Sky and the Right to Observe Stars [http://www.starlight2007.net/], in particular the educational, scientific, cultural, health and recreational importance of preserving access to an unpolluted night sky for all humankind.

and further resolves that

5. Protection of the astronomical quality of areas suitable for scientific observation of the Universe should be taken into account when developing and evaluating national and international scientific and environmental policies, with due regard to local cultural and natural values.

3. Résolutions Approuvées

RESOLUTION B1

Plan Stratégique de l'UAI: Astronomie pour les pays en voie de développement

La XXVIIme Assemblée Générale de l'Union Astronomique Internationale,

reconnaissant:

1. que l'UAI a pour objectif d'encourager le développement de l'astronomie et de promouvoir une meilleure compréhension de l'univers,

2. que les activités entreprises durant l'Année Mondiale de l'Astronomie ont grandement contribué à l'avancement des connaissances astronomiques auprès des citoyens de toutes les nations, et à une meilleure appréciation de la valeur de l'astronomie pour la société,

3. que l'éducation scientifique et la recherche sont des ingrédients essentiels au développement scientifique et technologique,

recommande que l'Union Astronomique Internationale

1. donne plus d'importance aux programmes qui contribuent à l'éducation astronomique dans les pays en voie de développement,

2. approuve les objectifs décrits dans le Plan Stratégique: ' Astronomie pour les pays en voie de développement ' et les adopte comme objectifs de l'UAI pour la prochaine décennie

3. évalue les programmes entrepris durant l'Année Mondiale de l'Astronomie pour déterminer lesquels permettront de promouvoir l'astronomie de manière la plus efficace.

RESOLUTION A1

Mise en œuvre du Plan Stratégique de l'UAI

La XXVIIme Assemblée Générale de l'Union Astronomique Internationale,

reconnaissant:

1. que l'UAI a pour objectif d'encourager le développement de l'astronomie et de promouvoir une meilleure compréhension de l'univers,

2. que les activités entreprises durant l'Année Mondiale de l'Astronomie ont grandement contribué à l'avancement des connaissances astronomiques auprès des citoyens de toutes les nations, et à une meilleure appréciation de la valeur de l'astronomie pour la société,

3. que l'éducation scientifique et la recherche sont des ingrédients essentiels au développement scientifique et technologique,

4. que la résolution B1, adoptant le Plan Stratégique, a été ratifiée par la XXVII Assemblée Générale,

recommande que l'Union Astronomique Internationale

1. accorde une haute priorité au soutien du développement d'infrastructures astronomiques dans les pays émergents,

2. procède à la mise en œuvre de ce plan au travers de la création d'un Bureau Mondial de Développement et cherche à mobiliser des ressources additionnelles appropriées à cet effet.

RESOLUTION B2

sur les constantes astronomiques UAI 2009

La XXVIIme Assemblée Générale de l'Union Astronomique Internationale,

Considérant:

1. le besoin de disposer d'un ensemble cohérent de valeurs numériques de haute précision pour les constantes astronomiques,

2. que des valeurs améliorées des constantes astronomiques ont été déduites d'observations récentes et publiées dans des revues à comité de lecture, et

3. que des valeurs conventionnelles ont été adoptées pour un certain nombre de quantités astronomiques par les résolutions des Assemblées générales de l'UAI en 2000 et 2006,

Reconnaissant

1. le besoin permanent de disposer des meilleures valeurs numériques disponibles (CBEs) pour les constantes astronomiques, et

2. le besoin d'un service opérationnel destiné à la communauté astronomique, pour la maintenance de ces valeurs,

Recommande

1. que la liste des valeurs des constantes, préalablement publiées, qui ont été rassemblées dans le rapport du Groupe de travail 'Numerical Standards of Fundamental Astronomy' de la Division I de l'UAI (voir http://maia.usno.navy.mil/NSFA/CBE.html), soit adoptée comme Système UAI (2009) de constantes astronomiques,

2. que les meilleures valeurs numériques disponibles pour les constantes astronomiques fassent l'objet d'une maintenance permanente sous forme d'un document électronique,

3. qu'afin de garantir l'intégrité de ces valeurs, la Division I mette en place une procédure formelle pour l'adoption de nouvelles valeurs et l'archivage des anciennes versions, et

4. que l'UAI établisse au sein de la Division I un organe permanent chargé de la maintenance des meilleures valeurs numériques disponibles pour les constantes de l'astronomie fondamentale.

RESOLUTION B3

La seconde réalisation du repère céleste international de référence (ICRF)

La XXVIIme Assemblée Générale de l'Union Astronomique Internationale,

Notant:

1. que la résolution B2 de la XXIIIme Assemblée générale (1997) a décidé *'qu'à compter du 1er Janvier 1998, le systme céleste de référence sera le Systme Céleste International de Référence (ICRS)'*,

2. que la résolution B2 de la XXIIIme Assemblée générale (1997) a décidé que '*le repère fondamental correspondant sera le Repère Céleste International de Référence (ICRF) construit par le Groupe de travail de l'UAI sur les repères de référence*',

3. que la résolution B2 de la XXIIIème Assemblée générale (1997) a décidé que '*l'IERS devrait prendre des mesures appropriées, conjointement avec le Groupe de travail de l'UAI sur les repères de référence, pour la maintenance de l'ICRF et de ses liens avec les autres repères de référence d'autres longueurs d'onde*',

4. que la résolution B7 de la XXIIIème Assemblée générale (1997) a recommandé que '*des programmes d'observations astronomiques de haute précision soient organisés de sorte que la maintenance des systèmes de référence astronomiques puissent être assurée au meilleur niveau de précision pour les hémisphères Nord et Sud*',

5. que la résolution B1.1 de la XXIVème Assemblée générale (2000) a reconnu '*l'importance de la poursuite des observations opérationnelles à l'aide de l'interférométrie très longue base (VLBI) pour maintenir l'ICRF*',

Reconnaissant

1. que depuis la réalisation de l'ICRF, des observations VLBI continues de sources ICRF ont permis de plus que tripler le nombre d'observations de sources,

2. que depuis la réalisation de l'ICRF, des observations VLBI continues de sources extragalactiques ont permis d'augmenter de façon significative le nombre de sources dont les positions sont connues avec un haut degré d'exactitude,

3. que depuis la réalisation de l'ICRF, une instrumentation améliorée, des stratégies d'observations, et l'utilisation des modèles astrophysiques et géophysiques les plus récents, ont amélioré de façon significative à la fois la qualité des données et l'analyse de l'ensemble complet des données VLBI astrométriques et géodésiques concernées,

4. qu'un groupe de travail sur l'ICRF, établi par le Service International de la Rotation de la Terre et des Systèmes de référence (IERS) et le Service international VLBI pour la géodésie et l'astrométrie (IVS), conjointement avec le Groupe de travail "Second Realization of the International Celestial Reference Frame" de la Division 1 de l'UAI, a produit une seconde réalisation possible de

l'ICRF dans un repère de cordonnées qui a été aligné sur celui de l'ICRF dans la limite de tolérance des incertitudes de ce dernier (voir Note 1),

5. que cette seconde réalisation possible de l'ICRF, telle qu'elle a été présentée par le Groupe de travail 'Second Realization of the International Celestial Reference Frame' de la Division 1 de l'UAI, représente une amélioration importante en termes de sélection des sources, exactitude des coordonnées et nombre total de sources, et représente ainsi une amélioration significative de la réalisation du repère de référence fondamental de l'ICRS par rapport à l'ICRF adopté par la XXIIIème Assemblée générale (1997).

Décide

1. qu'à compter du 1er Janvier 2010, la réalisation astrométrique fondamentale du Système de Référence Céleste International (ICRS) sera la Seconde réalisation (ICRF2) du Repère Céleste International de Référence, telle qu'elle a été produite par le Groupe de travail IERS/IVS sur l'ICRF conjointement avec le Groupe de travail 'Second Realization of the International Celestial Reference Frame' de la Division 1 de l'UAI (voir note 1),

2. que les organisations responsables des programmes d'observations VLBI astrométriques et géodésiques (c.a.d. l'IERS et l'IVS) prennent des mesures appropriées afin de poursuivre et d'améliorer les programmes d'observations et d'analyse VLBI pour maintenir et améliorer l'ICRF2,

3. que l'IERS, avec les autres organisations concernées, continue les efforts d'amélioration et de densification des repères de référence de grande exactitude définis à d'autres longueurs d'onde et continuent à améliorer le raccordement de ces repères de référence avec l'ICRF2.

Note 1: The Second Realization of the International Celestial Reference Frame by Very Long Baseline Interferometry, Presented on behalf of the IERS / IVS Working Group, Alan Fey and David Gordon (eds.). (IERS Technical Note ; 35) Frankfurt am Main: Verlag des Bundesamts für Kartographie und Geodäsie, 2009. See http://www.iers.org/MainDisp.csl?pid=46-25772 or http://hpiers.obspm.fr/icrs-pc/.

RESOLUTION B4

Soutenir les femmes en Astronomie

La XXVIIème Assemblée Générale de l'Union Astronomique Internationale,

Rappelle:

1. l'objectif numèro 3 de développement pour le millénaire aux Nations Unies : promouvoir l'égalité des sexes et donner plus de responsabilités aux femmes

2. l'objectif numéro 7 de l'UAI/UNESCO pour l'Année Mondiale de l'Astronomie 2009 : améliorer l'équilibre des sexes dans la représentation des scientifiques à tous les niveaux et, promouvoir une meilleure participation des minorités sous-représentées dans les carrières de scientifiques et d'ingénieurs,

Reconnassant

1. que l'excellence individuelle en Science et en Astronomie est indépendante du sexe

2. que l'égalité des sexes est un principe fondamental des droits humains,

considèrant

1. le rôle du groupe de travail de l'UAI pour les Femmes en Astronomie

2. le rôle du projet de Pierre Angulaire de l'AMA 'Elle est une Astronome',

décide

1. que les membres de l'UAI devraient encourager et soutenir les femmes Astronomes dans leurs communautés,

2. que les membres de l'UAI et les représentants nationaux devraient encourager les organisations nationales à éliminer les barrières et à s'assurer qu'hommes et femmes aient des chances égales pour réussir leur carrière en Astronomie, à tous les niveaux et à toutes les étapes de celle-ci.

RESOLUTION B5

sur la défense du ciel nocturne et le droit à la lumière des étoiles

La XXVIIème Assemblée Générale de l'Union Astronomique Internationale,

Rappellant:

1. L'objectif no 8 de l'Année Mondiale de l'Astronomie 2009: faciliter la préservation et la protection du patrimoine culturel et naturel mondial que constitue un ciel sombre dans des endroits tels que les oasis urbaines, les parcs nationaux et les sites astronomiques,

2. la déclaration approuvée lors du Colloque international sur la défense de la qualité du ciel nocturne et du droit observer les étoiles (La Palma, Iles canaries, 2007),

Reconnaissant

1. que le ciel nocturne a été et continue d'être une source d'inspiration pour l'humanité et que sa contemplation représente un élément essentiel dans le développement de la pensée scientifique dans toutes les civilisations,
2. que la diffusion de l'astronomie et des valeurs scientifiques et culturelles associées doit être considérée comme un contenu de base pour les activités éducatives,
3. que la vue du ciel nocturne sur la majeure partie des zones peuplées de la terre est déja compromise par la pollution lumineuse qui menace de s'accrotre dans le futur
4. que l'utilisation intelligente d'un éclairage artificiel non agressif qui minimise la diffusion atmosphérique implique également une utilisation plus efficace de l'énergie, ce qui va dans le sens des engagements plus généraux pris par la communauté internationale sur le changement climatique et pour la protection de l'environnement,
5. que le tourisme, parmi les autres acteurs peut devenir un levier majeur pour une nouvelle alliance en faveur de la défense de la qualité du ciel nocturne.

Considérant

1. le rôle de la commission 50 de la division X de l'UAI et de son GT Contrôle de la pollution lumineuse,

2. le rôle du projet de pierre angulaire Sensibilisation aux Ciels Sombres de l'AMA2009,

Decide

1. Un ciel nocturne non pollué autorisant la jouissance et la contemplation du firmament doit être considéré comme un droit fondamental socio-culturel et environnemental, et la dégradation progressive du ciel nocturne doit être considérée comme une perte fondamentale.

2. Le contrôle d'éclairages agressifs ou qui augmentent la diffusion atmosphérique doit être un élément fondamental des politiques de conservation de nature, car il a des incidences négatives sur les êtres humains et la faune, les habitats, les écosystèmes et les paysages.

3. Le tourisme responsable, sous toutes ses formes, doit être encouragé à prendre en considération le ciel nocturne comme une ressource à protéger et à apprécier, quelles que soient les destinations.

4. Les membres de l'UAI sont encouragés à prendre toutes les mesures nécessaires pour impliquer les parties liées à la protection du ciel nocturne en vue de la sensibilisation du public -que ce soit au niveau local, régional, national ou international- sur le contenu et les objectifs du Colloque international sur la défense de la qualité du ciel nocturne et du droit à observer les étoiles [http://www.starlight2007.net/],

et plus particulièrement sur l'importance éducative, scientifique, culturelle, de santé et de loisir, de préserver l'accès à un ciel nocturne non pollué pour tout l'humanité.

ainsi que

5. La protection de la qualité astronomique des zones convenables pour l'observation scientifique de l'univers doit être prise en compte au cours de la mise en uvre des politiques nationales et internationales scientifiques et environnementales, et lors de leur évaluation, tout en tenant compte de valeurs culturelles et naturelles locales.

Transactions IAU, Volume XXVIIB
Proc. XXVII IAU General Assembly, August 2009
Ian F. Corbett, ed.

doi:10.1017/S1743921310004849

CHAPTER V

REPORT OF THE EXECUTIVE COMMITTEE

1. Executive Committee 2006 - 2009

1.1. *Composition of the Executive Committee*

During the triennium 2006 - 2009, the Executive Committee was composed as follows:

Catherine J.	Cesarsky	President
Robert	Williams	President-Elect
Karel A.	van der Hucht	General Secretary
Ian F.	Corbett	Assistant General Secretary
Beatriz	Barbuy	Vice-President
Chen	Fang	Vice-President
Martha P.	Haynes	Vice-President
George K.	Miley	Vice-President
Giancarlo	Setti	Vice-President
Brian	Warner	Vice-President
Ronald D.	Ekers	Adviser
Oddbjørn	Engvold	Adviser

1.2. *Executive Committee meetings*

The IAU Executive Committee had the following meetings during the triennium:
- EC 82, 25 August 2006, at the IAU XXVI General Assembly, Prague, Czech Republic;
- EC 83, 15 - 17 May 2007, at Cape Observatory, Cape Town, South Africa;
- EC 84, 28 - 30 May 2008, at the Norwegian Academy of Science and Letters, Oslo, Norway;
- EC 85, 7 - 8 April 2009, at the Institute d'Astrophysique de Paris, France;
- EC 86, 2, 6, and 12 August 2009, during the IAU XXVII General Assembly in Rio de Janeiro, Brazil.

The business conducted by the Executive Committee is recorded in the minutes of the EC meetings. Brief reports of the minutes of the EC meetings have been published in the IAU *Information Bulletin*: EC 82 in IB 99, p. 22; EC 83 in IB 100, p. 24; EC 84 in IB 102, p. 43; and EC 85 in IB 104, p. 18.

1.3. *Officers' Meetings*

Between the meetings of the Executive Committee, the IAU Officers (President, President-Elect, General Secretary and Assistant General Secretary) convened at the IAU Secretariat in Paris, France, for their annual Officers' Meetings: OM-2007 on 30 January - 1 February 2007, OM-2008 on 29 - 31 January 2008, OM-2009-1 on 12 January 2009, and OM-2009-2 on 6 April 2009.

The business conducted by the Officers is recorded in the minutes of the Officers' meetings. Brief reports of the minutes of the Officers' Meetings have been published in the IAU *Information Bulletin*: OM-2007 in IB 100, p. 23; OM-2008 in IB 102, p. 43; OM-2009-1 in IB 103, p. 80; and OM-2009-2 in IB 104, p. 18.

1.4. *Executive Committee Working Groups*

1.4.1. *EC Working Group on the International Year of Astronomy 2009*

At its General Assembly in Sydney in July 2003, the International Astronomical Union voted unanimously in favor of a resolution asking that the year 2009 be declared the *Year of Astronomy* by the United Nations, in recognition of the significance of Galileo's introduction of the astronomical telescope in 1609. The proclamation was subsequently prepared by the EC Working Group 'International Year of Astronomy 2009' (EC-WG-IYA2009), and forwarded to the Executive Board of UNESCO.

The UNESCO General Conference, in October 2005, recommended that the United Nations General Assembly would adopt a resolution declaring 2009 as the *International Year of Astronomy*. In its recommendations to the UN General Assembly, the General Conference of UNESCO recognizes *"that the study of the Universe has led to numerous scientific discoveries that have great influence not only on humankind's understanding of the Universe but also on the technological, social and economic development of society"* and *"that astronomy proves to have great implications in the study of science, philosophy, religion and culture"*.

The United Nations General Assembly adopted a Resolution on 19 December 2007 declaring the year 2009 the *International Year of Astronomy* (UN-A/RES/62/200). The full UN Resolution is presented in Appendix I.

The *International Year of Astronomy* had its UNESCO-IAU Opening Ceremony on 15 and 16 January 2009 at UNESCO Headquarters in Paris, France.

The EC-WG-IYA 2009, chaired by Catherine J. Cesarsky, started its work during the IAU XXVI General Assembly in Prague in August 2006, and created a dedicated IYA 2009 secretariat hosted by ESO (Garching, Germany) and staffed by Lars Lindberg Christensen (WG secretary), Pedro Russo, Mariana Barrosa, and Lars Holm Nielsen.

As of July 2009, 141 national contacts and numerous organizational contacts have signed up to participate in national and international IYA2009 activities.

Progress reports on the activities of the EC-WG-IYA-2009 have been published in the IAU *Information Bulletin*: IB 100, p. 72, IB 101, p. 51, IB 102, p. 57, IB 103, p. 88, and IB 104, p. 20.

IAU Press Release IAU0913, dated 1 July 2009 summarizes the mid-year accomplishments as follows:

International Year of Astronomy 2009 raises millions of eyes to the skies.

1 July 2009, Paris: As the *International Year of Astronomy 2009* (IYA2009) reaches its six-month milestone, already over a million people have looked at the sky through a telescope for the first time, and even more have newly engaged in astronomy. This is only the tip of iceberg, though, as countless ongoing projects and planned initiatives indicate that IYA2009 is well on the way toward achieving many of its goals.

UNESCO and the International Astronomical Union (IAU) launched 2009 as the *International Year of Astronomy* (IYA2009) under the theme *The Universe, Yours to Discover*. IYA2009 is a global celebration of astronomy and its contributions to society and culture, with events at national, regional and global levels throughout the whole of 2009. Now halfway through 2009, a lot has been achieved and even more can be expected in the future.

Headlining the IYA2009 effort is the *Galileoscope* project. With the aim of providing low-cost telescopes that offer views far better than those obtained by Galileo Galilei some 400 years ago, the venture has picked up significant pace since IYA2009 began. By the end of July, the first 60 000 Galileoscopes will have shipped, and additional of 100 000 are currently in production. More than 4000 Galileoscopes have been generously donated by IYA2009 and individuals to organisations and schools in developing countries. This gesture aptly demonstrates astronomy enthusiasts' commitment to the IYA2009 goal of making the skies accessible to all.

But perhaps the most impressive figures for IYA2009 come from the *national* activities that have brought together hundreds of thousands of people in many countries for astronomy-themed events. For example, more than 400 000 people gathered for the Sunrise Event on New Year's Day in Busan City, South Korea. In Brazil, the 2009 Brazilian Olympiad of Astronomy and Astronautics saw more than 750 000 students participate from 32 500 schools. In Paraguay, the launch of IYA2009 featured a concert with more than 1600 musicians and over 15 000 attendees. In Norway, every single student in the country from grade levels 5 - 11 will soon receive a free astronomy kit, including a Galileoscope and an educational guide. For the first time in postal service history, and in just six months, more than 70 postal agencies around the world have issued over 140 new stamps inspired by astronomy.

The many other signature events of IYA2009 have further enabled astronomy enthusiasts to share their excitement. In April, the highly anticipated *100 Hours of Astronomy* extravaganza kicked off. This planet-wide celebration involved over 100 countries holding thousands of events, with more than two million people taking part in observing events. Widely regarded as an outstanding success, 100 Hours of Astronomy brought people from all seven continents together with the help of a live 24-hour webcast called "Around the World in 80 Telescopes". This groundbreaking broadcast was watched by over 150 000 individuals.

Astronomical images have the power to inspire people to think about our place in the Universe, a fact used by *From Earth To The Universe* (FETTU), a project to run exhibitions in unusual locations around the world, like train stations and shopping malls. So far, over 60 countries around the world have signed up to host FETTU exhibitions in more than 200 separate locations.

Dark Skies Awareness is an ongoing initiative to combat light pollution and raise awareness of the importance of deep darkness for appreciating and studying the cosmos. As part of this effort, the GLOBE at night project encourages members of the public to become citizen scientists by performing star-counts and reporting their findings. The 2009 campaign, held this March, garnered 15 700 measurements, nearly 80% more than the previous record in 2007.

The *Cosmic Diary* cornerstone project continues to flourish. Professional scientists are blogging about their lives and work, giving the public insight into what it is really like to be a researcher. Since its launch on 1 January 2009, the Cosmic Diary has recruited 60 professional astronomers from 28 countries. There have been over 1000 blog posts, attracting more than 97 000 visitors.

Further impressive web statistics are provided by the *Portal To The Universe* (PTTU), a global, one-stop clearing house for on-line astronomy content. PTTU serves as an index and aggregator for astronomy content for laypeople, press, educators, decision-makers, scientists and more. During its first two months of operation, PTTU received almost 100 000 visitors.

Universe Awareness (UNAWE) is an international outreach initiative that uses the beauty and scale of the Universe to inspire very young children in underprivileged environments. To date, programmes have been organised in 30 countries, producing many hundreds of educational resources. For example, in Tunisia more than 40 000 children have been participating in UNAWE activities since January 2009.

The cornerstone project *Developing Astronomy Globally* (DAG) has surveyed the status of astronomy research and education in more than 45 countries. To support projects and activities in developing regions, IYA2009 provided seed funding for development initiatives coordinated via DAG. Proposals from the following countries have been selected: Nepal, Uganda, Mongolia, Nicaragua, Nigeria, Kenya, Ethiopia, Gabon, Rwanda, Uruguay, the Former Yugoslav Republic of Macedonia and Tajikistan. More global educational activities come courtesy of the Galileo Teacher Training Program, which is running workshops in 25 countries.

Astronomical Heritage is a strong theme running throughout IYA2009. The IAU's Working Group on *Astronomy and World Heritage*, along with the International Council on Monuments and Sites, has begun the first Thematic Study in any field of science heritage. The results of the study will be reported to UNESCO's World Heritage Committee, bringing the protection and preservation of important astronomy sites to the world's attention. The Thematic Study's findings will also form the basis for developing specific guidelines for UNESCO member states on the inscription of astronomical properties.

It is fitting that cutting-edge astronomical research is reaching new heights in 2009. In May ESA launched *Herschel* and *Planck* – two of the most ambitious space missions ever attempted, ready to break new ground in astronomy. The IYA2009 logo and motto was proudly displayed on the Ariane 5 rocket that sent them to space. Also in May astronauts performed repairs and equipped the NASA/ESA *Hubble Space Telescope* with the latest in instrument technology. To honour IYA2009, astronaut Mike Massimino took on-board with him a replica of Galileo's telescope as well as an IYA2009 flag.

Although IYA2009's achievements to date are certainly impressive, it has only reached its halfway point and many new initiatives are in the works. For example, 23 - 24 October will see the launch of *Galilean Nights*, the follow-up to the highly successful *100 Hours of Astronomy* presentation.

For further details see the IAU IYA-2009 web site `<www.astronomy2009.org/>`.

1.4.2. *Category name for transneptunian dwarf planets*

In the aftermath of the planet definition resolutions debate at the IAU XXVI General Assembly in Prague, 2006, the EC asked IAU Division III to fulfill the task asked for in the Footnote 2 to Prague Resolution 5: "An IAU process will be established to assign borderline objects to the dwarf planet or to another category".

At its EC84 meeting, 28-30 May 2008 in Oslo, Norway, the EC accepted the recommendation of Division III to use the name *plutoid* for transneptunian dwarf planets similar to Pluto.

1.4.3. *EC Working Group on Women in Astronomy*

The Working Group on Women in Astronomy was formed during the IAU XXV General Assembly in Sydney, Australia, July 2003. The aims of the Working Group are to evaluate the status of women in astronomy through the collection of statistics over all countries where astronomy research is carried out; and to establish strategies and actions that can help women to attain true equality as research astronomers, which will add enormous value to all of astronomy.

1.4.4. *EC Working Group on Publishing*

In Spring 2009, the GS received notice from the chair that the EC-WG on Publishing had been disbanded.

1.4.5. *EC Working Group on Future Large Scale Facilities*

This WG is a small activity dedicated to ensuring that proposed Future Large Scale Facilities are widely reported to the IAU community at appropriate meetings. Multi-wavelength meetings are proposed to ensure project complementarity, and that there is the widest feasible community involvement in planning. Additionally, national and international facility plans are reported at the relevant meetings, again in the interests of improved communication.

The WG role is largely in maximising the breadth of information transfer. Facility plans are undertaken by dedicated national and international groups, such as NASA, ESA, AstroNet, and the US NAS/NSF Decadal Review. The relevant projects are presumed to manage their own information to their own communities.

1.4.6. *EC-WG on General Assemblies*

The Working Group stands by for questions on the organization of General Assemblies.

1.4.7. *EC Advisory Committee on Hazards of Near-Earth Objects*

During the triennium the Advisory Committee did not identify cases of Near Earth Objects justifying to alert the EC on potentially hazardous NEOs.

2. Membership of the Union

2.1. *National Membership*

During the IAU XXVI General Assembly in Prague, 2006, upon recommendation of the Executive Committee, the General Assembly welcomed Lebanon, Mongolia and Thailand as new National Members of the IAU. This brought the total number of National Members to 63.

2.2. *Individual Membership*

The number of new Individual Members welcomed by the IAU XXVI General Assembly in Prague, 2006, upon recommendation of the Nominating Committee, was 925.

The IAU Executive is saddened to report the death of the following 159 members of the Union, which have been reported to the IAU Secretariat since the IAU XXVI General Assembly in 2006:

Abhyankar, K.D.	Alpher, Ralph A.	Andrillat, Henri
Anteres K., Oliveira G.	Bateson, Frank M.	Behr, Alfred
Beurle, Kevin	Bisiacchi Giraldi, Gianfranco	Blunck, Jürgen
Boldt, Elihu	Bonifazi, Angelo	Boydag-Y., Fatma S.
Bracewell, Ronald N.	Brahde, Rolf	Brault, James W.
Brown, Ronald D.	Bruck, Mary T.	Bunclark, Peter S.
Cacciani, Alessandro	Cheffler, Helmut	Chen, Kwan-Yu
Chen, Meidong	Christiansen, Wilbur N.	Clarke, Arthur C.
Code, Arthur D.	Codina Ladanberry, Sayd J.	Cohen, Raymond J.
Cudaback, David D.	Damle, S.V.	Davis, Sumner P.
de Araújo, Francisco X.	Debehogne, Henri	Delbouille, Luc
Douglas, James N.	Edmondson, Frank K.	Ehlers, Jürgen
Fairall, Anthony P.	Fehrenbach, Charles	Fenton, Keith B.
Ferrari DO., Konradin	Firor, John W.	Fitton, Brian
Forti, Giuseppe	Friend, David B.	Giclas, Henri L.

Greisen, Kenneth I.	Haddock, Fred T.	Hagfors, Tor
Harrison, Edward R.	Hassan, S.M.	Hauge, Øivind
Hazen, Martha L.	Heintz, Wulff-Dieter	Helin, Eleanor F.
Henrard, Jacques	Hoffleit, E. Dorrit	House, Franklin C.
Isobe, Syuzo	Jarrett, Alan H.	Jennison, Roger C.
Johansson, Lars E.B.	Johansson, Sveneric	Jovanovic, Bozidar
Kadla, Zdenka I.	Kazš, Ilya	Kiang, Tao
Klarmann, Joseph	Kopecky, Miloslav	Kosugi, Takeo
Kourganoff, Vladimir	Krivsky, Ladislav	Kuznetsov, Oleg A.
Labs, Dietrich	Lanning, Howard H.	Lantos, Pierre
Little, Leslie T.	Low, Frank J.	Lyuty, Victor M.
Macrae, Donald A.	Maffei, Paolo	Mahra, H.S.
Makarov, Valentin I.	Marsh, Julian C.D.	Massevich, Alla G.
Mendoza-Briceno, César	Mulholland, John D.	Naumov, Vitalij A.
Ne'eman, Yuval	Niemela, Virpi S.	North, John D.
Novick, Robert	Novoselov, Viktor S.	Odgers, Graham J.
Oliveira, Grijo A.K.	Osterbrock, Donald E.	Ostro, Steven J.
Paczynski, Bohdan	Pagel, Bernard E.J.	Pal, Arpad
Pallavicini, Roberto	Pasinetti, Laura E.	Perinotto, Mario
Petropoulos, Basil Ch.	Plavec, Mirek J.	Pokorny, Zdenek
Popelar, Josef	Popov, Viktor S.	Potter, Heino I.
Poulakos, Constantine	Praderie, Françoise	Predeanu, Irina
Prodan, Yurij I.	Pustylnik, Izold	Raghavan, Nirupama
Reeves, Edmond M.	Refsdal, Sjur	Rivenq, Claude
Rudkjøbing, Mogens	Rumsey, Norman J.	Sadzakov, Sofia
Salpeter, Edwin E.	Schimmins, Albert J.	Schueker, Peter
Schwarz, Hugo E.	Sears, Richard L.	Seaton, Michael J.
Seitter, Waltraut C.	Shapley, Alan H.	Shitov, Yurij P.
Shmiedler, Felix	Sholomitsky, Gennady B.	Shulman, Leonid M.
Simovljevic, Jovan	Slysh, Vyacheslav I.	Smirnov, Mikhail A.
Smith, Harding E.	Sochilina, Alla S.	Solomon, Philip M.
Tifrea, Emilia	Toner, Clifford G.	Traving, Gerhard
Treder, Hans Juergen	Turner, Barry E.	Valiron, Pierre
Van Allen, James A.	Van Flandern, Tom	Voglis, Nikos
von Weizsäcker, Carl Friedrich	Walraven, Theo	Wendker, Heinrich J.
Westerlund, Bengt E.	Wetherhill, George W.	Wheeler, John A.
Whitrow, Gerald J.	Wild, John P.	Wilson, Andrew S.
Wilson, James R.	Xi, Zezong	Yuan, Chi

As of 24 July 2009 the total number of IAU Individual Members was 9 260. The IAU Secretariat received a number of 889 nominations for new Individual Membership, for consideration by the Nominating Committee at the IAU XXVII General Assembly in Rio de Janeiro, August 2009. The number of Consultants listed in the data base is 45.

3. Divisions, Commissions, Working/Program Groups, Services

3.1. *Working Groups of Divisions and Commissions*

Discontinuation of a Commission and of Working Groups, going into effect at the IAU XXVI General Assembly in Prague, 2006, has been published in IAU *Information*

Bulletin No. 99, section 4.2. Creation of new Commissions and Working Groups, going into effect at the IAU XXVI General Assembly in Prague, 2006, has been published in IAU *Information Bulletin* No. 99, section 4.3.

Reviews of all IAU scientific bodies will be presented by the IAU Division Presidents in the Executive Committee meeting EC86 at the IAU General Assembly in Rio de Janeiro, 2009. The Divison Presidents will present to the IAU Executive Committee their recommendations about necessary changes in the current division and commission structure, in order to meet the demands and challenges for the next triennium and beyond.

4. IAU scientific meetings

4.1. *IAU Symposia*

The IAU supported nine IAU Symposia in each of the years 2006, 2007, 2008, 2009. In General Assembly years, six of the nine Symposia are held at the General Assembly. The venues and dates of these Symposia are given in the list of IAU Proceedings presented below.

4.2. *Regional IAU Meetings*

In the triennium, the following Regional IAU Meetings (RIMs) have been supported:
- the 12th Latin-American Regional IAU Meeting, 22 - 26 October 2007, Isla de Margarita, Venezuela;
- the 10th Asian-Pacific Regional IAU Meeting, 3 - 6 August 2008, Kunming, China P.R.

As a new intitiative, a Regional IAU Meeting in the Middle-East was supported, reaching also out to Africa:
- the 1st Middle-East Africa Regional IAU Meeting, 5 - 10 April 2008, Cairo, Egypt.

5. IAU publications

During the triennium the following volumes have been published.

5.1. *IAU Highlights of Astronomy*

Highlights of Astronomy, Volume 13
AS PRESENTED AT THE XXVth GENERAL ASSEMBLY OF THE IAU
Sydney, Australia, 13-26 July 2003
Ed. Oddbjørn Engvold
(San Francisco: ASP) ISBN 1-58381-189-3, August 2006

Highlights of Astronomy, Volume 14
AS PRESENTED AT THE XXVIth GENERAL ASSEMBLY OF THE IAU
Prague, Czech Republic, 14-25 August 2006
Ed. Karel A. van der Hucht
(Cambridge: CUP) ISBN 978-0-521-89683-2, December 2007
URL: <journals.cambridge.org/action/displayIssue?jid=IAU&volumeId=2&issueId=14>

5.2. *IAU Transactions*

Proceedings of the XXVth General Assembly of the IAU
Transactions of the IAU, Volume XXVB
Sydney, Australia, 13 - 26 July 2003
Ed. Oddbjørn Engvold
(San Francisco: ASP) ISBN 978-1-58381-647-9, 2007

Reports on Astronomy 2003 - 2006
Transactions of the IAU, Volume XXVIA
Ed. Oddbjørn Engvold
(Cambridge: CUP) ISBN 0-521-85604-3, 2007
URL: `<journals.cambridge.org/action/displayIssue?jid=IAU&volumeId=1&issueId=T26A>`

Proceedings of the XXVIth General Assembly of the IAU
Transactions of the IAU, Volume XXVIB
Prague, Czech Republic, 14 - 25 August 2006
Ed.: Karel A. van der Hucht
(Cambridge: CUP) ISBN 978-0-521-85606-5, November 2008
URL: `<journals.cambridge.org/action/displayIssue?jid=IAU&volumeId=3&issueId=T26B>`

Reports on Astronomy 2006 - 2009
Transactions of the IAU, Volume XXVIIA
Ed. Karel A. van der Hucht
(Cambridge: CUP) ISBN 978-0-521-85605-8, January 2009
URL: `<journals.cambridge.org/action/displayIssue?jid=IAU&volumeId=4&issueId=T27A&iid=3578468>`

5.3. *IAU Symposium Proceedings*

IAU S233 Solar Activity and its Magnetic Origin
31 March - 4 April 2006, Cairo, Egypt
Eds. Volker Bothmer & Ahmed A. Hady
(Cambridge: CUP) ISBN 978-0-521-86342-1, November 2006

IAU S234 Planetary Nebulae in our Galaxy and Beyond
3 - 7 April 2006, Waikoloa Beach, HI, USA
Eds. Michael J. Barlow & Roberto H. Mendez
(Cambridge: CUP) ISBN 978-0-521-86343-8, December 2006

IAU S235 Galaxy Evolution across the Hubble Time
14 - 17 August 2006, Prague, Czech Republic
Eds. Françoise Combes & Jan Palouš
(Cambridge: CUP) ISBN 978-0-521-86344-5, March 2007

IAU S236 Near Earth Objects, our Celestial Neighbors: Opportunity and Risk
14 - 18 August 2006, Prague, Czech Republic
Eds. Andrea Milani, Giovanni B. Valsecchi & David Vokrouhlický
(Cambridge: CUP) ISBN 978-0-521-86345-2, May 2007

IAU S237 Triggered Star Formation in a Turbulent Interstellar Medium
14-18 August 2006, Prague, Czech Republic
Eds. Bruce G. Elmegreen & Jan Palouš
(Cambridge: CUP) ISBN 978-0-521-86346-9, April 2007

IAU S238 Black Holes: from Stars to Galaxies – across the Range of Masses 21 - 25 August 2006, Prague, Czech Republic
Eds. Vladimír Karas & Giorgio Matt
(Cambridge: CUP) ISBN 978-0-521-86347-6, May 2007

IAU S239 Convection in Astrophysics
21 - 25 August 2006, Prague, Czech Republic
Eds. Friedrich Kupka, Ian W. Roxburgh & Kwing Lam Chan
(Cambridge: CUP) ISBN 978-0-521-86349-0, May 2007

IAU S240 Binary Stars as Critical Tools and Tests in Contemporary Astrophysics
22 - 25 August 2006, Prague, Czech Republic
Eds. William I. Hartkopf, Edward F. Guinan & Petr Harmanec
(Cambridge: CUP) ISBN 978-0-521-86348-3, July 2007

IAU S241 Stellar Populations as Building Blocks of Galaxies
10 - 16 December 2006, La Palma, Tenerife, Spain
Eds. Alexandre Vazdekis & Reinier F. Peletier
(Cambridge: CUP) ISBN 978-0-521-86350-6, August 2007

IAU S242 Astrophysical Masers and their Environments
12 - 16 March 2007, Alice Springs, Australia
Eds. Jessica M. Chapman & Willem A. Baan
(Cambridge: CUP) ISBN 978-0-521-87464-9, January 2008

IAU S243 Star-Disk Interaction in Young Stars
21 - 25 May 2007, Grenoble, France
Eds. Jérôme Bouvier & Immo Appenzeller
(Cambridge: CUP) ISBN 978-0-521-87465-6, November 2007

IAU S244 Dark Galaxies and Lost Baryons
25- - 29 June 2007, Cardiff, Wales, UK
Eds. Jonathan I. Davies & Michael J. Disney
(Cambridge: CUP) ISBN 978-0-521-87466-3, February 2008

IAU S245 Formation and Evolution of Galaxy Bulges
16 - 20 July 2007, Oxford, UK
Eds. Martin G. Bureau, Evangelia Athanassoula & Beatriz Barbuy
(Cambridge: CUP) ISBN 978-0-521-87467-0, July 2008

IAU S246 Dynamical Evolution of Dense Stellar Systems
5 - 9 September 2007, Capri, Italy
Eds. Enrico Vesperini, Mirek Giersz & Alison I. Sills
(Cambridge: CUP) ISBN 978-0-521-87468-7, May 2008

IAU S247 Waves and Oscillations in the Solar Atmosphere: Heating and Magneto-Seismology
17 - 22 September 2007, Porlamar, Isla de Margarita, Venezuela
Eds. Robert Erdélyi & César A. Mendoza-Briceño
(Cambridge: CUP) ISBN 978-0-521-87469-4, May 2008

IAU S248 A Giant Step: from Milli- to Micro-arcsecond Astrometry
15 - 19 October 2007, Shanghai, China P.R.
Eds. Wenjing Jin, Imants Platais & Michael A.C. Perryman
(Cambridge: CUP) ISBN 978-0-521-87470-0, July 2008

IAU S249 Exoplanets: Detection, Formation and Dynamics
22 - 26 October 2007, Suzhou, China P.R.
Eds. Yi-Sui Sun, Sylvio Ferraz Mello & Ji-Lin Zhou
(Cambridge: CUP) ISBN 978-0-521-87471-7, May 2008

IAU S250 Massive Stars as Cosmic Engines
10 - 14 December 2007, Kauai, Hawaii, USA
Eds. Fabio Bresolin, Paul A. Crowther & Joachim Puls
(Cambridge: CUP) ISBN 978-0-521-87472-4, June 2008

IAU S251 Organic Matter in Space
18 - 22 February 2008, Hong Kong, China P.R.
Eds. Sun Kwok & Scott A. Sandford
(Cambridge: CUP) ISBN 978-0-521-88982-7, October 2008

IAU S252 The Art of Modelling Stars in the 21st Century
6 - 11 April 2008, Sanya, Hainan Island, China P.R.
Eds. Licai Deng & Kwing-Lam Chan
(Cambridge: CUP) ISBN 978-0-521-88983-4, October 2008

IAU S253 Transiting Planets
19 - 23 May 2008, Cambridge, MA, USA
Eds. Frédéric Pont, Dimitar D. Sasselov & Matthews J. Holman
(Cambridge: CUP) ISBN 978-0-521-88984-1, February 2009

IAU S254 The Galaxy Disk in Cosmological Context
9 - 13 June 2008, Copenhagen, Denmark
Eds. Johannes Andersen, Joss Bland-Hawthorn & Birgitta Nordström
(Cambridge: CUP) ISBN 978-0-521-88985-8, 2009

IAU S255 Low-Metallicity Star Formation: from the First Stars to Dwarf Galaxies
16 - 20 June 2008, Rapallo, Genova, Italy
Eds. Leslie K. Hunt, Suzanne C. Madden & Raffaella Schneider
(Cambridge: CUP) ISBN 978-0-521-88986-5, December 2008

IAU S256 The Magellanic System: Stars, Gas, and Galaxies
28 July - 1 August 2008, Keele University, Stafffordshire, UK

Eds. Jacco Th. van Loon & Joana M. Oliveira
(Cambridge: CUP) ISBN 978-0-521-88987-2, March 2009

IAU S257 Universal Heliophysical Processes
15 - 19 September 2008, Ioannina, Greece
Eds. Natchimuthukonar Gopalswamy & David F. Webb
(Cambridge: CUP) ISBN: 987-0-521-88988-9, March 2009

IAU S258 The Ages of Stars
13 - 17 October 2008, Baltimore, MD, USA
Eds. Eric E. Mamajek, David R. Soderblom & Rosemary F.G. Wyse
(Cambridge: CUP) ISBN: 978-0-521-88989-6, June 2009

IAU S259 Cosmic Magnetic Fields: from Planets, to Stars and Galaxies
3 - 7 November 2008, Puerto Santiago, Tenerife, Spain
Eds. Klaus G. Strassmeier, Alexander G. Kosovichev & John E. Beckman
(Cambridge: CUP) ISBN: 978-0-521-88990-2, April 2009

For a complete list of IAU Symposium Proceedings, see
<www.iau.org/science/publications/iau/symposium/>.
For e-versions, see: <journals.cambridge.org/action/displayJournal?jid=IAU>.
For printed volumes, see: <www.cambridge.org/uk/series/sSeries.asp?code=IAUP>.

5.4. *Proceedings of Regional IAU Meetings*

Proc. 11th Latin-American Regional IAU Meeting (LARIM 2005)
12 16 December 2005, Pucón, Chile
Eds. Leopoldo Infante, Mónica Rubio & Silvia Torres-Peimbert
Revista Mexicana de Astronomía y Astrofísica - Serie de Conferencias, Vol. 26, August 2006

Proc. 12th Latin-American Regional IAU Meeting (LARIM 2007)
22 - 26 October 2007, Isla Margarita, Venezuela
Eds. Gladis Magris, Gustavo A. Bruzual & Leticia Carigi
Revista Mexicana de Astronomía y Astrofísica - Serie de Conferencias, Vol. 35, June 2009

Proc. 1st Middle East and Africa Regional IAU Meeting (MEARIM 2008)
5 - 10 April 2008, Cairo, Egypt
Eds. Athem W. Alsabti, Ahmed Abdel Hady & Volker Bothmer (Cairo: University of Cairo Press), in preparation, 2009

Proc. 10th Asian-Pacific Regional IAU Meeting (APRIM 2008)
3 - 6 August 2008, Kunming, China P.R.
Eds. Shuang-Nan Zhang, Yan Li, Qingjuan Yu & Guo-Qing Liu
(Beijing: China Science & Technology Press), June 2009

For a complete list of Proceedings Regional IAU Meetings, see:
<www.iau.org/science/publications/iau/regional_meeting>.

5.5. *IAU Information Bulletin*

During the triennium, the IAU Secretariat published six IAU *Information Bulletins*: IB No. 99 (January 2007), IB No. 100 (July 2007), IB No. 101 (January 2008), IB No. 102 (July 2008), IB No. 103 (January 2009), and IB No. 104 (June 2009). The on-line version of the IAU *Information Bulletin* is available at
<www.iau.org/science/publications/iau/information_bulletin/>.

5.6. *IAU e-Newsletter*

During the triennium, the IAU Secretariat informed the IAU membership 13 times on timely matters by e-mail in its *e-Newsletters*. The on-line version of the IAU *e-Newsletter* is available at <www.iau.org/science/publications/iau/newsletters/>.

5.7. *IAU Press Releases*

During the triennium, the IAU Press Officer, Lars Lindberg Christensen, issued 6 Press Releases in 2006, 3 in 2007, 10 in 2008, and 14 in 2009. IAU Press Releases are available on-line at <www.iau.org/public_press/news/>.

5.8. *Other IAU related publications*

Astronomy for the Developing World
Proc. IAU XXVI General Assembly Special Session 5
21 - 22 August 2006, Prague, Czech Republic
Eds. John Hearnshaw & Peter Martinez 2007
(Cambridge: CUP), ISBN 978-0-521-87657-5, August 2007

5.9. *IAU Publisher*

The contract between the IAU and Cambridge University Press for publishing the IAU *Symposium Proceedings* series, the IAU *Highlights of Astronomy* series, and the IAU *Transactions-A and -B* series was renewed by January 2009 and runs till December 2013.

5.10. *IAU Editorial Board*

The IAU Editorial Board ensures good communication among the Editors of the individual Proceedings. The Editorial Board is chaired by the IAU Assistant General Secretary. In the past triennium the average production for publishing of individual IAU Proceedings was eight months (five months for the editors and three months for the publisher).

5.11. *History of the IAU*

At the occasion of the 100th IAU *Information Bulletin*, reminiscences of twelve past IAU General Secretaries have been published in IB No. 100. At the occasion of the 90th anniversary of the IAU, reminiscences of six past IAU Presidents have been published in IB No. 104. Histories of the Minor Planet Center, the Committee on Small Bodies Nomenclature, the Working Group on Planetary System Nomenclature, and the Central Bureau for Astronomical Telegrams have been publishes in IB No. 104.

6. Educational activities

Educational programmes under IAU Commission 46 on *Astronomy Education and Development* represent a high priority activity and responsibility of the IAU. The Executive

Committee recognizes the many successful and promising results obtained by the Commission 46 Program Groups.

On the initiative of the IAU Executive Committee, in particular of Vice-President George K. Miley a discussion meeting of parties associated with IAU educational activities took place in Paris, 27-29 January 2008. Subsequently a comprehensive document was prepared: *Astronomy for the Developing World. Strategic Plan 2010 - 2020.* This document was approved by the EC at its meeting EC85 and EC86, and will be submitted for implementation to the IAU XXVII General Assembly in Rio de Janeiro, Brazil, August 2009.

6.1. *Commission 46 PG International Schools for Young Astronomers*

The Commission 46 Program Group on *International Schools for Young Astronomers* (PG-ISYA) arranged ISYAs in the years between General Assemblies:
- 29th ISYA, Kuala Lumpur and Langkawi, Malaysia, 5 - 24 March 2007,
- 30th ISYA, Istanbul, Turkey, 1 - 22 July 2008.

Information and reports of these ISYAs have been published in IAU *Information Bulletins* No. 100, p. 46, No. 101, p. 38, and No. 102, p. 76.

A report on IAU *International Schools for Young Astronomers* over the period 1967 - 2007 has been published by PG chair Prof. Michèle Gerbaldi in: van der Hucht, K.A. (ed.), 2008, Proceedings of the XXVIth General Assembly of the IAU, Prague, August 2006, *Transactions IAU*, Vol. XXVIB (Cambridge: CUP), p. 238. A comprehensive review of the same has been published by past PG chair Prof. Michèle Gerbaldi in: Hearnshaw, J. & Martinez, P. (eds.), 2007, *Astronomy for the Developing World*, Proc. IAU IAU XXVI GA Special Session No. 5, Prague, August 2006 (Cambridge: CUP), p. 221.

During the past triennium, the IAU was co-sponsor of COSPAR Capacity Building Workshops related to education in astrophysics.

6.2. *Commission 46 PG World Wide Development of Astronomy*

In agreement with its mission, members of the Commission 46 Program Group on *World Wide Development of Astronomy* (PG-WWDA) has made exploratory visits to countries developing in astronomy. Reports of visits to Bangladesh, Ecuador, Laos, Mozambique, Peru, Thailand, Uruguay, and Uzbekistan,have been published in IAU *Information Bulletins* No. 100, p. 50, and in: van der Hucht, K.A. (ed.), 2008, Reports of Astronomy, *Transactions IAU*, Vol. XXVIIA, (Cambridge: CUP), p. 429.

6.3. *Commission 46 PG on Teaching for Astronomy Development*

The Commission 46 Program Group on *Teaching for Astronomy Development* (PG-TAD) has continued and initiated educational programs in Kazakhstan, Kenya, Democratic People's Republic of Korea, Mongolia, Morocco, Nepal, Nicaragua, Philippines, Trinidad & Tobago, Vietnam, and Uzbekistan. Reports on these have been published in IAU *Information Bulletins* No. 100, p. 47, No. 101, p. 40, and No. 102, p. 73.

7. Relations with international scientific organizations

The IAU maintains relations with a number of international scientific organizations. All those organizations, as well as the IAU, are members of the International Council for Science (ICSU). ICSU <www.icsu.org> sets out its 'mission' as "to strengthen international science for the benefit of society" and its 'vision' as "a world where science is used for the benefit of all, excellence in science is valued and scientific knowledge is effectively linked to policy-making. In such a world, universal and equitable access to high quality scientific data and information is a reality and all countries have the scientific capacity to use these and to contribute to generating the new knowledge that is necessary to establish their own development pathways in a sustainable manner."

In the triennium the IAU was represented in the following international scientific organizations:

	Organization	*IAU Representative*
ICSU	International Council for Science	Ian F. Corbett
BIPM	Bureau International des Poids et Mesures	
— CCTF	Consultative Committee for Time and Frequency	Toshio Fukushima
— CCU	Consultative Committee for Units	Nicole Capitaine
— CIE	Compagnie Internationale de l'Eclairage	*vacant*
CODATA	Committee on Data for Science and Technology	Raymond P. Norris
COSPAR	Committee on Space Research	
— Council		Jean-Claude Vial
— SC B	Space Studies of the Earth-Moon System, Planets and Small Bodies of the Solar System	Mikhail Ya. Marov
— SC D	Space Plasmas in the Solar System, including Planetary Magnetospheres	Marek Vandas
— SC E	Research in Astrophysics from Space	Mariano Mendez
— SC E1	Galactic and Extragalactic Astrophysics	Ganesan Srinivasan
— SC E2	The Sun as a Star	Arnold O. Benz
FAGS	Federation of Astronomical and Geophysical Services	Nicole Capitaine
IAF	International Astronautical Federation	Bernhard H. Foing
IERS	International Earth Rotation and Reference Systems Service	Nicole Capitaine
IGBP	International Geosphere-Biosphere Programme	Richard G. Strom
IHY	International Heliophysical Year	David F. Webb
ISES	International Space Environment Service	Helen E. Coffey
ITU	International Telecommunication Union	William J. Klepczynski Tomas E. Gergely
— ITU-R	Radiocommunication Bureau	Masatoshi Ohishi
IUCAF	Scientific Committee on Frequency Allocations	Wim van Driel
IUPAP	International Union of Pure and Applied Physics	Ian F. Corbett
— C4	Commission on Cosmic Rays	Heinrich J. Voelk
— C19	Commission on Astrophysics	Virginia L. Trimble
IVS	International VLBI Service for Geodesy and Astrometry	Patrick Charlot
SCAR	Scientific Committee on Antarctic Research	John W.V. Storey
SCOPE	Scientific Committee on Problems of the Environment	Derek McNally
SCOSTEP	Scientific Committee on Solar-Terrestrial Physics	Nat Gopalswamy
UN-COPUOS	UN Committee on the Peaceful Uses of Outer Space	Karel A. van der Hucht
URSI	Union Radio-Scientifique Internationale	Luis F. Rodriguez

8. Financial matters

The IAU budget for 2006 - 2009 had been approved by the IAU XXVI General Assembly in Prague, August 2006. Appendix II of this chapter presents the IAU budget, income and expenditure for the years 2006, 2007 and 2008, as well as the IAU budget for 2009.

The General Secretary is very pleased to report that the following IAU National Members stepped up their Category of Adherence to the IAU during the triennium:
- from Category I to II: Chile, China Taipei, New Zealand and Norway
- from Category II to III: Czech Republic, Israel, Mexico and Ukraine
- from Category IV to V: India, the Netherlands
- from Category V to VI: China Nanjing
- from Category V to VII: Italy
- from Category IX to X: USA.

The accounts for the year 2006 have been certified by the IAU Auditor, Msr. Arnaud de Boisanger; the accounts for the years 2007 and 2008 have been certified by the new IAU Auditor, Msr. Henri Grillet. The accounts for 2006 and 2007 were complicated by financial fraud, as explained in the section on the IAU Secretariat, given below.

The Finance Sub-Committee, chaired by Dr. Paul G. Murdin, provided most helpful advice and guidance to the General Secretary on all financial matters of the Union during the triennium.

8.1. *Comments on budget 2010 - 2012*

The proposed budget for the years 2010 - 2012 is presented at the end of chapter II of these Transactions. The budget for the years 2010 - 2012 was approved by the Executive Committee at its meeting EC85, 7 - 8 April 2009 in Paris. The Executive Committee recommends an increase of the IAU educational activities, and the hiring of a third full time staff member in the IAU Secretariat in Paris, France.

9. Administrative matters

9.1. *Revision of Statutes and Bye-Laws*

The IAU XXVI General Assembly adopted a revision of the Statutes and Bye-Laws. See: van der Hucht, K.A. (ed.), 2008, *IAU Transactions*, Vol. XXVIB, p. 251.

Proposed changes in the Statutes and Bye-Laws up for voting at the IAU XXVII General Assembly in Rio de Janeiro have been presented in IAU *Information Bulletin* No. 104, p. 27.

9.2. *Upgrading of IAU data base and web site*

During the previous triennium, the IAU data base and web site had been upgraded thoroughly under the guidance of past General Secretary Oddbjørn Engvold and with help of a French contractor, AMEOS. Because of the ever increasing demands placed on the web site from both astronomers and the public at large, it appeared desirably to overhaul data base and website again, with as contractor a software group at ESO, overseen by Lars Lindberg Christensen. The new IAU data base and web site became operational in May 2008. Our contractor at ESO continuous to maintain and upgrade the IAU web site and data base. The IAU greatly appreciates the level of support it has received under its agreements with ESO.

9.3. *Secretariat*

The IAU Secretariat, based in Paris, has undergone profound changes in the past three years. The triennium started with two full-time staff members, Mme Monique Orine, Executive Assistant, and Mme Claire Vidonne, data base and website assistant, plus two part-time assistants. When Mme Vidonne left in October 2006, her work was partly outsourced, eventually leading to a complete overhaul of both the database and the website by a contractor at ESO. Part-time office assistant Mme Mary Noel Giraud left in January 2008. Office assistant Mme Maitena Mitschler left in May 2009 and was replaced in July 2009 by Mme Jana Žilová.

Mme Orine, who had served the IAU for over twenty years, was due to retire in December 2008, while continuing as a consultant through the General Assembly in Rio de Janeiro. Her sudden demise in January 2008 left the IAU Secretariat without a transition of knowledge to a successor.

The General Secretary, working full time for the IAU, usually spends on-third of the time physically at the IAU Secretariat. During the triennium 2006 - 2009 he did spend an average of 8.6 working days per month at the IAU Secretariat in Paris.

In March 2008, Mme Vivien A. Reuter accepted the job of IAU Executive Assistant. General Secretary and Executive Assistant together embarked upon the re-invention of the IAU Secretariat.

While perusing numerous documents and computer files, we found ample evidence that Mme Orine had been much appreciated and praised by the IAU community. There is no doubt that she invested much energy and enthusiasm in her work and identified with the IAU. However, we also found clear evidence that, through a clever scheme of financial fraud, involving fake and inflated invoices and unauthorised signatures, Mme Orine had, since 1998 and possibly from the very beginning of her employment in 1987, systematically taken money from the IAU. Disappointing and upsetting as this may be, two points should be emphasized: the fraud was restricted to inflating the administrative costs of the Secretariat, and none of the programmes or activities of the IAU is implicated in any way. All the money stolen went through the books, so no unforeseen deficits have emerged as a consequence. In spite of this illegal activity, I am pleased to report that the current finances of the IAU are healthy.

Designed to be virtually undetectable by the General Secretaries, by the Finance Subcommittees and by the external auditor, this fraud was clearly facilitated by the practices of our bank Le Crédit Lyonnais (LCL) in Paris. Having recovered and analysed all bank cheques written out on behalf of the IAU and cashed over the period mid-1998 – 2007 (older cheques having been routinely destroyed by the bank), the IAU lawyer has filed a claim for about EUR 350 000 against the bank LCL in Paris on behalf of the IAU.

The General Secretary and the Executive Assistant have taken the following measures, among others, in the IAU Secretariat to reduce the risk of recurrence of fraud:

- A sophisticated financial software package (SAGE) has been installed to increase transparency of the IAU book-keeping and to render posterior changes impossible.
- The preparation of the annual accounts is supervised by a professional accountant clerk.
- A new auditor has been contracted for a period of seven years at maximum.
- Payments are made through electronic transfer, whenever feasible, thus reducing the use of bank cheques to a minimum.

10. Relationship with the Peter and Patricia Gruber Foundation

The relationship between the IAU and the Gruber Foundation has been reconfirmed in a First Amended Agreement on Collaboration. See Appendix III.

10.1. *The Gruber Cosmology Prize*

During the triennium, the IAU has provided the following members for the Gruber Foundation Cosmology Prize Advisory Board: Jocelyn S. Bell Burnell (UK), James E. Peebles (USA), Ronald D. Ekers (Australia), and Andrei Linde (USA). Gruber Foundation

Cosmology Prize laureates are:
2006: John C. Mather (USA) and the Cosmic Background Explorer Team.
2007: Saul Perlmutter (USA) and the members of the Supernova Cosmology Project; and Brian P. Schmidt and the members of the High-z Supernova Search Team.
2008: J. Richard Bond (Canada).
2009: Wendy L. Freeman (USA), Robert C. Kennicutt (UK) and Jeremy R. Mould (Australia).

10.2. *The Gruber Fellowships*

During the triennium the Gruber Foundation provided funds for post-doc Fellowships. The Gruber Fellows, selected from proposals reviewed by an IAU selection committee, are:
2006: Inma Martinez-Valpuesta (Spain, working in France) and
Hum Chand (India, working in France).
2007: Krzysztof Blejko (Poland, working in Australia).
2008: Karen L. Masters (USA, working in UK).
2009: Mattheus B.N. Kouwenhoven (UK, working in Beijing).

11. Relationship with the Norwegian Academy of Science and Letters. The Kavli Prize in Astrophysics. The ISYAs.

11.1. *Kavli Prizes*

The Kavli Prizes are awarded every other year to one or more leaders in the areas of astrophysics, nano science and neuro science. Winners receive a US $1 000 000 prize as well as a medal and a diploma in recognition of their cutting-edge research. The Prizes have been established in order to recognise outstanding scientific research, honour highly creative scientists, promote public understanding of scientists and their work, and foster international cooperation among scientists. The first Kavli Prizes were given in 2008.

The *Kavli Prize in Astrophysics* is awarded for outstanding achievement in advancing our knowledge and understanding of the origin, evolution, and properties of the universe. It will include the fields of cosmology, astrophysics, astronomy, planetary science, solar physics, space science, astrobiology, astronomical and astrophysical instrumentation, and particle astrophysics.

The Norwegian Academy of Science and Letters and the International Astronomical Union agreed to cooperate on future *Kavli Prizes in Astrophysics.*

11.2. *Kavli Prize in Astrophysics 2010*

Beginning in 2009, the IAU and the NASL will collaborate on choosing the members of the *Kavli Prize in Astrophysics* selection committee. The Academy will remain responsible for the announcement of the Prize and the nomination of the winners; however, they will seek advice from the Kavli Foundation and the IAU. The IAU will also announce the names of the winners through its IAU *Information Bulletin* and on its web site.

The agreement between the Norwegian Academy of Science and Letters and the IAU and was signed on 29 May 2008 by the President of the NASL, Prof. Ole Didrik Laerum, and the President of the IAU, Dr. Catherine J. Cesarsky. The text of the Agreement is presented in Appendix IV.

11.3. *The IAU International Schools for Young Astronomers and the Norwegian Academy of Science and Letters*

For the benefit of countries and regions that need support in development of astrophysical education and research, the NASL will co-sponsor one IAU International School for Astronomy (ISYA) every year, beginning in 2009. The financial support by the NASL per year is US$ 30 000. The curriculum and venue for an IAU ISYA will be determined by the IAU Division XII / Commission 46 Program Group for ISYA, together with the Local Organizer of the ISYA, and will be overseen by the IAU Vice-President in charge of IAU educational programs. Appropriate procedures for the announcement of the IAU ISYAs will be developed by the NASL and the IAU. Announcement and application information for the IAU ISYAs will be published in the IAU Information Bulletin and made available on the IAU web site.

Each IAU ISYA will invite earlier Kavli Laureates as one main speaker at each International Schools for Young Astronomers. For all information material about the IAU ISYA, it will be stated that the school is sponsored by the Kavli Prize/the NASL.

12. UNESCO and the IAU on Astronomy and World Heritage. A new IAU Working Group

On 30 October 2008 a formal Memorandum of Understanding (MoU) was signed between the IAU and UNESCO at the UNESCO Headquarters in Paris, France, agreeing a number of ways in which the two organisations will work together to advance UNESCO's Astronomy and World Heritage Initiative <whc.unesco.org/en/activities/19> and ensure its full implementation. This initiative aims to ensure the recognition, promotion and preservation of achievements in science through the nomination to the World Heritage List (WHL) of properties whose outstanding significance to humankind derives in significant part from their connection with astronomy.

Following the signing, a new Working Group of Division XII / Commission 41 (*History of Astronomy*) has been set up, which is charged with fulfilling the IAU's commitments under the MoU.

The WG Terms of Reference are:

1. To work on behalf of the IAU to help ensure that cultural properties and artefacts significant in the development of astronomy, together with the intangible heritage of astronomy, are duly studied, protected and maintained, both for the greater benefit of humankind and to the potential benefit of future historical research.
(The range of properties and objects in question includes ancient sites and monuments with demonstrable links to the sky (such as Stonehenge), instruments of all ages, archives, and historical observatories.)

2. To fulfill, on behalf of the IAU, its commitments under the Memorandum of Understanding with UNESCO on Astronomy and World Heritage.

3. To liaise with other international and national bodies concerned with astronomical history and heritage, in so far as their interests and activities impinge on these aims, to help achieve these aims.

4. To work, in conjunction with IAU C41 (History of Astronomy), IAU C46 (Education) and other Commissions and Working Groups within the IAU as appropriate, to enhance public interest, understanding, and support in the field of astronomical heritage.

The first main task for the IAU WG is to work with the International Committee on Monuments and Sites (ICOMOS) to produce a global Thematic Study on astronomical heritage. This will provide the basis upon which UNESCO will produce criteria for judging WHL nominations relating to astronomy. Detailed work plan are being prepared with UNESCO and ICOMOS.

The Memorandum of Understanding is presented in Appendix V.

Karel A. van der Hucht
IAU General Secretary 2006 - 2009
Utrecht / Paris / Rio de Janeiro, 11 August 2009

APPENDIX I (ad Section 1.4.1)

United Nations General Assembly, sixty-second session, agenda item 56 (b)

Resolution adopted by the General Assembly

[*on the report of the Second Committee (A/62/421/Add.2)*]

62/200. International Year of Astronomy, 2009

The General Assembly,

Recalling its resolution 61/185 of 20 December 2006 on the proclamation of international years,

Aware that astronomy is one of the oldest basic sciences and that it has contributed and still contributes fundamentally to the evolution of other sciences and applications in a wide range of fields,

Recognizing that astronomical observations have profound implications for the development of science, philosophy, culture and the general conception of the universe,

Noting that, although there is a general interest in astronomy, it is often difficult for the general public to gain access to information and knowledge on the subject,

Conscious that each society has developed legends, myths and traditions concerning the sky, the planets and the stars which form part of its cultural heritage,

Welcoming resolution 33 C/25 adopted by the General Conference of the United Nations Educational, Scientific and Cultural Organization on 19 October 2005† to express its support for the declaration of 2009 as the International Year of Astronomy, with a view to highlighting the importance of astronomical sciences and their contribution to knowledge and development,

Noting that the International Astronomical Union has been supporting the initiative since 2003 and that it will act to grant the project the widest impact,

Convinced that the Year could play a crucial role, inter alia, in raising public awareness of the importance of astronomy and basic sciences for sustainable development, promoting access to the universal knowledge of fundamental science through the excitement generated by the subject of astronomy, supporting formal and informal science education in schools as well as through science centres and museums and other relevant means, stimulating a long-term increase in student enrolment in the fields of science and technology, and supporting scientific literacy,

1. *Decides* to declare 2009 the International Year of Astronomy;

2. *Designates* the United Nations Educational, Scientific and Cultural Organization as the lead agency and focal point for the Year, and invites it to organize, in this capacity, activities to be realized during the Year, in collaboration with other relevant entities of the United Nations system, the International Astronomical Union, the European Southern Observatory and astronomical societies and groups throughout the world, and, in this regard, notes that the activities of the Year will be funded from voluntary contributions, including from the private sector;

3. *Encourages* all Member States, the United Nations system and all other actors to take advantage of the Year to promote actions at all levels aimed at increasing awareness among the public of the importance of astronomical sciences and promoting widespread access to new knowledge and experiences of astronomical observation.

78th plenary meeting, 19 December 2007

† United Nations Educational, Scientific and Cultural Organization, Records of the General Conference, 33rd session, Paris, 3-21 October 2005, vol. 1: Resolutions, chap. V.

APPENDIX II (ad Section 8)

I. IAU – STATEMENT OF INCOME 2006 - 2008

	I INCOME *BUDGET*	2006 (CHF)	2007 (CHF)	2008 (CHF)	2006 - '08 (CHF)	2009 (CHF)
	unit of contribution	3 580	3 685	3 800		3 910
	adjustment for inflation	3.4%	3.0%	3.0%		3.0%
	number of units	262	262	262		301.5
A	ADHERING ORGANIZATIONS					
A1	National Member dues	**843 269**	**1 066 347**	**775 425**	**2 685 041**	
		937 960	*965 470*	*995 600*	*2 899 030*	*1 178 865*
B	GRANTS & REFUNDS					
B1	GA related refunds		21 660		21 660	
B3	Symposia related refunds		772	26 649	27 421	
B4	ESO, grant for GA 2006	79 710			79 710	
B5	UN	6 050			6 050	
B6	Registration GA 2006	2 117			2 117	
	sub-total GRANTS & REFUNDS	**87 877**	**22 432**	**26 649**	**136 958**	
C	ROYALTIES					
C2	CUP	...	70 762	55 638		
C3	Springer	...	646	538		
C4	other (e.g., National Geographic)			588		
	sub-total ROYALTIES	**54 281**	**71 407**	**56 764**	**182 452**	
		15 000	*20 000*	*20 000*	*55 000*	*20 000*
D	FUNDS IN TRANSIT					
D1	IYA2009		74 050	281 329	355 379	
D2	NAS for GA2009 WiA Lunch			9 685	9 685	
D3	PPGF Fellowships	90 750	57 000	53 804	201 554	
	sub-total FUNDS IN TRANSIT	**90 750**	**131 050**	**344 818**	**566 618**	
E	ADMINISTRATIVE REFUNDS					
	Government Organizations		3 913		3 913	
	ACCOR			443	443	
	EROM			5 862	5 862	
	Cash in Safe IAP			13 561	13 561	
	Other			3 517	3 517	
	sub-total ADMINISTRATIVE REFUNDS		**3 913**	**23 383**	**27 296**	
F	BANK REVENUE					
F2	Strategy Fund	16 619	2 712	2 675	22 006	
	sub-total BANK REVENUE	**16 619**	**2 712**	**2 675**	**22 006**	
		25 000	*10 000*	*10 000*	*45 000*	*10 000*
	total INCOME	**1 092 796**	**1 297 862**	**1 229 714**	**3 620 372**	
G	**total INCOME minus 'In Transit'**	**1 002 046**	**1 166 812**	**884 896**	**3 053 754**	
		977 960	*995 470*	*1 025 600*	*2 999 030*	*1 208 865*

II. IAU – STATEMENT OF EXPENDITURE 2006 - 2008

II	EXPENDITURE *BUDGET*	2006 (CHF)	2007 (CHF)	2008 (CHF)	2006 - '08 (CHF)	2009 (CHF)
M	GENERAL ASSEMBLIES (GAs)					
M1	*Preparation costs*	20 318	25 083	7 747	53 148	
		45 000	*4 000*	*8 000*	*57 000*	*50 000*
M2	*Grants GA incl. 6 Symposia*	523 316	--	--	523 316	
		440 000	--	--	*440 000*	*385 000*
	total GENERAL ASSEMBLIES	**543 634**	**25 083**	**7 747**	**576 464**	
		485 000	*4 000*	*8 000*	*497 000*	*435 000*
N	SCIENTIFIC ACTIVITIES					
N1	*Sponsored meetings*					
N1.1	Grants IAU Symposia outside GA	78 200	250 527	172 668	501 395	
		75 000	*225 000*	*225 000*	*525 000*	*75 000*
N1.2	Grants Regional IAU Meetings (RIMs)	--	29 000	52 477	81 477	
		--	*30 000*	*30 000*	*60 000*	--
N1.3	Co-sponsored meetings	--	27 644	2 092	29 736	
	sub-total Sponsored meetings	78 200	307 171	227 236	612 607	
		75 000	*255 000*	*255 000*	*585 000*	*75 000*
N2	*Working Groups*					
N2.1	EC Working Groups	8 484	- -	--	8 484	
		7 000	*5 000*	*5 000*	*17 000*	*5 000*
N2.1.2	HNEOs – book (Rickman)	2 617	--	--	2 617	
N2.1.3	IYA2009 sponsoring and UN lobbying	--	50 363	49 125	99 488	
		--	*2 000*	*2 000*	*4 000*	*8 000*
N2.2	Commission Working Groups	3 594	--	--	3 594	
		7 000	*5 000*	*5 000*	*17 000*	*5 000*
N2.2.1	CB for Astronomical Telegrams (C6)	4 000	--	4 000	8 000	
		4 000	*4 000*	*4 000*	*12 000*	*4 000*
N2.2.2	Minor Planet Center (DIII)	--	16 000	12 000	28 000	
		8 000	*12 000*	*12 000*	*32 000*	*12 000*
N2.2.3	Meteor Data Center (C22)	8 000	--	--	8 000	
		1 500	*2 000*	*2 000*	*5 500*	*2 000*
N2.2.4	Astronomy & World Heritage (C41)	--	2 088	2 524	4 612	
N2.2.5	Planetary System Nomenclature (DIII)	--	--	1 756	1 756	
	sub-total Working Groups	26 695	68 451	68 451	164 551	
		27 500	*30 000*	*30 000*	*87 500*	*36 000*
	total SCIENTIFIC ACTIVITIES	**104 895**	**375 622**	**296 641**	**777 158**	
		102 500	*285 000*	*285 000*	*672 500*	*111 000*

	Expenditure (cont'd)	2006 (CHF)	2007 (CHF)	2008 (CHF)	2006 - '08 (CHF)	2009 (CHF)
O	EDUCATIONAL ACTIVITIES (C46)					
O5	*PG Intern. Schools for Young Astronomers*	--	43 269	31 540	74 809	
		--	*45 000*	*45 000*	*90 000*	
O1	*PG Exchange of Astronomers*	11 537		8 066		
		15 000	*15 000*	*15 000*	*45 000*	*15 000*
O8	*PG Teaching for Astronomy Development*	20 134		33 439		
		45 000	*45 000*	*45 000*	*135 000*	*45 000*
O9	*PG World Wide Development of Astronomy*	4 598		5 138		
		10 000	*15 000*	*15 000*	*40 000*	*15 000*
	sub-total PGs EA, TAD, WWDA	36 269	55 512	46 643	138 424	
		70 000	*75 000*	*75 000*	*220 000*	*75 000*
O2	*PG Exchange of Books, Journals*	--	1 931	674	2 605	
O10	*C46 Brainstorm in Paris*	--	--	7 651	7 651	
O11	*co-sponsoring COSPAR CBWs*	--	1 693	5 000	6 693	
	total EDUCATIONAL ACTIVITIES	**36 269**	**102 405**	**91 508**	**230 182**	
		70 000	*120 000*	*120 000*	*310 000*	*75 000*
P	FUNDS IN TRANSIT					
P1	*IYA2009 secretariat finances*	--	41 183	241 465	282 648	
P2	*PPGF Fellowships*	23 451	60 617	79 896	163 964	
P3	*NAS for GA2009 WiA Lunch*	--	--	11 388	11 388	
	total FUNDS IN TRANSIT	**23 451**	**101 801**	**332 749**	**458 001**	
Q	COOPERATION WITH OTHER UNIONS					
Q1	*Delegates' travel*	10 978	8 606	5 564	25 148	
		10 000	*12 000*	*12 000*	*34 000*	*12 000*
Q2	*Dues to other Unions*					
Q2.1	ICSU	14 331	18 317	19 434	52 082	
		20 000	*7 500*	*7 500*	*35 000*	*7 500*
Q2.2	IERS/FAGS	--	2 500	3 138	5 638	
		7 500	*10 000*	*10 000*	*27 500*	*10 000*
Q2.3	IUCAF	--	7 500	3 332	10 832	
		7 500	*7 500*	*7 500*	*22 500*	*7 500*
	sub-total Dues to other Unions	14 331	28 317	25 904	68 552	
		35 000	*25 000*	*25 000*	*85 000*	*25 000*
	total COOP. WITH OTHER UNIONS	**25 309**	**36 924**	**31 467**	**93 700**	
		45 000	*37 000*	*37 000*	*119 000*	*37 000*

	Expenditure (cont'd)	**2006** (CHF)	**2007** (CHF)	**2008** (CHF)	**2006 - '08** (CHF)	**2009** (CHF)
R	EXECUTIVE COMMITTEE					
R1	*Executive Committee Meetings*	87 961 *60 000*	54 458 *45 000*	53 505 *80 000*	195 924 *185 000*	 *80 000*
R2	*Officers' Meetings*	19 418 *5 500*	17 900 *12 000*	11 859 *12 000*	49 177 *29 500*	 *12 000*
R3	*Officers' expenditure (other)*					
R3.1	General Secretary expenditure					
R3.1.1	General Secretary – Paris		37 410	41 072		
R3.1.2	General Secretary – other		5 455	620		
	sub-sub-total GS expenditure	44 093 *30 000*	42 865 *30 000*	41 692 *30 000*	128 650 *90 000*	 *30 000*
R3.2	President	-- *2 000*	668 --	4 126 --	4 794 *2 000*	 --
R3.3	President-elect	--	--	800	800	
R3.4	Assistant General Secretary	740 *2 000*	598 *2 000*	4 557 *2 000*	5 895 *6 000*	 *2 000*
R3.5	FSC-chair visit to IAU Secretariat	--	--	1 191	1 191	
	sub-total Officer's expenditure (other)	44 833 *34 000*	44 131 *32 000*	52 365 *32 000*	141 329 *98 000*	 *32 000*
R4	*Press-Office*		--	4 447	4 447	
	total EXECUTIVE COMMITTEE	**152 212** *99 500*	**116 489** *89 000*	**122 176** *124 000*	**390 877** *312 500*	 *124 000*
S	PUBLICATIONS					
S1	*IAU Information Bulletins*	20 497 *20 000*	10 022 *10 000*	15 704 *10 000*	46 223 *40 000*	 *10 000*
	total PUBLICATIONS	**20 497** *20 000*	**10 022** *10 000*	**15 704** *10 000*	**46 223** *40 000*	 *10 000*

	Expenditure (cont'd)	2006 (CHF)	2007 (CHF)	2008 (CHF)	2006 - '08 (CHF)	2009 (CHF)
T	ADMINISTRATION / SECRETARIAT					
T1	*Salaries and Charges*					
T1.1	Salaries		93 893	128 897		
T1.2	Charges		34 922	68 704		
T1.3	Salary M.L.-O. CNRS		137 802	– 25 059		
	sub-total Salaries and Charges	239 709 *175 000*	266 618 *250 000*	172 542 *250 000*	678 869 *675 000*	*250 000*
T2	*Training Courses*	- - *5 000*	- - *5 000*	1 755 *5 000*	1 755 *15 000*	*5 000*
T3	*Outsourced Tasks*					
T3.1	Web/DB Development at ESO	- -	37 122	61 598	98 720	
T3.2	Web Management at ESO	- -	10 502	34 270	44 772	
T3.3	IT Assistance in Paris	22 433	– 1 617	1 087	21 903	
T3.4	Personnel Administration	- -	1 789	2 385	4 174	
T3.5	Accounting	- -	1 391	8 638	10 029	
T3.6	Auditing	11 726	1 969	1 744	15 439	
	sub-total Outsourced Tasks	34 159 *2 500*	51 155 *2 500*	109 722 *2 500*	195 036 *7 500*	*2 500*
T4	*General Office expenditure*					
T4.1	Post	11 187	3 855	6 496	21 538	
T4.2	Telephone and Internet	11 282	6 771	5 245	23 298	
T4.3	Rent (INSU / IAP)	7 967	5 963	5 693	19 623	
T4.4	IT Software	5 720	739	3 626	10 085	
T4.5	IT Hardware	2 814	9 900	2 365	15 079	
T4.6	Office Supplies	17 883	912	1 166	19 961	
T4.7	Books and CDs	- -	222	- -	222	
T4.8	Furniture and Locks	- -	2 238	- -	2 238	
T4.9	Contributions F&D IAP	- -	1 443	685	2 128	
T4.10	Miscellaneous (flowers, gifts tips)	- -	468	1 075	1 543	
	sub-total General Office expenditure	56 853 *86 000*	32 510 *86 000*	26 351 *89 000*	115 714 *281 000*	*94 000*
T5	*Bank Expenses*					
T5.1	Transaction Charges		7 561	7 917		
T5.2	Debit Interest		2 715	76		
	sub-total Bank Expenses	8 813 *4 000*	10 275 *4 000*	7 993 *4 000*	27 081 *4 000*	*4 000*
T6	*Other*	[appr. 55 000]	55 772	- -	55 772	
	total ADMIN./SECRETARIAT	**339 534** *268 500*	**416 329** *343 500*	**318 362** *346 500*	**1 074 226** *958 500*	*351 500*

	Expenditure (cont'd)	2006 (CHF)	2007 (CHF)	2008 (CHF)	2006 - '08 (CHF)	2009 (CHF)
	total EXPENDITURE	**1 246 267**	**1 184 676**	**1 216 354**	**3 647 297**	
V	**total EXP. minus 'In Transit'**	**1 222 816**	**1 082 875**	**883 605**	**3 189 296**	
		1 090 500	*888 500*	*930 500*	*2 909 500*	*1 143 500*
	INCOME OVER EXPENDITURE	**– 220 770**	**+ 83 937**	**+ 1 291**	**– 135 542**	
		– 112 540	*+ 106 970*	*+ 226 200*	*+ 220 630*	*+ 65 365*
	TOTAL ASSETS	**1 041 788**	**1 247 590**	**1 395 828**		

APPENDIX III (ad Section 10)

First Amended Agreement on Collaboration Between

The Peter Gruber Foundation and The International Astronomical Union

Preamble

In pursuance of their common goal to promote the public awareness and recognition of the achievements in the field of Cosmology in a broad sense and further the development of the field in the future, the Peter Gruber Foundation (PGF), located in St. Thomas, U.S. Virgin Islands (P.O. Box 503210, St. Thomas, VI 00805), and the International Astronomical Union (IAU), headquartered in Paris, France (98 bis Boulevard Arago, 75014 Paris, France), hereby agree to collaborate on the announcement and award of the Cosmology Prize of the Peter Gruber Foundation. The content and form of this collaboration are defined in the following.

The Cosmology Prize

The Cosmology Prize of the PGF is awarded annually for scientists or institutional groups of any nationality to honor the most distinguished work in the fields of astronomy, physics, mathematics, and philosophy. The Prize consists of a medal and a cash sum of USD $500,000.00. Nominations may be submitted by individuals, organizations, and institutions that have experience in and an appreciation for the study of cosmology and modern cosmological research. Nominations are received and prize recipients are selected by an international Advisory Board (composed as described below), with the advice and consent of the directors of the PGF, so as, among other things, to insure compliance with the PGF Articles of Incorporation and U.S. laws (including, but not limited to, the Internal Revenue Code) and regulations relating to foundation activities.

Advisory Board

The Advisory Board for the Cosmology Prize of the PGF will be composed of seven distinguished scientists representing the field of cosmology. Members are appointed for three-years periods, renewable once, and serve on the Board in their personal capacity only. Three members will be nominated by the IAU and one each by International Union of Pure and Applied Physics (IUPAP), the International Mathematical Union (IMU), and the International Union of the History and Philosophy of Science (IUHPS). The seventh member will be selected by these six individuals.

Award of the Prize

In the years of the IAU General Assembly, beginning in 2003, the Cosmology Prize will be presented at the IAU General Assembly. In non-General Assembly years, the Prize will be awarded in locations of importance in the history of cosmology, and where the event could have a positive impact on the further development of the field. Representatives of the four scientific Unions will be invited to all events.

The Peter Gruber Foundation Fellowships

In order to allow promising young scientists of any nationality to pursue education and research at a center of excellence in their field, the PGF will support fellowships each three years. The total amount of support per three-year period will be US $150,000 beginning in 2007. Appropriate procedures for the announcement and award of the fellowships will be developed by the IAU and the PGF, including provision for the advice and consent of the directors of the PGF, so as to insure compliance with the PGF Articles of Incorporation and U.S. laws (including, but not limited to, the Internal Revenue Code) and regulations relating to foundation activities.

Amendment of Termination of the Agreement

Either party to this Agreement may propose amendments to it or announce their intention to terminate it by written notification to the other party at any time. For a proposed amendment to be adopted, both parties must so agree. Failure of the parties to reach agreement on a proposed amendment or termination within one year of its notification will force the termination of this Agreement.

Implementation of the Agreement

This Agreement will enter into force when signed by the duly authorized representatives of the PGF and the IAU. This agreement supersedes all prior agreements. Announcements and application information for the Prize and the Fellowships will be published in the IAU Information bulletin for January of each year, beginning in January 2007, and be made available on-line at the IAU web site.

Collaborating Unions:

The parties to this Agreement are pleased to invite the international unions IUPAP, IMU, and IUHPS to join their collaboration on the award of the Cosmology Prize as described above. These Unions are invited to add their signatures below to confirm their interest and willingness to join this collaboration.

International Union of Pure and Applied Physics
c/o American Physical Society, 1 Physics Ellipse, College Park, MD 20740, USA

International Mathematical Union
Institute for Advanced Study, Olden Lane, Princeton, NJ 08540-0631, USA

International Union of the History and Philosophy of Science
Centre d'Histoire des Sciences, Université de Liège, 5 Avenues des Tilleuls, B-4000 Liège, Belgium

Signed by Karel A. van der Hucht, IAU General Secretary
Paris, February 2007

APPENDIX IV (ad Section 11)

Agreement between the Norwegian Academy of Science and Letters and the International Astronomical Union

In pursuance of their common goal to promote the public awareness and recognition of outstanding achievements in the science of astrophysics in the broader sense, and to promote the development of astrophysical research in the future,

- the Norwegian Academy of Science and Letters (The Academy, residing at Drammensveien 78, N-0271 Oslo, Norway), and
- the International Astronomical Union (IAU, headquartered at 98-bis, Boulevard Arago, F-75014 Paris, France)

hereby agree to collaborate on selection of the Prize Committee, the announcement of the Kavli Prize in Astrophysics and the IAU International Schools for Young Astronomers. The content and form of this cooperation are defined in the following.

The Kavli Prize in Astrophysics

The Kavli Prize in Astrophysics is awarded every second year to one or more scientists for outstanding achievement in advancing our knowledge and understanding of the origin, evolution, and properties of the universe, including the fields of cosmology, astrophysics, astronomy, planetary science, solar physics, space science, astrobiology, astronomical and astrophysical instrumentation, and particle astrophysics. The Kavli Prize consists of a gold medal, a cash sum of US$ 1 000 000.-, and a scroll. Nominations may be submitted by all who wish to nominate candidates, with the exception that self-nominations are not accepted. Nominations are received by the Academy. The Kavli Prize recipients are selected by an international Prize Committee, composed as described below.

The Prize Committee

The Prize Committee is composed of five distinguished scientists representing different fields in astrophysics. The Norwegian Academy of Science and Letters appoints the Committee on the basis of suggestions from leading international academies and other equivalent scientific organizations. These organisations are:

The Max Planck Society (Germany)
The National Academy of Sciences (US)
The Norwegian Academy of Science and Letters
The Royal Society (UK)

After receiving suggestions from these organisations, The Norwegian Academy of Science and Letters will seek the advice by the International Astronomical Union in order to establish a balanced committee with respect to the various fields of Astrophysics.

The Officers of the International Astronomical Union together with the Chair of the Prize Committee will work out a proposal to the Academies Board.

Announcement and award of the Prize

Announcement and nomination information of the Prize are the Academies responsibility, and the Academy with seek advice from the Kavli Foundation and the IAU. IAU will publish the Announcement in its IAU Information Bulletin and on its web site. The Prize will be awarded in September in Oslo, Norway. Representatives of the IAU will be invited to the Kavli Prize in Astrophysics Award Ceremony.

IAU International Schools for Young Astronomers

For the benefit of countries and their regions that need support in development of astrophysical education and research, the Academy will support one IAU International School for Astronomy (ISYA) every year, beginning in 2009. The total amount of financial support by the Academy per year is US$ 30 000.-. The curriculum for an IAU ISYA will be determined by the IAU Division XII / Commission 46 Program Group for a ISYAs, together with the Local Organizer of the ISYA, and will be overseen by the IAU Vice-President in charge of IAU educational programs. Appropriate procedures for the announcement of the IAU ISYAs will be developed by the Academy and the IAU. Announcement and application information for the IAU ISYAs will be published in the IAU Information Bulletin and will be made available on the IAU web site.

The IAU ISYA should invite earlier Kavli Laureates as one main speaker at each International Schools for Young Astronomers, and for all information material about the IAU ISYA it should be stated the school is sponsored by the Kavli Prize/the Academy.

Amendment of termination of the Agreement

Parties to this Agreement may propose amendments to it, or announce their intention to terminate it by written notification to the other party at any time. For a proposed amendment to be adopted, all parties must so agree. Failure of the parties to reach agreement on a proposed amendment or termination within one year of its notification will force the termination of this Agreement.

Implementation of the Agreement

This Agreement will enter into force when signed by the authorized representatives of the Academy and the IAU. This agreement supersedes all prior agreements. This Agreement is not intended to create legal obligations on any of the signatories. In the event of a dispute over the application of the terms of this Agreement, the matter shall be put to the President of the Academy, and the President of the IAU, who shall determine the process for resolution of the dispute. This Agreement is subject to Norwegian law.

For the Norwegian Academy of Sciences and Letters
Prof. Ole Didrik Laerum, president

For the International Astronomical Union
Dr. Catherine J. Cesarsky, president

Signed: Oslo, 29 May 2008

APPENDIX V (ad Section 12)

Memorandum of Understanding between the IAU and UNESCO on Astronomy and World Heritage

The International Astronomical Union, hereinafter referred to as "IAU",

and

The United Nations Educational, Scientific and Cultural Organization, hereinafter referred to as "UNESCO"

UNESCO and IAU being referred to hereinafter as "the Parties"

PREAMBLE

Whereas

UNESCO's mission, in implementing the 1972 *Convention concerning the protection of the world cultural and natural heritage* (the *World Heritage Convention)*, is to identify, protect and transmit to future generations the natural and cultural heritage properties inscribed on the World Heritage List and List of World Heritage in Danger,

Whereas

UNESCO is seeking support from civil society for the achievement of its strategic objectives and programme priorities,

Whereas

IAU's mission is to promote and safeguard the science of astronomy in all its aspects through international cooperation that includes promotion of its history, engaging people worldwide in scientific field research and educational activities,

In consideration of the foregoing premises, the Partners hereby agree to co-operate as follows:

Article 1 – Objectives of the Initiative

By the present Memorandum of Understanding, the Parties confirm their commitment to promote the *World Heritage Convention* and in particular to explore and to implement the

Astronomy and World Heritage" Initiative. This Initiative aims to the recognition, promotion and preservation of achievements in science through the nomination of the properties whose outstanding significance derives in significant part from their connection with science, in particular astronomy.

The Parties hereby agree on the activities which they will respectively and jointly undertake, in order to explore and develop the feasibility, viability and impact of this Initiative, as well as on the principal conditions and modalities whereby these activities would be carried out.

Article 2 – Development of the Content of the Initiative

2.1. The Parties shall work together to develop and implement a joint programme of activities focused on the following three areas of cooperation:

1. research: to provide experts who will assist with sites identification and inventory, as well as for given World Heritage properties included on the State Parties' Tentative Lists, will provide assistance for monitoring activities, data collection, the preparation of nomination files, the development of management plans, and related activities;

2. education: to build the capacities of World Heritage property managers and enlarge their experience through fellowship programmes;

3. promotion: to provide institutional partners who will assist with development of the data base, as well as public web pages created within the framework of the Initiative;

4. partnerships: to create tri-partite partnerships between the Parties and others, including private sector entities, in order to gather enough financial support to implement the Initiative.

2.2. The Parties shall decide on the nature of the projects which will be implemented within the framework of the Thematic Initiative "Astronomy and World Heritage" upon mutual agreement.

Article 3 – Responsibilities

3.1. Responsibilities of UNESCO

UNESCO commits itself to ensure the implementation of the Thematic Initiative "Astronomy and World Heritage" worldwide, in agreement and co-operation with the national authorities dealing with natural and cultural heritage in the States Parties to the World heritage Convention;

UNESCO commits itself to respect the architectural, cultural, natural and historic authenticity of properties in accordance with the World Heritage Convention, the Operational Guidelines and the relevant expert reports;

There will be no financial obligation for UNESCO in respect of any given Project developed within the framework of the Initiative, until the corresponding funds for such Project have been identified and received by UNESCO from an appropriate donor.

3.2. Responsibilities of IAU

IAU commits itself to assist UNESCO in the implementation of the Thematic Initiative "Astronomy and World Heritage" worldwide;

IAU commits itself to provide, through its bodies, the scientific expertise in the field of Astronomy required for the implementation of the Thematic Initiative "Astronomy and World Heritage" worldwide;

IAU commits itself into ensuring, through its bodies, the development of the DataBase of the Thematic Initiative "Astronomy and World Heritage" created on the website of the UNESCO World Heritage Centre within the support of the Royal Astronomical Society of the United Kingdom. All rights to the website and its content, including intellectual property rights, shall belong to UNESCO;

IAU commits itself into respecting the historic and scientific authenticity of properties in accordance with the relevant expert reports and into bringing the scientific expertise to the Thematic Initiative "Astronomy and World Heritage" worldwide;

There will be no financial obligation placed on IAU for a given Project to be developed within the framework of the Initiative.

Article 4 – Conditions and Modalities

4.1. IAU shall keep UNESCO informed of any promotional and fund-raising opportunities and shall not develop any part of this Initiative without the involvement and prior written approval of UNESCO. Such written approval shall be promptly considered and not to be unreasonably withheld.

4.2. In the event that projects are identified with corresponding funding and the Parties agree to develop it, the donors identified as contributors for such projects will need to be signatory to any Agreements developed in this context. The recognized donors shall not be granted the right to use the names and emblems of UNESCO and World Heritage until appropriate modalities of cooperation between UNESCO and potential contributors are established. Granting of the right to use the name and emblems of UNESCO World Heritage shall, in any case, be done in accordance with UNESCO rules and through specific subsequent instruments that set out the purpose and scope of the right to use the name and logo.

4.3. The Parties will closely consult with relevant governmental authorities at all phases of the Initiative.

4.4. The Parties agree to confer and develop working methods and formal processes for the implementation of these activities and to present their proposals for review and approval by both Parties no later than 6 months after the entry into force of this Memorandum of Understanding.

4.5. The present Memorandum of Understanding can be amended, at any time, upon the formal request by one of the Parties and written confirmation of the other party.

4.6. UNESCO's undertakings and activities with regard to the Initiative shall be governed by its relevant rules and procedures.

Article 5 – Scope

Territory: Worldwide

Article 6 – Duration

The present Memorandum of Understanding will take effect upon signature by both Parties. The Memorandum of Understanding is effective for the term of two (2) years from the date of signature by the Parties, and thereafter may be extended by written consent of the Parties for two (2) additional years on the basis of an evaluation undertaken by the Parties at the end of the initial term.

Article 7 – Termination

Either Party shall be entitled to terminate, with immediate effect, the Memorandum of Understanding, by giving notice in writing to that effect to the other Party in any of the following circumstances:

a) if either Party is unable, due to circumstances beyond its control, to continue to carry out its obligations in terms of the Memorandum of Understanding;

b) if the other Party breaches the terms of the Memorandum of Understanding.

The present Memorandum of Understanding may not be waived, modified or changed in any manner except by a written amendment signed by each of the Parties hereto.

The present Memorandum of Understanding is concluded *intuitu personae*. Therefore, it may not be assigned nor transferred in any case.

It is understood that this Memorandum of Understanding does not confer IAU any exclusivity regarding activities such as those covered by this Memorandum of Understanding and IAU accepts that UNESCO is currently collaborating on similar activities worldwide with other partners.

Article 8 - Notification

Any notification addressed to one of the Partners will be made preliminarily by fax and confirmed by mail to the following address:

For UNESCO	7, place de Fontenoy F-75352 Paris 07 SP France
To:	Director of the World Heritage Centre
Fax:	(331) 45 68 55 70
For IAU	98bis, boulevard Arago F-75014 Paris France
To:	IAU General Secretary
Fax.	(331) 43 25 26 16

Article 9 – Use of UNESCO and WHC Names, Emblems, Logos or Official Seals

IAU is entitled to be considered an 'official partner of UNESCO's World Heritage Centre'. However, IAU shall not be granted the use of the names, abbreviations, emblems, logos and/or official seals of UNESCO and the World Heritage in the Initiative unless UNESCO has approved a written request from IAU. Each request, including a narrative and visual proposal, shall be made by IAU and UNESCO shall do its utmost to respond within 10 working days. Proof of all materials produced with UNESCO's name(s) and/or emblems(s) shall be provided, once authorized by UNESCO, and produced by IAU to UNESCO for reporting purposes.

Concluded in two original copies in the English language, on 30 October 2008, at UNESCO Headquarters, Paris

For the International Astronomical Union	For the United Nations Educational, Scientific and Cultural Organization
Karel A. VAN DER HUCHT General Secretary	Kïchiro MATSUURA Director-General

Transactions IAU, Volume XXVIIB
Proc. XXVII IAU General Assembly, August 2009
Ian F. Corbett, ed.

doi:10.1017/S1743921310004850

CHAPTER VI
REPORTS ON DIVISION, COMMISSION and WORKING GROUP MEETINGS

DIVISION I — FUNDAMENTAL ASTRONOMY
(ASTRONOMIE FONDAMENTALE)

PRESIDENT	**Jan Vondrak**
VICE-PRESIDENT	**Dennis D. McCarthy**
PAST PRESIDENT	**Toshio Fukushima**
BOARD	**Toshio Fukushima, George H. Kaplan, Joseph A. Burns, Zoran Knezevic, Irina I. Kumkova, Daffyd W. Evans, Aleksander Brzezinski, Chopo Ma, Pascale Defraigne, Richard N. Manchester, Sergei A. Klioner, Gerard Petit**

PARTICIPATING COMMISSIONS

Commission 4	**Ephemerides**
Commission 7	**Celestial Mechanics and Dynamical Astronomy**
Commission 8	**Astrometry**
Commission 19	**Rotation of the Earth**
Commission 31	**Time**
Commission 52	**Relativity in Fundamental Astronomy**

DIVISION WORKING GROUPS
Second Realization of International Celestial Reference System
Numerical Standards of Fundamental Astronomy
Astrometry with Small Ground-based telescopes

INTER DIVISION WORKING GROUPS

Division I / Division III	**Cartographic Coordinates and Rotational Elements of Planets and Satellites**
Division I / Division III	**Natural Satellites**

PROCEEDINGS BUSINESS SESSIONS, 3 August and 4 August 2009

1. Introduction

There were four 1.5-hour sessions of Division I business meetings during the XXVIIth IAU General Assembly. The first three were devoted to the reports of Commissions, Working Groups and services associated with the Division, discussion about plans for the next triennium and future structure of the Division. Scientific presentations on the future space astrometric mission Gaia were made at the fourth session.

2. Business Meeting, Monday 3 August 2009, 11:00 hr

This session, chaired by Dennis McCarthy, was devoted to the reports of all Commissions pertaining to Division I on activities 2006–2009 and plans for 2009–2012.

The report of **Commission 4 (Ephemerides)** was presented by its President, Toshio Fukushima.

The report of **Commission 7 (Celestial Mechanics)** was presented by its President, Joseph Burns. He gave a summary of the activities of the Commission during the last 3 years and of the current state of the Commission, including a list of recently deceased members, introduced the incoming officers, and emphasized the following issues:
- organization of IAU Symposium 236: Near Earth Objects, our Celestial Neighbors: Opportunity and Risk, in Prague (Czech Republic), in 2006;
- organization of IAU Symposium 249: Exoplanets, Detection, Formation and Dynamics, in Suzhou (China), in 2007;
- preparation of the Triennial Report.

The journal *Celestial Mechanics and Dynamical Astronomy* is in a good state. Commission 7 also spent some time considering whether, and what, symposia the commission might wish to organize. The most specific of these was a proposal to organize a 2012 symposium in Finland on the occasion of the 100th anniversary of the publication of Karl Sundman's pivotal work on regularization methods in the three-body problem. This will be considered further. Improvements to the Commission's web page were also suggested, and many have already been implemented.

The report of **Commission 8 (Astrometry)** was presented by its President, Irina Kumkova. She touched the following topics in which the Commission was active:

Instrumentation and Reduction Methods: The activities in Brazil (CCD heliometer), China (NEOST – Near Earth Objects Space Telescope and LAMOST – Large Sky Area Multi-object Fiber Spectroscope Telescope), France (the reduction of mosaic CCD images obtained at the 3.6m CFHT in order to derive coordinates of pulsars and QSOs in the optical at the mas level), Russia (modernized automatic telescope MTM-500M for Solar System bodies astrometric and photometric observations), Ukraine (artificial satellites and space debris, small solar system bodies), UK (new reduction of the Hipparcos data) and USA (StarScan plate measuring machine completed at USNO, measuring all applicable Black Birch, AGK2, and Hamburg Zone astrograph plates) were reported.

Space Astrometry: The European Space Agency's astrometry mission Gaia is very important because of its ability to obtain accurate astrometric measurements for very large samples of stellar, extragalactic and solar-system objects, and the matching collection of synoptic, multi-epoch photometric and radial-velocity data as well as to address an extremely broad range of topics in galactic and stellar astrophysics, solar-system astronomy, reference frame and fundamental physics. JASMINE and Nano-JASMINE projects being developed in Japan. HST FGS data were used for mas-precision of astrophysically interesting stars, to establish perturbation orbits due to planetary-mass companions of nearby stars. The observations continue for RR Lyr and Pop II Cepheid stars to calibrate a Pop II Period-Luminosity relationship.

Reference Frames: Investigations and studies are active in Brazil (work on the extragalactic frame concentrates on the reconciliation between optical and radio positions). A list of 173 candidate stable sources was proposed; the precession and equinox motion correction were obtained from various samples of PPM and ACRS proper motion data in China, the Large Quasar Astrometric Catalogue was derived in France; relative positions of reference stars around ICRF radio sources were improved in Rumania.

Positions and Proper Motions: Extensive studies were made in Brazil, France, Germany, Italy, Japan, Russia, Spain, Ukraine, USA and China.

Open and Globular Clusters were studied in Brazil, China, Italy, Ukraine and USA.

Education in Astrometry: The first Chinese-French Spring School on Astrometry 'Observational campaign of solar system bodies', was held in Beijing, April 7-13, 2008. A new web-page 'Infrared Astrometry' (Kharin and Vedenicheva) was launched at the web-site of the MAO NAS of Ukraine http://www.mao.kiev.ua/IR. New trends of research oriented towards perspective usage of plate archives of MAO NAS of Ukraine as an element of Virtual Observatory was designed. A database of CCD, photographic observations and astrometric catalogues of NAO, Ukraine was made (Protsyuk) within the framework of IVOA. The databank of photographic observations of NAO is included in the WFPDB (Bulgaria).

Symposia, Colloquia, Conferences:
- Journées 2007 The Celestial Reference Frame for the Future, Meudon, France, September 17–19, 2007
- IAU Symposium 248 A Giant Step: From Milli- To Micro-Arcsecond Astrometry, Shanghai, China, October 15–19, 2007
- ADELA 2008 – IV Meeting on Dynamical Astronomy in Latin-America, Mexico City, February 12–16, 2008.

The report of **Commission 19 (Rotation of the Earth)** was presented by its President, Aleksander Brzeziński. Members of Commission 19 concentrated on different problems concerning Earth rotation. This phenomenon is conventionally described by the Earth Orientation Parameters (EOP) which are determined by the space geodetic techniques with continuously improving accuracy and temporal resolution. New observation techniques have been recently developed, such as the ring laser gyroscope and the new generation system VLBI2010 with small antennas. The advances in monitoring variations in Earth rotation are closely connected to the establishment of the IAG Global Geodetic Observing System (GGOS). Besides tidal influences of the external bodies, the principal causes of EOP variation are changing motions and mass distribution of the fluid parts of the planet. Therefore the development of observations and modeling of geophysical fluids, the atmosphere, oceans, and land hydrosphere, has been considered as an important mean for understanding Earth rotation variability.

Considerable progress could be achieved thanks to the observations of mass redistribution done by the gravity recovery and climate satellite experiment GRACE. Commission 19 closely cooperated with the International Earth Rotation and Reference Systems Service (IERS) on updating the IERS Conventions which provide a set of astronomical and geodetic constants and fundamental procedures. Besides determination of the EOP and the related geophysical excitation parameters, there were also many efforts to improve forecasts of these quantities, which are needed for spacecraft navigation and other practical purposes. Further refinement of the theoretical description of Earth rotation is considered as an important task for the commission activity during the next term. It is planned to organize for this purpose a common IAU/IAG Working Group on 'Theory of Earth rotation'.

President of **Commission 31 (Time)**, P. Defraigne, presented the main results achieved by the commission members during the last triennium. Concerning the relative frequency stability and accuracy of the atomic time scales, the 1-month stability of the International Atomic Time (TAI) is currently close to 3×10^{-16}, which is a limit of present clock performance. However, the long term stability will be improved in the near future thanks to improvements of the algorithms. The last version of TT(BIPM) was also presented with its present uncertainty of 5×10^{-16}. The recommendations of the Consultative Committee for Time and Frequency having an impact on the Astronomical Activities have also been reported. The pulsar time scales were shown to achieve a stability of 10^{-15} at 7 years.

The meetings relating to the scope of the Commission that were held during the triennium, in addition to the usual meetings of the time and frequency community, comprised the Fifth International Time Scale Algorithm (San Fernando, April 2008), IAU Symposium 261 'Relativity in Fundamental Astronomy: Dynamics, Reference Frames, and Data Analysis', (Virginia Beach, April 2009) and the Joint Discussion 'Time and Astronomy' during the 27th IAU General Assembly in Rio.

The activities of **Commission 52 (Relativity in Fundamental Astronomy)** were presented by its President, Sergei Klioner. The general goals of this newest Commission, created in 2006, are to clarify geometrical and dynamical concepts of Fundamental Astronomy within a relativistic framework; to provide adequate mathematical and physical formulations to be used in Fundamental Astronomy; to deepen the understanding of the obtained results among astronomers and students of astronomy, and to promote research needed to accomplish these tasks.

The projects on which the Commission works are the following:
- Compile a list of unresolved problems;
- Frequently asked questions;
- Relativistic glossary for astronomers;
- 'Task teams', *i.e.* ad hoc discussion groups for well posed issues (e.g., for TDB units).

The Commission understands the 'Applied Relativity' as a multidisciplinary research topic, with important social and educational problems:
a) People doing practical work have limited knowledge of relativity and often can not understand the details of the suggested relativistic models.
b) People working in relativity have limited experience with real data and often can not judge if what they suggest is at all relevant.
So, basic education in relativity should be a part of astronomical education and discussions between the two above mentioned groups are necessary.

The above mentioned problems were discussed at IAU Symposium 261 'Relativity in Fundamental Astronomy: Dynamics, Reference Frames, and Data Analysis' (Virginia Beach, USA, 27 April – 1 May, 2009), with 92 participants from 18 countries.

The topics on which the work will be mainly concentrated in the future are the definition of the ecliptic (or 'ecliptic image') in the Geocentric Celestial Reference System (GCRS), and the system of astronomical units.

3. Business Meeting, Monday 3 August 2009, 14:00 hr

This session, chaired by Jan Vondrák, was devoted to the reports of Division I Working Groups and associated Services.

WG **Second Realization of the International Celestial Reference Frame** (ICRF2), led by Chopo Ma, prepared the solution that is based on VLBI data in S/X bands through March 2009. 4540 sessions were used with about 6.5 million observations of 3414 different radio sources. Out of these, 295 sources are defining, position uncertainty of individual sources is $\geqslant 40\mu$as, the accuracy of the orientation of the axes is 10μas. The orientation is independent of the equator, ecliptic and equinox. The solution was accepted by the WG for ICRF2, and the Directing Boards of the IERS and IVS. All details of the solution are given in IERS Technical Note 35 'The second realization of the International Celestial Reference Frame by Very Long Baseline Interferometry', published by Bundesamt für Kartografie und Geodäsie, Frankfurt a.M., Germany, and is also available in PDF format at `http://iers.org/MainDisp.csl?pid=46-25772`. The WG proposed the IAU resolution on ICRF2, and if it is approved by the IAU General Assembly, the WG is to be disbanded.

Brian Luzum presented the report on the WG **Numerical Standards of Fundamental Astronomy** (NSFA) that he chairs. The WG was established at the IAU 2006 General Assembly to update the IAU Current Best Estimates (CBE) of astronomical constants while conforming with existing resolutions and conventions where possible. In order to ensure consistency, the Working Group coordinated with IAU Commissions 4 and 52, the IERS, and the BIPM Consultative Committee for Units. Following extensive discussions by e-mail as well as a few meetings, a new set of CBE was proposed for use as the IAU 2009 System of Astronomical Constants, and the corresponding IAU resolution was prepared, to be approved by the IAU General Assembly. In the 2009-2012 triennium, the WG will establish the procedures for a CBE electronic document and resolve the differences regarding the future of the astronomical unit. It is anticipated that IAU Division I will, in the 2009-2012 triennium, find a permanent home for the responsibilities of the WG.

Report on the activities of the WG **Astrometry by Small Ground-based Telescopes** was presented by William Thuillot, chairman of the WG, which includes 29 people from 12 countries. The goal of the WG is to maintain information on the astrometric activities performed with telescopes having diameters up to 2 meters, to facilitate the collaboration between the teams, and to encourage coordinated activities. These instruments are numerous, worldwide, generally easy to access and are therefore well adapted to medium and long-term programs and to observations in networks. They are well adapted to the teaching of practical astronomy and astrometry. Thus the WG is also interested in teaching astrometry. A web site located at `www.imcce.fr/astrom/` disseminates information on the WG activity. Several research teams are involved in the WG, several groups are already collaborating and all are very active in the WG activity domain.

The following activities are performed: Zacharias *et al.* in USA, densification of reference frames; Van Altena *et al.*, stellar astrometry and the providing of the Yale/San Juan Southern Proper Motion Catalog; Muinos *et al.* in Spain, stellar astrometry and the providing of the Carlsberg Meridian Catalog; Wenjing *et al.* in China, asteroids astrometry and natural satellites studies; Vieira-Martins, Assafin *et al.* in Rio, natural satellites astrometry and optical counterpart of extragalactic sources useful for the celestial reference frame; Pinigin *et al.* in Ukraine collaboration with teams in Kazan, Istanbul and Shanghai, space debris and Near-Earth asteroids; Pauwels *et al.* in Belgium, asteroids astrometry; Souchay *et al.* in Paris, astrometry of extragalactic sources and GBOT project of astrometry of the Gaia probe; Thuillot, Arlot *et al.* in Paris, mutual events of natural satellites campaigns and a ground-based follow-up network for the Solar System objects detected by Gaia.

Two specific actions were organized by members of the Working Group:

• In order to gather information on astrometry teaching, a census has been performed in March 2008 and several interesting answers were received (see the WG website).

• A spring school of astrometry was organized and held on 7-12 April 2008 in Beijing, China, with the topic 'Observational campaigns for Solar System bodies'. 28 PhD and postdoc students attended this training.

The report of WG **Cartographic Coordinates and Rotational Elements** was presented by its chairman, Brent Archinal. The WG, now consisting of 18 members from 10 countries, was established in 1976, and belongs to Divisions I and III. So far it prepared and published nine reports. Certain concerns were expressed about possible 6-year limits on WG chairs and limits on WG existence. The WG is now finishing the discussion on new issues for the triennial report. The present status of the current WG progress/issues is the following:
Mercury model: New radar results (J. Margo); continued discussions about feature-based vs. axis-of-figure system.
Moon: DE 421 ephemeris and mean Earth/polar axis system (recommendation from NASA Lunar Geodesy and Cartography WG).
Mars: No changes (recommendation from NASA Mars Geodesy and Cartography WG).
Saturnian moons: New results from Cassini; Titan, icy Satellites, Saturn rotation.
Jupiter/Saturn/Neptune rotation? Expertise needed - help wanted!
Small bodies: Subgroup to decide which to include.
Category for dwarf planets: Separate category for now; will possibly treat same as small bodies.
Long term issues (not resolved now): Work on re-establishing International Association of Geodesy affiliation; develop recommendations for extra solar planets.
Jean-Eudes Arlot reported on WG **Natural Satellites** that he is chairing. It belongs to Divisions I and III because of its interest both in dynamics and physics. The working group coordinates the study of the families of satellites:
- main satellites such as Galileans, eight Saturnians, five Uranians, Triton. They are small solar systems interesting because of resonances, Laplacian libration of the Galileans, evolution of the systems, internal dissipation of energy.
- outer irregular satellites: formation and evolution
- inner satellites: relationship with rings, evolution.
Some of these objects are goals for the space probes: Jupiter-Europa: project 'Laplace'; Saturn-Titan: project 'Tandem'; Pluto-Charon: mission 'New Horizons'. The topics of interest of the working group are: Celestial mechanics, Astrometry; Physics of the surfaces and internal structures; Formation and evolution of the solar system; Astrometry of the giant planets reachable through their satellites. The studies of the working group are of interest for the planets: the giant planets are difficult to observe for astrometric purpose, but the satellites are easier to observe and their ephemerides provide the position of the center of mass of the planet.

Works performed during the last triennium: Observations:

• USNO (Flagstaff): transit circle observations,
• USNO/IMCCE/Brussels/Pulkovo: scan of old plates,
• CNPq Brazil: CCD observations of all systems,
• IMCCE: mutual events campaigns (Uranus, Saturn, Jupiter),
• MPC: gathering outer satellites observations.

Ephemerides:

• JPL: all satellites (R.A. Jacobson),
• IMCCE and SAI: all satellites (J.E. Arlot, N. Emelianov),
• IAA: Galilean satellites (G. Krasinsky),
• MPC: outer irregular satellites,
• IMCCE and SAI: extrapolation of errors in ephemerides.

Note that observations are needed each year for all satellites! Some faint outer satellites were observed only once by large telescopes. So, they are lost soon. The observations have been stopped after 2003 or 2006, depending on the satellites.

The astrometric database NSDC is maintained through the working group. The main interest of the working group is to gather all available astrometric observations and to encourage the observational campaigns needed to be international, *i.e.* organized worldwide. The members of the working group help keep the database exhaustive (at 90%). It is maintained thanks to IMCCE (Paris) and SAI (Moscow).

The WG encourages the improvement of astrometric accuracy by means of:
- evaluation of the distance center of mass photocenter (need to know the surfaces);
- taking into account the changing absorption of the sky for moving objects (need monitoring of a photometric source);
- the star catalogue used for the reduction (corrections of zonal errors, use of a new catalogue UCAC3, GAIA,);
- re-reducing the scanned old photographic plates with new methods and new star catalogues.

The activities of the **International Earth Rotation and Reference Systems Service** (IERS) were described by the chairman of its Directing Board, Chopo Ma. The news from the IERS Product Centers is the following:

• Earth Orientation: Web service for Earth Orientation Parameters (EOP) and Earth orientation matrix; new EOP series IERS05 C04 consistent with the new terrestrial system ITRF2005 (see below); revision of Bulletin B, following community survey.

• Rapid Service / Prediction: Bulletin A changed to match IERS05 C04; initiated WG on prediction; web site moved and revised considerably.

• International Terrestrial Reference System (ITRS): Released ITRF2005, based on time series of SINEX files, containing site positions and EOP as combined by the Technique Centers, where available; issued call for participation for ITRF2008, now in progress; participation in some local surveys of co-located sites.

• International Celestial Reference System (ICRS): Increased VLBI observations in the southern hemisphere; edited IERS Technical Note 35 on ICRF2; drafted resolution on ICRF2 for IAU GA.

• IERS Conventions: Updated various chapters in online version IERS Conventions (2003); preparing for new Conventions (2009).

• Global Geophysical Fluids Center: There are eight Special Bureaus (Atmosphere, Core, Gravity/geocenter, Hydrology, Loading, Mantle, Oceans, Tides) that were reviewed by the IERS Directing Board and will be re-constituted. A call for proposals was released in May 2009, with the goal to establish operational products and include new products.

Combination Research Centers are to be replaced by Research Centers, responsible for specific topics related to Product Centers. Working Groups are established for:
- Prediction (studying EOP prediction at various time scales);
- Site Survey and Co-location (reconstituted with the new charter, chair and membership);
- Combination (will be ended on completion of Combination Pilot Project);
- Combination at the Observation Level (newly established);
- ICRF2.

The Central Bureau in Frankfurt further developed the IERS Data and Information System (DIS), based on modern technologies for internet-based exchange of data and information using XML and the generation and administration of ISO metadata. It also provides access to and visualization of IERS data products, and links to IERS components.

There were several workshops organized:

• Workshop on Global geophysical Fluids (San Francisco, CA, USA, December 6–7, 2006);

• Workshop on Conventions (Sèvres, France, September 20–21, 2007);

• FFOS Unified Analysis Workshop (Monterey, CA, USA, December 5–7, 2007);

• GGOS Unified Analysis Workshop, Follow-Up Meeting (Vienna, Austria, April 15, 2008),

and several IERS publications produced (IERS Technical Notes 34 and 35, IERS Annual reports 2005 and 2006, IERS Bulletins A, B, C and D – weekly to semi-annually, IERS Messages No. 92 trough 149).

The work done at the **International VLBI Service for Geodesy and Astrometry** (IVS) was described by Harald Schuh. During the report period the IVS held two General Meetings, one in Concepción, Chile in January 2006 and the other in Saint Petersburg, Russia in March 2008. Further, two Technical Operations Workshops were held at MIT Haystack Observatory in Westford, MA in April/May 2007 and April 2009, respectively. Another important meeting was the VLBI2010 Workshop on Future Radio Frequencies and Feeds (FRFF) in Wettzell, Germany in March 2009. The FRFF workshop resulted in recommendations for the VLBI2010 frequency range and a demonstration campaign in 2012, which were subsequently endorsed by the IVS Directing Board.

The IVS completed its first ten years of being a service for geodetic and astrometric VLBI on March 1, 2009. To commemorate the first decade a 10th Anniversary Celebration event was held in Bordeaux, France on March 25, 2009. The event included a symposium featuring the history of VLBI and the IVS, the interrelation of the IVS with the other space geodetic services (IGS, ILRS, IDS), and IVS' place among the other VLBI networks (EVN, VLBA, Asian networks). The event was live broadcast over the Internet. A recording of the various presentations is available at `http://canalc2.u-strasbg.fr/video.asp?idvideo=855`.

In August 2008, a 15-day continuous VLBI observation campaign called CONT08 was observed. The network consisted of eleven IVS stations. Unlike the CONT05 campaign, CONT08 was observed on the basis of UT days, *i.e.*, an observing day was run from 0 UT to 24 UT. Observational gaps between the single observation days (30 min in the CONT05 case) were avoided by performing the daily station checks (*e.g.*, pointing) not at the change of schedules but at well-coordinated, staggered times for all stations (*i.e.*, different daily check times for each station). A special issue of Journal of Geodesy on CONT08 is in the planning stage.

The IVS VLBI2010 Committee (V2C) submitted a Progress Report on the status of the development of the next generation geodetic VLBI system (VLBI2010 system), which summarizes the progress made in the development of the new system up to the end of 2008. The report was published as a NASA Technical Memorandum TM-2009-214180 and is available online on the IVS Web site.

The report on **Standards of Fundamental Astronomy** (SOFA) was presented by Patrick Wallace. SOFA is a service operated by Division I to provide authoritative fundamental-astronomy algorithms, for example, precession-nutation models. The activity is carried out by an international board that includes both astronomers and software experts, and has cross-membership with the IERS Directing Board.

During the period 2006-2009, the SOFA software collection has grown from 121 routines to 160, the increase being due to the introduction of the IAU 2006 precession. The software has a core of 41 canonical models, used by 67 astronomy-related routines and supported by 52 vector-matrix utilities.

The triennium saw the development of a second version of the SOFA software, this time in the C programming language and supported in parallel with the existing Fortran version. At the same time, the SOFA licensing conditions have been made more liberal, to harmonize with IAU/IERS practices in general and to encourage the widest possible use. SOFA's products are disseminated through a website: `http://www.iau-sofa.rl.ac.uk/`.

Considerable thought has been given to SOFA's formal position within the IAU organization, with the object of raising its profile and increasing user confidence while at the same time complying with IAU statutes, bye-laws and working rules. These discussions are still in progress.

4. Business Meeting, Monday 3 August 2009, 16:00 hr

The session was devoted to general problems of the Division and discussion on its future structure, and was chaired by Sergei Klioner and Dennis McCarthy.

Nicole Capitaine, IAU Representative to the **Comité Consultatif des Unités** (CCU), reported on the topics discussed during the recent CCU meeting 2009, relevant to Division I.

1) The CCU was consulted on the proposal by Capitaine & Guinot (2008) to the IAU NSFA Working Group about the astronomical unit (ua) of length. According to that proposal, the ua would be re-defined trough a fixed relation to the SI meter by a defining number. The ua would then no more be determined experimentally, which would directly permit the possible variation of the mass of the Sun, and/or of G. This would limit the role of the ua to that of a unit of length of 'convenient' size for solar system applications. The CCU declared its support to move to a fixed relationship to the SI meter through a defining number determined by continuity with past determinations. The CCU, also consulted on the symbol to be used ('ua' or 'au'), suggested

that the astronomers should decide upon an appropriate symbol for this unit independently of the language.

2) The CCU discussed a Note by Capitaine & Guinot suggesting improvements in the definition of the second, namely (i) that the definition should distinguish the concept and its realization and (ii) that the SI brochure make clear that the SI units (s and m) should be used for coordinate quantities in GR (*e.g.* the second for the coordinate times TDB and TT). This is in agreement with the conclusion of a paper by Klioner *et al.* (2009) within IAU Commission 52 on Relativity in Fundamental astronomy. This was acknowledged by the CCU.

3) Following a problem raised by the International Union of Pure and Applied Chemistry (IUPAC) about the fact that there exists no official definition of the year, and that the symbols 'yr', 'y' and 'a' are all used, the CCU established a small group in order to propose a formulation for the definition of the year, and an official symbol. That group would include a representation from IAU (N. Capitaine), IUPAP and ICRU (International Commission on Radiation Units and Measurements) and should report at the next CCU meeting.

Jan Vondrák then reported on the activities of **Division I (Fundamental Astronomy)**. He reminded the audience of the present structure (6 Commissions, 5 Working Groups, including the two common with Division III, and 3 associated Services), and stressed that most of the scientific activities are made through individual Commissions and Working Groups that reported during the first two sessions. In the preceding triennium, Division I coordinated and/or supported the following IAU meetings:

- IAU Symposium 248 A Giant Step: from Milli- to Microsecond Astrometry, October 15–19, 2007, Shanghai, China;
- IAU Symposium 249: Exoplanets, Detection, Formation and Dynamics, October 22–26, 2007, Suzhou, China;
- IAU Symposium 261 Relativity in Fundamental Astronomy, April 27 – May 1, 2009, Virginia Beach, USA;
- IAU Joint Discussion 6 Time in Astronomy, August 5–6, 2009, Rio de Janeiro, Brazil;
- IAU Special Session 9 Marking the 400th Anniversary of Kepler's *Astronomia Nova*, August 11–14, 2009, Rio de Janeiro, Brazil;
- Mathematics and Astronomy, a Joint Long Journey, November 23–27, Madrid, Spain.

Organizational activities of the Division President consisted mainly in regular annual evaluation of proposals for IAU Symposia (early 2007 – Symposia in 2008; May 2008, common meeting of Division Presidents with IAU Executive Committee in Oslo – Symposia, Joint Discussions, Special Sessions at the 27th IAU General Assembly in 2009; early 2009 – Symposia in 2010). The recommendations of DP were mostly accepted by IAU EC, with only a few exceptions.

Elections of the new President (P), Vice-President (VP) and Organizing Committee (OC) for the next triennium 2009-2012. The former VP, Dennis McCarthy agreed to serve as P for the next period. Next VP was elected by all members of the Division I (about 700) in e-mail ballot from the two candidates (Aleksander Brzeziński and Sergei Klioner), proposed by the outgoing OC. Sergei Klioner received a majority of votes and becomes the next VP. For the selection of the next OC membership the Division followed the same practice as introduced at the 26th IAU GA in Prague: OC consists of Presidents and Vice-Presidents of all six Commissions pertaining to Division I.

During the last year or so, an electronic discussion on the future structure of the Division took place, among OC members and some members of WG's. The main goal was to find a more permanent place for the people taking care of the astronomical constants, numerical standards, software *etc.*.. than a Working Group. There were several possibilities open, *e.g.*, to create a new Commission, or to widen the scope of an existing Commission (C4?). However, none of the proposed solutions was acceptable for all concerned, and no consensus was found. Therefore it was decided that a discussion will be organized at this meeting with a goal to find a solution for the future.

The following discussion was led by Dennis McCarthy. He began by outlining the potential issues for the future:
- Astronomical constants

- Gaussian gravitational constant, Astronomical Unit, GMSun, geodesic precession-nutation
- Astronomical software? - Solar System Ephemerides?
- Pulsar research - Comparison of dynamical reference frames
- Future Optical Reference Frame - Future Radio Reference Frame
- Exoplanets: Detection, Dynamics - Predictions of Earth orientation
- Pulsar timing
- "Units of measurements" for astronomical quantities in relativistic context
- "Astronomical units" in the relativistic framework
- Time-dependent ecliptic in the GCRS - Asteroid masses
- Review of space missions - Detection of gravitational waves
- VLBI on the Moon - Real time electronic access to UT1-UTC.

After the discussion it was decided that an ad hoc sub-committee will be formed to prepare the proposals for the changes in Division I structure, to be adopted at the next IAU GA.

5. Scientific Meeting on Gaia, Tuesday 4 August 2009, 11:00 hr

The session was chaired by Jan Vondrák; four scientific papers, describing the ESA project Gaia from different points of view, were presented.

Gaia project: current status from ESA perspective was presented by Jos de Bruijne, working in the Gaia Project Scientist Support Team of the European Space Agency. He presented the status of the Gaia spacecraft development. The main message was that Gaia is moving from the paper and design to the hardware and testing phase. The torus (optical bench) has been brazed successfully, the various mirrors are being polished, and the CCD production is close to finalization. The next major milestone is the spacecraft Critical Design Review (CDR), which is due mid 2010. With a launch foreseen in 2012, the final Catalogue is expected in the 2020-2021 timeframe.

Dimitri Pourbaix presented the paper **Gaia astrometry: accuracy and processing for different classes of objects**, in which he described in detail how Gaia will scan the sky, the expected standard errors of the positions in terms of the magnitudes (for single stars ranging from magnitude 6 to 20 from 6 to 200μas), and how non-single stars will be treated. An important part of the presentation was devoted to exoplanet detection, and also to the Solar System Astrometry (orbits, masses from close encounters, diameters, photometric data, testing of General Relativity from perihelion precessions *etc...*).

The paper **Gaia photometry and spectroscopy: accuracy, data processing and applications** was presented by Laurent Eyer. He described the payload of the satellite, and concentrated on G-band photometry with Sky Mapper and Astrometric Field. The expected precision decreases with the growing magnitude, roughly from 1 mmag to 20 mmag for magnitudes between 10 and 20. It is estimated that between 50 and 150 million variable objects will be detected. Low resolution spectro-photometry with Blue and Red Photometer (BP, RP) CCDs will be made with the goals to determine (i) color to be used for the chromatic astrometric corrections, and (ii) astrophysical parameters of the observed sources; about one billion stars will be observed from magnitude 6 to 20. Radial Velocity Spectrometer (RVS) CCDs will be used to determine radial velocities of about 150 million stars up to magnitude 17.

Sergei Klioner then concluded the session with his talk **Astrophysical applications of Gaia astrometry**, where he touched on many astrophysical applications that can be expected from the Gaia mission. Gaia, being an astrometric project, is expected to have a deep impact on many areas of astronomy and astrophysics. This impact comes from the extremely powerful combination of three distinct qualities in a single instrument:
- very accurate, global and absolute astrometric measurements,
- large and complete (flux limited) sample of objects,
- the matching collection of multi-epoch spectro-photometric and radial-velocity measurements.

Giving a short overview of the expected results, one can mention (1) galactic structure and dynamics (large, volume-complete samples allow the determination of spatially resolved statistics in phase space); (2) stellar astrophysics (more than a million stars with distances known at about 1%, calibration of the standard candles, *etc.*); (3) reference frame with more than 1500 object per square degree, direct link to the quasars, final accuracy of about 0.4μas/yr; (4) fundamental physics (light deflection to about 10^{-6}; low frequency gravitational waves; *etc.*).

Jan Vondrák *President of the Division*

Transactions IAU, Volume XXVIIB
Proc. XXVII IAU General Assembly, August 2009
Ian F. Corbett, ed.

doi:10.1017/S1743921310004862

COMMISSION 4 EPHEMERIDES (*EPHEMERIDES*)

PRESIDENT	**Toshio Fukushima**
VICE-PRESIDENT	**George H. Kaplan**
PAST PRESIDENT	**George A. Krasinsky**
ORGANIZING COMMITTEE	**Jean-Eudes Arlot, John A. Bangert, Catherine Hohenkerk, Martin Lara, Elena V. Pitjeva, Sean E. Urban, Jan Vondrak**

PROCEEDINGS BUSINESS SESSIONS, Session 4 of August 7th, 2009

1. Introduction

Dr. George Kaplan, the current Vice-President of the Commission was nominated to be the new President. Dr. Catherine Hohenkerk was elected to be the next Vice-President of the Commission. As for the Membership of the Organizing Committee, Dr. Vondrak stepped down and Drs William Folkner of JPL and Steve Bell of HMNAO have been added. In the below, we present summaries of the reports from various institutions presented at the business session.

Toshio Fukushima

2. United States Naval Observatory, U.S.A.

This report covers activity in the Astronomical Applications (AA) Department since the XXVIth General Assembly in Prague. The AA Department employs 13 scientists in three divisions: The Nautical Almanac Office (NAO), the Software Products Division (SPD), and the Science Support Division (SSD). During the reporting period, A. Monet, formerly at USNO's Flagstaff Station, was appointed chief of the SPD and J. Bartlett transferred to the SPD staff from USNO's Astrometry Department. G. Kaplan became a part-time contractor to USNO, working within the AA Department. M. Murison transferred to the Flagstaff Station.

Kaplan served as vice president of Commission 4, and J. Bangert and S. Urban served on the organizing committee. Bangert also served as a member of the Standards of Fundamental Astronomy (SOFA) reviewing board. J. Hilton served as a member of the inter-division Working Group on Cartographic Coordinates and Rotational Elements. M. Efroimsky chaired the local organizing committee for IAU Symposium 261, held in April 2009 at Virginia Beach USA.

Publication of *The Astronomical Almanac* and *The Astronomical Almanac Online*, *The Nautical Almanac*, *The* (U.S.) *Air Almanac*, and *Astronomical Phenomena* continued as a joint activity between Her Majesty's Nautical Almanac Office of the United Kingdom and the NAO. A new memorandum of understanding between the parent organizations, governing the collaboration, became effective in August 2008. *The Astronomical Almanac* for 2009, released in January 2008, was the first edition to incorporate the resolutions adopted by the IAU in 2006. *The Air Almanac* transitioned from a paper publication to an electronic (CD-ROM) publication effective with the 2009 edition.

Significant progress was made on a major revision of *The Explanatory Supplement to the Astronomical Almanac*, in collaboration with P.K. Seidelmann (Univ. of Virginia) and numerous contributors. The book is expected to be sent to the printer by the end of 2009.

Version 3.0 of the Naval Observatory Vector Astrometry Subroutines (NOVAS), which implements relevant IAU resolutions adopted from 1997 through 2006, will be released by the end of 2009. The software will be available in both Fortran and C editions.

Version 2.2 of the *Multiyear Interactive Computer Almanac* (MICA), which incorporates NOVAS 3.0, was in the final stages of testing as of August 2009. MICA is available for computers running Microsoft Windows and Apple Mac OS operating systems.

All USNO departmental Web sites were consolidated into a single Web portal accessible at *http://www.usno.navy.mil/USNO*. Prior to the consolidation, usage of the AA Department Web site varied from about 0.5 to 2.8 million visits per month.

A modest research program in positional astronomy, dynamical astronomy, and navigation continued within the department. Research topics included the spin evolution of Iapetus, the theory of bodily tides, determination of asteroid masses, and new methods of celestial navigation.

Other projects underway at USNO and of interest to Commission 4 include the USNO CCD Astrograph Catalog (UCAC), and observations of solar system bodies made with the Flagstaff Astrometric Scanning Transit Telescope (FASTT). Additional information on these projects can be found at *http://www.usno.navy.mil/USNO/astrometry/*.

John A. Bangert

3. Her Majesty's Nautical Almanac Office, U.K.

In the reporting period, Her Majesty's Nautical Almanac Office (HMNAO) has been operating within the UK Hydrographic Office (UKHO), a trading fund of the United Kingdom's Ministry of Defence. HMNAO has three staff, two based at UKHO in Taunton, Somerset and one at Rutherford Appleton Laboratory in Didcot, Oxfordshire. As two of the staff will reach retirement age within the next reporting period, a recruitment process has been initiated. The first interview cycle was unsuccessful and the second is now under way. If successful, it is hoped to have a new member of staff in place before the end of 2009. It is planned that HMNAO's final complement will be four people. Despite operating within the UKHO for the past three years, HMNAO does not yet have a business plan, nor does it have appropriate means to sell its astronomical products. Our nautical products are sold successfully through the UKHO distributor network but our astronomical customers do not associate this distributor chain with HMNAO's astronomical products. New methods of distributing this material still have to be found. Concerns also exist about the peripheral nature of HMNAO within UKHO, its profile and the difficulty in attracting new staff to the group.

The Astronomical Almanac, *The Nautical Almanac* and *Astronomical Phenomena* continue as joint publications with the US Naval Observatory. *The Star Almanac* and *the UK Air Almanac* and the quinquennial products Rapid Sight Reduction Tables and *NavPac and Compact Data* continue as UKHO products. Indeed, the 2011-2015 edition of *NavPac and Compact Data* will be available at the beginning of 2010 which will include updates to the *NavPac* PC software. *The Astronomical Almanac* is fully compatible with all IAU resolutions including those of IAU GA 2006. Starting with the 2011 edition, lunar librations will be based on the rotation ephemeris of the Moon in DE403/LE403. *The Astronomical Almanac Online*, the web companion of the book, continues to be updated and expanded including material such as lunar occultation maps of the planets, Pluto and minor planets. HMNAO staff have also contributed to the new edition of *The Explanatory Supplement to The Astronomical Almanac* due for publication in early 2010. They have also participate in other research projects such as a study of the migratory patterns of the Sooty Shearwater.

HMNAO continues to operate six web sites; http://www.hmnao.com, its general web site, http://websurf.hmnao.com, offering "dynamic" data for the general public, http://asa.hmnao.com, a mirror of *The Astronomical Almanac Online*, http://www.eclipse.org.uk, solar and lunar eclipse data for 1501 to 2100CE inclusive, http://www.crescentmoonwatch.org, a public participation project and http://iau-sofa.hmnao.com, the SOFA web site. Further developments of these sites are planned in the next reporting period.

Steve Bell

4. Institute of Applied Astronomy, Russia

(*a*) Fundamental ephemerides. During the 2006–2009 the regular publication of *The Russian Astronomical Yearbook* is continued. Planetary and lunar ephemerides are based on numerical model EPM2004 available to outside users via `ftp://quasar.ipa.nw.ru/incoming/EPM2004`. Ephemerides for planetary configurations, eclipses and occultations, as well as the ephemerides of the Moon (as Tchebyshov polynomials) and the mutual phenomena in the system of the Galilean satellites of Jupiter are updated and located at `http://quasar.ipa.nw.ru/PAGE/EDITION/RUS/rusnew.htm`. The P03 precession and IAU2000A nutation theories have been introduced into practice. The sidereal time is calculated from Earth rotation angle and by new formula of Equation of the Equinoxes accepted by "IERS Conventions 2003". Fundamental catalogues FK6 and HIPPARCOS have been used for calculation of star ephemerides. The parameters concerned the new concept of CIO and elements of the matrix for conversion from ICRS to CIRS are given.

(*b*) Special ephemerides. *The Naval Astronomical Yearbook* (annual issues for 2006–2009) and biennial *The Nautical Astronomical Almanac* (issues 2007–2008, 2009–2010) have been published. The basic purpose of producing the Almanac is to increase its applicability without essential increase of its volume and to give the same accuracy as NAY does. The explanation and part of auxiliary tables are given in both Russian and English versions.

(*c*) Software. Constructing numerical dynamical models, fitting the ephemerides to observations, as well as preparation of the ephemerides for publishing are carried out in the framework of the universal program package ERA (Ephemerides for Research in Astronomy). The Windows and DOS versions are available via anonymous FTP `ftp://quasar.ipa.nw.ru/incoming/ERA`. The Linux version is developed. The first electronic version of *The Personal Astronomical Yearbook (PersAY)* has been constructed. It is intended for calculation of the ephemerides published in the Astronomical Yearbook, including the topocentric ephemerides for any observer. The system PersAY is implemented as the Win32 application on the basis of the package ERA. The first version of PersAY for interval 2000–2015 based the fundamental ephemerides DE405/LE405 and EPM2004 is available on `ftp://quasar.ipa.nw.ru/pub/PERSAY/persay.zip`. The electronic system the *Navigator* for solution of basic naval astronavigating problems by the mode of remote access is in progress.

(*d*) Research work. The updated Ephemerides of Planets and the Moon — EPM2008 (Pitjeva 2009, Proc. IAU Symp. 261) have been constructed by the simultaneous numerical integration of the equations of motion of the major planets, the Moon, the Sun, 301 biggest asteroids, 21 trans-Neptunian objects and the lunar physical libration accounting for the perturbations due to the solar oblateness and the massive ring of small asteroids. Some tests have been made for estimating influence other TNO on the motion of planets. Their perturbations have been modeled by the perturbation from a circular ring having a radius 43 AU and the estimated mass $M_{TNOring}=(498 \pm 14)\cdot 10^{-10} M_{\odot}$ (5σ). The parameters of EPM2008 (65 ones for the lunar part and about 260 ones for the planet part) have been fitted to lunar laser ranging measurements 1970–2008, as well as to 550000 planet and spacecraft observations 1913–2008 of different types. EPM2008 have been oriented to ICRF by including into the total solution the 118 ICRF-base VLBI measurements of spacecraft 1989–2007 near Venus and Mars. The numerical ephemerides of the main satellites of the outer planets have been constructed and fitted to modern photographic and CCD observations. These ephemerides are used for improving the ephemerides of their parent planets and publication in the Russian Astronomical Yearbook.

Elena V. Pitjeva

5. Institut de Mécanique Céleste et de Calcul des Éphémérides, France

The ephemerides service of IMCCE has three missions:

- performing research activities on the motions of the solar system objects,
- making the French official ephemerides on behalf of Bureau des longitudes,
- providing calculations on request for professionnal, space agencies and general public.

For the planets, the Sun and the Moon, we use:

- VSOP model(Secular Variations of Planetary Orbits)
- INPOP model (Planetary Numerical Integration of Paris Observatory)

The latter is a brand new 4-D theory based upon a high precision model for planets and the Moon, fitted on space observations, support of the next Gaia mission. INPOP06 has been published (Fienga *et al.* 2008, A&A, 477, 315) and INPOP08 is submitted. Ephemerides "Miriade" are provided on a web server.

For the natural satellites, the NOE model (Numerical Orbit and Ephemerides) is used for the Martian satellites, the Galileans and the main Uranians (Lainey *et al.*, A&A 2006, 456, 783, Arlot *et al.* A&A 2006, 456, 1173). An estimation of the propagation of the ephemerides error is provided (Desmars *et al.* A&A 2009, 499, 321) and the ephemerides web server (Multi-sat ephemerides) is made in collaboration with the Sternberg Astronomical Institute in Moscow.

Yearly printed publications are:

- Connaissance des temps since 1679 for high precision ephemerides
- Annuaire du Bureau des longitudes since 1795 for the general public
- Ephemerides nautiques, the french nautical almanac for the Navy

An electronic version is available for Connaissance des temps.

Electronic ephemerides are provided through Internet (5 Mhits per month, i.e. 70 000 users per month) at www.imcce.fr. Specific web services are provided in the Virtual observatory framework using VO standard protocols, metadata and VOTable for exchange of self-defined information. The Sky Body Tracker facility identifies any solar system object in any field and is implemented in CDS/Aladin. Asteroid search based upon pre-calculated ephemerides funded on Astorb database (from Lowell observatory), daily updated on the time span 1949-2009 (¿450000 asteroids) interfaced with Aladin Sky Atlas V3.6. More objects (all satellites, comets) will be added soon. The time period is going to be enlarged.

William Thuillot, and Jean-Eudes Arlot

6. Jet Propulsion Laboratory, California Institute of Technology, U.S.A.

The development of planetary ephemerides at JPL continues to be driven by needs for planetary missions and improved by observations of spacecraft and other astrometric observations. Current missions with need of improved ephemerides include the MESSENGER mission to Mercury and the New Horizons mission to Pluto and the upcoming Mars Science Laboratory mission.

In the past three years the accuracy of the ephemerides of Venus and Saturn have improved dramatically due to observations from the ESA Venus Express mission and the NASA Cassini mission. The Mars ephemeris accuracy is being maintained by continuing observations of Mars orbiting spacecraft, such observations needed to compensate for the disturbance by asteroids that would otherwise cause the ephemeris accuracy to degrade. The outer planet ephemerides are being slowly improved by continuing ground-based astrometric observations, and some older observations being re-processed against modern ICRF-based star catalogs.

The most recent JPL ephemeris released in DE421. The next release is planned for September 2009, primarily to include data from two encounters of Mercury by the MESSENGER spacecraft and in preparation for its third Mercury encounter in November 2009. Following that the Mars spacecraft VLBI data set will be re-reduced against the recently released ICRF 2.0, followed by an new ephemeris fit in spring 2010.

William M. Folkner

7. National Astronomical Observatory of Japan and Japan Hydrographic and Oceanographic Department, Japan

In National Astronomical Observatory of Japan (NAOJ), Project of enhancing "Calendar and Ephemeris", which is a basic almanac designed for astronomical observers, teachers and citizens is currently underway. As the first step, NAOJ not only implemented new precession formula adopted by IAU in 2006, but also enhanced its volume from 2009 edition, for example, almost doubled its size and its number of pages. In addition, NAOJ has added more tools to its web site http://www.nao.ac.jp/koyomi/, while establishing a web site for mobile phones as well as English version for foreigners living in Japan.

Japan Hydrographic and Oceanographic Department (JHOD) finished occultation observations at the hydrographic observatory at the end of March, 2008 and completely withdrew from astronomical observations. JHOD discontinued the international Lunar Occultation Centre (ILOC) service on March, 2009 and the International Occultation Timing Association (IOTA) took over this role. JHOD delivered the 2010 edition of the Japanese Ephemeris as the final issue, and will publish only Nautical Almanac and Abridged Nautical Almanac hereafter.

Masato Katayama

Transactions IAU, Volume XXVIIB
Proc. XXVII IAU General Assembly, August 2009
Ian F. Corbett, ed.

doi:10.1017/S1743921310004874

COMMISSION 7 — CELESTIAL MECHANICS AND DYNAMICAL ASTRONOMY
(MÉCANIQUE CÉLESTE ET ASTRONOMIE DYNAMIQUE)

PRESIDENT	**Joseph A. Burns**
VICE-PRESIDENT	**Zoran Knežević**
PAST PRESIDENT	**Andrea Milani**
ORGANIZING COMMITTEE	**Evangelia Athanassoula, Cristián Beaugé, Bálint Érdi, Anne Lematre, Andrzej Maciejewski, Renu Malhotra, Alessandro Morbidelli, Stanton J. Peale, Miloš Sidlichovský, David Vokrouhlický, Ji-Lin Zhou**

PROCEEDINGS BUSINESS SESSION, 5 August 2009

1. Introduction

The meeting began at 11:00 am with a brief address by outgoing president Burns highlighting the most relevant advances in Celestial Mechanics that occurred in the last 3 years.

2. Membership status

Then, Burns presented a summary of the Commission membership in quantitative terms. There are 275 members presently active and 38 new applicants. He asked for a minute of silence in memory of the 6 deceased members during the last triennium: Roger Broucke (1932–2005), Jacques Henrard (1940–2008), Božidar Jovanović (1932–2008), J. Derral Mulholland (1934–2008), Viktor Novoselov (1925–2007), and Arpad Pal (1925–2006).

3. Past activities

The president gave a summary of the activities of the Commission during the last 3 years, emphasizing the following issues:

- organization of the Symposium 236: Near Earth Objects, our Celestial Neighbors: Opportunity and Risk, in Prague (Czech Republic), in 2006;
- preparation of the Triennial Report.

Ferraz Mello recalled that the Commission also proposed and organized Symposium 249: EXOPLANETS, Detection, Formation and Dynamics, in Suzhou (China), in 2007.

4. New officers

President Burns introduced the Commission Officers for the next triennium: Zoran Knežević (president), Alessandro Morbidelli (vice-president), Evangelia Athanassoula, Renu Malhotra, and Stanton Peale (second term), Jacques Laskar, Seppo Mikkola, and Fernando Roig (first term). Roig will act as the Commission's secretary.

Burns stated that the first activity for the new Organizing Committee (OC) will be to prepare the Terms of Reference of the Commission.

5. Report on the journal *Celestial Mechanics and Dynamical Astronomy*

Ferraz Mello, editor-in-chief of the journal, presented his report for the last triennium (see Appendix). He pointed out that:

- the time to reach final decision on submitted papers has decreased,
- the ratio between accepted and refused papers has stayed constant at approximately 50%,
- the journal adopted a policy of publication of selected papers from conferences, but not proceedings,
- the present impact factor (IF) computed over 2 years is above 1.5,
- the IF computed over 5 years is more stable than the IF over 2 years, which shows large fluctuations,
- the number of citations of an article becomes relevant only after the 2nd and 3rd years from publication.

All this information is also available at `http://www.astro.iag.usp.br/~sylvio/celmech.html`

6. Address by incoming president

The outgoing president then passed the word to the incoming president Knežević, who congratulated Burns for his services to the community in the past triennium, smooth running of the Commission, and for successful representation of Commission with Division I and the IAU.

7. Future activities

Then, Knežević introduced the plans for the next triennium:

(*a*) To organize another symposium on Celestial Mechanics. It is necessary to define the subject, the organizers, and the dates. Knežević pointed out that it could be organized only in 2011 or 2012, because the proposal must be submitted to the IAU in September of the previous year, and there is no time to apply for 2010. Moreover, the last symposium organized by the Commission, S 249, was in 2007, and it might be difficult to have another symposium after a short period of time, especially if held in China where S 249 was held. Ferraz Mello said that the next symposium should preferably not be held during the next GA in Beijing, in 2012, due to the difficulties and big expenses for traveling to China. Mikkola proposed to organize such symposium in Finland, in 2012, on the occasion of the 100th anniversary of the work on regularization methods in the Three Body Problem by Karl Sundman (*Mémoire sur le problème des trois corps*, Sundman 1912). This proposal might be a good option, but to try to organize a symposium the same year as a GA may be a problem because only 3 symposia can be organized that year outside the GA. In any case, Mikkola should send a letter of intention to the Commission OC.

(*b*) To write the Terms of Reference in order to justify the continuity of the Commission to the IAU Division I. Ferraz Mello offered to send the manuscript of a similar document written some years ago.

(*c*) To improve the web page of the Commission, since the present one is largely outdated. It is desirable to have a more dynamic web page, including some items on top of the already existing information like highlighted publications in the area. It is also desirable to implement a system for updating the members list periodically and a member search engine. The new web page is currently under construction and its latest version is at `http://staff.on.br/iaucom7`.

(*d*) To search for and to propose hot topics to be included in the next Triennial Report.

(*e*) To prepare for the elections before the next GA.

(*f*) To propose how the Commission may contribute to improve each area of the field.

8. Closing remarks

After this, and having no more issues to discuss, president Burns closed the session at 11:55 am.

Joseph A. Burns
President of the Commission

9. Appendix

Celestial Mechanics and Dynamical Astronomy (CDMA)

Report for the period 2005–2008

CMDA has been published since 2007 with a new cover showing an artistic representation of a low- thrust trajectory between the Earth and the Moon, including an halo orbit around L2 and using a lunar gravity assist to direct the flight towards a distant target. This new cover reminds readers that mission analysis and other aspects of astrodynamics, such as determination, prediction, computation and selection of orbits are among the aims of CMDA, as defined in the first issue of the journal, in 1969. The number of published papers on astrodynamics is increasing and a large special issue is scheduled for publication later in 2009.

The impact factor calculated by the Institute for Scientific Information and published in the Journal Citation Reports (JCR) reached 1.56 in 2008, largely compensating for the low of 0.844 in 2007. The JCR impact factor measures the number of citations, in a given year, of papers published in a journal in the 2 previous calendar years and is impaired by large statistical fluctuations. The new more robust 5-yr factor, published in JCR since 2007, reached 1.27 in 2008. The analysis of previous years shows that the 5-yr impact factor of CMDA was close to or above 1.0 since 2004 (q.v. `http://www.astro.iag.usp.br/~sylvio/celmech.html`).

Recently about 50 percent of the submitted papers are being published. The average times between the submission of one (accepted) paper and the final decision is being kept within 6 months. Some extremely long times are still recorded, but they generally correspond to papers demanding time- consuming, thorough revisions.

The transients from the merger of Kluwer and Springer are over and the journal is being published rigorously according to the schedule. Because of continuing efforts, CMDA is currently reaching, through electronic and hardcopy subscriptions, more than 5,000 institutions. A few recent papers are being offered by Springer on a free-access basis, but it is important that authors use the rights allowed by current copyright rules to make easily available e-preprints of their articles. The old issues (before 1997) are available for free download through the NASA-ADS site, but Springer will not extend this allowance to more recent issues.

Sylvio Ferraz Mello
Editor-in-Chief

References

Sundman, K. F. 1912, *Acta Mathematica* 36, 105

Transactions IAU, Volume XXVIIB
Proc. XXVII IAU General Assembly, August 2009
Ian F. Corbett, ed.

doi:10.1017/S1743921310004886

COMMISSION 8 ASTROMETRY (ASTROMETRY)

PRESIDENT	**Irina Kumkova**
VICE-PRESIDENT	**Dafydd Wyn Evans**
PAST PRESIDENT	**Imants Platais**
ORGANIZING COMMITTEE	**Alexandre Andrei, Alain Fresneau, Petre Popescu, Ralf-Dieter Scholz, Mitsuru Sôma, Norbert Zacharias, Zi Zhu.**

PROCEEDINGS BUSINESS SESSION, 10 August 2009

1. Business session (Chair I. Kumkova)

The business meeting was opened by the President, Irina Kumkova. She presented the agenda, which was approved. This session was attended by 42 participants. The meeting approved Dafydd Evans as secretary of minutes. A minute of silence was dedicated to the memory of the Commission members who had passed away since Prague GA, namely:

Gérard Billaud (Grass Observatory, France),
Peter Stephen Bunclark (Institute of Astronomy University of Cambridge, UK),
Dmitry Polozhentsev (Pulkovo Observatory, Russia).

1.1. *Commission Activities 2006–2009*

Kumkova reported about the main work done during the triennium. Most of the information dissemination was conducted via the Commission's WWW home page at `http://www.ast.cam.ac.uk/iau_comm8/`. A total of five electronic newsletters were circulated during the 2006–2009 period. Unfortunately, several Commission members have not reported their changed and new e-mail addresses which resulted in a loss of communication. All members are strongly encouraged to check and update their personal data in the IAU membership directory. The difficult task undertaken was the compilation of the Triennial Report. Kumkova thanked the national organizers of the job in collecting the individual reports. The science highlights in this triennium were:

(*a*) A new reduction of Hipparcos data van Leeuwen (2007), which resulted in a higher accuracy of parallaxes, improved by up to a factor 5 for the brightest stars.

(*b*) Much effort was dedicated to the preparation of the Gaia mission, now approved to be launched in 2011.

(*c*) JASMINE – a Japanese infrared astrometric satellite – Nano-JASMINE will be launched in 2010.

(*d*) IAU Symposium 248 "A Giant Step: From Milli- To Micro-Arcsecond Astrometry" was very successfully held in Shanghai, China PR, October 2007.

1.2. *New Commission Members*

The following new Commission members were confirmed by the meeting:

1.2.1. *IAU members requesting membership of Commission 8*

Sergei Klioner (Russia)
Timo Prusti (Netherlands)
Jan Souchay (France)

1.2.2. *New IAU members requesting membership of Commission 8*

Ummi Abbas (USA)
Eswar Bacham (India)
Antoine Bouchard (South Africa)
Alfred Chen (Taiwan)
Lan Du (China PR)
David Hobbs (Ireland)
Belinda Kalomeni (Turkey)
Nadiia Maigurova (Ukraine)
Yury Nefed'ev (Russian Federation)
Sabine Reffert (Germany)
Alessandro Sozzetti (Italy)
Andrea Zacchei (Italy)
Yong Zheng (China PR)
Nakagawa Akiharu (Japan)
Octavian Badescu (Romania)
Yavor Chapanov (Bulgaria)
Melania Del Santo (Italy)
Octavi Fors (Spain)
Anatoliy Ivantsov (Ukraine)
Sebastien Lepine (Canada)
Claudio Mallamaci (Argentina)
Jinsong Ping (China PR)
Zhiqiang Shen (China PR)
Fuping Sun (China PR)
Wei Zhang (China PR)

The total membership in the Commission has now reached 259, representing 36 countries.

1.3. *Election Results and New Commission Officers*

Kumkova thanked cordially the outgoing members of the Organizing Committee: Imants Platais, Alain Fresneau, Ralf-Dieter Scholz, Mitsuru Sôma for their dedicated service. Elections were held for the positions of President, Vice-President and four vacant positions in the Organizing Committee. By IAU tradition the outgoing Vice-President, Dafydd Evans, becomes the new Commission President. Out of two nominations for Vice-President only one agreed to remain on the ballot list. Norbert Zacharias was elected as Vice-President unanimously by the Organizing Committee. Seven scientists were nominated to the new Organizing Committee. Four among these have been elected (taking into account geographical distribution): A. Brown (Netherlands), N. Gouda (Japan), J. Souchay (France), S. Unwin (USA). The new Organizing Committee was approved unanimously.

The Commission officers for 2009–2012 are
Dafydd W. Evans (UK) President
Norbert Zacharias (USA) Vice-President

Organizing Committee:
A. Andrei (Brazil)
A. Brown (Netherlands)
N. Gouda (Japan)
I.Kumkova (Russia)
P. Popesku (Romania)
J. Souchay (France)
S. Unwin (USA)
Z. Zhu (China)

1.4. *Report of the WG on Densification of the Optical Reference Frame*

Zacharias (USNO) presented the final report of this WG, who suggested to terminate the WG and received no objections from the members. More work in this area will continue by a few dedicated colleagues but it was felt that no IAU organizing structure is needed to support this activity.

The web site of the WG was moved to `www.astro.yale.edu/astrom/dens_wg/` where all 3 newsletters of the past triennium can be found together with a general overview about optical astrometric surveys and their future.

The achievements of the WG up to 2008 have been summarized in Zacharias (2009). Highlights for this year are the almost completed Southern Proper Motion (SPM) 4th catalog and the release of the third US Naval Observatory CCD Astrograph Catalog (UCAC3) at the XXVII GA. The Yale San-Juan first-epoch catalog (YSJ1) forms the basis for both the early epoch SPM4 and the UCAC3 southern proper motions.

New observing programs were briefly mentioned. Within the next few years large amounts of sky survey data are expected from projects like PanSTARRS (Hawaii), SkyMapper (Siding Springs), and URAT/U-mouse (USNO). Several astrometric space missions are in preparation as well: Jasmine (Japan), J-MAPS (USNO), and Gaia (ESA).

1.5. *General discussion and AOB*

Kumkova started the discussion by asking how do we make the Commission more active? Pourbaix thought that we didn't have to worry too much and that there was nothing to change. Look at the triennial report which lists many Commission 8 activities. Most of the rest of the discussion focussed on organizing more astrometric meetings.

van Altena thought that it was good to reinstate the science sessions at the General Assemblies, Souchay wondered if there is a big astrometry meeting planned? Kumkova said that we have had difficulty to get IAU support for these. We tried 3 or 4 times and the very good Shanghai meeting was the result. Small meetings are probably more likely. Also we need to wait for new results.

Klioner pointed out that next year there will be a Gaia meeting in Paris. It is important to stress the new face of astrometry. It is a tool for astronomy in general. We need to attract new people. Commission 8 should help with this.

Kumkova said that the China-France school was good for this and Capitaine pointed out that there will also be regular Journée meetings. Evans asked that any details about future meetings be sent to him to put on the Commission 8 website.

Popescu suggested that next year or so might be a good opportunity to have a small meeting on QSOs and the reference frame. He asked Souchay if he would be willing to organize this. Ma thought that this would be a good idea since the optical-radio link will still be important before Gaia catalogues get released.

Souchay pointed out that in Paris they have not had problems attracting young astronomers to do astrometry. Gaia will have a magnitude limit of 20, so transfering the reference frame to denser magnitudes using CCD observations will be important.

Kumkova finished by pointing out that we should not consider astrometry as separate from astrophysics. We will need to draw attention of young astronomers to the topic and perhaps organize groups on specific problems.

Two short presentations were given by van Altena on SPM4 and his new astrometry book.

2. **Science Sessions** (chairs N.Zacharias & D.W.Evans)

The following presentations were given during the meeting. Summaries are given here. Many of the presentations can be found on the meeting website:
`http://www.ast.cam.ac.uk/iau_comm8/iau27/`

2.1. *400 Years of Astrometry: From Tycho Brahe to Hipparcos*
Erik Høg (Niels Bohr Institute, Copenhagen)

The four centuries of techniques and results were reviewed, from the pre-telescopic era until the use of photoelectric astrometry and space technology in the first astrometry satellite, Hipparcos, launched by ESA in 1989. Galileo Galilei's use of the newly invented telescope for astronomical observation resulted immediately in epochal discoveries about the physical nature of celestial bodies, but the advantage for astrometry came much later. The quadrant and sextant were pre-telescopic instruments for measurement of large angles between stars, improved by Tycho Brahe in the years 1570–1590. Fitted with telescopic sights after 1660, such instruments were quite successful, especially in the hands of John Flamsteed. The meridian circle was a new type of astrometric instrument, already invented and used by Ole Roemer in about 1705, but it took a hundred years before it could fully take over. The centuries-long evolution of techniques was reviewed, including the use of photoelectric astrometry and space technology in the first astrometry satellite, Hipparcos, launched by ESA in 1989. Hipparcos made accurate measurement of large angles a million times more efficiently than could be done in about 1950 from the ground, and it will soon be followed by Gaia which is expected to be another one million times more efficient for optical astrometry.

2.2. *Series of JASMINE projects – Exploration of the Galactic Bulge*

Naoteru Gouda and JASMINE Working Group (National Astoronomical Observatory of Japan)

An introduction was given for the following series of JASMINE projects in Japan:

Nano-JASMINE project: Nano-JASMINE project is planned to demonstrate the first space astrometry in Japan and to perform experiments for verification of some techniques and operations for JASMINE. Nano-JASMINE uses a nano-satellite whose size and weight are about 50 cm^3 and 25 kg, respectively. The targeted accuracy of parallaxes is about 3 mas at z=7.5 mag. Moreover, proper motions with high accuracies (0.1 mas/year) can be achieved by combining the Nano-JASMINE catalogue with the Hipparcos catalogue. There is a high possibility that Nano-JASMINE will be launched by a Cyclone-4 rocket in July 2010.

Small-JASMINE project: Small-JASMINE is an astrometric mission that observes in an infrared band (Kw-band). The central wavelength is 2.0 micron. Small-JASMINE will determine positions and parallaxes accurate to 10 micro-arcseconds for stars in the Galactic bulge, brighter than Kw=11 mag, and proper motion errors of 10 micro-arcseconds/yr. It will observe small areas of the Galactic bulge with a single beam telescope with a diameter of the primary mirror of around 30cm. The target launch date is around 2015. The main science objective of small-JASMINE is to clarify the formation history of the Galactic bulge and also determine the moderate model of bulge structure formation.

JASMINE project: JASMINE is an extension of the small-JASMINE mission. It is designed to perform a survey towards the whole Galactic bulge region with a single-beam telescope with a primary mirror diameter of around 80cm, determining positions and parallaxes accurate to 10 micro-arc seconds for stars brighter than Kw=11 mag, and proper motion errors of 10 micro-arc seconds/yr. The target launch date is around 2010–2015

2.3. *A Very Small Satellite for Space Astrometry: Nano-JASMINE*

Yoichi Hatsutori, Naoteru Gouda, Yukiyasu Kobayashi, Taihei Yano, Yoshiyuki Yamada and the Nano-JASMINE team, National Astoronomical Observatory of Japan

The outline and the current status of the Nano-JASMINE project was presented. The objective of this project is a scientific, astrometric and technical demonstration for JASMINE and a first experience of space astrometry in Japan. Nano-JASMINE is a very small satellite for space astrometry. It is only 25 kg and aims to carry out astrometric measurements of nearby bright stars (z<7.5mag) with an accuracy of 3 milli-arcseconds. This satellite adopts the same observation technique used by the HIPPARCOS satellite. In this technique, two different fields of view are observed by a beam-combiner simultaneously. The Nano-JASMINE telescope is based on a standard Ritchey-Chretien type optical system and has a beam-combiner, a 5 cm effective aperture, a 167 cm focal length and a field of view of 0.5×0.5 degree. The major technical difference between Nano-JASMINE and HIPPARCOS is the CCD sensor. A full depletion CCD will be used in a time delay integration (TDI) mode in order to efficiently survey the whole sky in wavelengths, including the near infrared. By using TDI mode, Nano-JASMINE will achieve astrometric accuracy comparable to that achieved by HIPPARCOS but with a small satellite. From a scientific viewpoint, Nano-JASMINE will measure the same stars that were observed by HIPPARCOS with the same accuracy, then a significant improvement on the accuracy of proper motions can be made and correct the degradation of the HIPPARCOS catalogues. The current status of Nano-JASMINE is that it is in the process of production as an engineering model. Thermal tests and vibration tests have already been conducted with a Structure-Thermal Model (STM) last summer, and the design validation of the satellite was confirmed. Moreover, it is confirmed that the telescope can achieve the diffraction limit during the performance test. The plan is to launch Nano-JASMINE in 2010. National Astronomical Observatory of Japan, The University of Tokyo, Alcantara Cyclone Space and SDO Yuzhnoye have reached a consensus to launch Nano-JASMINE with the Cyclone-4 rocket in the Federal Republic of Brazil.

2.4. *The Second Realization of the International Celestial Reference Frame: ICRF-2*

Alan L. Fey, USNO

Construction of a second realization of the International Celestial Reference Frame, ICRF-2, has been underway for the last several years. The work was carried out by two working groups: the ICRF-2 Working Group of the International Earth Rotation and Reference System Service (IERS) in cooperation with the International VLBI Service for Geodesy and Astrometry

(IVS) and the ICRF-2 Working Group of the International Astronomical Union. The task of the IERS/IVS Working Group was to generate ICRF-2 from Very Long Baseline Interferometry observations of extragalactic radio sources consistent with the current realization of the International Terrestrial Reference Frame and Earth Orientation Parameter data products with oversight from the IAU Working Group. A brief summary of the results were presented.

2.5. *Strengthening the ICRS optical link in the northern hemisphere*

P. Popescu, A. Nedelcu, O. Badescu & P. Paraschiv, Astronomical Institute of the Romanian Academy

In 2005 the Astronomical Institute of the Romanian Academy has started an observational program, using the Belogradchik Zeiss Telescope (Bulgaria), to investigate the link between the International Celestial Reference Frame (ICRF) and its representation at optical wavelengths. 59 astrometric positions of ICRF optical counterparts were obtained with average values of the optical-radio offsets of +6 mas and +7 mas in R.A. and Declination and standard deviation of 51 mas and 57 mas respectively. The radio-stars astrometry program is in progress and it will be extended to include sources from the VLBA Calibrator Survey – the largest high resolution radio survey available.

2.6. *UCAC, NOMAD, URAT - star catalogs for astrometry*

Norbert Zacharias, USNO

Properties of the final release of the USNO CCD Astrograph Catalog were presented. This all-sky astrometric catalog supersedes UCAC2. The USNO Robotic Astrometric Telescope is a new observing program which will begin in late 2009. Plans for updates of the Naval Observatory Merged Astrometric Dataset (NOMAD) were discussed and recommendations given about the "best" star catalog to be used for astrometric reference stars for the general astronomer.

2.7. *Determination of L Dwarf Distances and Objects in the L/T Transition*

Richard Smart, Jucira Lousada Penna, Alexandre H. Andrei, Ramachrisna Teixeira, Beatrice Bucciarelli, Victor A. d'vila, Julio Ignácio Bueno de Camargo, Dario N. da Silva Neto, Mario Lattanzi, Kátia M.L. da Cunha

L and T dwarfs are ultracool objects, cooler than M dwarfs, which are fundamental to the understanding of the star/planet transition. They have spectra dominated by molecular absorption due to water, methane and pressure-induced molecular hydrogen. Since the first defining L dwarfs, GD165B was known in 1997, there have been nearly 500 discovered. These come primarily from the Sloan Digital Sky Survey and from 2MASS. Model atmosphere analyses indicate temperatures of 2500 to 750 K. To understand the intrinsic properties of ultra cool dwarfs and ultimately massive Jupiter-like exoplanets, it is essential to determine their absolute luminosities. The only direct method to achieve this is with astrometric parallaxes, yet to date less than 40 have measured parallaxes. In this project a systematic determination is undertaken of L and T dwarf parallaxes. While the sequence of subdwarf luminosities is already reasonably defined by the objects with known parallaxes, this program allows a substantial improvement on that calibration and allow for direct confrontation with the structure models for sub-stellar objects. The observations are being made at the WFI ESO2.2m, La Silla. The program started in April 2007 and has secured time at least to the end of 2009. It contains 140 objects, all of which already with four or more observations. Typically the observations are made every other month, and so far there are at least four observations for each object, up to ten observations. This has enabled a first determination of parallaxes to some objects and a comprehensive study of the 2MASS referred proper motion field. The astrometric repeatability is at 10mas. At this level there is a significant reduction on the length and number of observations usually required for this type of program.

2.8. *The LQRF - An Optical Representation of the ICRS*

A.H. Andrei, J. Souchay, N. Zacharias, R.L. Smart, R. Vieira Martins, D.N. da Silva Neto, J.I.B. Camargo, M. Assafin, C. Barache, S. Bouquillon, J.L. Penna

The large number and all sky repartition of quasars from different surveys combined with their presence in large, deep astrometric catalogues, enables us to build an optical materialization of the ICRS following its defining principles - namely, kinematically non-rotating with respect to the ensemble of distant extragalactic objects, aligned to the mean equator and dynamical equinox of J2000, and realized by a list of adopted coordinates of extragalatic sources. The LQRF (Large Quasar Reference Frame) was built with the care of avoiding misrepresentation of its constituent quasars, of homogenizing the astrometry from the different catalogues and lists from which the constituent quasars are gathered, and of attaining the milli-arcsecond global alignment to the ICRF, as well as typical individual source position accuracies even to better than 100 milli-arcsecond. Starting from the updated and presumably complete LQAC (Large Quasar Astrometric Catalog) list of QSOs, initial optical positions for those quasars are found in the USNO B1.0 and GSC2.3 catalogues and from the SDSS Data Release 5. The initial positions are next placed onto the UCAC2-based reference frames, followed by an alignment to the ICRF, as well as of the most precise sources from the VLBA calibrator list and from the VLA calibrator list - in the three cases under the proviso that reliable optical counterparts also exist. Finally, the LQRF axis are surveyed through spherical harmonics, considering right ascension, declination and magnitude terms. The LQRF contains 100,165 quasars, well represented on an all-sky basis, from -83.5 to +88.5 degrees of declination, and with 10 arcmin as the average distance between adjacent elements. The global alignment to the ICRF is of 1.5 mas, and the individual position accuracies are represented by a Poisson distribution peaking at 139 mas on right ascension and at 130 mas on declination. As a by product, significant equatorial corrections appear for all the used catalogues (except the SDSS DR5), an empirical magnitude correction can be discussed for the GSC2.3 intermediate and faint regimens, both the 2MASS and the preliminary northernmost UCAC2 positions show consistent astrometric accuracy, and the harmonic terms come out small always. The LQRF contains J2000 referred equatorial coordinates and is completed by redshift and photometry information from the LQAC. It is aimed to be an astrometric frame, but it is also the basis for the Gaia mission initial quasars list and can be used as a testbench for quasar space distribution and luminosity function studies. The LQRF will be updated when there is a release of new quasar identifications and newer versions of the used astrometric frames. In the later case, it can itself be used to examine the interrelations between those frames.

2.9. *Astrometry of ICRF Sources: The Influence of Radio Extended Structures on Offsets between the Optical and VLBI Positions*

J.I.B. Camargo, M. Assafin, A.H. Andrei, R. Vieira-Martins & D.N. Da Silva Neto

The International Celestial Reference Frame – ICRF – is the currently adopted IAU celestial reference frame. Its coordinate axes are materialized by the positions of 212 extragalactic radio sources unevenly distributed over the entire sky. Such positions are determined by VLBI techniques and have a median uncertainty better than 0.5 milliarcsecond. In addition to these so called defining sources, other 505 extragalactic radio sources are also listed in the ICRF. Their VLBI positions are consistent with the ICRF and serve to densify the frame. All of them, no matter whether defining or not, are in practice used to directly access the ICRF and may present spatially extended structures as seen from their high resolution S/X-band images. In this work, optical positions are obtained of 14 compact and extended ICRF sources with the ESO/MPG 2.2m telescope and compared to their VLBI counterparts. The intrinsic radio structure of the extragalactic sources is one of the limiting factors in defining the ICRF. It may also lead to the noncoincidence between the VLBI and optical positions. This question of noncoincidence, already addressed and verified by da Silva Neto (2002), is revisited here. In particular, two sources are identified for which this noncoincidence may have been motivated by the presence of the extended radio structure. As given by the Bordeaux VLBI Image Database, the structure indices of these sources in the X band are 3 and 4, indicating that they are probably not very good reference frame objects. From their high resolution images, as given by the USNO Radio Reference Frame Image Database, one may infer a possible correlation between the VLBI/optical offset and the plane of the sky orientation of the extended radio structures. One implication is that the relationship between the radio and optical frames should take into consideration structure effects in the future.

2.10. *The Joint Milli-Arcsecond Pathfinder Survey (J-MAPS) Mission: Introduction and Science Goals*

Ralph Gaume, USNO

J-MAPS is a small, US-funded, space-based, all-sky visible wavelength astrometric and photometric survey mission for 0th through 14th V-band magnitude stars with a 2012 launch. The primary objective of the J-MAPS mission is the generation of an astrometric star catalog with better than 1 milliarcsecond positional accuracy and photometry to the 1% accuracy level, or better at 1st to 12th mag. A 1-mas all-sky survey will have a significant impact on our current understanding of galactic and stellar astrophysics. J-MAPS will improve our understanding of the origins of nearby young stars, provide insight into the dynamics of star formation regions and associations, investigate the dynamics and membership of nearby open clusters, and discover the smallest brown dwarfs at distances up to 5 pc after a 2-year mission, and Jupiter-like planets out to 3 pc after 4 years. J-MAPS will provide critical milliarcsecond-level parallaxes of tens of millions of stars in the difficult 8-14th magnitude range, which when combined with stellar spectroscopy and relative radii determined from exoplanet transit surveys, allows a determination of stellar radii and exoplanet densities. In addition, the 20-year baseline between the groundbreaking Hipparcos mission and the J-MAPS mission allows a combination of the J-MAPS and Hipparcos catalogs to produce common proper motions on the order of 50–100 microarcseconds per year.

2.11. *The Wavelet Search for Stellar Clusters in NOMAD*

Veniamin Vitayzev, Alex Tsvwtkov and Irina Kumkova

The wavelet technique was presented for searching the heterogeneities of stellar density in the data of the NOMAD (Naval Observatory Merged Astrometric Dataset) catalogue which contains more than a billion stars. Many globular and open clusters have been detected in various photometric bands up to V=18. A lot of artifacts in NOMAD data were found in addition. This technique can be used for future catalogues including the products of the Gaia mission.

2.12. *Posters*

Significant radio-optical reference frame offsets from CTIO data
Zacharias, M. I. et al., USNO
Astrometric surveys 2000 to 2020
Erik Høg, Niels Bohr Institute, Copenhagen

Irina Kumkova
President of the Commission

References

van Leeuwen, F. 2007, *Hipparcos, the new reduction of the raw data* (Dordrecht: Springer)
da Silva Neto *et al.* 2002, *AJ* 124, 612
Zacharias, N. 2009, in *Trans. IAU XXVIIA*, Ed. K. A. van der Hucht, Cambridge Univ. Press, p. 34

Transactions IAU, Volume XXVIIB
Proc. XXVII IAU General Assembly, August 2009
Ian F. Corbett, ed.

doi:10.1017/S1743921310004898

Commission 19 — ROTATION OF THE EARTH (*ROTATION DE LA TERRE*)

PRESIDENT	**Aleksander Brzeziński**
VICE-PRESIDENT	**Chopo Ma**
PAST PRESIDENT	**Véronique Dehant**
ORGANIZING COMMITTEE	**Pascale Defraigne, Jean O. Dickey, Cheng-Li Huang, Jean Souchay, Jan Vondrák, Patrick Charlot** (IVS representative), **Bernd Richter** (IERS representative), **Harald Schuh** (IAG representative)

PROCEEDINGS BUSINESS AND SCIENCE SESSIONS, 5 August 2009

1. Introduction

The IAU Commission 19 meeting during the XXVII IAU General Assembly in Rio de Janeiro was held on Wednesday 5 August 2009, sessions 3 (14:00–15:30) and 4 (16:00–17:30). It was attended by about 40 participants. The meeting was split into three sessions.

Session I (14:00–14:30) was the Commission business meeting. It began with a report on the C19 activities during the triennium 2006–2009. Then the composition of new Organizing Committee for the triennium 2009-2012 was announced. Next, the President proposed some modifications to the Commission Terms of Reference and briefly discussed the membership issues. Finally, the upcoming President presented his plans for the Commission's activities in the next triennium. A summary of Session 1 is given in Section 2 below.

Session II (14:30–15:30) was devoted to the reports of the following Services and Working Groups: International Earth Rotation and Reference Systems Service (IERS), International VLBI Service for Geodesy and Geodynamics (IVS), Standards Of Fundamental Astronomy (SOFA), IERS Conventions Center, IERS Working Group on Prediction, IAG's Global Geodetic Observing System (GGOS). A summary of the reports can be found in Section 3.

Session III (16:00–17:30) was the science session on themes defined in the C19 Terms of Reference. It was composed of four invited presentations concerning the important developments in monitoring and modeling variations in Earth rotation, and two contributed reports on the recent research projects and scientific achievements in Germany and PR China. Section 4 below contains summaries of those scientific presentations.

2. Business Session

2.1. *IAU Commission 19 triennial report 2006–2009*

The most important scientific developments related to the objectives of the Commission were described in the C19 triennial report 2006–2009 (Brzeziński *et al.*, 2008). Some important issues were also discussed in the services / WG's reports and scientific presentations during this meeting; see sections 3 and 4 below for summaries.

Commission 19 was one of the participating commissions in organizing the Joint Discussion 6 "Time and Astronomy" during this Assembly. In addition, members of C19 were involved in organization of several meetings devoted to the subject of Earth rotation and reference systems:

• conference series "Journées Systèmes de Référence Spatio-Temporels", in Paris (2007) and in Dresden (2008);

• the 6th Orlov Conference "The study of the Earth as a planet by methods of geophysics, geodesy and astronomy" in Kiev, Ukraine, 2009;

- sessions during the worldwide scientific meetings:
  - IUGG General Assembly 2007 in Peruggia,
  - EGU General Assemblies 2007, 2008 and 2009 in Vienna,
  - AGU Fall Meetings 2006, 2007, 2008 in San Francisco,
  - Earth Tide Symposium 2008 "New Challenges in Earth's Dynamics" in Jena;
- IERS Workshops on Conventions (2007 in Paris), and on Earth Orientation Parameters (EOP) Combination and Prediction (to be held in October 2009 in Warsaw);
- IVS 2008 General Meeting in St Petersburg, International Workshop on Laser Ranging 2008 in Poznań.

Commission 19 closely cooperated in the last triennium with the following scientific bodies:

- IAG Commission 3 "Geodynamics and Earth Rotation" and its sub-commissions
  - 3.1: Earth Rotation and Earth Tides,
  - 3.3: Geophysical Fluids;
- IAU Div. I Commission 31 "Time";
- IAU Div. I Commission 52 "Relativity in Fundamental Astronomy";
- IAU Div. I WG on Second Realization of International Celestial Reference Frame (ICRF2);
- IAU Div. I WG on Numerical Standards in Fundamental Astronomy (NSFA).

The following two scientific campaigns, closely related to the objectives of C19, were organized recently:

- IERS EOP Prediction Comparison Campaign (EOP-PCC), between July 2005 and March 2008;
- CONT08: two-week campaign of continuous VLBI sessions, scheduled for observing during the second half of August 2008.

The website of Commission 19 http://iau-comm19.cbk.waw.pl/ was an important form of contact with the members of C19 and other researchers working in related fields. All decisions or actions of C19 were taken after email discussion with either the Commission Organizing Committee or the entire membership. The bibliography received from the Commission members during the preparation of the triennial report has been regularly updated and further extended by adding the references found in the worldwide internet data-bases such as SAO/NASA ADS. The current list which is available from the Commission website consists of more than 200 papers published between 2006 and 2009.

2.2. *Organization issues: elections, ToR and membership*

After elections conducted by email in February 2009 the following new OC of C19 for the term 2009–2012 was proposed for approval by the IAU General Assembly:

- President: Harald Schuh (Austria),
- Vice-President: Cheng-Li Huang (PR China),
- Past-President: Aleksander Brzeziński (Poland),
- OC members at large:
  - Christian Bizouard (France),
  - Benjamin Chao (Taiwan),
  - Richard Gross (USA),
  - Wiesław Kosek (Poland),
  - David Salstein (USA).

The representatives of the IAG, IERS and IVS to the C19 OC will be announced soon.

The organizing Committee of C19 discussed before the IAU General Assembly modifications to the C19 Terms of Reference, taking into account critical remarks of the IAU members. All changes adopted by the OC and accepted by the participants of this meeting concern the last section of the ToR, entitled "Organization". We will write this section below with all new changes in italics; the whole document is available from the C19 website.

Terms of Reference of C19, section "Organization":

The organization of the Commission is governed by the IAU statutes, by-laws and working rules which are available from the IAU website: http://www.iau.org/administration/statutes_rules/.

The Commission consists of its members and is chaired by the President. To coordinate its activity, the Commission forms the Organizing Committee (OC).

Each IAU member who is interested in the participation in the Commission activity may be a member of the Commission. No election procedure for a membership is established; only recommendation from the Commission 19 OC is needed.

The IAU General Assembly elects the President and Vice-President of the Commission for a 3-year term coordinated with the IAU General Assemblies. This term is not renewable. Candidates are proposed by the Commission from its members through voting and are elected by the IAU General Assembly. The Vice President is *normally* proposed to the General Assembly as the President for the next term.

The Organizing Committee includes ex-officio members (the present Commission President, the present Commission Vice-President, the past Commission President, and representatives from the IAG, the IERS, the IVS) and up to 5 additional members at large. *The representatives of the IAG, IERS and IVS are supposed to represent also the other Services that are not in the OC: the IGS, IDS and ILRS.* At-large members are nominated by the OC, selected by the Commission members through the voting and elected by the IAU General Assembly. As a rule, an OC member cannot serve more than two successive terms unless (s)he is elected as the President or Vice-President.

The Commission may propose Working Groups to fulfil specific tasks at the Commission level as well as at the Division level.

Representatives of the Services providing Earth orientation parameters (IERS, IVS, IGS, ILRS, IDS) should be invited to the Commission 19 Business Meetings and other events and report there.

The last organizational issue discussed during the meeting concerned the membership of C19. According to the IAU website, Commission 19 consists of 155 members. However, among them about 50 never responded to the email inquiries. Paragraph 42 of the IAU Working Rules, states "Before each General Assembly the Organizing Committee may (...) decide to terminate the Commission membership of persons who have not been active in the work of the Commission". The President and OC of C19 tried to clarify the status of the inactive members of C19. Unfortunately, this action could not be finished before the IAU GA in Rio: therefore it should be continued by the new OC of C19. But an even more important task for the OC will be to attract to C19 young scientists who are active in the related research fields. But first they must become members of the IAU. The procedure, which is described in p.III of the IAU Working Rules http://www.iau.org/administration/statutes_rules/working_rules/, is quite long therefore should be started sufficiently early to be completed before the next IAU GA.

2.3. *Plan of the IAU Commission 19 activities 2009–2012*

Harald Schuh, Vienna University of Technology, Vienna, Austria
President-Elect, IAU Commission 19

What are the recent developments and new challenges in Earth rotation research?

- Scientific Services (IVS, IGS, ILRS, IDS, ...) provide the highest quality EOPs, which are combined by the IERS in the frame of GGOS, the IAG Global Geodetic Observing System.
- Hourly time resolution and high precision allow investigation of a lot of small (second order) effects that have been hidden by the measurement noise in the past.
- Global models of geophysical fluids have been considerably improved in the last years (atmosphere, oceans, hydrology, ...) which allow detailed studies of EOP excitation.
- New instruments and technological developments w.r.t. Earth rotation (e.g. ring-laser, excellent clocks, ...).

Possible activities of IAU C19 for the next term:

- Organize a dedicated Workshop or Symposium on Earth Rotation (jointly with IAG Commission 3?).
- Organize a special issue on Earth rotation of an international journal (e.g. Journal of Geodesy).
- Follow new technological achievements with relevance for EOP research (e.g. optical clocks, ...) and make proposals on applications.
- Establish even better collaboration with neighboring disciplines (oceanography, meteorology, hydrology, ...).
- Encourage interdisciplinary endeavors with the aims of improving the understanding of processes and interactions in the Earth system in view of global change.
- Establish a new IAU/IAG Working Group on "Theory of Earth rotation".

3. Working Groups and Services report

3.1. *Report of the International Earth Rotation and Reference Systems Service*

Chopo Ma, NASA/Goddard Space Flight Center, Greenbelt MD, USA
Chair, IERS Directing Board

The Earth Orientation PC established a web service for EOP values and the Earth orientation rotation matrix. A new EOP series IERS 05 C 04 was issued consistent with ITRF2005. The Rapid Service / Prediction PC changed Bulletin A to match IERS 05 C04. It also initiated a Working Group on Prediction. The ITRS PC released ITRF2005 based on time series of SINEX files containing site positions and EOP as combined by the Technique Centers, where available. It issued a call for participation for ITRF2008, now in progress. The ICRS PC increased VLBI observations in the southern hemisphere, edited the IERS Technical Note 35 on ICRF2, and drafted resolution B3 on the ICRF2 for the IAU General Assembly. The Conventions PC updated various chapters in the online version of the IERS Conventions 2003 (McCarthy & Petit, 2003) and is preparing for new Conventions (2009).

The Global Geophysical Fluids Center was reviewed by IERS Directing Board and will be reconstituted. A Call for Proposals was released in May 2009 emphasizing the renewal of existing operational products and inclusion of new operational products.

A Working Group was established to study EOP prediction at various time scales. The Site Survey and Co-location WG was reconstituted with new charter, chair and membership. The Combination WG was ended on completion of Combination Pilot Project, and a WG on Combination at the Observation Level was begun. The ICRF2 WG compiled its work in IERS Technical Note No. 35 for presentation to the IAU.

The Central Bureau further developed the IERS Data and Information System (DIS) based on modern technologies for internet-based exchange of data and information using XML and the generation and administration of ISO metadata. The DIS provides access to and visualization of IERS data products as well as links to IERS components. The Central Bureau organized a number of workshops with the abstracts and presentations available on the IERS web site. Publications included two Technical Notes, No. 34 "The International Celestial Reference System and Frame" (Souchay & Feissel-Vernier, 2006), and No. 35 "The Second Realization of the International Celestial Reference Frame by Very Long Baseline Interferometry" (Fey & Gordon, 2009). The online version of those IERS Technical Notes is available from www.iers.org.

3.2. *The IVS on its way to the next generation VLBI system VLBI2010*

Harald Schuh, Vienna University of Technology, Vienna, Austria
Chair, IVS Directing Board

The International VLBI Service for Geodesy and Astrometry (IVS) is well on the way to fully defining a next generation VLBI system, called VLBI2010 (Petrachenko *et al.*, 2009). The goals of the new system are to achieve 1-mm position accuracy over a 24-hour observing session and to carry out continuous observations, i.e. observing seven days per week, with initial results shall to be delivered within 24 hours of taking the data. These goals require a completely new technical and conceptual design of VLBI measurements. Based on extensive simulation studies, strategies have been developed by the IVS to significantly improve its product accuracy through the use of a network of small ($\sim$ 12-m) fast-slewing antennas, a new method for generating high precision delay measurements, and improved methods for handling biases related to system electronics, deformations of the antenna structures, and radio source structure. As of June 2009, the construction of ten new VLBI2010 sites has already been funded, which will improve the geographical distribution of geodetic VLBI sites on Earth and provide an important step towards a global VLBI2010 network.

3.3. *SOFA 2006-2009*

Patrick Wallace, Rutherford Appleton Laboratory, Chilton, UK
Chair, SOFA Review Board

SOFA (Standards Of Fundamental Astronomy) is a service operated by Division I to provide authoritative fundamental-astronomy algorithms, for example precession-nutation models. The activity is carried out by an international board that includes both astronomers and software experts, and has cross-membership with the IERS Directing Board. SOFA's products are disseminated through a website: http://www.iau-sofa.rl.ac.uk/.

During the period 2006-2009, the SOFA software collection has grown from 121 routines to 160, the increase being due to the introduction of the IAU 2006 precession. The software has a core of 41 canonical models, used by 67 astronomy-related routines and supported by 52 vector-matrix utilities.

The triennium saw the development of a second version of the SOFA software, this time in the C programming language and supported in parallel with the existing Fortran version. At the same time, the SOFA licensing conditions have been made more liberal, to harmonize with IAU/IERS practices in general and to encourage the widest possible use.

Considerable thought has been given to SOFA's formal position within the IAU organization, with the object of raising its profile and increasing user confidence while at the same time complying with IAU statutes, bye-laws and working rules. These discussions are still in progress.

3.4. *Report of the IERS Conventions center*

Gérard Petit, Bureau International des Poids et Mesures, Sèvres Cedex, France
and
Brian Luzum, U.S. Naval Observatory, Washington DC, USA
Principal scientists, IERS Conventions Center

The IERS Conventions Center is provided jointly by the BIPM and the USNO. The Conventions Center provides updated versions of the Conventions in electronic form, after approval of the IERS Directing Board. In the mean time, work on interim versions is also available by electronic means. The Center works with the help of an Advisory board for the IERS Conventions update. All updates since the last registered edition, the IERS Conventions 2003 (McCarthy & Petit, 2003) are provided in the web page http://tai.bipm.org/iers/convupdt/convupdt.html.

In the triennium 2006–2009, the main achievements have concerned the following chapters of the Conventions. Chapter 4 (Terrestrial reference systems and frames) has been rewritten to account for ITRF2005. Chapter 5 (Transformation between the ITRF and GCRS) has been completely rewritten to implement the IAU 2000–2006 resolutions and corresponding terminology. In Chapter 6 (Geopotential), a section on the ocean pole tide has been added. In Chapter 7 (Displacement of reference points), sections have been added or rewritten for the conventional ocean tide loading, for the ocean pole tide loading, and for technique-dependent effects. Chapter 9 (Atmospheric propagation) has been completely rewritten for optical signals and for radio signals. Chapter 10 (General relativistic models for space-time coordinates and equations of motion) has been modified in line with IAU recommendations and a new section has been added.

The IERS Workshop on Conventions was held in September 2007 to plan a new registered edition of the IERS Conventions. At the workshop, relevant models for inclusion in the Conventions and long-term technical and institutional issues were discussed as well as determining milestones for achieving the next registered edition. For an executive summary, see http://www.bipm.org/utils/en/events/iers/workshop_summary.pdf.

In an effort to make the IERS Conventions more efficient to maintain and more user-friendly, a series of changes have been initiated. The Conventions Update page has been modified to not only include information and links to past updates, but to also provide information and links to planned and possible changes. A new web page of additional, supporting material has been created. In an effort to work towards providing standardized software, a software template was revised to be analogous to the new IAU Standards of Fundamental Astronomy (SOFA) software template. Implementation of this template on existing draft software and additional standardization of code was begun in 2008.

It is planned that the next registered edition of the Conventions be issued at the end of 2009.

3.5. *Activities of the IERS Working Group on Prediction (WGP)*

William Wooden, U.S. Naval Observatory, Washington DC, USA
WG Chair

The IERS Working Group on Prediction (WGP) was tasked to determine what Earth orientation parameter prediction products are needed by the user community and to examine the fundamental properties of different input data sets and algorithms with a goal of improving the prediction process. The WGP establishment grew out of IERS Rapid Service/Prediction Center (RS/PC) concerns about the continued relevance of current products, new accuracy requirements, the impact of new data sets, viable new prediction methodologies, and the desire to build on the interest generated by the EOP Prediction Comparison Campaign organized under the umbrella of the IERS by the Vienna University of Technology and the Space Research Center, Warsaw, and the efforts of the IERS Combination Pilot Project. The WGP expectations include definitive user requirements, a comprehensive look at prediction methods, a comprehensive look at new data sets, and a definitive assessment of the current state-of-the-art in EOP prediction.

The first task of the WGP was to determine whether the current RS/PC products, which were developed more than 15 years ago, are adequate or whether modifications and/or improvements are necessary to meet more stringent requirements. A short EOP user survey was developed by the WGP and posted on the RS/PC website www.iers.org/MainDisp.csl?pid=71-49. The IERS invited participation from those on the IERS mailing lists, those who receive IERS RS/PC products, and any others thought to have an interest in EOP predictions (see IERS Message No. 104, www.iers.org). The survey confirmed that there is a large class of operational users that need daily predictions up to 30 days in advance in tabular format with one-day spacing. Other classes of users have different needs. Some users would like to improve long term predictions and although it is an interesting problem, the emphasis on shorter term forecasts as described by the terms of reference under which the IERS RS/PC currently operates has been re-validated by the survey results. However, there is a need for increased accuracy, and the efforts of the WGP to examine algorithms and incorporate potential new sources of data appear to address that need. In addition there seems to be a growing interest in daily and sub-daily predictions which will require more timely measurements of EOP quantities and some increased processing capability.

The second task of the WGP was to examine the fundamental properties of various input data sets and algorithms. A repository of test data sets was established containing a general test set from 2000-2006, a test set for polar motion loops, large amplitude annual/Chandler polar motion, and large and small UT1 changes. Analysis of the test data sets confirms that the polar motion loop periods are the most challenging. The activities of the WGP members have been reported at recent EGU and AGU meetings. In addition a special session of the 2007 Journées meeting, "Prediction, Combination and Geophysical Interpretation of Earth Orientation Parameters," contained an overview of WGP activities, a panel discussion on the prediction of EOP, and numerous papers devoted to all aspects of the problem (see Proc. Journées Systemes de Référence Spatio-Temporels 2007, N. Capitaine (ed.), pp. 143–229, Paris Observatory). An IERS Workshop on EOP Combination and Prediction will be held at the Space Research Centre in Warsaw, Poland from 19-21 October 2009 to provide specific recommendations to improve current techniques/IERS products based on the analysis done to date. The final product of the WGP will be an IERS Technical Note summarizing the activities of the WGP and providing the current state-of-the-art in EOP prediction.

3.6. *Global Geodetic Observing System (GGOS) and its Relation to IAU*

Harald Schuh, Vienna University of Technology, Vienna, Austria

GGOS is the Global Geodetic Observing System of the International Association of Geodesy (IAG); see (H-P. Plag & M. Pearlman, 2009) and the website www.ggos.org. It provides observations of the three fundamental geodetic observables and their variations, that is, the Earth's shape, the Earth's gravity field and the Earth's rotational motion. GGOS integrates different geodetic techniques, different models and approaches in order to ensure a long-term, precise monitoring of the geodetic observables in particular by bringing together geometrical and physical measurements. GGOS provides the observational basis to maintain a stable, accurate and global reference frame and in this function is crucial for all Earth observation and many practical applications. GGOS contributes to the emerging Global Earth Observing System of Systems (GEOSS) not only with the accurate reference frame required for many components of GEOSS but also with observations related to the global hydrological cycle, the dynamics of atmosphere and oceans, and natural hazards and disasters. GGOS acts as the interface between the geodetic Services of the IAG and the IAU and external users such as GEOSS and United Nations authorities. With this the geodetic community can provide the global geosciences with a powerful tool consisting mainly of high quality products, standards and references, and of theoretical and observational innovations. In the presentation special emphasis was given to the relation between GGOS and IAU Commission 19.

4. Scientific presentations

4.1. *Optical reference frames for monitoring Earth rotation (invited)*

Jan Vondrák, Astronomical Institute Prague, Czech Republic

Some years ago, we collected about 4.5 million optical astrometric observations of latitude / universal time / altitude variations, made at 33 observatories in 1899.7 - 2002.6. These observations, brought into the system of the Hipparcos Catalogue, served to determine Earth Orientation Parameters. When doing so, we found that a large proportion of observed stars were double or multiple, which caused problems when extrapolating their positions far from the mean epoch of Hipparcos Catalogue, 1991.25. Therefore, we started improving the positions, and in particular proper motions, by combining the observations with their positions in ARIHIP, Tycho-2 and Hipparcos catalogues to derive an Earth Orientation Catalog (EOC). The main goal was to create a new and improved reference frame for long-term Earth rotation studies, by using the best available star catalogues and their combination with the rich observational material of the existing program of measuring Earth orientation in the 20th century. So far, three versions of EOC have been produced : EOC-1 (based only on meridian observations, Vondrák & Ron, 2003), EOC-2 (based on all observations, Vondrák, 2004), EOC-3 (improved EOC-2 by adding periodic motions of 586 stars, Vondrák & Štefka, 2007) and EOC-4 (Vondrák & Štefka, 2009). The last version, EOC-4, was constructed by a new procedure that removed disadvantages of its predecessors. The main improvements were more precise amplitudes of periodic parts, more double stars detected (599) and consistency with space-based observations at the Hipparcos mean epoch. The catalog EOC-4, containing nearly 5 thousand stars, has already been used to derive a new solution of EOP from optical astrometry in the 20th century (JD6, this IAU GA).

4.2. *Long-period tidal effects on Earth rotation (invited)*

Richard S. Gross, Jet Propulsion Laboratory, Pasadena CA, USA

Tidal forces due to the gravitational attraction of the Sun, Moon, and planets deform the solid and fluid regions of the Earth causing the Earth's inertia tensor to change and thus causing the Earth's rotation to change. Here, available models of the (1) elastic response of the solid Earth, (2) inelastic response of the solid Earth, and (3) dynamic response of the oceans to the tide raising potential at periods greater than a day are compared to length-of-day (lod) observations from which atmospheric and nontidal oceanic effects have been removed. Spectra of the residual lod observations exhibit prominent peaks at the tidal frequencies, most but not all of which are eliminated when an elastic body tide model is removed. While inelastic effects are only a few percent of the elastic, it is shown that it is important to include both the in-phase and out-of-phase terms when removing them from lod observations. Of the oceanic tide models evaluated, models constrained by data are shown to perform better than those not constrained by data. The sum of the Yoder *et al.* (1981) elastic body tide model, Wahr & Bergen (1986) inelastic body tide model, and Kantha *et al.* (1998) ocean tide model was found to do the best job of removing tidal effects from lod observations, particularly at the fortnightly tidal frequency. The model recommended by the IERS Conventions 2003 (McCarthy & Petit, 2003) was found to do the worst job.

4.3. *Recent progress in modeling precession-nutation (invited)*

Nicole Capitaine, Observatoire de Paris, Paris, France

The IAU 2006/2000 precession-nutation has been adopted in two stages. The first stage (IAU 2000 Resolution B1.6) was the adoption of the IAU 2000 precession-nutation, which has been implemented in the IERS Conventions 2003 (McCarthy & Petit, 2003). The second stage (IAU 2006 Resolution B1) was the adoption of the P03 Precession (Capitaine *et al.*, 2003; Hilton *et al.*, 2006) (i.e. the IAU 2006 precession) as a replacement for the precession part of the IAU 2000A precession-nutation, beginning on 1 January 2009. Expressions and procedures for implementing the new precession-nutation have been provided by Capitaine & Wallace (2006) and Wallace & Capitaine (2006); they have been implemented in Chapter 5 of the IERS Conventions updated in 2009 (http://tai.bipm.org/iers/convupdt/convupdt_c5.html) and the Standards Of Fundamental Astronomy (SOFA). The difference between IAU 2006 and IAU 2000 lies essentially in the precession part, though very small changes are needed in a few of the IAU 2000A nutation amplitudes in order to ensure compatibility with the IAU 2006 values for ϵ_0 and the J_2 rate. Whenever these small adjustments are included in the periodic terms (e.g. in the SOFA and IERS implementations), the notation "IAU $2000A_{R06}$" can be used to indicate that the nutation has been revised for use with the IAU 2006 precession.

Comparisons of the IAU 2006/2000 precession-nutation model with VLBI observations, once corrected for an empirical model for the FCN show residuals with a w.r.m.s. of about 130 μas (Capitaine *et al.*, 2009). The residuals would be compatible with corrections of a few tens of μas to the 18.6-yr nutation, which would be absorbed in small corrections to the estimates for a couple of the BEP of the MHB model. This result is consistent with independent fits (Zerhouni *et al.*, 2009) to the LLR celestial pole offsets with respect to the same IAU precession-nutation model and a comparison over 400 years with the INPOP06 numerical integration of Fienga *et al.* (2008) of the GCRS motion of the axis of angular momentum.

Future improvements of the precession-nutation model would require a rigorous consideration in the GR framework, an improvement in the model for the electromagnetic couplings and a specific account for the second order torque. Potential for improvement lies in solving the rotational equations in the X, Y variables (Capitaine *et al.*, 2006) or estimating the BEP parameters in the time domain with a non-linear Bayesian inversion method (see Koot *et al.*, 2008).

4.4. *Monitoring Earth Orientation variations, state of the art and prospective (invited)*

Daniel Gambis, Observatoire de Paris, France

One task of the Earth Orientation Centre of the IERS is to monitor Earth orientation parameters (C04 series) including long term consistency with both celestial and terrestrial reference frames. The combined series C04 is computed and made available to a broad community of users in Astronomy, Geodesy, Geophysics, Space sciences and Time (Bizouard and Gambis, 2009). In addition to the improvement of the precision, it is as well essential to monitor the consistency of EOP with the ITRF. This is now achieved by a yearly comparison between the current C04 series and the EOP solution based on the stacking of IGS and IVS SINEX files using the CATREF software (Altamimi *et al.*, 2006). The current accuracy of the current C04 solution is about 50 μas for pole components, 10 μs for UT1 and 60 μas for nutation offsets. We expect that in a near future there will be a strong evolution in that field. EOP solutions and the terrestrial reference frame will be simultaneously derived in a global combination at the normal equation level (Gambis *et al.*, 2009). The method is based on the fact that techniques have their own strengths and weaknesses. Their combination should benefit from their mutual constraints. A program within the Groupe de Recherches de Géodésie spatiale, GRGS, has been carrying out for several years. Observations derived from the various techniques are processed using unique software (GINS) with same conventional models and constants and inverted using DYNAMO developed by the GRGS. The main difficulties of the procedure lie in the combination strategy to be applied, in particular the way to ensure stability of reference frames over successive weeks and in addition the weighting of the various techniques in the combination. It is likely that future generation of EOP and TRF products will be routinely based on this rigorous combination. The method seems as well promising to study short term variations of EOP.

4.5. *Earth Rotation and Global Dynamic Processes: Research activities in the framework of the German DFG-Research Unit FOR584/2*

Florian Seitz, Technical University Munich, Munich, Germany
and
Harald Schuh, Jurgen Müller, Hermann Drewes, Hansjörg Kutterer, Michael Soffel, Maik Thomas

A coordinated research initiative on the fields of Earth Rotation and Global Dynamic Processes has been organized in a joint program with partners from Germany, Austria and Switzerland. For a period of six years (2006-2012), the program integrates the expertise and competencies on these topics available in 11 participating research and university institutes from various disciplines (geodesy, oceanography, meteorology, geophysics). The group consists of 10 closely inter-related sub-projects with 12.5 funded scientific positions, mainly filled with PhD students. The initiative's main objective is the comprehensive and consistent analysis, modeling, and interpretation of all facets of Earth rotation in interdisciplinary co-operation. Principal research tasks comprehend the processing and analysis of observations (determination of EOP, combined analysis of heterogeneous observation types, integration of Earth rotation, gravity field, and surface deformations) as well as theoretical and numerical modeling (analysis, description, and explanation of relevant physical phenomena and couplings in the Earth system). Through the integral treatment of Earth rotation and related physical phenomena on all relevant time scales, the research also means a valuable contribution to IAG's Global Geodetic Observing System (GGOS). Research highlights in the field of relativity and nutation theory are the extension of the post-Newtonian nutation/precession theory from a rigid to an elastic Earth model and the assessment of precise nutation angles and relativistic parameters from LLR. Activities in the field of consistent modeling of Earth system processes comprehend the assimilation of Earth rotation parameters into an ocean model, the improvement of a forward model for core/mantle interaction, the analysis of climate variability from fully coupled atmosphere-hydrosphere models, and the inversion of a dynamic Earth system model for the estimation of physical Earth parameters from EOP. Research associated with the analysis and interpretation of observations of ring laser gyroscopes has been promoted in order to confirm sub-daily and episodic variations of Earth rotation and improve the sensor orientation using a 3-D FEM topographic model for surface deformation. Advances in the field of combined processing and analysis of both techniques and parameters cover the simultaneous estimation of consistent high-quality time series of EOP and TRF parameters from integrated VLBI, SLR, and GNSS analysis as well as the combination of EOP and 2nd-degree harmonic gravity field coefficients for the separation of mass and motion effects. More detailed information on projects and participants as well as numerical data and results are provided on the research unit's web portal ERIS (Earth Rotation Information System) at www.erdrotation.de.

4.6. *Research activities on Earth rotation and reference systems in China*

Cheng-Li Huang, Shanghai Astronomical Observatory, Shanghai, P.R. China

Some of the research activities on Earth rotation and reference systems in China during 2005–2008 were reported in the IAU Commission 19 Triennial Report (Brzeziński *et al.*, 2008). They include 1) EOP and interior geophysics; 2) EOP and surface fluids; 3) EOP prediction; 4) ITRF and ICRF; 5) Data analysis (VLBI, SLR, GPS); 6) observations and instruments; and 7) miscellaneous. Most of them were achieved in Shanghai Astron. Obs. (SHAO), Wuhan Inst. Geodesy and Geophys. (WHIGG), and National Astron. Obs. of China (NAOC). In addition, there were some other related contributions which are briefly described below.

Possible connection between global significant earthquakes and Earth's variable rotation rate was investigated (Ma *et al.*, 2007). The observation of plumb line variation from gravimetry in China and its relation with earthquakes was also studied (Li & Li, 2008).

Research concerning the ITRF2005 showed that this conventional frame does not satisfy the need of monitoring mm-level geodynamic change in construction of mm-level TRF (Zhu *et al.*, 2008). Geocenter motion was also studied using the wavelet transform, and was estimated from the SLR data.

In the Chinese first lunar mission, Chang'E-1, launched on 24 October 2007 and impacting the Moon on 1 March 2009, the Chinese VLBI network (CVN) provided rapid Earth rotation service, tracking and navigation. A new 65-m VLBI antenna is under construction in SHAO and will work from 2012 for astrophysics, geodesy and DSN. A group in SHAO has been working to develop their own software for processing VLBI data for both astrometry and geodesy, including model of high frequency EOP variation. Co-location gauge for the VLBI, SLR and GPS stations in She-shan station, west of Shanghai, was made successfully in 2008 and will be made again in the second half of 2009.

Some experiments in time synchronization between the Earth and satellites were conducted in recent years with the purpose of application for satellite navigation and positioning. For example, the laser time transfer (LTT) payload onboard the Chinese experimental navigation satellite COMPASS-M1 with an orbital altitude of 21500 km was launched in April 13, 2007. After 17 months orbital flight, the LTT payload has maintained its good performance.

5. Closing remarks

The President of Commission 19 thanks the members of the current Organizing Committee of C19 for their active work during the past triennium. He also congratulates the new President, Vice-President, and the OC members on their election and wishes them successful activity during the term 2009–2012.

Aleksander Brzeziński
President of the Commission

References

Altamimi, Z., Boucher, C., & Gambis, D. 2006, *Adv. Space Res* 33, 342

Bizouard, C. & Gambis, D. 2009, in: H. Drewes (ed.), *Geodetic Reference Frames*, IAG Symposia Series vol. 134, Springer-Verlag, New York, p. 3, DOI 10.1007/978-3-642-00860-3

Brzeziński, A., Ma, C., Dehant, V., Defraigne, P., Dickey, J. O., Huang, C.-L., Souchay, J., Vondrák, J., Charlot, P., Richter, B., & Schuh, H. 2008, in: K. van der Hucht (ed.), *IAU Transactions, Vol. 4, Iss. 27A, Reports on Astronomy 2006-2009*, Cambridge University Press, p. 37, DOI: 10.1017/S1743921308025271

Capitaine, N. & Wallace, P. T. 2006, *A&A* 450, 855

Capitaine, N., Folgueira, M., & Souchay, J. 2006, *A&A* 445, 347

Capitaine, N., Mathews, P. M., Dehant, V., Wallace, P. T., & Lambert, S. B. 2009, *Celest. Mech. Dyn. Astr.* 103, 179

Capitaine, N., Wallace, P. T., & Chapront, J. 2003, *A&A* 412, 567

Fey, A. & Gordon, D. (eds.) 2009, *IERS Technical Note* 35, Verlag des Bundesamts für Kartographie und Geodäsie, Frankfurt am Main

Fienga, A., Manche, H., Laskar, J., & Gastineau, M., 2008, *A&A* 477, 315

Gambis, D., Biancale, R., Carlucci, T., Lemoine, J. M., Marty, J. C., Bourda, G., Charlot, P., Loyer, S., Lalanne, L., & Soudarin, L. 2009, in: H. Drewes (ed.), *Geodetic Reference Frames*, IAG Symposia Seri es vol. 134, Springer-Verlag, New York, p. 265, DOI 10.1007/978-3-642-00860-3

Hilton, J. L., Capitaine, N., Chapront, J. J., Ferrandiz, J. M., Fienga, A., Fukushima, T., Getino, J., Mathews, P., Simon, J.-L., Soffel, M., Vondrák, J., Wallace, P., & Williams, J. 2006, *Celest. Mech. Dyn. Astron.* 94, 351

Kantha, L. H., Stewart, J. S., & Desai, S. D. 1998, *J. Geophys. Res.* 103, 12, 639

Koot, L., Rivoldini, A., de Viron, O., & Dehant, V. 2008, *J. Geophys. Res.* 113, B08414, doi: 10.1029/2007JB005409

Li, Z. X. & Li, H. 2008, *Science in China* 38(4), 432

Ma, L. H., Han, Y. B., & Yin, Z. Q. 2007, *Astronomical Research & Technology* 4(4), 406

McCarthy, D. & Petit, G. (eds.) 2003, *IERS Technical Note* 32, Verlag des Bundesamts für Kartographie und Geodäsie, Frankfurt am Main

Petrachenko, B., Niell, A., Behrend, D., Corey, B., Böhm, J., Charlot, P., Collioud, A., Gipson, J., Haas, R., Hobiger, T., Koyama, Y, MacMillan, D., Malkin, Z., Nilsson, T., Pany, A., Tuccari, G., Whitney, A., & Wresnik, J. 2009, in: D. Behrend & K. Baver (eds.), *International VLBI Service for Geodesy and Astrometry 2008 Annual Report*, NASA/TP-2009-214183, p. 13

Plag, H.-P. & Pearlman, M. (eds.) 2009, *Global Geodetic Observing System: Meeting the Requirements of a Global Society on an Changing Planet in 2020*, Springer-Verlag, New York

Souchay, J. & Feissel-Vernier, M. (eds.) 2006, *IERS Technical Note* 34, Verlag des Bundesamts für Kartographie und Geodäsie, Frankfurt am Main

Vondrák, J. 2004, *Serbian Astron. J.*, 168, 1

Vondrák, J. & Ron, C. 2003, in: N. Capitaine & M. Stavinschi (eds.), *Proc. Journées 2002 Systèmes de référence spatio-temporels*, Observatoire de Paris, p. 49

Vondrák, J. & Štefka, V. 2007, *A&A* 463, 783

Vondrák, J. & Štefka, V. 2009, *A&A* in press

Wahr, J. M. & Bergen, Z. 1986, *Geophys. J. R. astr. Soc.* 87, 633

Wallace, P. T. & Capitaine, N. 2006, *A&A* 459, 981

Yoder, C. F., Williams, J. G., & Parke, M. E. 1981, *J. Geophys. Res.* 86(B2), 881

Zerhouni, W., Capitaine, N., & Francou, G. 2009, in: M. Soffel & N. Capitaine (eds.), *Proc. Journées 2008 Systèmes de référence spatio-temporels*, Lohrmann-Observatorium Dresden and Observatoire de Paris, p. 186

Zhu, W. Y., Xiong, F. W., & Song, S. L. 2008, *Progress in Astronomy* 26(1), 1

Transactions IAU, Volume XXVIIB
Proc. XXVII IAU General Assembly, August 2009
Ian F. Corbett, ed.

doi:10.1017/S1743921310004904

COMMISSION 31 TIME
(TEMPS)

PRESIDENT: Pascale Defraigne
VICE-PRESIDENT: Richard Manchester
PAST PRESIDENT: Demetrios Matsakis
ORGANIZING COMMITTEE: Gerard Petit, Mizuhiko Hosokawa, Sigfrid Leschiutta, Zao-Cheng Zhai

PROCEEDINGS OF THE BUSINESS SESSION, August 7, 2009

1. Appointment of new officials

Dr R. N. Manchester and Dr M. Hosokawa have been elected as, respectively, the new President and Vice-President of the Commission for the next term, 2009-2012. Concerning the Organizing Committee (OC) Members, we welcome as new members: Felicitas Aris (BIPM, France), Philip Tuckey (LNE-SYRTE, France), Vladimir Zharov (MSU, Russia), and Shougang Zhang (NTSC, China) and we acknowledge the outgoing members: Demetrios Matsakis, Gerard Petit, Mizuhiko Hosokawa, Sigfrid Leschiutta and Zao-Cheng Zhai. P. Defraigne remains on the OC as immediate Past-President.

2. Report of the triennium

P. Defraigne presented the main results achieved by the commission members during the last triennium. Concerning the relative frequency stability and accuracy of the atomic time scales, the 1-month stability of the International Atomic Time (TAI) is currently close to 3×10^{-16}, which is a limit with present clock performances. However, the long term stability will be improved in the near future thanks to improvements of the algorithms. The last version of TT(BIPM) was also presented with its present uncertainy of 5×10^{-16}. The recommendations of the Consultative Committee for Time and Frequency having an impact on the Astronomical Activities have also been reported. The evolution of the pulsar time scales during the triennium was shown with a stability of 10^{-15} at 7 years currently being achieved. The presentation of the report ended with the list of meetings relating to the scope of the Commission that were held during the triennium, including in addition to the usual meetings of the time and frequency community, the Fifth International Time Scale Algorithm, April 2008 in San Fernando, the IAU Symposium 261 "Relativity in Fundamental Astronomy: Dynamics, Reference Frames, and Data Analysis", April 2009 in Virginia Beach and the Joint Discussion "Time and Astronomy" during the IAU General Assembly.

3. New activities

The future scope and role of the Commission was then discussed. The main function is to provide a forum and a list of contacts for discussion of issues related to time in the scientific community. The Commission brings together people from the time-standard and the astronomical communities. A focus of this effort will be to develop the Commission 31 website with information on: the relationship between UTC and UT1 (with a link to IERS), status of the discussions about leap seconds, references on atomic time scales and atomic clocks and on pulsar timing. A link to the BIPM pages on UTC (CircularT) and TT(BIPM) was additionally proposed. The role of the Commission in providing information about time to the general public was also discussed. This is also best done through the website.

4. New members

Four new members have been proposed to join the Commission. R. N. Manchester will try to increase the membership, in particular, by inviting more people from the pulsar community to become members.

5. Closing remarks

R. N. Manchester thanked P. Defraigne and the outgoing OC members for their efforts over the past three years and said that he looked forward to helping to develop the role of the Commission during the next three years.

Pascale Defraigne
President of the Commission

Transactions IAU, Volume XXVIIB
Proc. XXVII IAU General Assembly, August 2009
Ian F. Corbett, ed.

doi:10.1017/S1743921310004916

COMMISSION 52 — RELATIVITY IN FUNDAMENTAL ASTRONOMY

RELATIVITÉ DANS ASTRONOMIE FONDAMENTALE

PRESIDENT — Sergei A. Klioner
VICE-PRESIDENT — Gérard Petit
PAST PRESIDENT — —
ORGANIZING COMMITTEE — Victor A. Brumberg, Nicole Capitaine, Agnès Fienga, Toshio Fukushima, Bernard Guinot, Cheng Huang, François Mignard, Ken Seidelmann, Michael Soffel, Patrick Wallace

COMMISSION 52 WORKING GROUPS

none

PROCEEDINGS BUSINESS SESSION, 7 August 2009

1. The Commission and its membership

The IAU Commission 52 "Relativity in Fundamental Astronomy" (RIFA) has been established during the 26th General Assembly of the IAU (Prague, 2006) to centralize the efforts in the field of Applied Relativity and to provide an official forum for corresponding discussions. The general scientific goals of the Commission are:

- clarify geometrical and dynamical concepts of Fundamental Astronomy within a relativistic framework,
- provide adequate mathematical and physical formulations to be used in Fundamental Astronomy,
- deepen the understanding of the above results among astronomers and students in astronomy,
- promote research needed to accomplish these tasks.

In the period 2006–2009 the Commission has actively invited people to participate in the activities of the Commission. In July 2009 the Commission had 39 full members from 10 countries (USA and France having more than 60% of the members). A considerable problem for the Commission was the rule that only IAU members can become members of an IAU commission. In the particular case of Commission 52, which shares the research field with gravitational physics, this has led to the situation when many people actively working in the research field were not allowed to officially join the Commission. In several cases the Commission has undertaken certain efforts in order to make certain individuals new members of the IAU. During General Assembly in Rio de Janeiro many new members have joined the Commission so that the total number of members is now about 70.

2. Activities of the Commission

A web page containing all the information concerning the work of the Commission has been created and activated: `http://astro.geo.tu-dresden.de/RIFA`. It has been decided that this web page will be maintained at the same place. The content of the web page will be obviously determined by the President of the Commission and by its Organizing Committee.

Several scientific and educational projects have been initiated in 2007. The educational projects include compilation of a list of open problems in the field of applied relativity as well as a list of frequently asked questions. The Relativistic Glossary for astronomers has been also created and is available on the web page of the Commission. These three tools are intended to serve the broad astronomical community. The corresponding documents are expected to be updated and enriched in the future by the Commission itself and by future Working Groups of the Commission.

The following three scientific topics were identified by the Commission as important to discuss in the years to come:

1. units of measurements for astronomical quantities in the relativistic context,
2. astronomical units in the relativistic framework,
3. time-dependent ecliptic in the GCRS.

2.1. *Task team "Units of measurements"*

During the period 2006–2009 only the first topic has been discussed in detail. The content of this topic can be summarized as follows. In the literature (including very recent papers) one can find different units used in precise work: "TDB units", "TCB units", "TT units" along with "SI units". The co-existence of these units is related to the relativistic scaling of time and space coordinates. On the other hand, the IAU 1991 resolutions clearly state that only SI units without any additional relativistic scaling should be used for all astronomical quantities (astronomical units like AU are not meant here). Besides the non-SI units lead to certain logical contradictions. For example, time scale TAI cannot be considered as being expressed in the units of SI, if TT is considered to be expressed in non-SI units. A balanced approach to this issue had to be suggested and discussed. Such an approach helps also to unify the notations and numerical values of astronomical constants throughout the literature. As a material for discussions (Klioner 2008) has published a concise review of the problem of relativistic scaling of astronomical quantities.

For this discussion a dedicated discussion forum – task team – has been established. The task team was open for any member of the Commission. All members of a task team were expected to participate *actively* in the discussion. A total of 12 members of the Commission have decided to participate in the task team. The discussion was organized in the form of e-mail exchange. Several documents circulated as a basis for the discussion. The task team had also a number of short workshops during scientific meetings in Paris, Dresden, and in Virginia Beach, VA, USA.

Although the topic of the discussion did not seem to be complicated a priori, the discussion had encountered several difficulties which have also substantially slowed down the elaboration of the common point of view and recommendations for the nomenclature. The work of the task team has finally resulted in a joint paper (Klioner *et al.* 2009) containing the summary of the discussion and the recommendation to use the units of SI with any time coordinates – TCG, TCB, TT, and TDB. This should be considered as a crucial step forward in this area.

2.2. *IAU Symposium 261*

A fundamental problem for the field of Commission 52 is its multidisciplinary nature. Experts in general relativity usually have limited experience with real data and often cannot judge if what they suggest is really relevant for practical purposes. Experts in astronomical data processing usually have limited knowledge of relativity and often try to apply Newtonian way of thinking to general-relativistic concepts. One remedy of this situation is to organize dedicated scientific meeting where both kind of researchers come together and try to understand each other.

The Commission has organized IAU Symposium 261 "Relativity in Fundamental Astronomy: Dynamics, Reference Systems and Data Analysis". The Symposium was held in Virginia Beach, VA, USA from 27 April to 1 May 2009 and gathered about 100 participants from 18 countries. The IAU Symposium 261 had the goals 1) to summarize the advances in Applied Relativity in the past quarter of a century, 2) to highlight the astonishing achievements in testing General Relativity and to elucidate the tests to be expected in the near future, 3) to facilitate the communication and collaboration between scientists working with high-accuracy data of different kinds by providing a chance to meet at a common scientific meeting, and 4) to consider the future developments of Applied Relativity.

3. Main scientific achievements in Applied Relativity

During Business Session an attempt to summarize the most important developments in the field of Applied Relativity in the period from 2006 to 2009 have been undertaken. The following list does not pretend to be exhaustive.

Brumberg (2007) has demonstrated that in accordance to the old idea of Infeld the variational principle of the Einstein field equations may be used to derive the commonly employed Einstein-Infeld-Hoffman equations of motion from the linearized metric.

Recent re-definition of TDB adopted by the 26th General Assembly of the IAU has created the possibility to define the time scales of new solar system ephemerides in full consistency with General Relativity. The ephemeris group in Paris Observatory (Fienga *et al.* 2009) has used this possibility and created the first version of the "4-dimensional ephemeris" which contains not only the spatial positions and velocities of the solar system objects, but also the numerical transformation between the time parameter of the ephemeris (TDB) to the time scale TT which can be easily used to compute the moments of observations in TAI, UTC or similar time scales generally available to an observer on the Earth.

Earth rotation is the only astronomical phenomenon which is observed with very high accuracy, but still modelled in a Newtonian way. Although a number of attempts to estimate and calculate the relativistic effects in Earth rotation have been undertaken, no consistent theory has appeared until now. At least two projects have been recently started to improve the situation. Brumberg & Simon (2007) consider the formally Newtonian equations of rotational motion with all quantities relativistically transformed into dynamically non-rotating version of the GCRS. Klioner, Gerlach & Soffel (2009) have developed the fully post-Newtonian theory of Earth rotation using numerical integration of the post-Newtonian equations of rotational motion.

Subtle effects in the GPS model are important to improve the accuracy of GPS observations and to model future higher-accuracy navigation systems. Recent papers (Kouba 2004; Larson *et al.* 2007) consider such additional relativistic effects in the GPS model. A detailed review on Relativity in geodesy has been published by Müller *et al.* (2007).

Several new results in the models for light propagation have been published (Kopeikin & Makarov 2007; Le Poncin-Lafitte & Teyssandier 2008; Klioner & Zschocke 2009) aimed at improving the consistency and accuracy of the practical models for high-accuracy positional and ranging observations. These refined models will be used e.g., in the space missions like Bepi-Colombo, Gaia and SIM.

Another research area where substantial progress has been reached is planning and verifying astronomical tests of General Relativity and alternative theories of gravity. In particular, the influence of translational motion of the gravitating bodies on experiments involving light propagation has been considered in detail. Several authors (Bertotti, Ashby & Iess 2008; Kopeikin 2009a,b) have discussed the effect of the motion of the Sun on the light travel time in experiments such as Cassini relativity experiment. The authors finally confirm the claimed accuracy of the Cassini experiment (Bertotti, Iess & Tortora 2003). Considering the anticipated accuracy of the future relativistic experiments it is important to provide realistic errors of the achieved estimates of relativistic parameters. A sophisticated statistical analysis of the relativistic experiments with the ESA mission BepiColombo (Milani *et al.* 2002) has been performed by Ashby, Bender & Wahr (2007).

The ESA project ACES has been recently selected and is aimed at having very accurate clock (stable and accurate at the level of $10^{-16} - 10^{-17}$) in space in the near future. Special care must be taken in practical relativistic modelling of such clocks. Duchayne, Mercier & Wolf (2009) have investigated the relativistic modelling of high-accuracy clock on board of an Earth satellite in full detail.

Finally, the work on the improvement of the relativistic formulations and semantics of the IERS Conventions has been continued. As mentioned above substantial progress has been achieved in developing a consistent nomenclature on units of measurements in the relativistic framework (Klioner 2008; Klioner *et al.* 2009).

4. Future work

The subject of astronomical units mentioned at the beginning of Section 2 above will be treated by a special Working Group of Division I. The participants of Business Session considered that it is appropriate for Commission 52 to discuss the definition of an ecliptic in the

GCRS. The future directions of the work of the Commission 52 will be decided by the future President and the Organizing Committee of the Commission.

Sergei A. Klioner
President of the Commission

References

Ashby, N., Bender, P. L., & Wahr, J. M. 2007, *Phys. Rev. D*, 75, 022001

Bertotti, B., Ashby, N., & Iess, L. 2008, *Class. Quan. Grav.*, 25, 045013

Bertotti, B., Iess, L., & Tortora, P. 2003, *Nature*, 425 374

Brumberg, V. A. 2007, *Cel. Mech. Dyn. Astr.*, 99, 245

Brumberg, V. A. & Simon, J.-L. 2007, *N*otes scientifique et techniques de l'insitut de méchanique céleste, S088

Duchayne, L., Mercier, F., & Wolf, P. 2009, *A&A*, 504, 653

Fienga, A., Laskar, J., Morley, T., Manche, H., Kuchynka, P., Le Poncin-Lafitte, C., Budnik, F., Gastineau, M., & Somenzi, L., 2009, *arXiv:0906.2860*

Kopeikin, S. M. 2009a, *Phys. Lett. A*, 373, 2605

Kopeikin, S. M. 2009b, *MNRAS*, 399, 1539

Kopeikin, S. M. & Makarov, V. V. 2007, *Phys. Rev. D*, 75, 062002

Klioner, S. A. 2008, *A&A*, 478, 951

Klioner, S. A., Capitaine, N., Folkner, W., Guinot, B., Huang, T.-Y., Kopeikin, S. Pitjeva, E., Seidelmann, P. K., & Soffel, M. 2009, in "Relativity in Fundamental Astronomy: Dynamics, Reference Frames, and Data Analysis", S. Klioner, P. K. Seidelmann & M. Soffel (eds.), Cambridge Univesity Press, 79

Klioner, S. A., Gerlach, E., & Soffel, M. 2009, in "Relativity in Fundamental Astronomy: Dynamics, Reference Frames, and Data Analysis", S. Klioner, P. K. Seidelmann & M. Soffel (eds.), Cambridge Univesity Press, 112

Klioner, S. A. & Zschocke, S. 2009, "Numerical versus analytical accuracy of light propagation solution in the Schwarzschild field", *Clas. Quan. Grav.*, submitted (see *arXiv:0902.4206*, *arXiv:0904.3704*, and *arXiv:0911.2170*)

Kouba, J. 2004, *GPS Solutions*, 8, 170

Milani, A. *et al.* 2002 *Phys. Rev. D*, 66, 1

Larson, K. M., Ashby, N., Hackman, C., & Bertiger, W. 2007, *Metrologia*, 44, 484

Le Poncin-Lafitte, Chr. & Teyssandier, P. 2008, *Phys. Rev. D*, 77, 044029

Müller, J., Soffel, M., & Klioner, S. A. 2007, *J. Geod.*, 82, 133

Transactions IAU, Volume XXVIIB
Proc. XXVII IAU General Assembly, August 2009
Ian F. Corbett, ed.

doi:10.1017/S1743921310004928

DIVISION II — SUN and HELIOSPHERE

SOLEIL et HELIOSPHERE

PRESIDENT	**Donald B. Melrose**
VICE-PRESIDENT	**Valentin Martinez Pillet**
PAST PRESIDENT	**David F. Webb**
BOARD	**Jean-Louis Bougeret, James A. Klimchuk, Alexander Kosovichev, Lidia van Driel-Gesztelyi and Rudolf von Steiger**

PARTICIPATING COMMISSIONS

Commission 10	**Solar Activity**
Commission 12	**Solar Radiation & Structure**
Commission 49	**Interplanetary Plasma & Heliosphere**

DIVISION WORKING GROUPS

Solar Eclipses
Solar and Interplanetary Nomenclature
International Solar Data Access
International Collaboration on Space Weather

PROCEEDINGS BUSINESS SESSION

1. Introduction

This report is on activities of the Division at the General Assembly in Rio de Janeiro. Summaries of scientific activities over the past triennium have been published in Transactions A, see Melrose *et al.* (2008), Klimchuk *et al.* (2008), Martinez Pillet *et al.* (2008) and Bougeret *et al.* (2008). The business meeting of the three Commissions were incorporated into the business meeting of the Division. This report is based in part on minutes of the business meeting, provided by the Secretary of the Division, Lidia van Driel-Gesztelyi, and it also includes reports provided by the Presidents of the Commissions (C10, C12, C49) and of the Working Groups (WGs) in the Division.

2. Report on the Business Meeting held on 7 August 2009

The Division II Business Meeting at the General Assembly in Rio de Janeiro was held on Friday 7 August 2009, chaired by the outgoing President of Division II, Don Melrose. An agenda, containing draft motions to be discussed at the meeting had been circulated electronically to all members of the Division prior to the General Assembly.

The meeting was in two parts due to a conflict with Symposium 264. The first part started at 9am, when it was agreed to defer some agenda items to later in the day, after the last talk at the Symposium. These items included the Business Meetings for C10 and C12.

2.1. *Items from the Business meeting in Prague*

The Chair reported on the following items:
1. Outcomes following the business meeting at the last General Assembly in Prague in 2006. The minutes of the Prague meeting were shown and briefly discussed. Members were advised that links to reports published by the IAU on the activities of the Division, its Commissions and WGs are available on the Division website.

2. The chair reported briefly on the three IAU Executive meetings to which the Division Presidents were invited during the triennium. Recommendations concerning the scientific sessions at the General Assembly were delegated to the Division Presidents at the Executive meeting in Oslo in May 2008. At the Executive Meeting on 2 August (EC86-1) the Secretary, Lidia van Driel-Gesztelyi, represented the Division.

2.2. *Future of Commissions and WGs*

The IAU requires that the Division make recommendations on the continuation of existing Commissions and WGs, and on the creation of new Commissions and WGs.

An explanatory note on the structure of the Division had been distributed to all members of the Division prior to the General Assembly. The Chair summarized the status quo by remarking that the Division operates as a federation of the three Commissions. The following motion was moved from the Chair:

Motion 1: Division II recommends that its three Commissions continue: C10 Solar Activity, C12 Solar Radiation & Structure, and C49 Interplanetary Plasma & Heliosphere.

After a brief discussion, this motion was carried unanimously.

The Chair reported that the Presidents of three of the four WGs had recommended that their respective WGs continue, and one President recommended that his WG not continue. The following motion was moved from the Chair:

Motion 2: Division II recommends that three of its four WGs continue: WG Solar Eclipses, WG International Solar Data Access, and WG International Collaboration on Space Weather; and that the fourth WG not continue: WG Solar & Interplanetary Nomenclature.

It was agreed that Division II WG Solar Eclipses, and Division II WG International Collaboration on Space Weather continue, and that WG Solar & Interplanetary Nomenclature not continue. After a vigorous discussion it was recommended that WG International Solar Data Access formally expire and that discussions begin on replacing it by a new WG after consideration of possible overlap and synergy with the IAU Virtual Observatory group.

With this change, Motion 2 was carried unanimously.

2.3. *Election Procedure*

The IAU requires that the Executive (President, Vice President, Secretary) and other members of the Organizing Committee (OC) of the Division for the following triennium be elected at the General Assembly. The Division decided on guidelines for the election process at the General Assembly in Sydney in 2003. The following explanatory note was distributed electronically by the Chair prior to the General Assembly.

Election Procedure for Division II

The current policy for election of the OC for Division II was decided at the Business Meeting at the General Assembly in Sydney 2003. The policy is that the OC consist of the Presidents and Vice Presidents of C10, C12 & C49. The Vice President is regarded as the President-elect. There was no decision to appoint a Secretary at the General Assembly in Sydney; in Prague it was decided to appoint a Secretary from within the OC, and one was chosen from amongst the Vice President of the Commissions. The Chair reported that there has been no move within the Division to change these arrangements.

The important choice that needs to be made at this General Assembly is the election of the Vice President, to take over as President at the following General Assembly. In Sydney, it was decided that the Vice President of the Division should be chosen from among the incoming Commission Presidents. David Webb took over as President of the Division at the Sydney General Assembly, after being President of C49 in the previous triennium, Don Melrose took over as President of the Division at the Prague General Assembly, after being President of C10 in the previous triennium, and Valentin Martinez Pillet takes over as President of the Division 2009–2012 after being President of C12 for 2006–2009. Strict rotation suggests that the Vice President of the Division for 2009–2012 should be from C49.

This scheme is too restrictive, as the situation now facing us shows.† It is important that the Vice President of the Division have appropriate experience, either as the current or immediate

† The VP of C49, who would have been the VP of the Division under this procedure, declined to accept the appointment.

past President of one of the Commissions. This year the OC of the Division decided to broaden the group from whom the incoming Vice President of the Division could be chosen, specifically to include the current Presidents and Vice Presidents of the Commissions. A proviso was that the President and Vice President of the Division should not be from the same Commission. This left four possible candidates, and the OC chose from those of these who were willing to be nominated. The nominee is James Klimchuk, the outgoing President of C10.

The following motion was moved from the Chair.

Motion 3: The incoming Vice President of the Division is to be chosen from amongst the current members of the OC of the Division, subject to the proviso that the President and Vice President are from different Commissions.

After a brief discussion, this motion was carried unanimously.

The position of Secretary of the Division was discussed, and it was agreed that the Secretary be chosen by the incoming OC from amongst its members. It was agreed that it is desirable (but not necessary) that the Secretary be from the Commission not represented by the President or Vice President of the Division.

2.4. *Incoming OC for the Division*

The membership of the incoming OC was noted:
President: Valentin Martinez Pillet (Spain)
Vice President: James A. Klimchuk (USA)
Secretary: to be decided
Past President: Donald B. Melrose (Australia)
Board: Gianna Cauzzi (Italy), Natchimuthuk Gopalswamy (USA), Alexander Kosovichev (USA), Ingrid Mann (Japan), Carolus J. Schrijver (USA) and Lidia van Driel-Gesztelyi (France, UK, Hungary).
None of the VPs of the Commissions was present, and the appointment of the Secretary, from amongst them, was deferred until they could be consulted.

2.5. *Electronic contact with Division members*

An e-mail list <iaudivii@iac.es> of all members of the Division is available, allowing effective electronic communication between the OC and members of the Division. There was discussion on how this new possibility should be used. Valentin Martinez-Pillet, who has oversight of the list, emphasized that we need to restrict the number of emails, avoiding any that could be regarded as SPAM. At present only a few messages have been sent to this list, and the Division President is the only person who has permission to send messages. Nevertheless, a few people have asked to be removed from the list.

The use of the list for news was discussed. The consensus was that any news items should be specifically IAU-related. There is no intention that this will replace Solar News and a source of more general news for the solar community. Jean-Louis Bougeret pointed out that many heliosphere/magnetosphere people do not read Solar News.

One specific use of the list is for election of the OCs for the three Commissions. For this purpose, it is desirable to have separate lists for the three Commissions. As most of the work is done through the Commissions, such lists would be useful for other purposes.

2.6. *Divisional website*

The Division has a website at
http://www.iac.es/proyecto/iau_divii/IAU-DivII/main/index.php
Suggestions relating to the website and how it might be used will be welcome.

2.7. *C49 Business Meeting*

The President of C49, Jean-Louis Bougeret took the chair for the business meeting of C49. He reviewed the broad scope of C49 and proposed that it cover the solar wind and heliosphere, including the magnetospheres. It was questioned up to what point the Earth's magnetospheric community (that has strong links with C49) is involved in IAU matters such as Symposia. While it is acknowledged that this community is more linked with other international organizations, it is recommended that the Commission OC take measures to encourage this community to participate more actively in the IAU. Their input can be very relevant for important astronomical topics such as exo-planet magnetospheres.

The President thanked the outgoing OC for their work. The outgoing OC comprised 13 members including P and VP, covering most topics in the field of C49. (P: J.-L. Bougeret, VP: R. von Steiger, Board: S. Ananthakrishnan, H. Cane, N. Gopalswamy, S. Kahler, K. Shibata, R. Lallement, B. Sanahuja, M. Vanda, F. Verheest, D. Webb). During the last triennium, OC members have been much involved in special activities in the context of the International Heliophysical Year (IHY 2007-2008). He expressed the view that C49 is the best context to pursue these efforts of international coordination.

The President referred to the Triennial Report by C49, cf. Bougeret *et al.* (2008). A big effort was made by contributors who produced a very nice summary of progress made in the last three years. In particular, Richard Marsden, the *Ulysses* Project Scientist, wrote a very nice review of the main results of the *Ulysses* mission –a landmark in this field– that was recently terminated after 18 years (five IAU triennia!) of successful operation.

The election for the new OC was discussed and conducted within the OC. It was expected that a C49 e-mail list will be available in the future, so that all regular members can participate in the election. The OC acknowledged that the outgoing VP, R. von Steiger declined to continue as President, being already much involved in COSPAR and ISSI activities. J.-L. Bougeret proposed as new President Nat Gopalswamy, who has been much involved in the organization of the IHY as "International Coordinator", and as Vice-President Ingrid Mann, who is much involved in the hot topics of interplanetary dust and nano-particles. For the Board, it was agreed to enforce the IAU bye-laws for the maximum duration of OC membership (6 years), and 7–8 members maximum. The President reported the result of the election of the new OC,
President: Natchimuthuk Gopalswamy (USA)
Vice-President: Ingrid Mann (Japan)
Past President: Jean-Louis Bougeret (France)
Board: Carine Briand (France), Rosine Lallement (France), David Lario (USA), P.K. Manoharan (India), Kazunari Shibata (Japan), and David F. Webb (USA),
and he thanked them and wished C49 an excellent triennium.

2.8. *Afternoon session*

The meeting adjourned and reconvened after the conclusion of Symposium 264. The Chair reported on the items of business discussed before the adjournment. In response to a comment from Jean-Claude Pecker, the meeting was advised that triennial Commission science reviews are available on the IAU website.

2.9. *C10 Business Meeting*

The President of C10, James Klimchuk, took the chair for the business meeting of C10. He summarized the election of the incoming OC, and the activities of the C10 over the triennium.

Election: Because an e-mail distribution list for the 656 regular members of C10 is not yet available, the election was conducted within the OC. OC members were asked if they wished to serve another term, and they were reminded that serving more than two terms is discouraged. Nominations were also solicited for entirely new members. This produced a slate of candidates from which the incoming OC members were elected by anonymous ballot. Following tradition and with unanimous approval, the current Vice-President became the incoming President. A subsequent vote was taken to elect the incoming V-P from a list of returning OC members who wished to be considered. The outcome of the election is as follows:
President: Lidia van Driel-Gesztelyi (France, UK, Hungary)
Vice-President: Carolus J. Schrijver (USA)
Past President: James A. Klimchuk (USA)
Board: Paul Charbonneau (Canada), Lyndsay Fletcher (UK), S. Sirajul Hasan (India), Hugh S. Hudson (USA), Kanya Kusano (Japan), Cristina H. Mandrini (Argentina), Hardi Peter (Germany), Bojan Vršnak (Croatia), and Yihua Yan (China).
The question of whether to continue to have a Commission Secretary will be taken up by the new OC. It is expected that a Commission e-mail list will soon be available and that all regular members will participate in the next election of the OC.

Meeting Proposals: The OC solicited, reviewed, and endorsed proposals for Symposia and Joint Discussions. It also offered advice on how the proposals could be improved. Many OC members expressed a concern that Symposia tend to be too broad and suggested that more focused meetings would have greater scientific value and attract more leading scientists. It was recognized that focused proposals can have difficulty getting the support of multiple Division

Presidents and therefore are less likely to be selected. It is hoped that the IAU Executive Committee will consider this issue.

Triennial Report: A team effort involving the entire OC produced a very nice summary of science progress in Solar Activity over the prior three years. It was published in Transactions IAU, Vol. XXVIIA, cf. Klimchuk *et al.* (2008), and is available on the Commission website.

Website: A C10 website was created (http://www.mssl.ucl.ac.uk/iau_c10/index.html). Among other things, it includes recent reports of the Commission and a history of OC members and officers dating back to 1970 (Presidents dating back to 1925).

Solar Naming Convention: Carolus Schrijver, incoming V-P, proposed a standardized convention for identifying solar events (e.g., flares) to be included in publications so that search engines can easily identify other relevant publications in the on-line literature. C10 formally endorsed the proposal, and it has recently been adopted by *Solar Physics*, with other journals hopefully to follow. The proposal can be found on the Division II website.

At the end of the C10 business meeting, the President invited the audience to make suggestions on how the Commission could better serve its members. No significant response was received.

2.10. *C12 Business Meeting*

The President of C12, Valentin Martinez-Pillet, took the chair for the business meeting of C12. He thanked the members of the outgoing OC, described the election of the incoming OC, and summarizes the activities of C12 over the triennium.

The election was not carried out by polling the entire membership, due to the absence of a Commission specific e-mail list. It was commented that it would be ideal if IAU could provide such e-mail lists. In their absence, the Commission Presidents only have the option of creating such lists at their home institutions, but it would be difficult to justify the effort in creating, verifying and maintaining such lists. Only through Commission-wide e-mail lists) would it be possible to involve the full membership in future OC elections. The President noted the result of the election for the incoming OC:

President: Alexander Kosovichev (USA)
Vice-President: Gianna Cauzzi (Italy)
Past President: Valentin Martnez Pillet (Spain)
Board: Martin Asplund (Germany), Axel Brandenburg (Sweden), Dean-Yi Chou (China Taiwan), Jörgen Christensen-Dalsgaard (Denmark), Weiqun Gan (China PR), Vladimir D. Kuznetsov (Russian Federation), Marta G. Rovira (Argentina), Nataliya Shchukina (Ukraine), and P. Venkatakrishnan (India).

The President advised that the C12 website can be found at:
http://www.iac.es/proyecto/iau_divii/IAU-Com12/main/index.php
where information about related IAU Symposiums and links to appropriate web sites (e.g., IAU membership lits) can be found.

The President referred to the triennial report of C12, cf. Martinez Pillet *et al.* (2008), which includes contributions from scientists who are not OC members and who made relevant investigations to hot topics. Specific topics covered in this way were solar abundances, dynamo activity and solar-cycle minimum.

Sasha Kosovichev (incoming President) asked how members can change Commission/Division ascription within IAU. The President commented that IAU has already established a login facility at the Union web site where some personal information can be changed. It would be ideal if information about the ascription to specific Commissions/Divisions could be negotiated using this tool. As incoming President, Sasha noted that a specific request along these lines to the IAU Executive will be made.

2.11. *Proposals for IAU meetings*

The President of the Division resumed the chair. He opened the meeting for possible proposals for solar-sponsored IAU Symposia in 2011, 2012 and 2013. There are nine IAU Symposia in each calendar year. Two of the Symposia in 2010 are solar-related:
IAUS 273 "Physics of the Sun and star spots" Los Angeles, USA, 23–26 August
IAUS 274 "Advances in Plasma Astrophysics" Catania, Italy, 6–10 September

The schedule for proposing Symposia is that same each year. Specifically, for Symposia in 2011, the deadline for Letters of Intent is mid-September 2009, and complete proposals will be required by early December 2009. The successful proposals will be known by mid-2010.

2.12. *Other business*

Carine Briand suggested that Division II proposes to the General Secretary of the IAU that he send to the Italian government a letter in support of the reconstruction of Aquila University astronomically related sites. The incoming Division president, V. Martinez Pillet, suggested that he raise this issue at the Executive meeting on the last day of the GA.

Nat Gopalwamy presented a proposal from Sarah Gibson for a WG on comparative solar minima: characterizing the heliophysical "ground state". There was general support for the proposal. A more detailed report on this proposal is included in § 3.7 below.

Carine Briand raised the issue of the low visibility of Solar and Heliospheric sciences. A proposal was made to form a Division II WG whose main objective will be to start activities related to public outreach of Heliophysics sciences. It was pointed out that there would be overlap with the activities of C55 (Division XII), that links should be created with C55, and that eventually the activities of this WG are likely to be incorporated into those of C55. It was agreed that we should proceed with such a new WG for the next three years. A brief report is included in § 3.8 below.

Sasha Kosovichev presented a proposal from C12 for a WG on the coordination of synoptic observations of the Sun. In the subsequent discussion, Todd Hoeksema remarked on the importance of long-term continuous synoptic observations, and on the difficulty of maintaining a continuous series. It was agreed that C12 should decide what action to take on this matter.

Cristina Mandrini presented a brief report on solar physics research in Argentina. Solar physics in Argentina is young compared to other branches of astrophysics. There are about eight PhD students working in solar physics and/or heliospheric physics. Cristina emphasized the need for schools for these students. One such school was organized in association with the IHY in 2006, and she pointed out the difficulty of organizing of such meetings due mainly to financial problems. Two instruments provide data, that are freely available, the Hα Solar Telescope (HASTA, high temporal cadence increased during flares) and the Mirror Coronagraph (MICA, prototype of C1) both in Observatorio Felix Aguilar in San Juan. A third instrument is the Solar Submillimeter Telescospe (SST) in collaboration with the Brazilian group led by P. Kaufmann in Complejo Astronomico El Leoncito, San Juan.

3. Reports from WGs and IAU representatives

Chairs of WGs and IAU representatives on other international organizations (COSPAR, IHY, SCOSTEP, ISES) were invited to give reports. The following reports were received.

3.1. *WG on Solar Eclipses (Jay Pasachoff)*

During the 2006-9 triennium, members of the IAU WG on Solar Eclipses of IAU Division II worked on matters of professional liaison and public information for total solar eclipses in Siberia/China on 1 August 2008 and, just before the General Assembly, in India and China on 22 August 2009; for annular eclipses in French Guiana on 22 September 2006, New Zealand and the South Pacific on 7 February 2008 (annular in Antarctica), and Indonesia and Australia's Cocos Islands on 7 January 2009; and partial in eastern Asia on 19 March 2007 and in Argentina and Brazil on 11 September 2007. The WG Website, at
http://www.eclipses.info/>http://www.eclipses.info,
is a convenient reference for professional astronomers, amateur astronomers, and the general public. The WG is also in liaison with the Program Group on Public Education at the Times of Eclipses of IAU C46 on Astronomy Education and Development. WG Chair Pasachoff (2007) wrote "Observing solar eclipses in the developing world" for the proceedings of an IAU Special Session.

Members of the WG include Iraida Kim of Moscow State University (Russia), Hiroki Kurokawa of Kwasan Observatory (Japan), Jagdev Singh of Indian Institute of Astrophysics (India), Vojtech Rusin of the Astronomical Institute of the Slovak Academy of Sciences (Slovakia), Atila Ozguc of the Kandilli Observatory (Turkey), Fred Espenak of NASA's Goddard Space Flight Center (USA), Jay Anderson of University of Manitoba (Canada), Glenn Schneider of the University of Arizona's Steward Observatory (USA), Michael Gill of the Solar Eclipse Mailing List (U.K.) and Yihua Yan of the National Astronomical Observatories (China). Espenak is the world's major source of predictions of eclipse paths and circumstances and Anderson provides meteorological information; their work is available on the Web, linked to

http://www.eclipses.info/>http://www.eclipses.info
and directly at
http://eclipse.gsfc.nasa.gov/eclipse.html,
and as NASA Technical Publications, all of which acknowledge the support of the IAU WG on Eclipses. Espenak & Meeus (2007), Espenak & Meeus (2008) also published "Five Millennium Canon of Solar Eclipses: -1999 to +3000" and "Five Millennium Catalogue of Solar Eclipses: -1999 to +3000 (2000 BCE to 3000 CE)." Gill now runs the Solar Eclipse Mailing List at: SEML@yahoogroups.com, a valuable resource for many of the most dedicated eclipse observers.

Pasachoff and Anderson attended a Solar Eclipse Workshop in Delhi in January 2009 to help Indian students and others prepare for the 22 July 2009 eclipse. In the event, successful airplane observations were obtained and some ground-based observations were obtained from a few Indian sites, though the weather in India was, as predicted, overall not conducive to viewing.

Two eclipse review articles, Pasachoff (2009a) and Pasachoff (2009b), were published.

Yan was recently added to the WG because he was chief solar scientist for observations and planning for the 22 July 2009 eclipse, the longest in the 18 year 11 1/3 day Saros series, which was mainly observed from China, with totality approaching 6 minutes. Among the tasks of the WG on Solar Eclipses with which Yan helped were discussions about arrangements for the duty-free import of observing equipment for the eclipse observations and official invitations on behalf of the Chinese National Academy of Sciences to aid scientists in obtaining visas to China. Pasachoff had traveled with Yan, with Lin Lan of Hangzhou High School, and with Jin Zhu, director of the Beijing Planetarium, to reconnoitre near Shanghai for eclipse sites, and they chose three alternatives: Jinshanwei (on the coast south of Shanghai), Moganshan (at 300 m altitude north of Hangzhou), and Tianhuangping (at 900 m altitude west northwest of Hangzhou). The site at Tianhuangping was finally adopted by research teams from many nationalities, and was considered the IAU site. It included scientists from the United States, China, Australia, Greece, France, Russia, Georgia, Bulgaria, and elsewhere. For the actual eclipse, the weather was not ideal, but all aspects of the eclipse, including the corona, were viewed, though through clouds. Conditions at Moganshan (fog) and Jinshanwei (rain) were worse, so it is fortunate that most scientists were concentrated at Tianhuangping. Some preliminary images are available at www.williams.edu/astronomy/eclipse/eclipse2009.

Several ships observed the eclipse, including one that obtained the maximum of 6 min 42 sec under clear skies. Excellent observations were obtained at Enewetak atoll, Marshall Islands, by Miloslav Druckmüller, Vojtech Rusin, and colleagues. See
http://www.zam.fme.vutbr.cz/ druck/eclipse/Ecl2009e/0-info.htm

The good offices of the IAU WG on Solar Eclipses will continue during the next triennium with an annular eclipse on 15 January 2010 in India, Sri Lanka, Bangladesh, and Myanmar; with a total eclipse on 11 July 2010 in Easter Island and at sea and on atolls in French Polynesia; with four partial eclipses in 2011, and with an annular eclipse on 20 May 2012 whose path includes southeastern China, southeastern Japan, and the western United States, including major radio telescopes. Preparations will be made for the 13 November 2012 total solar eclipse that will cross Australia's Cape York peninsula and Cairns. A Solar Eclipse Workshop is under discussion for 2011, a year with no total or annular eclipses but with four partial eclipses, which would be the successor to similar Workshops held most recently at the Griffith Observatory, California, USA, in 2007, and prior to that in Milton Keynes, U.K., in 2004, and in Antwerp, Belgium, in 2000; such workshops have been held in years with no total or annular eclipses.

Xavier Jubier (France) has been added to the WG in recognition of his providing on-line eclipse maps in Google Maps and Google Eclipse:
http://xjubier.free.fr/en/site_pages/SolarEclipsesGoogleMaps.html
He has also provided Solar Eclipse Maestro for computer control of cameras during eclipses. With the end of the Chinese pair of total eclipses, Yihua Yan has rotated off the WG.

3.2. *WG on on Solar and Interplanetary Nomenclature (Ed Cliver)*

The WG on Solar and Interplanetary Nomenclature (formed in 2000) was chaired by Ed Cliver and included Jean-Louis Bougeret, Hilary Cane, Takeo Kosugi (deceased), Sara Martin, Reiner Schwenn, and Lidia van Driel-Gestelyi. With the help of the broader community, the WG identified terms used in solar and heliospheric physics that were thought to be in need of clarification and then commissioned topical experts to write essays reviewing the origins of these terms and their current usage/misusage. In all six terms were addressed and seven essays (listed below) were published. The first six essays, Burlaga (2001), Russell (2001), Cliver & Cane (2002), Daglis

(2003), Vršnak (2005) and St. Cyr (2005), and an introduction to the series, Cliver (2001), were published in *EOS*, the weekly newspaper of the American Geophysical Union, and the last, Švestka (2007), was published in *Solar Physics*.

It was decided by the WG that the backlog of significant solar/heliospheric terms "at risk" had been largely addressed and that the WG be terminated. As more terms arise, as will almost certainly be the case in a vigorous field of research such as solar-terrestrial physics, a new committee can be formed to address them.

The Chair thanks the WG members for their dedicated service during these past nine years as well as the essay authors for their insights on the science underlying the terms they considered.

3.3. *WG on International Collaboration on Space Weather (David Webb)*

The WG for International Collaboration on Space Weather has as its main goal to help coordinate the many activities related to space weather at an international level. It is chaired by David Webb and its website is at:
http://www.iac.es/proyecto/iau_divii/IAU-DivII/main/spaceweather.php
The site currently includes the international activities of the IHY, the International Living with a Star (ILWS) program, the CAWSES (Climate and Weather of the Sun-Earth System) WG on Sources of Geomagnetic Activity, and Space Weather studies in China.

The IHY is an international program of scientific collaboration that during the time period 2007-2009, centered on 2008, the 50th anniversary of the International Geophysical Year. The IHY was considered to be of sufficient importance to the IAU to have its own IAU scientific representative, currently David Webb. The physical realm of the IHY encompasses all of the solar system out to the interstellar medium, representing a direct connection between in-situ and remote observations. The IHY working group helped identify national leaders for the IHY program. The IHY organization has an International Advisory Committee and an International Steering Committee. Coordinators were appointed for eight regions of the world and ~60 countries have functioning national committees. Complete information on the IHY can be found at the main IHY site: http://ihy2007.org. The IHY has officially closed, but it is expected that the observatory legacy part of IHY will continue under UN auspices as the International Space Weather Initiative.

The CAWSES working group on Sources of Geomagnetic Activity, chaired by Nat Gopalswamy, has as its objectives to understand how solar events, such as CMEs and high speed streams, impact geospace by investigating the underlying science and developing prediction models and tools. CAWSES has been extended as CAWSES–II for 2009–2013. The website is at: http://www.cawses.org/

The WG on Space Weather Studies in China is chaired by Jingxiu Wang and is involved with many new initiatives on space weather. The working group will be adding information and websites on active space weather studies in other countries, such as in India, Russia and the Americas.

3.4. *WG on International Data Access (Bob Bentley)*

Website: http://www.mssl.ucl.ac.uk/grid/iau/index.php

The WG on International Data Access (Sun and Heliosphere) was originally formed as a group intended only to cover the solar part of Division II. However, it was extended to include heliospheric data sets needed to support Space Weather and related studies.

For several years, many of the members of the WG have discussed the idea of building a virtual observatory (VO) in heliophysics. In order to address science problems that span the disciplinary boundaries the idea was that this would provide enhanced access to solar and heliospheric data, and to magnetospheric and ionospheric data for planets with magnetic fields and/or atmospheres. Although initial attempts were unsuccessful, a proposal that was submitted in September 2008 to the Capacities thematic priority of the European Commission's Seventh Framework Programme (FP7) was accepted. The Heliophysics Integrated Observatory, HELIO, started in June 2009 and will last for three years; the HELIO Consortium has 13 members, many of whom are members of this WG. The Consortium includes experts in data from the different domains of heliophysics as well as computer scientists; NASA and ESA are also both involved.

One of the main objectives of HELIO virtual observatory is to create a collaborative environment where scientists can discover, understand and model the connections between solar phenomena, interplanetary disturbances and their effects on the planets. In order to achieve this the project has to address issues related to data quality and content, and manage the differences in the way data are stored and described in the different domains. It is in these areas that the activities of this WG are important.

With a few exceptions, data from space-based observatories are generally of good quality and well managed. However, data from ground-based observatories have more inconsistencies than their space-based counterparts that make them harder to use in the machine world of the virtual observatory. Within the WG we have been looking at two areas that should enhance data access for both groups:

- Metadata contents of the files headers
- The naming and organization of data in archives exposed to the Internet

There are many complexities associated with making changes to existing data sets and the principle objective of the former discussion is to improve the quality of new data sets by encouraging good practices. Virtual observatories such as HELIO must, as far as possible, be able to handle the issues related to accessing existing datasets within the infrastructure that they provide.

There is significant scope, however, for improving the status quo on the latter topic. Small changes to the way files are named and the directory structure under which they are stored can radically improve their accessibility from the Internet; the benefit of such changes are not necessarily obvious until they are explained to the providers. Also, the provision of observing logs allows virtual observatories to address issues of the continuity and completeness of data sets, and manage access where all the data are not necessarily on-line.

The WG is working with HELIO and with the SOTERIA project (also funded under FP7) to try to test our ideas for improving data access. In many ways, HELIO and SOTERIA complement each other: HELIO is concentrating on infrastructure issues while a significant part of the resources of SOTERIA are devoted to working with data providers. By using the groups involved in SOTERIA as test cases for the Working Groups guidelines, we should be able to iterate the guidelines to provide maximum benefit while causing minimal disruption to the data providers. HELIO should then be able to quickly demonstrate how any changes have improved access.

By validating the ideas of the WG through a collaboration with funded projects that we have influence over, we believe that we stand a chance of making real progress in our objective on improving access to solar and heliospheric data for the whole community. It is in this area that most of our effort will be focused over the next 12–18 months.

With HELIO we are also considering whether additional improvements are required to service the needs of other communities that could be interested in these data - astrophysics and planetary science being the most obvious examples. If necessary such changes will be incorporated into the guidance material provided by the WG.

Links to HELIO, SOTERIA and related projects can be found on the WG website.

3.5. *Report on Scientific Committee on Solar-Terrestrial Physics (Nat Gopalswamy)*

The Scientific Committee on Solar-Terrestrial Physics (SCOSTEP) has completed the successful Climate and Weather of the Sun-Earth System (CAWSES) program that ran during 2004–2008. Preparations are currently made to continue this program as CAWSES–II, which will be launched in 2010 and will end in 2014.

Based on the community input obtained during various meetings since 2007, four scientific topics have emerged that focus on some of the forefront scientific challenges facing the international solar terrestrial physics community:

1). Solar Influences on Earth's Climate
2). Geospace Response to Altered Climate
3). Short-term Solar Variability and Geospace
4). Geospace Response to Lower Atmospheric Waves

In addition, two working groups have been formed that will address the issues of Capacity Building and Escience and Informatics.

In collaboration with the Solar-Terrestrial Environment Laboratory of Nagoya University in Japan, a series of educational comic books was developed to communicate solar-terrestrial science to the general public in many languages. In addition to distributing the existing comic books, new ones that address CAWSES–II science topics will be developed. CAWSES–II will support scientists from developing nations as well as graduate students and young scientists to take part in SCOSTEP/CAWSES activities and will have access to data and research tools.

CAWSES–II is to establish a Virtual Institute in order to most effectively coordinate international collaborations among scientists around world, particularly those from developing countries as well as early career scientists and students. The Virtual Institute will take advantage of cyber-infrastructure technology and develop necessary software to facilitate cross-disciplinary research, and data and resource management. It will establish digital libraries and host virtual scientific conferences, which will benefit greatly young scientists. Public education and capacity building will continue to be a core part of CAWSES–II.

The twelfth Solar-Terrestrial Physics Symposium organized by SCOSTEP will be held in Berlin during 12–16 July 2010. The scientific deliberations will be organized along the following sessions reflecting the CAWSES–II activities: 1. Solar influences on climate, 2. Space weather: science and impacts, 3. Atmospheric coupling processes, and 4. Space climatology.

3.6. *Summary of Activities Related to the IHY Program (David Webb)*

The International Heliophysical Year (IHY) is an international program of scientific research and collaboration to understand the external drivers of the space environment and climate. Activities were centered in 2007 and 2008. The IHY involves utilizing the existing assets from space and ground as a distributed Great Observatory and the deployment of new instrumentation, new observations from the ground and in space, and public and student education. IHY officially closed in February 2009 at the United Nations in Vienna, Austria. Many IHY follow-on activities will continue over the next few years and will focus on the transition to the new International Space Weather Initiative (ISWI).

Within the IAU, coordination of IHY activities is within Division II. David Webb is the IAU representative to the IHY and will continue as the representative for the new ISWI. Hans Haubold has led the IHY effort for the United Nations under the auspices of COPUOS and the U.N. Basic Space Science (UNBSS) program.

IHY's focus on developing new and exciting EPO programs, led by Cristina Rabello-Soares, provided unique opportunities for the global community to increase the visibility and accessibility of heliophysics outreach programs. To address this focus, the IHY developed a Schools Program that developed a series of schools in 2007 and 2008 with the purpose of educating students about Universal Processes, the organizational principles and universal laws that underlie our understanding of the Universe. The IHY Schools Program was coordinated by David Webb and Nat Gopalswamy. Five main IHY schools were held: the North America school in Boulder, CO, USA, July-August 2007; the first Asia-Pacific school at Kodaikanal Observatory in Bangalore, India December 2007; the Latin America school in Sao Paulo, Brazil February 2008; the 2nd Asia-Pacific school in Beijing, China October 2008; and the Europe-Africa school in Nsukka, Nigeria November 2008.

The IHY Gold History initiative has had the goals of identifying and recognizing participants in the first IGY, preserving memoirs, etc. of historical significance for the IGY, making them available to historians and researchers, spreading awareness of the history of geophysics, and planning special events. About 150 IHY Gold members have been recognized.

Many scientific meetings and workshops related to IHY were held in 2007–2009 in many countries. In 2009 alone there were these IHY-related meetings: The IHY Africa 2009 Workshop in June in Livingstone, Zambia held in conjunction with the SCINDA 2009 workshop; an IHY session at the International Association of Geomagnetism and Aeronomy (IAGA) 11th Scientific Assembly in Sopron, Hungary in August; the Final IHY/ UNBSS Workshop in Daejeon, Korea in September; a Joint Discussion (JD16) on the results of the IHY Whole Heliosphere Interval August 12–14 at the IAU General Assembly in Rio de Janiero, Brazil; a second workshop following up on the successful 2008 IHY's Whole Heliosphere Interval (WHI) workshop will be held in Boulder, CO, USA November 10–13, 2009.

Several books and articles have or are now being published regarding IHY-related activities. These include:

1) Two versions of the Final Report of the International Heliophysical Year to the United Nations Committee on the Peaceful Uses of Outer Space (CUPOUS): a 50-page summary of IHY activities published in 2008 by the UN, and a complete report that is being published by Springer.

2) The Proceedings of IAU Symposium 257, "Universal Heliophysical Processes", held in Ioannina, Greece in September 2008, was published for the IAU by Cambridge University Press in 2009, edited by Nat Gopalswamy and David Webb.

3) Papers based on the IHY/UNBSS meeting held in Sozopol, Bulgaria in 2008 will be published in a special issue of the journal *Sun and Geosphere*.

4) A book on universal physical processes will be published by Springer based on the IHY School in Kodaikanal, India.

5) A summary of the papers presented at the IAU JD16 meeting on WHI, edited by David Webb and Sarah Gibson, will be published in the IAU Highlights of Astronomy, Volume 15.

6) Brief reports by David Webb summarizing the IHY activities as related to the IAU have been published in the IAU Information Bulletins over the past few years. The final one will appear in IB105.

3.7. *New WG on Comparative Solar Minima (David Webb, Sarah Gibson)*

Website: http://ihy2007.org/IAUWG/WEBPAGES/IAUWG.shtml

The solar activity cycle, as manifested by repeated increase and then decrease in the number of sunspots visible on the Sun, has been observed and analyzed for centuries. However, only for the past few ~11-year activity cycles have new capabilities in satellite and ground-based observations allowed us to consider how a broad range of solar, heliospheric, and geospace observables vary within and between cycles. These observations, in conjunction with theoretical and numerical modeling advances, enable an interdisciplinary, system-wide view on the origins and impacts of solar cycle variation.

Solar minimum represents the time of lowest solar activity and simplest heliospheric structure, and as such is a good place to begin putting together such a system-wide understanding. However, recent observations and analyses imply complexities in the variation within and between solar minima that have implications for analyzing and predicting space weather responses at the Earth during solar quiet intervals, and also for interpreting the Sun's past behavior as preserved in cosmogenic isotopes and historical sunspot and auroral records.

Determining the solar origins and net impacts at the Earth of solar minimum differences will require coordinated, interdisciplinary modeling efforts to bring the pieces together. The international observational and modeling coordinated campaigns known as the Whole Sun Month (WSM) and the Whole Heliosphere Interval (WHI) are examples of such efforts, for solar minimum periods in 1996 and 2008, respectively. The goals of these campaigns were to characterize the 3-D solar minimum heliosphere and to trace the effects of solar structures and activity through the solar wind to the Earth and other planetary systems, and beyond. A direct comparison of these two periods illustrates how very different solar minima may be.

The mission of this WG is to facilitate international and interdisciplinary research that focuses on the coupled Sun-Earth system during solar minimum periods. Such research seeks to characterize the system at its most basic "ground state" but also to understand the degree and nature of variations within and between solar minima.

The WG will build from the WSM and WHI legacy, with the goals of:

- Archiving observations, models, visualizations, and related publications from these periods and the solar minima that encompass them
- Coordinating ongoing research, via projects, workshops, Special Sessions/Joint Discussions, and Symposia (note: the WG has submitted a letter of intent for a 2011 IAU Symposium)
- Providing an infrastructure that could be extended to include observations and models that include both past and future solar cycles

The focus will be on variations between solar minima, but it will be essential to consider how such variations arise. This will require some degree of broader consideration of the solar cycle, both in the context of how a given solar minimum may depend upon the transport of solar magnetic flux in the years preceding it, and in the even greater context of long-term solar cycle variations. Solar activity in the past few decades has been very high compared to the past millenium. What was the heliospheric state during periods of the lowest sunspot activity, e.g., the Maunder minimum of the 17th century? Is there a minimum "ground state" for solar/heliospheric behavior? How might complexities in the solar magnetic configuration have influenced the Earth's response during such intervals? Finally, in the study the cyclic interactions of the heliophysical system, we will be mindful of insights gained into stellar variability over multiple time scales.

3.8. *New WG on Communicating Heliophysics (Carine Briand)*

There have been several valuable and successful efforts for communicating Heliophysics at national and international levels during the IHY 2007–8. These efforts are still in progress and a frame is needed to coordinate them, particularly with the activities of C55 of Division XII. As reported above, it was agreed that we should proceed with such a new WG for the next three years. A formal proposal has been prepared and forwarded to the incoming Division President.

4. Closing remarks

I thank all those who contributed to this report, specifically, Lidia van Driel-Gesztelyi, Valentin Martinez Pillet, Jim Klimchuk, Jean-Louis Bougeret, David Webb, Jay Pasachoff, Ed Cliver, Bob Bentley, Nat Gopalswamy and Sarah Gibson.

Don Melrose
Past President of the Division

References

Bougeret, J.-L., von Steiger, R., Webb, D. F., Ananthakrishnan, S., Cane, H. V., Gopalswamy, N., Kahler, S. W., Lallement, R., Sanahuja, B., Shibata, K., Vandas, M., & Verheest, F. 2008, *Commission 49: Interplanetary Plasma and Heliosphere*, in K. A. van der Hucht (ed.) *Transactions IAU, Vol. XXVIIA, Reports on Astronomy*, Cambridge University Press, pp. 124–144

Burlaga, L. 2001, *Terminology for Ejecta in the Solar Wind*, *EOS* 82, 433–435

Cliver, E. W. 2001, *The Last Word (Introduction)*, *EOS* 82, 433

Cliver, E. W. & Cane, H. V. 2002, *Gradual and Impulsive Solar Energetic Particle Events*, *EOS* 83, 61–68

Daglis, I. A. 2003, *Magnetic Storm – still an adequate name*, *EOS* 84, 207–208

Espenak, F. & Meeus, J. 2007, *Five Millennium Canon of Solar Eclipses: -1999 to +3000*, *NASA Technical Publication*
http://eclipse.gsfc.nasa.gov/SEpubs/5MCSE.html

Espenak, F. & Meeus, J. 2008, *Five Millennium Catalogue of Solar Eclipses: -1999 to +3000 (2000 BCE to 3000 CE)*, *NASA Technical Publication*
http://eclipse.gsfc.nasa.gov/SEpubs/5MKSE.html

Klimchuk, J. A., van Driel-Gesztelyi, L., Schrijver, C. J., Melrose, D. B., Fletcher, L., Gopalswamy, N., Harrison, R. A., Mandrini, C. H., Peter, H., Tsuneta, S., Vrsnak, B., & Wang, J. 2008, *Commission 10 Solar Activity: Triennial Report 2006-2009*, in K. A. van der Hucht (ed.) *Transactions IAU, Vol. XXVIIA, Reports on Astronomy*, Cambridge University Press, pp. 79–103

Martnez Pillet, V., Kosovichev, A., Mariska, J. T., Bogdan, T. J., Asplund, M., Cauzzi, G., Christensen-Dalsgaard, J., Cram, L. E., Gan, W., Gizon, L., Heinzl, P., Rovira, M. G., & Venkatakrishnan, P. 2008, *Commission 12: Solar Radiation and Structure*, in K. A. van der Hucht (ed.) *Transactions IAU, Vol. XXVIIA, Reports on Astronomy*, Cambridge University Press, pp. 104–123

Melrose, D. B., Martnez Pillet, V., Webb, D. F., van Driel-Gesztelyi, L., Bougeret, J.-L., Klimchuk, J. A., Kosovichev, A., & von Steiger, R. 2008, *Division II Sun and Heliosphere: Triennial Report 2006-2009*, in K. A. van der Hucht (ed.) *Transactions IAU, Vol. XXVIIA, Reports on Astronomy*, Cambridge University Press, pp. 73–78

Pasachoff, J. M. 2007, *Observing solar eclipses in the developing world*, in J. B. Hearnshaw, P. Martinez (eds.), *Astronomy in the Developing World*, Cambridge University Press, pp. 265–268

Pasachoff, J. M. 2009, *Solar Eclipses as an Astrophysical Laboratory*, *Nature* 459, 789–795, DOI 10.1038/nature07987
http://www.nature.com/nature/journal/v459/n7248/pdf/nature07987.pdf

Pasachoff, J. M. 2009, *Scientific Observations at Total Solar Eclipses*, *Research in Astronomy and Astrophysics* 9, 613–634
http://www.raa-journal.org/raa/index.php/raa/article/view/182

Russell, C. T. 2001, *In Defence of the Term ICME*, *EOS* 82, 434

St. Cyr, C. 2005, *The Last Word: The Definition of Halo Coronal Mass Ejections*, *EOS* 86, 281–282

Švestka, Z. 2007, *The Misnomer of "Post-Flare Loops"*, *Solar Physics* 246, 393

Vršnak, B. 2005, *Terminology of Large-Scale Waves in the Solar Atmosphere*, *EOS* 86, 112–113

Transactions IAU, Volume XXVIIB
Proc. XXVII IAU General Assembly, August 2009
Ian F. Corbett, ed.

doi:10.1017/S174392131000493X

DIVISION III PLANETARY SYSTEMS SCIENCE

PRESIDENT	EDWARD L. G. BOWELL
VICE-PRESIDENT	Karen J. Meech
PAST PRESIDENT	Iwan P. Williams
BOARD	Alan Boss, Guy J. Consolmagno, Régis Courtin, Julio A. Fernández, Bo Å S. Gustafson, Walter F. Huebner, Anny-Chantal Levasseur-Regourd, Mikhail Ya. Marov, Michel Mayor, Rita M. Schulz, Pavel Spurný Giovanni B. Valsecchi, Jun-ichi Watanabe, Adolf N. Witt.

PARTICIPATING COMMISSIONS

Commission 15	Physical Study of Comets and Minor Planets
Commission 16	Physical Study of Planets and Satellites
Commission 20	Positions and Motions of Minor Planets, Comets and Satellites
Commission 21	Light of the Night Sky
Commission 22	Meteors, Meteorites, and Interplanetary Dust
Commission 51	Bioastronomy
Commission 53	Extrasolar Planets

DIVISION WORKING GROUPS

Physical Study of Comets
Physical Study of Minor Planets
Motions of Comets
Distant Objects
Meteor Shower Nomenclature
Professional-Amateur Cooperation in Meteors
Small Bodies Nomenclature
Planetary Systems Nomenclature

SERVICES

Minor Planet Center
Minor Planet Center Advisory Committee

INTER DIVISION WORKING GROUPS

Division III / Division I	Near-Earth Objects (WGNEOs)
Division III / Division I	Cartographic Coordinates and Rotational Elements of Planets and Satellites

PROCEEDINGS BUSINESS SESSIONS, Monday 10 August 2009

1. Introduction

The meeting was opened by Ted Bowell, president, at 11 am. The 2006 Division III meetings were reviewed by Guy Consolmagno, secretary; as the minutes of those meetings have already been published, they were assumed to be approved.

2. Division III Structure

Bowell reviewed the new membership applications (by Commission): Commission 15 (Physical Study of Comets and Minor Planets) has 47 new applicants, Commission 16 (Physical Study of Planets and Satellites) 47, Commission 20 (Positions and Motions of Minor Planets, Comets and Satellites) 25, Commission 21 (Light of the Night Sky) 5, Commission 22 (Meteors, Meteorites and Interplanetary Dust) 22, Commission 51 (Bioastronomy) 27 , and Commission 53 (Extrasolar Planets) has 72 new applicants. As it is likely that applicants have joined more than one commission, all that can be concluded is that the total number of new members of Division III ranges from somewhere between 72 and 245 new members; probably the number is around 150, which would represent a 15% increase in Division III membership. At present, the officers of the Division include the Division Board, consisting in total of 17 members including the President, Vice President, Past President, Secretary, Organizing Committee. The IAU guidelines recommend that the board membership be between 8 and 12 members; however, this is just a guideline and the previous larger board was accepted by the Executive Committee. We are the third largest of the twelve divisions and we have an unusually large number of commissions; and so, to have presidents on the board we're bound to have a complex structure. There are seven Commissions, five Commission Working Groups (which are expected to last longer than a triennium), several Commission Task Force/Groups (which are expected to go away after a fixed time), one Division Service, one Division Advisory Committee, two Division Working Groups, and two Interdivisional Working Groups.

These are the commissions and working groups:

Commission 15: Physical Study of Comets and Minor Planets
WG Physical Study of Comets
WG Physical Study of Minor Planets
Commission 16: Physical Study of Planets and Satellites
Commission 20: Positions and Motions of Minor Planets, Comets, and Satellites
WG Motions of Comets
WG Distant Objects
Commission 21: Light of the Night Sky
Commission 22: Meteors, Meteorites, and Interplanetary Dust
TF Meteor Shower Nomenclature
WG Professional-Amateur Cooperation in Meteors
Commission 51: Bioastronomy
Commission 53: Extrasolar Planets
Service: Minor Planet Center
Advisory Committee: Minor Planet Center
WG: Committee on Small Bodies Nomenclature
WG: Planetary System Nomenclature
IWG: Cartographic Coordinates and Rotational Elements of Planets and Satellites (jointly with Division I: Fundamental Astronomy)
IWG: Natural Planetary Satellites (jointly with Division I)

The status of Commission 21, Light of the Night Sky, was discussed. (Note this commission is not to be confused with the commission on light pollution.) In 2006 it had been recommended that the future status of this commission be reviewed. As it happens, only about ten percent of the membership of C21 works in the field of planetary systems. Most C21 members are working on various aspects of diffuse or integrated galactic and extragalactic backgrounds. Thus C21, as currently constituted, does not really fit within DIII. Therefore it is proposed that the present Commission 21 "Light of the Night Sky", as it currently exists within IAU Division III "Planetary Systems Sciences", be dissolved; a new Commission 21 with the designation "Galactic and

Extragalactic Background Radiation" will be formed; the new Commission 21 will be located within IAU Division IX: "Optical and Infrared Techniques"; and these changes shall go into effect with the conclusion of the 2009 IAU General Assembly in Rio de Janeiro. Those members of C21 working on planetary systems science will be welcomed into C22 (Meteors, Meteorites and Interplanetary Dust), where their interests will be well served. The proposed action has been approved by the DIII and DIX Boards, and by the IAU Executive Committee.
Merging Commissions 15 and 20 was discussed. The merger was first suggested because of the apparent overlap in the science interests of C15 (Physical Study of Comets and Minor Planets) and C20 (Positions and Motions of Minor Planets, Comets and Satellites). However, further discussion between the Organizing Committees of C15 and C20 and with the DIII Board indicated that there is in fact little science overlap between the two Commissions. Only about 50 people, out of a total of some 600, are members of both C15 and C20. Consequently, the Executive Committee was asked not to merge C15 and C20, and they have accepted this position.

3. Executive Committee/Division III matters

Bowell reported on the three Executive Committee meetings he had attended in the last triennium; there include one in Prague, after the close of the General Assembly in 2006, in Oslo in May 2008 and in Rio de Janeiro, preceding the General Assembly in 2009. The following main areas of interest to Division III were discussed at these meetings: choosing IAU-sponsored Symposia and Joint Discussions, choosing a group name for trans-Neptunian dwarf planets, the naming of dwarf planets and starting mornings at the Rio General Assembly with plenary session reviews. Concerning the choosing IAU-sponsored Symposia, Joint Discussions, and Special Sessions, Bowell noted that in the IAU as a whole, the science interests of Division III are often overshadowed by those of stellar and galactic astronomers. Even though we are one of the larger divisions, we are represented by only one Division President. We make up 1200 members of a membership of ten thousand IAU membership total. Commission VIII has another 300 planetary people, plus others in other commissions, so we are about 15% of the total IAU. Therefore, in any vote among Division Presidents to choose among competing Symposia, Joint Discussions, and Special Sessions, Division III can be at a disadvantage. The Executive Committee recognizes this, and also recognizes that many planetary science astronomers will not attend the General Assembly if there is not a Symposium, Special Session, or Joint Discussion in their field. Consequently, Division III's interests are in fact well represented at GAs and during inter-GA years.
Sessions at the Rio GA of interest to Division III members and co-sponsored by Division III Commissions include:
Symposium 263: Icy bodies of the solar system (5 days)
Special Session 6: Planetary systems as potential sites for life (2 days)
Invited Discourse 2: Water on planets (James F. Bell III)
Plenary Review: Icy bodies of the solar system (David Jewitt)

Concerning dwarf planets, Bowell recalled that at the close of the Prague GA in 2006, we had a definition for planet, and we were using the term dwarf planet for bodies massive enough to be in hydrostatic equilibrium ("near round"), but not massive enough to clear their orbital zones. We were still lacking at least three items: A group name to replace dwarf planet; quantitative thresholds to define the boundaries between the different groups; and a group name for transneptunian dwarf planets.

The sense at Prague had been that "dwarf planet" was a term most people didn't like. Two main candidate names were discussed (nanoplanet and subplanet), but neither gained traction. Furthermore, since that time the term "dwarf planet" has entered the public consciousness and like it or not, it would be very difficult to change it now. It is now probably too late to reopen this discussion. At the time of the Prague GA, there were no refereed publications of direct relevance to the issue of defining quantitative thresholds for the boundaries between planets, dwarf planets, and small solar system objects, and so creating task groups to attack these problems would have been premature. Since then, Soter (Astron. J. 132, 2513) has discussed planetary orbit clearing by accretion and ejection of lesser bodies; and Tancredi and Favre (Icarus 195, 851) have quantified the lower diameter limits for icy and rocky dwarf planets. Thus, material for the basis of a discussion on quantitative thresholds is now available. That discussion has not yet taken place; whether it should is a matter for the incoming President. Concerning a "group name" for transneptunian dwarf planets, Bowell recalled that it was a resolution of the GA in

2006 that such a name be determined at some future time, and the Executive Committee wished Division III to act on this matter. The Committee on Small Bodies Nomenclature's preference was for the term plutoid; he transmitted this name to the Executive Committee in Oslo 2008, and the term was accepted. However, in part because of a communications failure, Division III's other naming group (the Working Group on Planetary System Nomenclature) was not consulted. A post facto vote of WGPSN members indicated that they would not have approved the term plutoid. All in all, the President took the blame for this slip up, and offered apologies for everyone. Even on the CSBN there were many against this name; the argument was that new names like this were unnecessary, and there was a possibility of confusing it with other terms. It is not clear that this name will be used by the community in the future. It was noted that most astronomical terms come from the community, and are not imposed from above.

Shortly after the Prague GA, there was urgency to name (136199) 2003 UB313 and its satellite. The CSBN and WGPSN joined forces to undertake the naming, and accepted the discoverer's suggestion of Eris for the primary body and Dysnomia for the secondary body. A press release and thematic web page were developed to explain this choice. Later, there was urgency to name two additional candidate dwarf planets [(136472) 2005 FY9 and (136108) 2003 EL61]. Still lacking quantitative thresholds for dwarf planethood, the Executive Committee decided to adopt the following recommendation (from the Division III Board and President Elect Bob Williams):

Any solar system body having (a) a semimajor axis greater than that of Neptune, and (b) absolute magnitude brighter than $H = +1$ *mag shall be considered for naming purposes to be a dwarf planet and named jointly by the WGPSN and CSBN. Name(s) proposed by the discoverer(s) will be given deference. Note that* $H = +1$ *mag implies diameter 850 km* ($p = 1$) ¡ *D* ¡ *5000 km* ($p = 0.03$).

Using this definition, (136472) was named Makemake (creator god from Rapa Nui mythology) and (136108) was named Haumea; its two satellites were named Hi'iaka and Namaka (Hawaiian goddess and children). Bowell noted that it is not critically important even if such bodies are in future deemed not to be dwarf planets; they still have appropriate names.

Keith Noll noted that the working definition for naming may be overly restrictive at present; other planets already named may turn out to be dwarf planets. We need a working definition and we need to establish the rules for these names, perhaps less restrictive than the Kuiper Belt naming conventions. A further discussion on nomenclature raised the question of whether the names of C15 and C20 need to be changed, given the preference in 2006 for the term "small solar system bodies" in place of "minor planets". The decision at Prague was to continue to use "minor planet" for these commissions, and the idea that "minor planet" is not to be used was dropped. As above, the usage of the community itself will eventually determine what terms are most likely to be used.

The idea of starting mornings at the Rio GA with plenary session review talks was suggested to the Executive Committee by Bowell, and discussed extensively at the Executive Committee meeting in Oslo (in 2008). The idea was to start the day in a session involving attendees from many disciplines and to provide reviews of the Symposia intelligible to all GA attendees. Possible negative consequences are that it may reduce time for Commission and other meetings, which may result in increased overlap of meetings.

This has actually occurred, as can be seen in the cases of the Division III business meeting occurring during Special Session 6, the simultaneous meetings of C15 and the CSBN, and a possible overlap of Division III and Commission 20 meetings. On the other hand, the plenary reviews have been very much appreciated. One attendee stated that they were "one of the best parts of the meeting." One suggestion was that division business meetings might be held during these plenary sessions, but it was noted that this would run counter to the desire for a plenary where everyone can participate.

It was noted that commission meetings are now much shorter than had been the tradition ten or twenty years ago. In the past, science was discussed at these commission meetings, but today they are almost entirely just business meetings.

The President was not consulted about the specifics of the General Assembly program; no draft program was given, and the final program was merely circulated three weeks before the meeting. He did request that the division meetings come after the commission meetings, so that the division could deal with such issues as might arise at the commission meetings; that didn't happen. The hope was expressed that an effort can be made to raise this issue and these complaints for the future; the incoming President will carry this forward.

Bowell reported on the most recent developments from the Executive Committee that impact Division III:
It is the desire of the Executive Committee that Working Group chairs serve for one triennium, or for a maximum of two triennia. This could in the future become an IAU Working Rule.
A new Executive Committee Working Group on Naming Dwarf Planets will be made up of all the members of CSBN and WGPSN.

There is ongoing discussion about whether CSBN and WGPSN will remain as Division III working groups or become Executive Committee working groups. The CSBN has not discussed this yet; the WGPSN would like to return to its original status as an Executive Committee working group. They argued that originally they reported to the Executive Committee because naming planetary features had a delicate political significance during the rivalry between the then Soviet Union and the United States, at the time when they were the only nations sending missions to other planets. By the 1990s, that issue had become moot. However, given the nature of space missions from many different countries and greater number of different national points of view today, we are returning to a situation where we need to accommodate a new complex set of political sensibilities. Furthermore, there have been communication errors ever since the WGPSN moved to Division III supervision, and it appears these have continued over several trienniums. They argued that would be more efficient if the WGPSN were informed directly by the Executive Committee about decisions that involve its work.

4. Appeals on suggested asteroid names rejected by the CSBN

The Committee for Small Body Nomenclature (CSBN) is, among other things, responsible for the naming of asteroids. Suggested names and brief explanatory "citations", mostly from asteroid discoverers, are distributed by the CSBN Secretary to CSBN members for review. Most are accepted, perhaps after modification, and are published approximately monthly in the Minor Planet Circulars. Some names are rejected by CSBN members, usually because they are thought to contravene the CSBN's guidelines. Examples of the CSBN's guidelines are that asteroid names will not be accepted if they are too similar to those of other asteroids or natural satellites, or are in questionable taste; names honoring persons or companies or products for no more than success in business are discouraged; names that resemble advertising will not be accepted. Individuals or events principally known for political or military activities are unsuitable until 100 years after the death of the individual or occurrence of the event. (A current list of these guidelines can be found at the web site http://www.ss.astro.umd.edu/IAU/csbn/mpnames.shtml). The rationale for these guidelines is to avoid value judgments not related to astronomy, and to avoid offending people or countries. These guidelines are not rules, and therefore can be overruled by majority decision of CSBN or Division III. The appeals process states: *Any decision of the CSBN with which a[n asteroid name] proposer disagrees may be appealed by the proposer. [T]hat appeal should be addressed, by electronic mail or by letter, to the President of Division III, for action by the [Division] membership at the following General Assembly.* In practice, appeals to Division III are quite rare. Three asteroid names and citations had been rejected by the CSBN on the ground that each violated the guideline against individual or events principally known for political or military activities within the past 100 years, and these rejections were appealed by the proposers of the names. In all three cases, it is noted, votes by CSBN members, for or against, were close. The three names and citations under appeal were:

(64000) Aungsansuukyi = 2001 SD115
Discovered 2001 Sept. 20 by W. K. Y. Yeung at the Desert Eagle Observatory.
Aung San Suu Kyi (b. 1945) is a pro-democracy activist in Burma and a noted prisoner of conscience and advocate of non-violent resistance. She won the Rafto Prize and the Sakharov Prize for Freedom of Thought in 1990 and the Nobel Peace Prize in 1991.

(169568) Asanaviciute = 2002 FN6
Discovered 2002 Mar. 16 by K. Cernis and J. Zdanavicius at the Moletai Astronomical Observatory.
Together with thirteen other defenders of Lithuania's independence, Loreta Asanaviciute (1967-1991) was killed at the TV tower in Vilnius on the tragic night of 1991 Jan. 13, when tanks ploughed into a crowd of unarmed demonstrators.

(192293) Georgelser = 1990 TA2
Discovered 1990 Oct. 10 by F. Brngen and L. D. Schmadel at Tautenburg.
Carpenter Georg Elser (1903-1945) was opposed to Nazism from the beginning of the regime, feeling that it would plunge Germany into a major war. In order to prevent greater bloodshed, he made an unsuccessful attempt on the life of Hitler in Nov. 1939. The name was suggested by the first discoverer.

A significant discussion, lasting for nearly an hour, focused on the issues of whether the names proposed, or the citations as written were political in nature, or offensive to people or countries. It was noted that there are some names with political overtones that were approved in the past, including anti-Nazi activists; the name Sakharov was noted as an example, though it was also noted that he also qualified for an asteroid name as a prominent scientist. It was argued that to honor a person for non-violent humanitarian activities is much easier than for these cases. There was concern that the IAU was being called to make a political decision, rather than one in our field of expertise. It was also noted that the first and third names are for people noted for very recent events.
It was decided to adjourn the meeting at this point and delay the vote until the afternoon, thus allowing further reflection on the issue.

5. Reviews of Activities of Commissions and Working Groups

Because this meeting was running at the same time as the Special Session, the vote on the CSBN appeal was delayed so that those attending the Special Session could participate, and so commission reports were presented before the vote on the CSNB appeal.
C53: Extrasolar Planets (Alan Boss for Michel Mayor): The commission currently has 120 IAU members plus 30 non-IAU associates. The highlights of progress in the field: more than 350 Extrasolar Planets have now been discovered. The number of planets now discovered is beginning to be large enough to put constraints between theory and observation. Many of these planets were detected by transits; when a planet goes behind a star, one can find the temperature of the planet's atmosphere and determine low resolution spectra, which have allowed certain constituents to be identified. The transit method also allows a determination of the spin of the star and the orbital plane; most recently we have started to discover a lot of objects with a strong inclination of the orbital plane compared to star spin axis, which is interesting for understanding the origin of hot Jupiters. We are now seeing a new population, "super Earths" with sizes ranging up to Neptune (20 Earth mass) sized bodies; it appears that one third of every solar type star has this kind of low mass object. The record smallest such body discovered is 1.9 Earth masses so far. Three different teams have done images of distant planets, and the Hubble and Spitzer space telescopes have played important roles. During the past triennium, the commission sponsored two important symposia, in China and in Boston; they were extremely successful, with more than 200 participants each. At this meeting we are a cosponsor of Special Session on Planetary Systems as Potential Sites for Life; as evidence of its success, the room is too small to accommodate all those interested in attending this session.
C16: Physical Study of Planets and Satellites (Rgis Courtin): Planetary exploration is in a golden age. As evidence, we note the in-depth exploration of Saturn by Cassini, the number of missions to Mars, both from orbit and on the surface, Venus Express, which is about to end its nominal mission with a good number of results on atmospheric dynamics, the Messenger mission to Mercury, which has done two flybys leading to an upcoming orbital mission, and whose discoveries so far include major basins and primary craters, including the second largest crater, Beethoven, discovered last year, the New Horizons mission en route to Pluto mission, several Lunar missions, and many results from the ground. The commission has had a stable membership. At the last AAS-DPS meeting we had lunchtime meeting where we presented to a roomful of more than 100 people the activities of the IAU, and strongly encouraged young planetary scientists to join the IAU. There was an attempt to put together an electronic news bulletin, but given the launch of a separate Planetary Exploration Newsletter we decided not to duplicate that effort; instead, we are restricting the news bulletin to the business of the commission. We also joined with C51 and C53 to sponsor the Special Session on Planetary Systems as Potential Sites for Life. As indicated by the number of posters (45) and registrants, this has

been very successful.
C15: Physical Study of Comets and Minor Planets (Walter Huebner): Over the past three years we have seen several missions to comets and asteroids, including the Japanese mission to Itokawa, missions to Temple I, Wild 2, and the ongoing Rosetta mission to Comet 67 P/Churyumov-Gerasimenko with encounters past two asteroids, 2867 Steins (already occurred, in 2008) and 21 Lutetia (in 2010). We have two working groups, one on the Physical Studies of Comets (chaired by D. C. Boice) and one on the Physical Studies of Minor Planets (chaired by R. A. Gil-Hutton) We also have four task groups: the Task Group on Cometary Magnitudes (chaired by G. Tancredi) and the Task Group on Asteroid Magnitudes (chaired by E. Tedesco) to investigate the quality and accuracy of their respective magnitudes, the Task Group on Asteroid Polarimetry and Albedo Calibration (chaired by A. Cellino) to examine the calibration of the polarimetric slope/albedo relation for asteroids and the Task Group on the Physical Properties of Near Earth Objects (chaired by Karri Muinonen) to inventory the geological and geophysical properties of NEOs for study and mitigation. This last task group was set up just last year, and its task is not yet finished; it will continue into the next triennium. Chairs of the appropriate working groups are co-chairs of the four task groups. Each has its own web site. A new task group is being proposed to discuss countermeasures for potentially hazardous objects.
C 20: Positions and Motions of Minor Planets, Comets and Satellites (Julio Fernandez). We discussed the proposed merger with C 15, but the opinion of the majority of our members was to keep separate commissions as it is now. In this discussion, we reviewed the more general questions about our role as a commission. What are the main goals of a commission? What are the reasonable duties of a commission? How can we help the astronomical community who are working in these topics? We discussed the issue of revising the rules for obtaining credit for an asteroid or comet discovery. Arguments were made in favor of giving credit to the original discoverers, to those making the essential follow-up observations, and to the dynamicists who contribute to making the orbit reliable and useful. Most members felt it was best to maintain the current system where the credit for discovery and the preference for suggesting names are left only to the original discoverers. The commission cosponsored Symposium 263 at this General Assembly, and other meetings related to our field were discussed. Finally, for the future, we plan to maintain the current organization of the commission. In our scientific work, we note the need to find a way to promote the improvement of orbits for the numerous objects likely to be discovered by the new sky surveys coming on line in the next few years.
C 21: Light of the Night Sky (no report). This commission is to be renamed and moved to Division IX.
C 22: Meteors, Meteorites and Interplanetary Dust (Jun-Ichi Watanabe for Pavel Spurn). Membership of this commission has increased to 132. The Working Group on Professional/Amateur Cooperation in Meteors has been a great success; Ryabova will continue as chair. The Task Group Meteor Shower Nomenclature, chaired by Jenniskens, has come up with official names of 64 meteor showers, and these were approved at the C22 business meeting. This Task Group still has a working list of further 264 candidate showers to consider; because they have this ongoing work, they have proposed to become a Working Group of the Commission for the next triennium, and this was approved in the C22 business meeting. Its task will be to adapt the new meteor shower nomenclature rules for newly discovered showers, arbitrate proposed names from new surveys, and determine if candidate meteor showers should be accepted and named. The Commission sponsored three conferences during the past triennium: Meteoroids 2007 in Barcelona, Asteroids-Comets-Meteors (ACM) 2008 in Baltimore, and the Workshop on Bolides and Meteorites in Prague, 2009. Future conferences include the 2008 TC3 Workshop in the Sudan, to be held during December 5-6, 2009, Meteoroids 2010, in Colorado, and ACM 2011, in Japan.
C 51: Bioastronomy (Alan Boss). The commission was renewed in Prague; with the addition of a dozen new members, it now has a membership of several hundred scientists. In the past triennium it has cosponsored Special Session 6 at this General Assembly, and sponsored the meeting Bioastronomy 2007 in San Juan, Puerto Rico, with an attendance of about 250. It was originally proposed to hold the next Bioastronomy meeting in 2010 but, following up on a long standing proposal it was decided instead to join forces with ISSOL (International Society for the Study of the Origin of Life) and hold a join meeting in France in 2011.
Service: Minor Planet Center (Ted Bowell for MPC). Minor Planet Circulars continued to be issued each lunation. About 20000 asteroids/year are numbered (the total of all numbered

asteroids is now 220,000). A "few score" comets, discovered with ground-based telescopes, and on the order of 100 SOHO comets are designated every year; names are published for about 1000 asteroids/year; about twenty periodic comets/year are numbered. The number of astrometric observations of asteroids in the database exceeds 50 millions, going back to the early 1800s. Hundreds of thousands of comet observations are now in the database. The Minor Planet Center has secured funding from NASA for the years 2008-2011; thus many MPC services, which previously were charged for, will be provided gratis after 1 October 2009; and there will be accelerated software and website development. We note that there is a commercial enterprise attempting to sell asteroid names; should the IAU do something about this? Perhaps the IAU should sell such names themselves and use the collected money to support research.

Division Working Group: Committee on Small Body Nomenclature (Pam Kilmartin for Jana Tich). As noted in the triennial report, 701 comets and 2228 minor planets received names between July 2005 and June 2008; there are now about 220,000 numbered asteroids, of which 14,574 are named. Among the notable names are 100000 Astronautka, named for the 50th anniversary of the start of the space age with the launch of the first satellite, Sputnik, in 1957; the significance of the number is the convention that "space" begins at 100,000 meters above the Earth. Names for dwarf planets and satellites have been assigned (with the WGPSN). A naming convention has been proposed for non-resonant objects with semimajor axes a ¿30.07: those with chaotic short lived orbits, like centaurs, will be named for mythical creatures that are hybrids or shape shifters, for example 42355 Typhon's satellite Icneda. An update of the minor planets dictionary is being published.

Division Working Group: Planetary System Nomenclature (Rita Schulz). We consist of 13 members, mostly task group chairs for various planets in solar system. Jrgen Blunck, our expert on Greek and Latin and classical history, died during the past triennium; he has been replaced in this role by Guy Consolmagno. The working group met in Ithaca, NY on 10 Oct 2008 (during the AAS-DPS meeting at Cornell University) and agreed to write a document to define clearly the roles and responsibilities of task groups; this document can be found on our web page now. Each task group now should have a minimum of 5, maximum 8, members; there was a need to renew task groups for Mercury and Venus which were suddenly needed after many years of being dormant. As of June 2008, there have been 255 features named on moons of Saturn, on Mars and Phobos, on Mercury, on Venus, and on Earth's Moon; in addition several newly discovered planetary satellites have been named.

Interdivisional (with Division I) Working Group: Cartographic Coordinates and Rotational Elements of Planets and Satellites (Brent Archinal). We have published a proposal in our triennial report to recommend coordinate systems for various bodies; nine have been so done since 1976. For Mercury we have discussed a model based on new radar results, and a new (observed) pole position. A question arose as to whether it would be better to use a system based on a particular feature (as is currently the case) or one based on a now much more accurately known axis of figure system? It was decided to continue the feature-based system but to publish the offsets between the two systems. We recommend the use of the Moon DE421 ephemeris, rotated to the mean earth polar axis system. For Mars we recommend no changes at this time, even as improvements continue; this may change in six years. For Saturn we note new results for its satellites, especially the librations of Titan, and Saturn's rotation; the committee has noted that it needs more expertise for understanding the issues of choosing rotation systems for the gas giant planets, especially Jupiter, Saturn, and Neptune. A subgroup has been formed to decide which small bodies to include in our data base (limited to those that have been directly imaged, not just those with good light curves). We are planning a separate category for dwarf planets; they will be treated as small bodies in terms of the default choices for coordinate systems. Long term issues include reestablish IAG affiliation and the issue of extrasolar planets where it is becoming possible to make temperature maps (for example). A uniform way for describing the location of these features also needs to be determined. Officers of the commission will continue as in the past triennium; the proposed six-year limit might be a problem for this group, where a particular technical expertise is required.

Interdivisional (with Division I) Working Group: Natural Planetary Satellites (Jean-Eudes Arlot). We are looking for new members to continue the important work of maintaining observations from which improved ephemerides can be derived; satellites are fast moving, and many observations are needed to provide good ephemerides including reasonable estimates of errors. We must avoid repeating the 1920-1970 gap where there were few good observations; the absence of such data impairs our ability to develop dynamical models for the orbits. Even in the past triennium it has happened that many newly discovered faint outer solar system irregular satellites have been lost due to the lack of follow-up observations, which are only possible at large telescopes. The main interest of the Working Group is maintaining the databases in Paris and Moscow, which include about 90% of all observations made (several centuries' worth) of great value for developing dynamic models; a web page provides further details. Mutual event campaigns are underway for Jupiter and Saturn in 2009; the hope is to provide an accuracy of 0.2 arcsecond, an improvement on the current 1 arcsecond accuracy. With such accuracy one can study the effects of tides in these satellites; for example, one can see accelerations in the motions of Io that can be related to its thermal state (see papers in Nature in 2009), thus constraining the internal structure of the Galilean satellites. The question arose as to whether this work would be better served by a new commission rather than as a working group; however, it was decided it would be more useful to remain as an interdivisional Working Group. Concerning the satellites of asteroids, we plan to produce a database for these objects, which currently is scattered in several locations. We are looking for more members for our Working Group because gathering such data is a lot of work; we need help. We note that our data and ephemerides at present are held at the MPC, at JPL, and at Moscow; it is good to have several sites maintaining these data.

6. Vote on the CSBN Appeals

Discussion on each of the three names was resumed, with a final five minutes provided for a summary discussion for each name, followed by a vote. There was concern in all cases that the problem might not be the selection of the names but rather with the citation describing the selection, so two votes were taken; those in favor as it stands, and those who would change their vote if the citation were altered.

Aungsansuukyi: There was great sympathy for the person in question but also a concern that this was clearly a political statement; any choice would reflect political opinions, which could be too subjective and in any case beyond the realm of the IAU's expertise. It was for such reasons that the IAU guidelines were adopted in the first place; suspending the 100-year rule in this case might lead to further complications later, if some future cases were accepted and other rejected. For Aungsansuukyi, the vote to overturn the decision of the CSBN was Yea: 4, Nay 22; if the citation wording were changed to remove the first sentence, 16 still would vote nay. The appeal was rejected.

Asanaviciute: It was noted that this was the most political case of the three, related to recent events, and it could reflect badly on the Russian community within Lithuania. There was also concern that the citation might be seen as provocative. Again, the utility of the 100-year rule allows one to recognize the significance of a given event with a better perspective; there does not appear to be an urgent need for such a naming at this time. For Asanaviciute, the vote was Yea: 1, Nay: 27; if the wording were changed, 22 indicated they would still vote against overturning the CSBN decision. The appeal was rejected.

Georgelser: He was clearly a brave person, and the IAU has already honored similar anti-Nazi dissidents; but they were peace activists, whereas he wanted to make a change via violence. For Georgelser, the vote was Yea: 6, Nay 20; the Nay votes stayed at 20 even if the citation were changed. The appeal was rejected.

7. Election of New Officers

Unlike in previous years, voting for the slate of new officers took place several months before the General Assembly, by e-mail. Constructing the ballot proved to be complicated, and the outcome was extremely close.

The Division III Board consists of President, Vice President, Secretary and Organizing Committee (OC) members. Currently there are 14 OC members, including past President Iwan Williams. The IAU desires broad geographic representation and gender balance. The Division III Board had extensive discussions about the voting process.

Following tradition, Vice President Karen Meech agreed to become President; incoming Commission Presidents, the outgoing Division III President, and the Chair of WGPSN were appointed to the OC without vote (8 people).

Division III members were asked to vote for Vice President, Secretary, and 6 OC members. Some outgoing Commission Presidents became candidates for incoming Vice President and Secretary. The current Board discussed additional nominees, and selected 3 candidates for each position. Outgoing Commission Presidents were invited to run as OC members. Division III members were asked to nominate themselves for OC membership (rather than the Division III membership being asked to nominate others). The current board voted to select a subset of nominees from this list. This list was then submitted to the membership for a vote by email with the following results:

President	Karen Meech (U.S.A.)	
Vice President	*Giovanni Valsecchi (Italy)	
Secretary	*Patrick Michel (France)	
Organizing Comm	*Dominique Bockele-Morvan (France)	Incoming D15 VP
	Alan Boss (U.S.A.)	Incoming C53 P
	Edward Bowell (U.S.A.)	DIII Past P
	Alberto Cellino (Italy)	Incoming C15 P
	*Guy Consolmagno (Vatican City State)	Outgoing DIII S
	*Julio Fernndez (Uruguay)	Outgoing C20 P
	William Irvine (U.S.A.)	Incoming C51 P
	*Uwe Keller (Germany)	
	*Daniela Lazzaro (Brazil)	
	Melissa McGrath (U.S.A.)	Incoming C16 P
	*Keith Noll (U.S.A.)	
	Rita Schulz (Germany)	Continuing WGPSN C
	Jun-Ichi Watanabe (Japan)	Incoming C22 P
	Makoto Yoshikawa (Japan)	Incoming C20 P

*Asterisks indicate members who were voted on by the Division III membership.
C = Chair. P = President. S = Secretary. VP = Vice President

The number of Board members at this point totaled 17, the same as the outgoing Board. Eight nations and four continents were represented, with the gender balance being five women, twelve men. This Board contained eight continuing members, and nine new members. In light of the desire for geographical balance, it was moved from the floor to add Mikhail Marov as a representative of Russia and Jin Zhu as a representative of China Nanjing. (Due to email problems, they had been left off the original ballot.) Both motions passed without dissent. It was moved to thank Ted Bowell for his service as President; this passed by acclimation.
The Division meeting closed at 3:50 pm.

Ted Bowell
President of the Division

Transactions IAU, Volume XXVIIB
Proc. XXVII IAU General Assembly, August 2009
Ian F. Corbett, ed.

doi:10.1017/S1743921310004941

COMMISSION 15 PHYSICAL STUDIES OF COMETS AND MINOR PLANETS

L'ETUDE PHYSIQUE DES COMETES ET DES PETITES PLANETES

PRESIDENT	**Alberto Cellino**
VICE-PRESIDENT	**Dominique Bockelée-Morvan**
PAST PRESIDENT	**Walter F. Huebner**
ORGANIZING COMMITTEE	**Bjrn J. Davidsson, Elisabetta Dotto, Alan Fitzsimmons, Petrus M. M. Jenniskens, Dmitrij Lupishko, Thais Mothé-Diniz, Gonzalo Tancredi, Diane H. Wooden**

PROCEEDINGS BUSINESS SESSIONS, 5 August and 11 August 2009

1. Introduction

This report of the business meeting of Commission 15 at the 2009 IAU GA is based on notes provided by Walter Huebner, past president, and on the minutes taken by Daniel Boice, secretary of Commission 15 in the triennium 2006 to 2009, with additional notes from the current secretary, Daniel Hestroffer. The business meeting was split into two sessions, the first held on 5 August and the second held on 11 August. This report presents the minutes of the two Commission 15 business-meeting sessions held during General Assembly XXVII.

Walter Huebner opened the Business Meeting, welcoming members and others in the audience interested in the commissions activities. The incoming President, Albert Cellino, the past Division III President, Ted Bowell, and the incoming Division III President, Karen Meech, were in attendance, as were approximately two-dozen commission members. The agenda as presented by Walter Huebner was adopted.

The first item of business was the justification for the continuation of Commission 15. It had been proposed at the Prague GA XXVI that Commissions 15 and 20 be merged at this GA. After much discussion prior to the GA, the Commission 15 membership could not support this proposed merged for several reasons, among them, lack of commission membership overlap and responsibilities, the large size that would result from the merger, etc. Ted Bowell, outgoing Division III president, was present and expressed his opposition to the proposed merger. Commissions 15 and 20 will continue to be separate during the next triennium and for the foreseeable future.

The next item of business was the justification for continuation of the Commission Working Groups and Task Groups. Since the issues with which they are concerned are broad and cannot be resolved within a given triennium, it was approved that the Comets and Minor Planets WGs should continue, as well as their respective TGs. Task Group deliverables should be updated for the next triennium.

The Commission decided to continue preparing the Triennial Report in the same format as the five preceding reports. The research descriptions will be included in the printed version, while the references will be made available on the Commissions website.

2. Commission 15 Elections

Walter Huebners term as President came to an end at this IAU GA. The membership extended their gratitude to him for his excellent leadership during the past triennium. It was decided to continue the tradition that the Vice President becomes President, insuring that the President has at least three years prior exposure to the operations of the Commission. As is also tradition, the new President selected the new commission secretary, Daniel Hestroffer. The position of Vice President is usually filled by the Chair of one of the two Working Groups on an alternating basis. Thus, the new Vice President would have been the Chair of the Comets Working Group of the previous triennium. However, Tetsuo Yamamoto (outgoing Chair of the Comets WG) had declined due to work commitments. As a result, Tetsuo strongly recommended Dominique Bockelée-Morvan for the position of Vice President. She was approached by the present and incoming presidents and accepted the nomination to fill this position. No additional nominations for the Vice President were submitted by the membership, so she was elected unanimously to this position. Tetsuo Yamamoto also completed his two-term limitation as Chair of the Comets WG so nominations to succeed him were solicited. Again, the appreciation of the membership for his service to the commission these past six years were extended to him. Nominations for the Chair of the Comets WG and the five open positions in the Organizing Committee to serve during the next triennium (2009-2012) were solicited via e-mail from the membership. Based on these nominations, an election was held prior to the IAU GA by e-mail with about one-fourth of the commission membership participating. The results of this election are summarized below.

Officers for Triennium 2009 2012:

PRESIDENT	Alberto Cellino (Italy)
VICE-PRESIDENT	Dominique Bockelée-Morvan (France)
SECRETARY	Daniel Hestroffer (France)
MINOR PLANETS WG CHAIR	Ricardo Gil-Hutton (Argentina)
COMETS WG CHAIR	Daniel Boice (USA)
ORGANIZING COMMITTEE	Björn J. Davidsson, Elisabetta Dotto, Alan Fitzsimmons, Petrus M. M. Jenniskens, Dmitrij Lupishko, Thais Mothé-Diniz, Gonzalo Tancredi, Diane H. Wooden

Thanks were given to the outgoing members of the OC who have completed their second term: Dominique Bockelée-Morvan (France), Yuehua Ma (China), Harold J. Reitsema (USA), and Rita M. Schulz (Netherlands). It was noted that more of an effort should be expanded to recruit members from the Asian continent.

3. Commission 15 Membership and Other Business:

Total memberships stands at 403 as of 5 August 2009.

New Members: The following 49 new members joined the Commission during the past triennium:

Baransky, Olexander	Ukraine
Barber, Robert	UK
Bendjoya, Philippe	France
Bhardwajl, Anil	India
Carruba,Valerio	Brazil
Chubko, Larysa	Ukraine
Conrad, Albert	United States
da Silveira, Enio	Brazil
Delsanti, Audrey	SF2A (France)
Doressoundiram, Alain	France
Duffard, Rene	Argentina
Filacchione, Gianrico	Italy
Fornasier, Sonia	SF2A (France)
Gajdoš, Stefan	Slovakia
Gounelle, Matthieu	France
Gulbis, Amanda	National Research Foundation (USA)
Jakubik, Marian	Slovakia
Karatekin, Özgür	Belgium
King, Sun-Kun	China Taipei
Koshkin, Nikolay	Ukraine
Küppers, Michael	Spain

Lara, Luisa	Spain
Licandro, Javier	Spain
Lo Curto, Gaspare	Chile
Marchi, Simone	Italy
Mazzotta Epifani, Elena	Italy
Mendillo, Michael	United States
Ortiz, Jose	Spain
Peixinho, Nuno	Portugal
Picazzio, Enos	Brazil
Roig, Fernando	Brazil
Roques, Françoise	France
Rosenbush, Vera	Ukraine
Schaller, Emily	United States
Sen, Asoke	India
Shinsuke, Abe	Japan
Sho, Sasaki	Japan
Tarashchuk, Vera	Ukraine
Tiscareno, Matthew	United States
Tosi, Federico	Italy
Trujillo, Chadwick	United States
Ugolnikov, Oleg	Russian Federation
Vzquez, Roberto	Mexico
Voelzke, Marcos	Brazil
Wang, Xiao-bin	China Nanjing CAS
Yang, Xiaohu	China Nanjing CAS
Zhang, Jun	China Nanjing CAS
Zhang, Xiaoxiang	China Nanjing CAS
Zhao, Haibin	China Nanjing CAS

Resignations: C.R. ODell (USA) resigned membership in the Commission, having retired and become inactive in the study of small solar system bodies.

Necrology: Five distinguished members of the Commission that had died during the preceding triennium were acknowledged and a moment of silence was observed in their memory.

Helin, Eleanor (Glo)	USA
Ostro, Steven	USA
Safaeinili, Ali	USA
van Flandern, Tom	USA
Zucconi, Jean-Marc	France

It was noted that the category of IAU consultant has been eliminated.

4. Summaries of Working Group Reports

4.1. *Working Group on Comets*

Outgoing WGC Chair: Tetsuo Yamamoto (Japan), presented by incoming WGC Chair: Daniel Boice (USA)

Three major accomplishments of the Comets WG were reported during the business meeting. The first was the submission of the triennial report in June 2008. Thanks were given to Walter Huebner and other contributors to the comet section. A website for the Comet WG was initiated by Tetsuo Yamamoto at Hokkaido University (www.lowtem.hokudai.ac.jp/iau-c15-wg/index.html). This website includes the web pages for the Task Group on Cometary Magnitudes (TGCM). Due to inactivity, the TG on Cometary X-Rays was abolished. Plans were made to transfer the website to Southwest Research Institute where the new Comets WG chair resides. The last major accomplishment was holding a mini-workshop on cometary missions at the Prague GA in August 2006. Speakers included Rita Schultz (Rosetta); Uwe Keller (proposed Fresh From the Fridge Mission); and Mike A'Hearn, Karen Meech and the Deep Impact team.

4.2. *Working Group on Minor Planets*

WGMP Chair: Ricardo Gil-Hutton (Argentina)

Several issues were advanced and summarized in the Minor Planets WG report. The first, was the usage of the words minor planets after the redefinition of a planet during the previous GA. There was no objection during the discussion of this issue so the name of the WG will not be changed. Another issue is that the WG consists of only one person, so the Chair will approach other Commission 15 members to expand the WG. The WG also prepared the part of the triennial report for asteroids. A website for the Minor Planets WG was initiated by Ricardo Gil-Hutton at Complejo Astronmico El Leoncito (Casleo) (www.casleo.gov.ar/c15-wg).

5. Summaries of Task Group Reports

5.1. *Task Group on Asteroid Magnitudes*

TGAM: Ed Tedesco (USA) and R. Gil-Hutton (Argentina), co-chairs, presented by Alberto Cellino (Italy), 9 members total

The TGAM was charged with investigating the quality of asteroid absolute magnitudes and the accuracy of the asteroid magnitude phase function and to suggest ways to improve both. The fundamental role played by the absolute magnitude in numerous areas of research is well known. An accurate asteroid magnitude phase function is needed to reduce these observations. Thus, the Task Group was impressed by the very poor state of the H values listed in the major databases, by the potential consequences on the results of a number of recent investigations concerning the size-frequency distribution of the asteroid population, and inferences on the most likely values of asteroid collisional impact strengths.

The TGAM have created alternate formulations of the H, G1, G2 magnitude phase function to replace the current (H, G) phase function which, if used on properly transformed and calibrated data will eventually result in a catalog of accurate H values. A paper in preparation will detail application of the new magnitude phase functions to asteroid magnitude phase data. In particular, it will demonstrate how the two-parameter versions of the H, G1, G2 phase function can be used to predict asteroid brightness from sparse observational data and will provide an estimate of the accuracies to be expected.

5.2. *Task Group for Albedo Polarimetric Albedo Calibration*

TGAPAC: Albert Cellino (Italy) and R. Gil-Hutton (Argentina), co-chairs, 8 members total

This Task Group is charged with recalibration of the albedo-polarization relation for asteroids; specifically to define improved procedures to determine the relation between the albedo at the polarization minimum (Pmin) and the slope of the albedo. The Task Group prepared progress reports to the Commission 15 OC and prepared a final report at the IAU General Assembly in 2009. Two meetings were held by TGAPAC during the previous triennium: Astronomical Polarimetry (Quebec, 2008) and the ISSI International Team on *Light Scattering Phenomena in Small Body Surfaces* (Bern, two meetings in 2008). In the future, the primary goals of TGAPAC are new data collection for albedo calibrations, aimed at obtaining very accurate polarization measurements of the Shevchenko-Tedesco target list, more theoretical work is needed, additional studies are needed for peculiar objects (Barbarians, F-class, etc), and the need of new instruments and optimal use of telescopes in the 2- to 4-m class.

5.3. *Task Group on Comet Magnitudes*

TGCM: Gonzalo Tancredi (Uruguay) and T. Yamamoto (Japan), co-chairs, 7 members total

This report was given by Gonzalo Tancredi during the August 11 business meeting. He reminded us that the problem of deriving a magnitude of a comet has increased with the use of CCD and visual devices. The discussions by members within the task group also included the community of amateur astronomers. The tasks and deliverables consist in a new derivation of the total and nuclear magnitude, new rules to report them with other valuable data, and production of a final report. The database on cometary magnitudes should be made more accessible. The history of the IAU (m1, m2) system was introduced in 1960 and is no longer used. A short report of the "5th International Workshop on Cometary Astronomy (August 8)" was given: CCD photometry and reduction with FoCAs (photometry with Astrometrica) mostly used in the community; a dispersion between CCD and visual magnitudes was noted; and a definition for the coma similar to the FWHM is proposed. D. Boice (USA) becomes the new co-chairman of the task group, replacing T. Yamamoto.

5.4. *Task Group for Physical Properties of Near-Earth Objects*

TGPPNEO: Karri Muinonen (Finland), 12 members total

The name was changed from physical properties of NEOs to geophysical and geological properties of asteroids and comets as originally proposed. The working group met by exchanging e-mails several times. The major goal is to collect knowledge and data on material properties (observed or simulated) of NEOs. There are many sources available on the web, journal publications, and in books. EuroPlanet will also be a potential source in the near future. Two points were raised: the introduction of spectroscopic properties into the database and whether one should reproduce the data in an internal database or (preferably) point to an external one. It is recommended to narrow the list of parameters, to prioritize those needed most, and to make systematic comparisons between specific asteroids and laboratory experiments on meteorites. A database has been established at: neodata.space.swri.edu.

6. New Task Groups

A TG for International Participation for Countermeasures Against Potentially Hazardous Objects (PHOs, astertoids and comet nuclei on a collision course with Earth) was proposed and discussed extensively. It was concluded that the new Division III President, Karen Meech, will approach the IAU Executive Committee to endorse international participation in research, planning, and execution of countermeasures against PHOs.

There was a call for other new Task Groups but no suggestions were put forth.

7. Proposals for IAU Symposia

It was suggested that a proposal be developed for an IAU Symposium for Countermeasures Against Potentially Hazardous Objects (PHOs) when the endorsement of the IAU Executive Committee mentioned above is obtained.

There was a call for other symposia sponsored by Commission 15 but no suggestions were put forth.

8. Closing Remarks

Outgoing President, Walter Huebner, closed the business meeting with a call to intensify the Commissions efforts to recruit new members, especially younger scientists, to keep the Commission vibrant and healthy. He then thanked all for their assistance during his tenure and wished the incoming Commission officers continued success. All present thanked the past President, Walter Huebner, and the past Secretary, Daniel Boice, for their untiring commitment and service to Commission 15.

Alberto Cellino
President of the Commission

Transactions IAU, Volume XXVIIB
Proc. XXVII IAU General Assembly, August 2009
Ian F. Corbett, ed.

doi:10.1017/S1743921310004953

COMMISSION 16 PHYSICAL STUDY OF PLANETS & SATELLITES *(ETUDE PHYSIQUE DES PLANETES & DES SATELLITES)*

PRESIDENT	**Melissa A. McGrath**
VICE-PRESIDENT	**Mark T. Lemmon**
PAST PRESIDENT	**Regis Courtin**
SECRETARY	**Luisa M. Lara**
ORGANIZING COMMITTEE	**Sang Joon Kim, Leonid V. Ksanfomality, David Morrison, Viktor G. Tejfel, Padma A. Yanamandra-Fisher**

PROCEEDINGS BUSINESS SESSION, 13 August 2009

1. Introduction

The Business Meeting of IAU Commission 16 was held at the General Assembly in Rio de Janeiro on August 13, 2009. The meeting was called to order at 9:15am by Commission President, Régis Courtin, with 12 members present. Commission 16 now has 312 members, up from 271 at the last General Assembly. It is now the third largest Commission in Division III, after Commission 15 and Commission 51.

A moment of silence was requested to honor departed colleagues during the 2006–2009 triennium: John Derral Mulholland (1934–2008); James A. van Allen (1914–2006); George W. Wetherill (1925–2006); Arthur C. Clarke (1917–2008); Gordon A. McKay (1945–2008); Steven J. Ostro (1916–2008). A brief rememberance that included slides highlighting each departed colleague was shown, and a few moments were spent discussing the careers of these colleagues.

The new officers and members of the Commission 16 Organizing Committee were then introduced. They are identified at the top of this report.

2. Report from the President and discussion

Régis Courtin summarized the activities of Commission 16 during the last triennium, which included an informal meeting of the Organizing Committee at the 2006 Division for Planetary Sciences meeting in Pasadena, California, in October 2006 to brain storm about the organization of a session for the Rio General Assembly. The idea of a special session devoted to astrobiology and extrasolar planets, in conjunction with Commissions 51 (Bio-Astronomy) and 53 (Extrasolar Planets), was identified. This resulted in the highly successful Special Session 6 "Planetary Systesm as Potential Sites for Life" held from August 10-11, 2009 at the General Assembly in Rio de Janerio. The session was highlighted on the front page of the Estrela D'Alva (the Morning Star), the daily newspaper of the General Assembly. The session consisted primarily of invited talks (22) and oral contributions (4), so many requests for oral talks were moved to the poster session (40).

Commission 16 was also at the origin of a lunch presentation of the IAU activities, specially those of Division III, during the Division for Planetary Sciences meeting held in Ithaca, NY, in October 2008. It was noted that the high number of new Commission members is probably a result of this initiative.

It was suggested during the discussion that in the future, it would be helpful and appropriate to give the posters some devoted time, perhaps either by having one person read the posters and review them during the oral session, or allowing each poster author to present one viewgraph during the oral session. The group also discussed ideas for a Special Session, Joint Discussion, or Symposium for the next General Assembly, which will be held August 20–31, 2012 in Beijing, China. It was noted that at this General Assembly the "Ices Bodies of the Solar System" was a 5 day Symposium sponsored by Division III. Ideas suggested included a theme of "Giant Planets to Exoplanets to Brown Dwarfs"; the Moon; or Binary Systems (including binary asteroids) in conjunction with Commission 15. The group endorsed the plan to hold another informal meeting of the Organizing Committee at the Division for Planetary Sciences meeting in Pasadena in 2010 to further flesh out potential ideas for a session in Beijing.

The group considered briefly a request from Professor R-M. Bonnet, the President of COSPAR, to Professor Karel A. van der Hucht, the General Secretary of the IAU, inviting IAU sponsorship of several special events (sessions) of the COSPAR Scientific Assembly and Associated Events to be held in Bremen, Germany, 18–25 July 2010.

A request was made during the meeting to post the Commission 15 membership list on the Commission web page.

The group also discussed possible actions that the Commission could undertake during the next triennium. Suggestions included the integration of the Working Group on Natural Satellites led by Jean-Eudes Arlot. It was noted that many small satellites that are discovered are subsequently lost; it would be helpful to support increased efforts not to lose them after discovery. A Working Group on Modeling Planets and Collecting Data, with a web page and links, was also suggested.

The meeting was adjourned at 10:15am.

Régis Courtin
President of the Commission

Transactions IAU, Volume XXVIIB
Proc. XXVII IAU General Assembly, August 2009
Ian F. Corbett, ed.

doi:10.1017/S1743921310004965

COMMISSION 20 — POSITIONS AND MOTIONS OF MINOR PLANETS, COMETS AND SATELLITES

(POSITIONS ET MOUVEMENTS DES PETITES PLANETES, DES COMETES ET DES SATELLITES)

PRESIDENT — Julio A. Fernandez
VICE-PRESIDENT — Makoto Yoshikawa
PAST PRESIDENT — Giovanni B. Valsecchi
ORGANIZING COMMITTEE — Steve Chesley, Yulia Chernetenko, Alan Gilmore, Daniela Lazzaro Karri Muinonen, Petr Pravec, Tim Spahr, Dave Tholen, Jana Ticha, Jin Zhu

PROCEEDINGS BUSINESS SESSION, 12 August 2009

1. Membership of the commission

We have for this General Assembly 25 new applications for C20 membership, which represents about 10% increase with respect to the total number of members. We are very satisfied with the increase rate of our commission membership, which reflects the high degree of activity in our field. Nevertheless we note that there are not many people involved in the commission business.

2. Past IAU meetings sponsored by the commission

During the last triennium C20 supported several proposals of meetings, the list includes: "Mathematics and Astronomy - A Long Journey" (Proposed to be held in Madrid, Spain in November 2009), "The Role of Astronomy in Society and Culture" (Proposed to be held in Paris in January 2009, and approved by the IAU Executive Committee as IAU Symposium 260), and "Icy Bodies in the Solar System" (Proposed as one of the symposia in connection with the IAU General Assembly, and approved by the IAU Executive Committee as IAU Symposium 263). In particular, the last proposal was considered for us key to attract the planetary community to the General Assembly.

3. Organizing committee

The C20 organizing committee had 13 members until the GA. Dr Giovanni B. Valsecchi, after ending his term as past President, stepped off, thus reducing the size of the organizing committee to 12 members. There were no proposals of new members for the OC, nor anybody from the OC renounced or reached the term-limit for stepping off. Dr. Makoto Yoshikawa (not present at the meeting) was appointed as the new President of the commission, Dr. Steve Chesley as the new vice President, whereas the undersigned will remain in the OC as past President. A discussion followed on the new Secretary. Dr. Daniela Lazzaro was considered a very suitable candidate for this post and her nomination finally approved by acclamation.

4. Permanence of Commission 20

There was next a discussion about the pertinence of keeping C20 as an independent IAU commission against other options, in particular the possibility of merging with Commission 15. There were arguments in favor of keeping C20 based on the little overlapping of the membership of C15 and C20 (about 10%), and also on the different science methods and interests.

5. Discovery credit rules

There was a discussion on whether we need to change to a lesser or greater degree the existing rules. There was a consensus that this issue deserves to be reviewed, whether or not we finally conclude that changes need to be introduced. A Task Group was accordingly set up, led by Dr. Giovanni Valsecchi, and constituted by a mixture of professionals and amateurs with the following criterium: to keep the people of a previous task group, namely Tim Spahr, Steve Chesley, Dan Green, Jana Ticha, Juri Medvedev, David Dixon, Peter Birtwhistle, Robert McNaught, and adding as new members Zoran Knezevic and Jin Zhu, whereas the undersigned, who was a member of the prevous task group, stepped down. It was given to the Task Group a two-year period (i.e. one year before the next GA) to present a report back to the OC.

6. Upcoming meetings

There was finally a discussion about future meetings of interest for C20 members and the need to keep an active role in sponsoring such meetings when appropriate, and even playing an active role in the organization of such meetings. It was informed that Drs. Hal Levison and Wing Ip have interest in organizing meetings in USA and Taiwan respectively, and that the next ACM meeting, that will be held in Japan in 2011, might become an IAU meeting.

Julio A. Fernández
President of the Commission

Transactions IAU, Volume XXVIIB
Proc. XXVII IAU General Assembly, August 2009
Ian F. Corbett, ed.

doi:10.1017/S1743921310004977

COMMISSION 22 — METERS, METEORITES AND INTERPLANETARY DUST

MÉTEORES, MÉTÉORITES ET POUSSIÈRE INTERPLANÉTAIRE

PRESIDENT	**Junichi Watanabe**
VICE-PRESIDENT	**Peter Jenniskens**
PAST PRESIDENT	**Pavel Spurný**
ORGANIZING COMMITTEE	**Jiří Borovička, Margaret Campbell-Brown, Guy Consolmagno, Tadeusz Jopek, Jeremie Vaubaillon, Iwan P. Williams, Jin Zhu (Secretary)**

PROCEEDINGS BUSINESS MEETING on 7 August 2009

1. Introduction

The business meeting of commission 22 was held at the room 5 in the SulAmerica Convention Center in Rio de Janeiro(14:00-15:30). Fifteen people attended at this meeting:J.Borovička, E.Bowell, G.Consolmagno, D.Green, P. Jenniskens, A. Pellinen-Wannberg, R. Rudawska, J. Watanabe, J. Zhu, P. H. A. Hasselmann, F. Ostroviski, D. A. Oszkiewicz, W. Thuillot, P. Mahajani, and A. Sule. This meeting was managed by Junichi Watanabe, the current C22 Vice-President. The summary of the meeting is described.

2. Membership

The membership of Commission 22 increased from 112 members at the end of the XXVI General Assembly in Prague (2006) to 132 members at the end of the current XXVII General Assembly.

There are 22 new members: Daniel Apai (Hungary, currently in United States), Pieter Dieleman (the Netherlands), Rene D. Duffard (Argentina, currently in Spain), Štefan Gajdoš (Slovakia), Matthieu Gounelle (France), Maria Hajduková, Jr. (Slovakia), Hiroko Nagahara (Japan), Marian Jakubík (Slovakia), Amanda I. Karakas (Australia), Maxim Y. Khovritchev (Russia), Svitlana V. Kolomiyets (Ukraine), Pavlo M. Kozak (Ukraine), Maria A. Lugaro (Italy, currently in Australia), Jesus Martinez-Frias (Spain), Romolo Politi (Italy), Shinsuke Abe (Japan, currently in Taiwan), Sho Sasaki (Japan), Toshihiro Kasuga (Japan), Federico Tosi (Italy), Juraj Tóth (Slovakia), Xiaoxiang Zhang (China), and Haibin Zhao (China).

Two members deceased since the last General Assembly: Eleanor Francis Helin (USA) and Koichiro Tomita (Japan). The meeting stood in silence in remembrance of the deceased C22 members.

3. New executive

The new executive of the Commission 22 for the next triennium has already been approved by Division III: President: Jun-ichi Watanabe (Japan), Vice-President: Peter Jenniskens (USA), Past President: Pavel Spurný (Czech Republic). The members of the organizing committee are Jiří Borovička(Czech Republic) Margaret Campbell-Brown (Canada), Guy Consolmagno (Vatican), Tadeusz Jopek (Poland), Jeremie Vaubaillon (France), and Iwan P. Williams (UK).

Following the suggestion of G. Consolmagno, Jin Zhu (China) was elected unanimously as the new secretary of Commission 22.

4. Activities since the last General Assembly

At the C22 business meeting held in Prague in 2006, several tasks were discussed:

4.1. *Stabilize C22 membership – attract new meteor scientists to the C22*

With 22 new members this task can be considered as successfully done. Nevertheless, A. Pellinen-Wannberg noted that many people investigating meteors with high power radars are not astronomers and do not know how to apply for IAU membership. Organizing committee of Commission 22 should give them an advice.

4.2. *Update list of C22 members including current email addresses*

The list of all C22 members is available at the web page of the Commission. The e-mail addresses of all members have not been collected. However, E. Bowell noted that it is important to have e-mail addresses, since e-mail is the only mean which can be used for communication with members. IAU Secretariat can be asked to provide e-mail addresses.

4.3. *Prepare and keep updated web pages of the Commission including its Task and Interest Groups*

The web pages of C22 were prepared by the former secretary J. Borovička at the address http://meteor.asu.cas.cz/IAU/. E. Bowell noted that commission web pages can be at any address. It is possible to include them into the official IAU web page, but this seems to be not very practical. From practical reasons (easy update), it would be preferable to move the C22 web pages to China under the supervision of the new secretary J. Zhu.

4.4. *Prepare merging of Commission 22 with Commission 21 in 2009*

The majority of members of C21 refused merging with C22 since they are not working in planetary sciences at all. E. Bowell informed about the solution of this matter as approved by the IAU Executive Committee. Commission 21 in the present form (Light of the night sky) will be dissolved. New Commission 21 will be formed with the name Galactic and Extragalactic Background Radiation and will be part of Division IX (Optical and Infrared Techniques). However, about 10% of current C21 members in fact work in Solar System astronomy (e.g. zodiacal light). They can easily become members of C22. The Organizing Committee of C22 should actively propose them the C22 membership.

5. Commission report

The five page report of Commission 22 was submitted by the president, P. Spurný in 2008, following the request of IAU Secretariat. The report was prepared in a shorter form, concentrating on Commission activities and not compiling all scientific results in the field. The report will be published in the triennial IAU Transactions XXVIIA, Reports on Astronomy 2006-2009. The reports of two C22 bodies, the Working Group on Professional-Amateur Cooperation in Meteors and the Task Group on Meteor Showers Nomenclature, have been included as well.

6. The list of established meteor showers

P. Jenniskens, on behalf of the Task Group on Meteor Shower Nomenclature, gave a brief history of the task assigned to the Task Group at the IAU General Assembly in Prague and its activities so far to get there. The Task Group was to come up with a list of established meteor showers that deserved to be officially named by the IAU at this Assembly. A two-step process was established, where all showers discussed in new literature are first added to the Working List of Meteor Showers, each being assigned a unique name, number, and three letter code. Tadeusz Jopek at Poznan University maintains this list on the Meteor Data Center website (http://www.astro.amu.edu.pl/%7Ejopek/MDC2007/index.php). A set of meteor shower nomenclature rules was established and published in the IAU Bulletin. At the May Bolides Meeting in Praque, the Task Group settled on a list of 64 showers, known to be real from either past meteor outbursts or from detection in two recent meteor orbit surveys. Comm! ission 22 was asked to approve officially the meteor shower names on this List of Established Showers.

After the queries of E. Bowell and D. Green, it was discussed what happens if an error is later discovered in the published list. P. Jenniskens confirmed that errors on occassion do creep into the IAU Working List of Meteor Showers, at the first stage of the process, and are later corrected. Green noted that such corrections should be reported on the CBETs (Central Bureau for Astronomical Telegrams) that announce when new showers have been added. Jenniskens pointed out that it is not difficult to make such corrections, even for the list of established showers, because the names are tied to numbers and codes. In case a future correction on an officially recognized name is needed, corrections need also be reported in a CBET, a procedure analogical to naming minor planets. Future additions to the list of established showers can be published at any time but they will become official only after the next General Assembly.

J. Borovička complained that the proposed list was not circulated among C22 members or at least the Organizing Committee. P. Jenniskens answered that the list had been posted on the website of IAU Meteor Data Center since May this year, where it could be accessed by all C22 members. He also noted that the C22 president is part of the Task Group and no direction was reveived to take further action. G. Consolmagno supported the authority of the Task Group to choose the list of established showers presented here for a vote in Commission 22.

Finally, the proposed list of 64 established meteor showers was approved by the commission, without changes (6 votes in favor, 1 abstention).

After consulting in Division III, E. Bowell later confirmed that no further action would be required within the IAU for the proposed names to be officially accepted. Green suggested that the list of established names be published in a CBET (Central Bureau for Astronomical Telegrams).

7. Working Groups

The present members of Commission 22 agreed unanimously that the Working Group on Professional-Amateur Cooperation in Meteors be continued in the next triennium. The chair will remain Galina Ryabova. E. Bowell noted that according to IAU Executive Committee, working group chairs should normally serve for one triennium and maximally for two triennia. There is no conflict since G. Ryabova just finished her first triennium.

The present members of Commission 22 further agreed unanimously that the Task Group on Meteor Shower Nomenclature be transformed into the Working Group on Meteor Shower Nomenclature for the next triennium. This proposal was approved by the commission (7 votes in favor). The working group will maintain the archive of meteor shower nomenclature, consider proposed shower names from new surveys, and identify what next set of showers to name officially in next IAU General Assembly. The chair will remain Peter Jenniskens, the vice-chair will be T. Jopek. Other members will be J. Watanabe (C22 president), V. Porubčan, J. Rendtel, S. Abe, and J. Baggaley. One member will be supplemented later instead of R. Hawkes, who offered to pass on the torch. A. Pellinen-Wannberg noted that it would be desirable to have at least one female member.

8. Future conferences

J. Watanabe informed about the future conferences of interest to C22 members:

- 2008 TC3 Workshop in University of Khartoum, Sudan, on December 5-6, 2009.
- METEOROIDS 2010 conference in Breckenridge, Colorado, USA, on May 24-28, 2010.
- Annual Meeting of the Meteoritical Society in New York City on July 26-30, 2010.
- Asteroids, Comets, Meteors conference (ACM 2011) in Niigata City, Japan, on July 18-22, 2011.

Junichi Watanabe
President of the Commission

Transactions IAU, Volume XXVIIB
Proc. XXVII IAU General Assembly, August 2009
Ian F. Corbett, ed.

doi:10.1017/S1743921310004989

COMMISSION 51 BIOASTRONOMY

BIOASTRONOMY

PRESIDENT	**William Irvine**
VICE PRESIDENT	**Pascale Ehrenfreund**
PAST PRESIDENT	**Alan Boss**
ORGANIZING COMMITTEE	**Cristiano Cosmovici**
	Sun Kwok
	Anny-Chantal Levasseur-Regourd
	David Morrison
	Stephane Udry

PROCEEDINGS BUSINESS SESSION, 12 August 2009

1. Introduction

Commission 51 met on August 12, 2009. Outgoing President Alan Boss chaired the meeting, and there were several dozen members present, including incoming President William Irvine, incoming Vice President Pascale Ehrenfreund, and outgoing Past President Karen Meech. Commission 51 (C51) was re-authorized for a term of six more years at the 2006 Prague General Assembly of the IAU, and hence comes up for renewal at the 2012 IAU General Assembly in Beijing, China.

2. Organizing Committee

The 2006-2009 C51 Organizing Committee (OC) had eight members, five from the USA, and three from Europe. Given that IAU rules state that the OC should not have more than eight members, we needed to allow several current members of the OC to rotate off in order to improve the geographical balance. Past President Karen Meech and OC member David Latham agreed to do so, allowing us to add two new OC members, Sun Kwok of Hong Kong and Anny-Chantal Levasseur-Regourd of France. Clearly C51 should also attempt to add OC members from other regions that are not currently represented on the OC, such as from Africa and South America, in time for the 2012 General Assembly.

3. Bioastronomy Meetings

The primary activity of C51 since its inception has been the approximately triennial meetings on bioastronomy which it holds. Past President Karen Meech presented a summary of the most recent bioastronomy meeting, held in San Juan, Puerto Rico on 16-20 July, 2007. The meeting was a complete success, with about 250 bioastronomers in attendance. A field trip was held to visit the Arecibo Observatory and the Camuy Caves. At the San Juan meeting, plans were made to hold a joint meeting with the International Society for the Study of the Origin of Life (ISSOL), which also holds triennial meetings on similar topics. It has been recognized for some time that in order to avoid conflicts between these two meetings, it would be desirable to try to hold joint meetings, for the ultimate benefit of both groups. This plan was put into action at the ISSOL meeting held in Florence, Italy in August, 2008, where the C51 OC met with ISSOL President Janet Siefert and her OC and jointly approved plans to hold the first combined meeting in Montpellier, France, on 3-8 July 2011. As incoming Vice President, Pascale Ehrenfreund agreed to co-chair the Scientific Organizing Committee for the 2011 meeting, representing C51, working with Muriel Gargaud and Robert Pascal of ISSOL and their colleagues from Montpellier on the Local Organizing Committee. Pascale Ehrenfreund presented a summary of the plans for the 2011 meeting at the Rio de Janeiro C51 business meeting, and it was clear that the plans for the joint meeting are moving along well and we can expect that the meeting will be a resounding success.

4. Symposium and Special Session Sponsorship

C51 has been asked to support various proposals to hold IAU Special Sessions and Symposia, either at the 2009 IAU GA, or on their own. Several of these proposals were judged to be appropriate for C51 support and were successful, such as the Special Session on "Planetary Systems as Potential Sites for Life" at the Rio de Janeiro General Assembly.

5. New Members

The following requests for membership in C51 were approved at the meeting:

DESPOIS, Didier (France)	GARGAUD, Muriel (France)
HAGHIGHIPOUR Nader (United States)	KUAN Yi-Jehng (Taiwan)
KWOK Sun (Hong Kong)	LAMONTAGNE Robert (Canada)
LAZIO Joseph (United States)	MELOTT Adrian (United States)
OHISHI Masatoshi (Japan)	VON HIPPEL Ted (United States)

The following requests for consultant status were also approved:

BRUCATO John R (Italy)	DENNING Kathryn (Canada)
LAREO Leonardo (Columbia)	MACCONE Claudio (Italy)

Consultants are bioastronomers who are not members of the IAU and so are not eligible for membership. Consultants are no longer recognized by the IAU, but C51 continues to recognize their interests in the field of bioastronomy and keeps them aware of C51 activities by virtue of their place on the e-mail distribution list.

C51 continues to seek astronomers who wish to be recognized as members. We encourage interested IAU members to ask to join C51, which can be accomplished simply by sending an e-mail to the incoming C51 President, William Irvine (irvine@fcrao1.astro.umass.edu).

6. Closing remarks

We expect that the experiment of holding Bioastronomy meetings in concert with our colleagues in ISSOL will be a success, and we hope that all C51 members will be able to participate in the Montpellier, France joint meeting. Bioastronomy is a relatively new field that continues to grow and expand, as we make continue to make progress toward understanding the prevalence of life in the universe, how it originates, and how it evolves.

Alan P. Boss
President of the Commission

Transactions IAU, Volume XXVIIB
Proc. XXVII IAU General Assembly, August 2009
Ian F. Corbett, ed.

doi:10.1017/S1743921310004990

COMMISSION 53 — EXTRA-SOLAR PLANETS
(Extra-Solar Planets)

PRESIDENT	**Alan Boss**
VICE-PRESIDENT	**Alain Lecavelier des Etangs**
PAST PRESIDENT	**Michel Mayor**
ORGANIZING COMMITTEE	**Peter Bodenheimer, Andrew Collier-Cameron, Eiichiro Kokubo, Rosemary Mardling, Dante Minniti, Didier Queloz**

PROCEEDINGS BUSINESS SESSIONS, 12 August 2009

1. Introduction

Commission 53 met in August 12, 2009. Outgoing President Michel Mayor chaired the meeting, and there were several dozen members present, including incoming President Alan Boss,incoming Vice President Alain Lecavelier des Etangs. Commission 53 (C53) was founded at the 2006 Prague General Assembly of the IAU. After a period of 6 years, C53 will come up for renewal at the 2012 IAU General Assembly in Beijing, China. For the moment, more than 150 IAU members have asked to be members of C53 and few dozen non-IAU members having asked to be informed of the commission activity.

2. Organizing Committee

The 2006-2009 C53 organizing Committee (OC) had 13 members, all of them being previously members of the IAU Working Group for Extra-Solar Planets (WGESP). The new OC includes 9 members of which only the President and past-president were members of the last triennum OC. The new OC has members with quite diverse competences and a broad geographical distribution.

3. Meetings

Meetings focussed on extra-solar planets are still extremely frequent, covering all of the aspects of that new chapter of astrophysics, from new instrumentation, detections, characterisation and theory. On the last triennum two IAU symposia have been devoted to extra-solar planetary systems. From October 22nd to 26th 2007 in Suzhou, the IAU Symp. 249, "Exoplanets: Detection, Formation and Dynamics" and from May 19th to 23rd 2008 the IAU Symposium 253, "Transiting Planets". Both conferences were attended by more than 200 participants.

At the IAU General Assembly in Rio de Janeiro, the SpS6 "Planetary Systems as Potential Sites for Life", co-organised by commissions 16, 51 and 53. The very large attendance to this special session has proven again the interest of the astronomical community for this cross-disciplinary topic.

4. Summary of discussions

Three topics have been discussed during the meeting :

• Definition of exoplanets and the upper limit for their mass. The present definition of extra-solar planets has been established by the IAU working group WGESP. It will be the responsability of the C53 to rediscuss that definition .

• Nomenclature of extrasolar-planets. The present rule for the nomenclature of extrasolar planets is simply to use the name of the star followed by a lower case letter given in the order of discovery, in case of multiplanetary systems, starting from "b". Examples: 51 Peg b, 55 CnC b, 55 CnC c, 55 CnC d ,...CoRoT-7 b, CoRoT-7 c ,... The discussion was focused on the possibility to add some specific names to exoplanets as it is the rule for asteroids. It was decided to let the Organising Committee of C53 to clarify that question of nomenclature having in mind the present very high rate of exoplanet discoveries.

• List of Exoplanets.

Michel Mayor
President of the Commission

Transactions IAU, Volume XXVIIB
Proc. XXVII IAU General Assembly, August 2009
Ian F. Corbett, ed.

doi:10.1017/S1743921310005004

DIVISION III
COMMITTEE ON SMALL BODY NOMENCLATURE

CHAIR	Jana Ticha (Czech Republic)
MPC Rep.	Brian G.Marsden (secretary) (USA)
Div III Rep.	Karen Meech (USA)
CBAT Rep.	Daniel Green (USA)
WGPSN Rep.	Rita Schulz (Netherlands)
MEMBERS	Michael F. A'Hearn (USA), Edward L. G. Bowell (USA), Julio Fernandez (Uruguay), Pam Kilmartin (New Zealand), Syuichi Nakano (Japan), Keith Noll (USA), Lutz Schmadel (editor of DMPN) (Germany), Viktor Shor (Russia), Gareth Williams (USA), Donald K. Yeomans (USA), Jin Zhu (China).

The CSBN meeting held in Rio de Janeiro on August 11 was attended by just six members, including Pam Kilmartin as the acting chair, and several visitors. Since there was not a quorum of members, it was not possible to make any decisions. But there was a good discussion on many topics, from which several points emerged that should be more fully discussed by the whole committee during the next few months:

1. Membership

The list of members was accepted, but those present thought that more specialized and geographically diverse members were needed. The total committee now numbers 16, but members recommended that consideration be given to co-opting more members who can join in the committee's work. More still needs to be done especially with regard to Japanese and Chinese names. The Executive Committee may soon have a rule (by-law) that chairpersons should not serve more than two three-year terms. Members agreed that it might be useful to have an assistant chairperson as there is a long learning curve in this type of committee. There could also be a role for the outgoing chair. In order to ensure continuity, however, it would be useful to discuss with the EC if the CSBN, like the WGPSN, could come directly under the EC in future and therefore not be subject to this rule.

2. Guidelines

Suggestions for consideration as guidelines included (a) some limit on the number of minor planets named for family members of the discoverer; (b) some limits on how closely names can resemble other names (one, two, three letters? how about pronunciation?); (c) is the limit of 16 characters still relevant? Other ideas mentioned include doubling up (even quadrupling up) names of discoverer family members into a single minor planet name, a practice that would surely help produce names that are dissimilar to existing names. A fixed limit of family names allowed for each discoverer could be difficult. Some members still want the 16-character maximum, but we should encourage shorter names. Keith Noll suggested that we should form a subcommittee (task group) to reconsider the guidelines, reporting back in about three months, and that this should be a regular event every three years, so that the guidelines are reviewed and (re)accepted before each General Assembly. This suggestion was generally acceptable to those present.

3. Number of name proposals

Even though the "Sydney 2003 guideline" requesting individual discoverers and teams to propose no more than two names for each two-monthly naming batch is regularly broken by a few proposers, these batches have been limited to no more than 100 name proposals. Considering the increasing number of minor planet discoveries and likely increase in the number of name proposals, the CSBN will work in close cooperation with the MPC on preparing a web-based system for computer automation of the name-approval process and the editing of citations.

4. Priorities of naming

Considering that proper names are an important part of solar system nomenclature, the CSBN must set priorities for dealing with naming proposals and give more emphasis to naming frequently cited objects. The CSBN priority will be the naming of NEAs, TNOs, binary bodies, satellites of minor planets, space mission and radar targets, objects of significance to physical studies, etc. Keith Noll remarked that TNOs are not being named, as the mythological convention is seen to be too restrictive. Compilation of a namebank of suitable names for objects in some of the categories listed above would be a good start, especially for TNOs and their satellites. We should also have a task group to do this.

5. Selling of minor planet names

Dan Green suggested that, as the list of numbered minor planets is now very large and only 7% had been named, the IAU should give some thought to the possibility of allowing the public to pay a fee and have a main belt (H 15) minor planet given a name of their choice, with the proceeds going to education and research. Most of the other members present could not agree with this suggestion, which conflicts with the IAU EC's views on organizations such as the "International Star Registry".

Jana Ticha
Chair of the CSBN

Transactions IAU, Volume XXVIIB
Proc. XXVII IAU General Assembly, August 2009
Ian F. Corbett, ed.

doi:10.1017/S1743921310005016

DIVISION III WORKING GROUP on PLANETARY SYSTEM NOMENCLATURE

PRESIDENT Rita Schulz
VICE-PRESIDENT
PAST PRESIDENT
MEMBERS K. Aksnes, J. Blue, E. Bowell
G. A. Burba, G. Consolmagno, R.Courtin
R. Lopes, M. Ya. Marov, B. G. Marsden
M. S. Robinson, V. V. Shevchenko, B. A. Smith

PROCEEDINGS BUSINESS SESSIONS, 7 August 2009

1. Introduction

The meeting was attended by 5 members of the WG (E. Bowell, G. Consolmagno, R. Courtain, R. Lopez, R. Schulz) one Task Group member (J. Watanabe), and several guests from the CSBN and CBAT. It was decided at the beginning of the meeting that the attending members of the WGPSN would discuss matters, provide their opinion or vote, and then ask the other 8 formal members to do the same via email. As a consequence the following discussed items have been agreed by majority vote of the WG members.

2. Discussion Items

In the first part of the meeting four specific items were addressed which needed discussion and agreement by the WG.
1) An information email shall be sent to CBAT every time a satellite has been officially named.
2) Categories of proper names shall be approved by the WG before being implemented by the TG.
3) If biographical information is only available in a language the WGPSN members cannot understand, a translation of this information into English should be made available, whereby it has to be ensured that this translation comes from a bona fide source. (The English language was chosen for practical reasons, because all members of the WG would speak it.) The respective TG chair is required to provide an English translation.
4) The need for relaxing rules about internet sources was discussed. The WG was of the opinion that it is not yet time for trying to put up a general rule on the use of internet sources. It was however noted that it should be allowed to use internet sources if no other sources are available and if the internet source is a bona fide source, of which the likelihood of staying available is not smaller that for non-internet sources (e.g. the Britannica). This issue should be discussed for the time being on a case by case basis.
If an internet page is used that page will be saved as a pdf file, linked to the sources page ("http://planetarynames.wr.usgs.gov/jsp/append4.jsp"), and stored at the USGS (in case the site becomes unavailable in the future).

3. Closing remarks

As the integration of the WGPSN from the Executive Committee into Division III has resulted in communication problems that affected the work, authority, and reputation of the WG, options for solving this problem were discussed. It was confirmed that the D-III president has the obligation to forward any information concerning the WGPSN to the WG chair. After discussion the WGPSN agrees that moving the WGPSN back to reporting directly to the EC provides the only effective solution to the communication problem. The IAU President has therefore been contacted and requested to suggest this topic as an agenda item at the next EC meeting.

Rita Schulz *President of the Working Group*

Transactions IAU, Volume XXVIIB
Proc. XXVII IAU General Assembly, August 2009
Ian F. Corbett, ed.

doi:10.1017/S1743921310005028

DIVISION IV STARS
(Etoiles)

PRESIDENT	**Monique Spite**
VICE-PRESIDENT	**Christopher Corbally**
PAST PRESIDENT	**Dainis Dravins**
BOARD	**Christine Allen, Francesca d'Antona, Sunetra Giridhar, John Landstreet Mudumba Parthasarathy**

PARTICIPATING COMMISSIONS

Commission 26	**Double and multiple Stars**
Commission 29	**Stellar spectra**
Commission 35	**Stellar constitution**
Commission 36	**Theory of stellar atmospheres**
Commission 45	**Stellar classification**

DIVISION WORKING GROUPS

Massive Stars
Abundances in red giants

INTER DIVISION WORKING GROUPS

Division IV-V WG	**Active B stars**
Division IV-V WG	**Ap and related stars**
Division IV-V WG	**Standard stars**

SCIENCE AND BUSINESS MEETING, 7 August 2006

1. Introduction

During the General Assembly in Rio de Janeiro the Division IV meeting, and the meetings of the participating working groups and commissions, were held on thursday 6th (session 1 and 2) and friday 7th (sessions 1, 2, 3, 4).
The meeting of the Division IV took place in one 1.5 hour session during the afternoon of 7 August 2009 and was attended by about 50 participants. Most of this meeting was devoted to a scientific session "**The Solar Composition**".
This scientific session was followed by a short business meeting where the composition of the new board of the Division for the triennium 2009-2012 was announced.

2. The solar composition

The chemical composition of the Sun is one of the most important yardsticks in astronomy with implications for almost all fields from planetary science, helioseismology, to the high-redshift Universe. It is the reference for the determination of stellar abundances. Recently with the introduction of new 3D atmospheric models the chemical composition of the Sun has been revised and it is a good time to survey the new results, and the new problems, in this field of research.

The three following talks were given and are available (pdf format) on the WEB site of the Div IV:

http://www.iau.org/science/scientific_bodies/divisions/IV/ and click on "Division WEB page" (documents).

Marc Pinsonneault (Ohio State University, Department of Astronomy, USA)
Absolute Solar Abundances from Helioseismology

Helioseismology permits the study of internal solar structure in exquisite detail, and in particular the internal temperature gradient. The temperature gradient in the core in turn depends on the opacity, which also depends on the composition. This opens the possibility of using stellar interiors studies to constrain the absolute solar abundances. Scalar solar features (in particular, the solar surface helium abundance and convection zone depth) are also sensitive to the solar abundances; the former is strongly affected by iron group element abundances because they retain electrons to high temperatures, and the latter by both iron and lighter metals such as CNONe. A two parameter abundance solution can be inferred from the scalar properties, with the relative heavier element abundance set to the meteoritic values and the lighter one set by photospheric abundance ratios. The scalar constraints alone do not distinguish between C, N, O, and Ne as the opacity sources, but a solution with high Ne and low CNO can be ruled out by the strong imprint of the Ne/O ratio on the sound speed profile in the solar core. The seismic solution for the solar mixture is similar to that of Grevesse & Sauval 1998, with modestly enhanced Ne/O. The inferred oxygen is compatible with the recent Caffau *et al.* 2008 value but not with the lower Asplund *et al.* 2004 oxygen. Convective overshoot may reduce the inferred seismic oxygen by a small amount, but not enough to bring the values into agreement with the lower scale. Independent seismic tests of the bulk metal abundance in the convection zone, based upon the equation of state, also favor a higher absolute oxygen abundance consistent with that derived from opacity calculations. Experimental efforts in progress will provide valuable constraints on the theoretical opacities for solar interiors conditions.

Martin Asplund (Max-Planck-Institut für Astrophysik, D-85741 Garching, Germany)
New determination of the abundances in the solar atmosphere

The solar chemical composition is an important ingredient in our understanding of the formation, structure and evolution of both the Sun and our solar system. Furthermore, it is an essential reference standard against which the elemental contents of other astronomical objects are compared. In this talk I evaluate the current understanding of the solar photospheric composition. In particular, a re-determination of the abundances of nearly all available elements is presented (Asplund, Grevesse, Sauval & Scott, 2009, ARAA, 47, 481).

The results are based on a realistic new 3-dimensional (3D), time-dependent hydrodynamical model of the solar atmosphere, which fulfills all key observational tests. We have carefully considered the atomic input data and selection of spectral lines, and accounted for departures from LTE whenever possible. The end result is a comprehensive and homogeneous compilation of the solar elemental abundances.

Particularly noteworthy findings are significantly lower abundances of carbon, nitrogen, oxygen and neon compared with the widely-used values of a decade ago. The new solar chemical composition is supported by a high degree of internal consistency between available abundance indicators, and by agreement with values obtained in the solar neighborhood and from the most pristine meteorites. There is, however, a stark conflict with standard models of the solar interior according to helioseismology, a discrepancy that has yet to find a satisfactory resolution.

Hans-Günter Ludwig (CIFIST, GEPI, Observatoire de Paris, 92195 Meudon Cedex, France)
Solar abundances from spectroscopy: what we know and what we need

We determined solar photospheric abundances from optical and near-infrared spectroscopy. The work was conducted in the broader context of spectroscopic studies in the CIFIST (Cosmological Impact of the FIrst STars) team hosted by Paris Observatory. The project constitutes a determination independent of other work applying several "self-made" tools: the 3D radiation hydrodynamics code CO^{5}BOLD, the 3D spectral synthesis code Linfor3D, the 3D NLTE code NLTE3D, and a 1D stellar atmosphere code called LHD sharing the micro-physics and in part numerics with CO^{5}BOLD. Also the standard 1D codes ATLAS, MARCS, and the Kiel NLTE code were applied. We completed the work on 12 elements (Li, C, N, O, P, S, Eu, Hf, Th, K, Fe, Os) using lines of atomic species. At the moment 14 people are involved in the project headed by E. Caffau bringing in all necessary expertise to handle code developments and the data analysis.

The main result might be summarized by the overall solar mass fraction of metals of $Z = 0.0154$; for the important CNO elements we obtained A(C)=8.50 ± 0.11, A(N)=7.86 ± 0.12, and A(O)=8.76 ± 0.07 on the usual spectroscopic scale A(H)=12. We took a rather conservative stand when estimating the uncertainties; in the case of N and C we quote the dispersion among abundances from individual lines, in the case of O it is a combination of individual errors and systematics due to NLTE effects. Our findings may be contrasted with the results of Asplund and collaborators who also applied 3D hydrodynamical model atmospheres in the abundance determination. While by standards of stellar spectroscopy the differences are modest (about 15%) they appear surprising when considering that all groups are basically using the same observational material and apply the same atomic parameters. What are the reasons for the sizable systematic differences?

Sources of systematic errors are related to i) line selection and blending problems; ii) the accuracy to which equivalent widths can be measured, here in particular the continuum placement and line profile shapes fitted to observations; iii) the accuracy of the 3D atmosphere models and spectral synthesis codes; iv) for NLTE calculations the efficiency of collisions with neutral hydrogen atoms; v) not-understood differences among high-quality solar atlases.

3D models atmosphere now provide an excellent match to the observed center-to-limb variation (including line blocking) of the solar radiation field. This indicates that the model's thermal structure closely resembles the actual solar conditions. This remains to be also shown for higher photospheric layers. Future analysis work would greatly benefit from a highest quality (signal-to-noise, correction for telluric absorption, wavelength calibration) solar atlas for several limb angles. Moreover, there remains a need for accurate atomic data: wavelength, oscillator strength, and particularly collisional cross sections needed in NLTE calculations.

3. The new Organising Committee members for the triennium 2009-2012

The election of a new Vice President for the Division had been organised before the General Assembly.
To chose the candidates we proceeded as it had been done last time: the candidates were the five current Presidents of the participating commissions (each was contacted with an invitation to be a candidate). In the case where the current President declined, the past president was invited to be candidate.
The electorate was defined as all members of the Division. More than 1000 e-mails were sent to those members (about 60 "could not been delivered"). 246 votes were received and Francesca d'Antona was elected as the new Vice President. (She will becomes President of the Division after the next General Assembly in China.)
The new board is formed by the President, the Vice President, the Past President of the Division and the five new Presidents of the participating Commissions. The new organising Committee of the Division IV for the triennium 2009-2012 is thus:
Christopher Corbally (President, Vatican), F. d'Antona (Vice President, Italy), M. Spite (Past President, France), J.A. Durantez Docobo (Comm 26, Spain), N.E. Piskunov (Comm 29, Sweden), C. Charbonnel (Comm 35, Switzerland), M. Asplund (Comm 36, Germany), R.O. Gray (Comm 45, USA)

The President of the division congratulates the new Vice President and the new SOC members on their election and she thanks the members of the current Organizing Committee of the Div IV for their very active work in ranking a large number of Symposia proposals, during the past triennium...

Monique Spite
President of the Division

Transactions IAU, Volume XXVIIB
Proc. XXVII IAU General Assembly, August 2009
Ian F. Corbett, ed.

doi:10.1017/S174392131000503X

COMMISSION 26 DOUBLE AND MULTIPLE STARS

PRESIDENT	**Christine Allen**
VICE-PRESIDENT	**Jose A. Docobo**
PAST PRESIDENT	**William I. Hartkopf**
ORGANIZING COMMITTEE	**Yuri I. Balega, John Davis, Brian D. Mason, Edouard Oblak, Terry D. Oswalt, Dimitri Pourbaix, Colin D. Scarfe**

PROCEEDINGS BUSINESS SESSION, 7 August 2009

1. Introduction

The business meeting of Commission 26 was held on Friday, 7 August 2009, and consisted mainly of scientific talks. At the end, incoming President J. A. Docobo gave a video presentation with his thoughts regarding future activities of Commission 26, including the comments of members of the OC. He also announced a December 2009 workshop and discussed potential meetings in the coming years.

2. The science talks

Brian Mason spoke about the present status and future plans for the USNO Double Star Catalog. During the ensuing discussion, the prospects of a joint commission 26/54 meeting on interferometric binaries were evaluated. Doubts were expressed as to whether a third double star CD was necessary. The consensus was that it was not yet needed.

Oleg Malkov talked about the Binary Star Data Base, set up and maintained in collaboration with Edouard Oblak and Bernard Debray. Although it may appear repetitious to the other binary star catalogs, it includes data from all observational categories. The database is currently available at bdb.obs-besancon.fr . It will be transferred to Moscow with a mirror at Besancon.

Dimitri Pourbaix gave a progress report on the 9th catalogue of spectroscopic binaries, which now contains 2946 systems and 3608 orbits, 1903 of which have radial velocities. To date, data from 591 papers have been added, with an estimated 80% completeness. More information, and the catalog itself, is available at http://sb9.astro.ulb.ac.be (but see also Pourbaix *et al.* 2004, A&A 424, 727). With adequate manpower the updated catalog could be completed in three years, but under the present circumstances, six years would be required. A very interesting proposal is to include in the USNO orbit catalog ORB6 a direct link to the SB9 page for spectroscopic binaries, when pertinent.

Christos Siopis presented a talk on Gaia eclipsing binaries. The expectation is that among 10^8 variable objects there will be $10^5 - 10^6$ eclipsing binaries. They will have 30-200 transits per object, with a mean of about 80.

Christine Allen (with A. Poveda) discussed a specific wide common proper motion companion to GJ 282AB. The probability of its being optical is 7.6×10^{-5}. All components have the same parallax, proper motion, radial velocity, and consistent spectral types. From their X ray luminosities similar ages are obtained. The larger than expected proper motion difference may be due to the dynamical disintegration of the wider C companion.

C. Allen (with M.A. Monroy) then presented a progress report on an improved list of wide halo binaries. They find the major semiaxes to follow Opik's distribution (a power law with exponent 1) rather than the power law with exponent 1.55 which is favored by Chaname and Gould. The widest systems follow Oepik's distribution up to separations of more than 60 000 AU.

In a video presentation José A. Docobo discussed work done at Santiago de Compostela. Their orbit catalog includes now 2061 orbits of 1618 orbits, and is available at http://www.usc.es/astro. This catalog can be considered to be complementary to ORB6.

3. Closing remarks

In the video sent by José A. Docobo he expressed his willingness to promote and host a C26 meeting, considering such meetings essential for better communication among C26 members, He discussed possible dates and subjects, which would need to be further evaluated among OC members.

My thanks are due to B. Mason, for making available his notes on the meeting.

Christine Allen
President of the Commission

Participants

Name	email	C26 member
Brian Mason	bdm@usno.navy.mil	yes
Markus Mugrauer	markus@astro.uni-jena.de	no
Theo ten Brummelaar	theo@chara-array.org	yes
Edward Weis	eweis@wesleyan.edu	yes
Thijs Kouwenhoven	t.kouwenhoven@sheffield.ac.uk	no
Christos Siopis	christos.siopis@ulb.ac.be	no
Frederic Arenou	frederic.arenou@obspm.fr	yes
Dimitri Pourbaix	pourbaix@astro.ulb.ac.be	yes
Oleg Malkov	malkov@inasan.ru	yes
Hans Zinecker	hzinnecker@aip.de	yes
Christine Allen	chris@astroscu.unam.mx	yes
Norbert Zacharias	nz@usno.navy.mil	no
Ralph Gaume	rgaume@usno.navy.mil	no

Transactions IAU, Volume XXVIIB
Proc. XXVII IAU General Assembly, August 2009
Ian F. Corbett, ed.

doi:10.1017/S1743921310005041

COMMISSION 29 — STELLAR SPECTRA (*STELLAR SPECTRA*)

PRESIDENT	Nikolai Piskunov
VICE-PRESIDENT	Katia Cunha
PAST PRESIDENT	Mudumba Parthasarathy
ORGANIZING COMMITTEE	Wako Aoki
	Martin Asplund
	David Bohlender
	Kenneth Carpenter
	Jorge Melendez
	Mudumba Parthasarathy
	Silvia Rossi
	Verne Smith
	David Soderblom
	Glenn Wahlgren

PROCEEDINGS BUSINESS SESSIONS, 7 August 2009

1. Past Activities

The business meeting was attended by 23 members of the Commission. The meeting started at 16:00 a short report of the activities during the triennium 2006-2009. The focus of the activities was the sharing of expertise between spectroscopic techniques in various areas of astronomical research. In particular, the progress in instrumentation, detectors, data reduction, data analysis and archiving. The second activity was the analysis of to IAU meeting proposals followed by recommendations for improvements and eventually support. The sponsored symposia included Sponsoring symposia The Ages of Stars and The Disk Galaxy Evolution in the Cosmological Context. The Commission was also disseminating information about the Commission activities and relevant meetings to the Commission members. In this respect the Commission web page is playing a crucial role.

At the time of the GA in Rio the Commission 29 included 393 members. The new president introduced the changes in the Organizing Committee. As the result of the election carried out via email. The new President thanked the outgoing members:

- Fiorella Castelli
- Philippe Eenens
- Ivan Hubeny
- Chris Sneden
- Masahide Takada-Hidai
- Werner Weiss

Special appreciation was expressed to the very important work by the past President of the Commission. The newly elected members are:

- Martin Asplund (Germany)
- Wako Aoki (Japan)
- David Soderblom (USA)
- Verne Smith (USA)

- David Bohlender (Canada)
- Jorge Melendez (Portugal)

Katia Cunha took over as the Vice-President.

2. Planning of activities for the next triennium

The report of the President was followed by a general discussion centered around 2 points: (1) the future of the Commission 29 and interaction with other IAU commissions and (2) the geographical representation of the Commission OC members. The first topic was stimulated by an obvious overlap between the areas of spectroscopy, atomic and molecular data, stellar atmospheres, interstellar medium, exoplanets etc. The general consensus was that regular revisions of commission specializations and commission sizes are necessary to maintain the efficiency and motivation of the commission members. This issue must be seriously discussed by the IAU EC. The 2nd topic was came of the analysis of the fast development of the astronomical research and stellar spectroscopy in particular outside the traditional centers. One excellent example of such change is the LAMOST project in China. These tendencies must be reflected in the membership of the organizing committees which will be beneficial for both sides.

Nikolai Piskunov
President of the Commission

Transactions IAU, Volume XXVIIB
Proc. XXVII IAU General Assembly, August 2009
Ian F. Corbett, ed.

doi:10.1017/S1743921310005053

COMMISSION 35 STELLAR CONSTITUTION
(CONSTITUTION DES ETOILES)

PRESIDENT Corinne Charbonnel
VICE-PRESIDENT Marco Limongi
PAST PRESIDENT Franca D'Antona
ORGANIZING COMMITTEE Gilles Fontaine, Jordi Isern, John Lattanzio, Claus Leitherer (Secretary), Jacco Van Loon, Achim Weiss, Lev Yungelson

PROCEEDINGS BUSINESS SESSION, 7 August 2009

1. Introduction

A business meeting of the IAU Commission 35 was held during the GA in Rio on Friday, August 7, 2009, with a few members of the Commission in attendance. Special care will be taken to have more members attending the BM at the next GA in Bejing in 2012. The points discussed during the BM are summarized below and are posted on the C35 website http://iau-c35.stsci.edu.

2. Presentation of the new Organizing Commitee

On behalf of the entire membership, C. Charbonnel thanked F. D'Antona for her important contribution to our Commission as former President and congratulated her for her election as Vice-President of Division IV (Stars). She also expressed the thanks of the Commission to N. Langer, R. Larson, J. Liebert, E. Mueller for their years of service on the OC, and welcomed the new OC members J. Isern, C. Leitherer, M. Limongi, J. Van Loon recently elected by Commission membership. The new Commission President and Vice-President for the term 2009–2012 are C. Charbonnel and M. Limongi. C. Leitherer is appointed Commission Secretary for the next three years.

The OC of Division IV STARS is composed by C. Corbally (President), F. D'Antona (Vice-President), M. Asplund (C36), C. Charbonnel (C35), J.A. Durantez Docobo (C26), R.O. Gray (C45), N.E. Piskunov (C.29).

3. Comments on the official tasks of the Organizing Commity

The main official tasks of the OC are to (1) provide a triennial report on the activities of the Commission and of its members in the field of "Stellar Constitution", and to (2) review the applications and establish a list of preference for the future IAU Symposia. It should also provide support for the animation of the community as a whole.

3.1. *The triennial report*

The triennial report should cover the organisational activities of the Commission and should highlight scientific accomplishments in the field of stellar structure and evolution. This document is thus especially important in order to demonstrate the vitality of C35 and of the research field it covers. The previous triennial reports can be found on http://iau-c35.stsci.edu/Reports/.

The next report will be due by the end of 2011. In order to improve the preparation process as well as the quality of the outcome, C. Charbonnel would like all the Commission members to send to the OC relevant information. The C35 members will all be contacted regularly (schedule to be decided) and will be asked to send a brief summary of their most important achievements in the past three years. This information will be made public and will be collected by OC members in order to prepare the triennial report.

3.2. *IAU Symposia*

The past triennium – apart for the first year – was a bit frustrating from the point of view of supervision of scientific IAU meetings. In particular, the number of meeting applications for 2010 were few enough that the selection was practically none. In the future we will encourage the C35 members to post more numerous strong proposals for IAU Symposia. This is crucial for the vitality of our scientific community.

4. Role of C35

During the BM, we raised the question of the role of the C35 (apart for the tasks described in § 2), and of the need for other activities and tools to be developed. The discussion focussed on the possible ways of stimulating exchanges among C35 members.

We decided to proceed to an evaluation of the needs, suggestions, and preferences of the C35, by performing a member survey. The main questions to be addressed will be the following: Do you want an electronic forum of discussion with round table about stellar structure problems, and open to the whole community? Should we add more information (what kind?) about the commission members on our website? Would you like to post on the C35 webpage a few lines about your work, together with references, whenever you publish results relevant to C35 (i.e., all along the triennal exercise)? Do you want/need a C35 electronic newsletter including scientific announcements?

The OC will first discuss the way to proceed during the Fall 2009, and a general consultation of all the C35 members will be carried out at the beginning of 2010.

5. C35 name and membership

In order to increase the visibility of our Commission, and to be in better adequation with the science it covers, C.Charbonnel proposes to change the name of C35 presently entitled "Stellar constitution" and suggests rather "Stellar structure and evolution". This suggestion was informally discussed with the new President of Division IV, C. Corbally, and will be addressed in the forthcoming electronic consultation of the Commission members.

Special care will be taken to look for "missing members" and to encourage both theoreticians/modellers and observers who are active in the field to adhere to C35. Any IAU scientist with research credentials who, for their own valid reasons wish to join our Commission and to contribute to its progress, should be admitted. Gender issues should also be addressed, in particular for the composition of the OC (only one woman out of nine OC members for the term 2009–2012 ...).

6. C35 webpage

The C35 webpage (launched in 2003) is maintained by our Secretary C.Leitherer. Its goal is to support the administrative and scientific needs of our Commission. It contains general information on the C35 structure and activities, membership directory, links to commission reports, conference announcements, bibliography related to stellar constitution, links to stellar structure resources that were made available by owners, links to main astronomical journals, and regularly updated news items. The ressources contain evolutionary tracks and isochrones from various groups, nuclear reactions, equations of state, opacity data, and astronomical tools. It will soon host the scientific results posted by the C35 members as described above.

http://iau-c35.stsci.edu

Corinne Charbonnel
President of the Commission

Transactions IAU, Volume XXVIIB
Proc. XXVII IAU General Assembly, August 2009
Ian F. Corbett, ed.

doi:10.1017/S1743921310005065

COMMISSION 36 THEORY OF STELLAR ATMOSPHERES (*THEORY OF STELLAR ATMOSPHERES*)

PRESIDENT	**Martin Asplund**
VICE-PRESIDENT	**Joachim Puls**
PAST PRESIDENT	**John Landstreet**
ORGANIZING COMMITTEE	**Carlos Allende Prieto**
	Thomas Ayres
	Svetlana Berdyugina
	Bengt Gustafsson
	Ivan Hubeny
	Hans Günter Ludwig
	Lyudmila Mashonkina
	Sofia Randich

PROCEEDINGS BUSINESS SESSION, August 7, 2009

The members of the Commission 36 Organizing Committee attending the IAU General Assembly in Rio de Janeiro met for a business session on August 7. Both members from the previous (2006–2009) and the new (2009–2012) Organizing Committee partook in the discussions. Past president John Landstreet described the work he had done over the past three years in terms of supporting proposed conferences on the topic. He has also spent significant amount of time establishing an updated mailing list of all > 350 members of the commission, which is unfortunately not provided automatically by the IAU. Such a list is critical for a rapid dissemination of information to the commission members and for a correct and smooth running of elections of IAU officials. Everyone present thanked John effusively for all of his hard work over the past three years to stimulate a high level of activity within the discipline.

Most of the discussion at the business meeting centered around the organization of conferences of interest for the commission members. At the 2009 General Assembly a large number of meetings were highly relevant for the field of stellar atmospheres. In particular the two-day Joint Discussion 10 *3D Views on Cool Stellar Atmospheres – Theory Meets Observation* organized by Hans Ludwig was entirely devoted to this topic and was a great success. In addition, several other symposia and joint discussions featured a large number of invited reviews and contributed talks related to stellar atmospheres, including JD4 *Progress in Understanding the Physics of Ap and Related Stars*, JD11 *New Advances in Helio- and Astero-Seismology*, Special Session 7 *Young Stars, Brown Dwarfs, and Protoplanetary Disks* and of course Symposium 265 *Chemical Abundances in the Universe – Connecting First Stars to Planets.* In addition, the scientific part of the business meeting of Division IV Stars on August 7 was devoted entirely to the solar chemical composition and the discrepancy with helioseismology; Marc Pinnsonneault, Hans Ludwig and Martin Asplund gave interesting talks describing their take on this hotly debated topic. In all respects, the work of the commission was very well represented at this General Assembly, which demonstrates that the field is still blooming and highly relevant for the broader field of astronomy.

The business meeting ended with a discussion about future meetings. Everyone agreed that we should plan for at least one larger meeting each triennia as more or less in the past. We should also try to ensure that the topic of stellar atmospheres are covered in other conferences, which should not be so difficult given that it plays a key role also in other areas such as exo-planet searches, Galactic and cosmic chemical evolution and stellar parameter estimations. One possibility is to organize a conference on modelling stellar and planetary atmospheres in Munich in 2011.

Martin Asplund
President of the Commission

Transactions IAU, Volume XXVIIB
Proc. XXVII IAU General Assembly, August 2009
Ian F. Corbett, ed.

doi:10.1017/S1743921310005077

COMMISSION 45 — STELLAR CLASSIFICATION (STELLAR CLASSIFICATION)

PRESIDENT	Sunetra Giridhar
VICE-PRESIDENT	Richard Gray
PAST PRESIDENT	Christopher Corbally
ORGANIZING COMMITTEE	Coryn Bailer-Jones
	Laurent Eyer
	Michael J. Irwin
	Davy Kirkpatrick
	Steve Majewski
	Dante Minitti
	Birgitta Nordström

PROCEEDINGS BUSINESS SESSION, 7 August 2009,

1. Business

The business meeting of Commission 45 was held on Friday, 7 August. It was attended by the Vice-President of the Commission (who chaired the meeting in the absence of the President) as well as nine other members of the Commission. Attendance was limited, as usual, by the unavoidable occurrence of parallel sessions.

The Vice President announced the names of the new organizing committee of the Commission and welcomed their participation. Some of those members are new members of the Commission and even of the IAU.

The Vice President also drew the attention of the meeting to the Commission 45's contribution to Reports on Astronomy 2006-2009, where activity in stellar classification over the last several years is cited, and he thanked all the contributers to this report.

One major development of interest to this Commission was the recent publication by Princeton University Press of the book "Stellar Spectral Classification", by Richard O. Gray and Christopher J. Corbally. Its publication was announced to the business meeting, and a sample copy passed around.

A major activity of Commission 45, in concert with other commissions under the aegis of Division IV, is the Working Group on Standard Stars. This Working Group was formally terminated as of this General Assembly. While standard stars continue to be of vital interest to Commission 45, Division IV and the IAU in general, it was felt that this working group had served its purpose. The archives of the Standard Star Newsletter will still be available on the Standard Star Working Group webpage, which will continue to be maintained by Richard Gray.

2. New Organizing Committee and President

The Organizing Committee (OC) for the next triennium was arranged by e-mail prior to the General Assembly. Nominations from the Commission membership at large were just sufficient to fill vacancies arising from the usual process of rotation, so it was not necessary to hold an election. Retiring members of the OC were sincerely thanked, and the incoming members and officers were acclaimed by those present.

President: Richard Gray (USA) grayro@appstate.edu
Vice-President: Birgitta Nordström (Denmark) birgitta@astro.ku.dk
Organizing Committee:

Sunetra Giridhar (ex-officio, India)
Adam J. Burgasser (USA)
Laurent Eyer (Switzerland)
Ranjan Gupta (India)
Margaret M. Hanson (USA)
Michael J. Irwin (UK)
Caroline Soubiran (France)

3. Science

The scientific program consisited of a talk by C. Corbally on "A Stellar Spectral Classification Encoding Scheme" recently devised by Myron Smith, co-workers at MAST (Multimission Archive at STScI), C. Corbally and R. Gray. The purpose of this encoding scheme is to enable sophisiticated searches of databases on the basis of spectral types. After the report, a discussion of various issues associated with the introduction of this scheme ensued, including how to select the "best" spectral type from the literature. Various ways to do this were proposed, and an informal straw poll was taken to help guide the authors of the encoding scheme.

4. Closing remarks

The vice president thanked those present for their participation in this meeting and for their support of Commission 45's work in the last three years, and he wished them every success in their work and projects during the coming triennium.

Richard O. Gray *Incoming President of the Commission*

Transactions IAU, Volume XXVIIB
Proc. XXVII IAU General Assembly, August 2009
Ian F. Corbett, ed.

doi:10.1017/S1743921310005089

DIVISION IV/V WORKING GROUP on ACTIVE B STARS

PRESIDENT	**Geraldine J. Peters**
VICE-PRESIDENT	**Carol E. Jones, Richard D. Townsend**
PAST PRESIDENT	**Juan Fabregat**
MEMBERS	**Karen S. Bjorkman, M. Virginia McSwain, Ronald E. Mennickent, Coralie Neiner, Philippe Stee, Juan Fabregat (non-voting)**

PROCEEDINGS BUSINESS SESSIONS, 6 August 2009

1. Introduction

The meeting of the Working Group on Active B Stars consisted of a business session followed by a scientific session containing nine talks. The titles of the talks and their presenters are listed below. We plan to publish a series of articles containing summaries of these talks in Issue No. 40 of the *Be Star Newsletter* †. This report contains an account of the announcements made during the business session, an update on a forthcoming IAU Symposium on active B stars, a report on the status of the *Be Star Newsletter*, the results of the 2009 election of the SOC for the Working Group for 2009-12, a listing of the Working Group bylaws that were recently adopted, and a list of the scientific talks that we presented at the meeting.

2. Business session

2.1. *Announcements*

1. A proposal to continue IAU recognition of the Working Group was submitted by G. Peters & J. Fabregat by the 2009 February 28 deadline.

2. The triennium report for the WG for 2006-09 was prepared and submitted by J. Fabregat & G. Peters by the 2008 July deadline. It was published in 2008 December in IAU Transactions XXVIIA and in Issue No. 40 of the Be Star Newsletter.

3. A proposal to hold a meeting on active B stars in Paris in the summer of 2010 was submitted by C. Neiner by the 2008 December 1 deadline. Our proposal entitled "Active OB stars: structure, evolution, mass loss, and critical limits" was accepted as one of the eight IAU Symposia to be held in 2010. It has been designated as IAU Symposium No. 272 and will take place from July 19-23.

2.2. *IAU Symposium No. 272*

C. Neiner, chair of the SOC for IAU S272 "Active OB stars: structure, evolution, mass loss, and critical limits" presented an update on the plans for the meeting. The venue for the meeting will be Eurosites-Rpublique, 8 bis, rue de la Fontaine au Roi, 75011 Paris. An overview of the scientific program, key dates, and social events can be found on the meeting's website http://iaus272.obspm.fr/.

† The official publication for the IAU Working Group on Active B Stars.

2.3. *Be Star Newsletter*

The *Be Star Newsletter*, which is published in hard copy at Georgia State University for the Working Group on Active B Stars, continues to be the main source of information on new discoveries, ideas, manuscripts, and meetings on active B stars. G. Peters, D. Gies, and D. McDavid continue, respectively, as Editor-in-Chief, Technical Editor, and Webmaster. Abstracts and announcements are usually posted on our website (http://www.astro.virginia.edu/dam3ma/benews/) within 48 hrs of being received. Articles submitted for publication have been refereed since 2000 leading to an improvement in the quality of the *Newsletter*. When we have accumulated about 50 pages of material we finalize an issue and print hard copies that are mostly distributed to libraries worldwide. We encourage researchers to submit material to our Community Comments section that we introduced in 2005 to allow Working Group members to voice opinions or ideas on which the community can submit rebuttal, similar to unedited discussion that is sometime published as part of meeting proceedings. During the past triennium we have published Issues Number 38 (March 2007) and 39 (June 2009). The latter contains the proceedings from the scientific session held during the meeting of the Working Group on Active B Stars at the 26th IAU General Assembly in Prague, Czech Republic.

2.4. *SOC Election Results*

The election to replace the four retiring members of the SOC was held in 2009 July. E-mail ballots were sent to all current IAU members of the Working Group on Active B Stars. The Scientific Organizing Committee (SOC) for the 2009-12 triennium is:

Term expiring in 2012: Karen Bjorkman, Coralie Neiner, Geraldine Peters, Philippe Stee
Term expiring in 2015: Carol Jones, Virginia McSwain, Ronald Mennickent, Richard Townsend
Non-voting: Christopher Corbally (President of IAU Division IV: Stars), Steven D. Kawaler (President of IAU Division V: Variable Stars), Juan FabregatOutgoing SOC Chair)

2.5. *Bylaws for the IAU Working Group on Active B Stars*

A key action item for the triennium 2006-09 was to establish a formal set of bylaws for the Working Group. These were finalized by the SOC prior to this businesses meeting and are presented verbatim below.

I. Nature and Goals of the Working Group

The Working Group on Active B Stars (formerly known as the Working Group on Be Stars) was re-established under IAU Commission No. 29 in 1979 at the 17th IAU General Assembly in Montreal, Canada, and has been in continuous operation to the present. Its main goal is to promote and stimulate research and international collaboration in the field of the active early-type (OB) stars. The focus of the WG was originally on the classical Be stars, but in recent years there has been an increasing contact and overlap with other research areas, particularly in closely aligned topics such as pulsating OB stars and B stars in interacting binaries. The Working Group on Active B Stars is an IAU Inter-Divisional Working Group sponsored by Divisions IV (Stars) and V (Variable Stars).

Our goal is to investigate active phenomena in B-type stars including mass loss and accretion, pulsations, rotation, magnetic fields, and binarity and determine the fundamental parameters for these objects and to promote collaboration and interaction between scientists specializing in these studies.

II. Membership

Membership is open to any scientist working in the field of active B stars, including amateur scientists. IAU membership is required to run for a place on the Scientific Organizing Committee (SOC) and to serve on the SOC. Members receive e-mail updates on news items published in the Be Star Newsletter, the official publication for the Working Group. One can join the Working Group by contacting the webmaster of the Be Star Newsletter.

III. Newsletter and Website

The Be Star Newsletter is the official publication for the Working Group. The Newsletter is published irregularly with a frequency governed by the number of items submitted. News items, including announcements for meetings of interest to the Working Group and abstracts of new papers, and longer articles are submitted to the Editor-in-Chief. All articles are reviewed by anonymous referees who advise the Editor-in-Chief on the suitability for publication. All accepted articles are published on the official website for the Working Group soon after they have been accepted by the Editor-in-Chief. The website is maintained by a web editor. The technical editor is responsible for arranging for the printing and mailing of a paper copy of the Newsletter.

IV. Election of the Scientific Organizing Committee for the Working Group

The duties of the Scientific Organizing Committee (SOC) are to establish scientific policy of the Working Group, set up mechanisms to stimulate the collaboration between its members, and help to prepare proposals for scientific meetings of interest to the Working Group. Each member of the Working Group who is also an IAU member is permitted to nominate four persons as candidates for the new SOC and send them to the Election Officer. The Election Officer selects 10 candidates with the highest number of votes. If several individuals have the same number of nominations for the last spot on the ballot, they are all accepted and the number of candidates will be higher than 10. According to the general IAU rules, only IAU members or new members pending approval at the forthcoming General Assembly can be accepted as candidates for SOC membership Geographically balanced representation should be taken into account in the nomination of the candidates. The Election Officer must verify that all nominees to be listed as candidates on the ballot are willing to serve if elected.

The four new SOC members are elected from the candidates by the members of the Working Group who are members of the IAU. Each IAU member may vote for up to four different persons. The four persons with the largest numbers of valid votes are elected. In the case of a tie, a runoff election will be held. The term of an SOC member lasts for a duration of six years, and begins at the conclusion of the IAU General Assembly in the year he/she is elected. A person must wait for six years, or two IAU General Assemblies, in order to run again for a position on the SOC. The four new and the four continuing SOC members determine among themselves the new Chairperson.

V. Ratification of the Bylaws

The bylaws are ratified by the SOC, and may be amended as per input from the Working Group membership. Ratification is by a simple majority, or a yes vote by at least 5/8 SOC members. Amendments become valid at the conclusion of each IAU General Assembly

3. Scientific program

Session 1 (G. Peters, Chair)

09:30 *CoRoT and the Be stars*
J. Fabregat

10:00 *Watching the growth of a disk: 37 days of H-alpha spectroscopy of HD 168797*
E. Grundstrom

10:15 *Influence of X-ray radiation on wind structure of hot stars*
J. Krticka

10:30 Coffee

Session 2 (J. Fabregat, Chair)

16:00 *Active B stars and the new class of gamma-ray binaries"?*
V. McSwain

16:15 *Magnetism in massive stars*
G. Wade

16:45 *On the incidence of magnetic fields in slowly-pulsating B, β Cephei and B-type emission line stars*
J. Silvester

17:00	*Be stars in the IPHAS and VPHAS+ galactic plane surveys* J. Fabregat
17:10	*Non-radial pulsations in the open cluster NGC 3766* R. Roettenbacher
17:20	*Analysis of B and Be star populations of the double cluster h and chi Persei* A. Marsh
12:30	Session Ends

4. Closing remarks

I would like to thank the speakers who presented a set of excellent talks and all who attended the Working Group meeting. Impressive new information on the nature of B stars and their activity is emerging that could barely be imagined three decades ago when this Working Group was re-established at the XVII IAU General Assembly in Montreal, Quebec. We heartedly thank outgoing SOC chair Juan Fabregat, for his skillful leadership and assistance in preparing the various reports and other documents that were submitted to the IAU General Secretary. We are looking forward to seeing you again in 2012 at the meeting of the Working Group on Active B Stars at the 28th IAU General Assembly in Beijing.

Geraldine J. Peters
Chair of the Working Group, 2009-12

Transactions IAU, Volume XXVIIB
Proc. XXVII IAU General Assembly, August 2009
Ian F. Corbett, ed.

doi:10.1017/S1743921310005090

INTER-DIVISION IV-V WORKING GROUP on Ap and Related Stars

CHAIR	**Margarida S. Cunha**
PAST CHAIR	**Werner Weiss**
BOARD	**Mike Dworetsky**
	Oleg Kochukhov
	Friedrich Kupka
	Francis Leblanc
	Richard Monier
	Ernst Paunzen
	Nikolai Piskunov
	Hiromoto Shibahashi
	Barry Smalley
	Jozef Ziznovsky

PROCEEDINGS BUSINESS SESSION, 7th of August 2009

1. Balance of Activities

The business meeting started at 11h00, in the presence of 18 members, with a brief summary of the activities and achievements of the Working group during the triennium 2006-2009.

Particular emphasis was given to the work done on the renovation of the ApN newsletter (http://ams.astro.univie.ac.at/apn), by the ApN editors, Stefano Bagnulo, Luca Fossati, and Gregg Wade, to respond more adequately to the community needs. The second activity to deserve particular attention was the organization of a *Wish List* on atomic and molecular data, to be iterated during the next triennium in close interaction with Commission 14. Finally, the Chair reported on the main events organized by members of the Working Group, in particular the *CP#Ap* Workshop, that took place in Vienna, Austria, in September 2007, and whose organization was led by Ernst Paunzen, and the Joint Discussion 4 *Progress in understanding the physics of Ap and related stars*, organized by the Working Group, that took place during the General Assembly.

2. Planning of activities for the next triennium

Following on the report by the Chair of the Working Group, there was a general discussion on specific activities to be carried out during the period 2009-2012.

No resolution was made with regards to the organization of future meetings, although the opinion of most members was that the Working Group should aim at organizing a large Symposium in the year following the next General Assembly (i.e., 2013). In the mean time, smaller meetings / workshops are expected to be organized on specific topics by members of the Working Group.

The members have also discussed the continuation of the efforts to strengthen the interaction between the Working Group and Commission 14, on Atomic and Molecular Data.

3. New composition of the Organizing Committee

During the Business Meeting elections were held to elect two new members for the Organizing Committee. The choice of the new Chair was postponed to a later date, having in the mean time been conducted by e-mail. The new composition of the organizing Committee is as follows:

- New Chair: Gautier Mathys (Chile).
- Outgoing OC members: Barry Smalley (UK); Werner Weiss (Austria).
- Incoming OC members: Gautier Mathys (Chile); Olga Pintado (Argentina).

The Chair expressed her gratitude to Barry Smalley and Werner Weiss for their endless contributions to the Working Group over at least three terms, during which they were part of the Organizing Committee. Thanks were also given to all members that helped organizing the activities of the Working Group over the past three years, with particular thanks to the members of the Organizing Committee and the editors of the ApN.

Margarida S. Cunha
Chair of the Working Group

Transactions IAU, Volume XXVIIB
Proc. XXVII IAU General Assembly, August 2009
Ian F. Corbett, ed.

doi:10.1017/S1743921310005107

DIVISION V — VARIABLE STARS (*ETOILES VARIABLES*)

PRESIDENT — **Alvaro Gimenez**
VICE-PRESIDENT — **Steven Kawaler**
PAST PRESIDENT — **Jørgen Christensen-Dalsgaard**
BOARD — **Michel Breger, Edward Guinan, Slavek Rucinski**

PARTICIPATING COMMISSIONS

Commission 27 — **Variable Stars**
Commission 42 — **Close Binary Stars**

DIVISION WORKING GROUPS

Spectroscopic Data Archiving

INTER DIVISION WORKING GROUPS

Division IV / Division V: Active B Stars, Ap & Related Stars

PROCEEDINGS BUSINESS SESSIONS, 5 and 6 August 2009

1. Introduction

Division V organized a brief Business meeting during the XXVIIth General Assembly, prior to Business meetings (reported separately) of Commissions 27 and 42. The Division V Business Meeting began at 11:00 on 5 August 2009.

The Division held a longer science meeting on 6 August 2009, which covered activities of both Commissions; the list of talks in this meeting is given below.

2. Overview of activities and events in the past triennium

The scientic activities in Division V predominantly take place through the two Commissions who both have long traditions and a continued high level of activity. We refer the reader to the triennial reports of the two commissions, prepared in advance of the General Assembly.

The Division and Commission web sites play an important role in the activities of these bodies. We are very grateful to Andras Holl of the Konkoly Observatory for setting up and maintaining these sites.

3. Election of New Officers

The elections for o?cers for the coming triennium resulted in the following Organizing Committee:

- President: Steven D. Kawaler (USA)
- Vice-President: Ignasi Ribas (Spain)
- Past President: Alvaro Giménez (Spain)
- Organizing Committee: Michel Breger (Austria), Edward F. Guinan (USA), Gerald Handler (Austria), and Slavek Rucinski (Canada)

4. Discussion Items

The contribution to the field of variable stars of space missions launched during the trienium (Corot and Kepler), as well as future databases to be provided with experiments such as LSST, was discussed at some length. This prompted many to speculate as to what the meaning of the term "variable star" might be when almost all stars show variability when observed with sufficiently high photometric precision.

It was discussed that, for the coming years, further attention is needed on the application of computer science advances in the exploration of data bases and data mining particularly with respect to automatic, unattended, classification of variable stars of all types. In this new data-intensive era, the limitations of the current GCVS naming of variable stars was pointed out and new alternatives should be explored and discussed sooner than later.

Gerald Handler presented the editorial policy of the *Information Bulletin on Variable Stars* as well as the composition of the new Editorial Board. Activities carried out by the GCVS group to ensure the archiving of all observations leading to the discovery of variable stars was mentioned. Finally, the activities of Michel Breger to restore an IAU archive of variable star data was acknowledged.

5. Division V Science Session

The Division hosted a session of science talks of interest and relevance to its constituent commissions on 6 August, beginning at 9AM. The talks were organized and presented as follows:

- Session 1 (chair: A. Giménez)
  - *The Baker Nunn Patrol Camera variable star survey done at Calgary*, Gene Milone
  - *Epsilon Aurigae this year*, Gene Milone
  - *Accurate stellar masses and radii*, Johannes Andersen
- Session 2 (chair: S. Rucinski)
  - *Probing into the deepest layers of solar-like pulsators*, Margarida Cunha
  - *The* Kepler *Kepler mission variable stars*, Douglas Caldwell
  - *Binary star database*, Oleg Malkov
  - *Activity of the IAU Commission 27 working group on the future of the GCVS*, Nikolai Samus

Alvaro Giménez and Steven Kawaler
President and Vice-President of the Division

Transactions IAU, Volume XXVIIB
Proc. XXVII IAU General Assembly, August 2009
Ian F. Corbett, ed.

doi:10.1017/S1743921310005119

COMMISSION 27 — VARIABLE STARS *(ETOILES VARIABLES)*

PRESIDENT — Steven D. Kawaler
VICE-PRESIDENT — Gerald Handler
PAST PRESIDENT — Conny Aerts
ORGANIZING COMMITTEE — Tim Bedding, Márcio Catelan, Margarida Cunha, Laurent Eyer, Simon Jeffery, Peter Martinez, Katalin Oláh, Karen Pollard, Seetha Somasundaram

PROCEEDINGS BUSINESS SESSIONS, 5 August 2009

1. Overview of the activities and achievements

The meeting started at 14h00. The president welcomed the participants to the business meeting of C27, and acknowledged the members of the Organizing Committee. Special thanks were expressed to the outgoing members: Conny Aerts, Peter Martinez, and Seetha Somasundaram.

After the approval of the agenda, he provided an overview of the activities of C27 of the past three years, rooted in the triennial report of the Commission submitted earlier in the year. The triennial report on Research in Variable Stars and C27 activities initially drafted by the president with voluminous input from the OC members. The OC and president then iterated to be as inclusive as possible while staying within the 6 page limit.

A total of 9 IAU meeting proposals was evaluated by the OC. As in previous years, the OC ranked the proposals as a group whenever possible. The occasional at-the-deadline proposals were treated with care to ensure fairness to those submitted earlier.

2. Election of New Officers

New officers were selected prior to the General Assembly following similar procedures to what had been done in past years. Nominations were solicited from the entire C27 membership, with selection of the final slate to replace outgoing Organizing Committee members and preserve representative balance.

The results were as follows:

- The new president is Gerald Handler (Austria)
- The new vice–president is Karen Pollard (New Zealand)
- Outgoing (retiring) OC members: Conny Aerts (Belgium), Peter Martinez (South Africa), and Seetha Somasundaram (India)
- Continuing OC members: Tim Bedding (Australia), Márcio Catelan (Chile), Margarida Cunha (Portugal), Laurent Eyer (Switzerland), Simon Jeffery (Northern Ireland), Steven Kawaler (ex officio, USA), and Katalin Oláh (Hungary).
- New OC members: S. O. Kepler (Brazil), Katrien Kolenberg (Austria), and David Mkrtichian (Ukraine)

3. Other matters

The meeting concluded with those in attendance anticipating further science discussions related to activities of the C27 members. These were discussed during the full-day science/business session of Division V jointly between C27 and C42 (forming the Division V) on 6 August 2009.

Steven D. Kawaler
President of the Commission

Transactions IAU, Volume XXVIIB
Proc. XXVII IAU General Assembly, August 2009
Ian F. Corbett, ed.

doi:10.1017/S1743921310005120

COMMISSION 42 — CLOSE BINARIES (*ETOILES BINAIRES*)

PRESIDENT	**Slavek M. Rucinski**
VICE-PRESIDENT	**Ignasi Ribas**
PAST PRESIDENT	**Alvaro Giménez**
ORGANIZING COMMITTEE	**Petr Harmanec, Ronald W. Hilditch, Janusz Kaluzny, Panayiotis Niarchos, Birgitta Nordström, Katalin Oláh, Mercedes T. Richards, Colin D. Scarfe, Edward M. Sion, Guillermo Torres, Sonja Vrielmann**

PROCEEDINGS BUSINESS SESSIONS, 5 August 2009

1. The Organizing Committee

During the commission business session, the past President presented the new Organizing Committee which was selected by the OC through a e-mail vote conducted during the months before the Rio de Janeiro General Assembly. The new OC will consist of Ignasi Ribas (President), Mercedes Richards (Vice President), and Slavek Rucinski (Past President) with the members: David Bradstreet, Petr Harmanec, Janusz Kaluzny, Joanna Mikolajewska, Ulisse Munari, Panos Niarchos, Katalin Olah, Theo Pribulla, Colin Scarfe and Guillermo Torres.

2. The matters discussed

Three subjects were discussed during the session:
(1) Ron Samec presented photometric results for W UMa-type binaries with shallow but apparently total eclipses. These properties are interpreted within the context of the contact "Lucy" model as due to extremely low mass ratios of these binaries. The interpretation is based solely on the basis of light curve modeling without any spectroscopic support. The binaries are important as they would exemplify best the wide applicability of the contact model even when the stellar components are very different.
(2) On the very related matter, the Past President summarized recently published results (Pribulla & Rucinski(2008)) for the particularly important and well-known, low mass-ratio system AW UMa. Photometry of this binary gives an excellent determination of the mass ratio in the spirit as above, yet spectroscopy (which resolves the spatial dimension of the radial velocity and thus the correct mass ratio) gives an entirely different picture: The mass ratio is indeed small, but larger than the photometric one by a margin much larger than any formal uncertainty but – more importantly – the binary does not seem to be a contact one! No W UMa-type system was as thoroughly spectroscopically analyzed as AW UMa so we should be prepared for further surprises with other W UMa binaries; its should be noted, however, that V566 Oph does agree with the Lucy model. The case of AW UMa is a serious warning on the mechanical application of light curve synthesis codes to derivation of physical parameters of close binary stars.
(3) The Past President expressed his view that – in general – the situation in the field of light curve solutions is far from satisfactory one: On one hand single – but frequently poorly observed – binaries are analyzed for multitude of parameters with seldom trustworthy determination of uncertainties, on the other hand, thousands – soon millions light curves from massive ground-based and space photometric surveys – are begging for analysis, even simplest characterization.

The Past President plans to prepare a memorandum on the current state of the affairs which will address also the over-production of low usefulness papers; a partial remedy of combining objects into large groups will be suggested.

Other science subjects related to activities of the C42 members were discussed during the full-day business session of the joint C27 and C42 (forming the Division V) during on 6 August 2009.

3. Other matters

The bibliographic notes on Close Binaries (BCB) are being produced under the coordination of Colin Scarfe. The Web pages of C42 and the sister commission C27 within the Division V are maintained at the Konkoly Observatory by Andras Hall.

We note that recently, because of the similarity of techniques and methods – a very large close binary specialists has "migrated" to the field of extras-solar planet detection. Yet, the close binary star expertise is and will be very much needed in interpretation of large amounts of data coming from large ground-based and satellite photometric variability surveys.

Slavek M. Rucinski
President of the Commission

Reference

Pribulla, T. & Rucinski, S. M. 2008, *Mon. Not. Roy. Astr. Soc.* 386, 377

Transactions IAU, Volume XXVIIB
Proc. XXVII IAU General Assembly, August 2009
Ian F. Corbett, ed.

doi:10.1017/S1743921310005132

DIVISION VI INTERSTELLAR MATTER (*MATIÈRE INTERSTELLAIRE*)

PRESIDENT **Tom Millar**
VICE-PRESIDENT **You-Hua Chu**
PAST PRESIDENT **John Dyson**
BOARD **Dieter Breitschwerdt, Mike Burton, Sylvie Cabrit, Paola Caselli, Elisabete de Gouveia Dal Pino, Gary Ferland, Mika Juvela, Bon-Chul Koo, Sun Kwok, Susana Lizano, Michal Rozyczka, Viktor Tóth, Masato Tsuboi, Ji Yang**

PARTICIPATING COMMISSIONS

Commission 34 **Interstellar Matter**

DIVISION WORKING GROUPS

Astrochemistry
Planetary Nebulae
Star Formation

BUSINESS MEETING 10 August 2009

1. Introduction

The business meeting of Division VI was held on Monday 10 October 2009. Apologies had been received in advance from D Breitschwerdt, P Caselli, G Ferland, M Juvela, S Lizano, M Rozyczka, V Tóth, M Tsuboi, J Yang and B-C Koo.

2. Election of Division Officers and New Organising Committee Members

An election was held among the members of the Organising Committee to elect new officers for the triennium 2009–2012. Following nominations, Sun Kwok (China, Hong Kong) was elected as Vice-President and Mike Burton (Australia) elected as Secretary. You-Hua Chu (USA) takes over as President of the Division following her term as Vice-President.

Two members of the Organising Committee, Gary Ferland (USA), and Susana Lizano (Mexico), finished their term of office in 2009 and an election was organised among the Division membership for their replacements. Following an electronic call to the membership for nominations, a list of 20 nominations from 7 countries was circulated by e-mail. Over 200 votes were cast with new members elected for the 2009-2012 triennium being Thomas Henning (Germany) and Neal Evans II (USA).

3. Possible New Working Group

Jacco van Loon (UK) had asked that the OC consider setting up a new Working Group on the Diffuse Interstellar Medium. In principle the OC felt that it would be useful to have such a WG but it broadened the discussion to consider whether the current structure of Division VI, one of the few Divisions that contained one Commission, was fit for purpose. The advantage of having all areas of interstellar medium science contained within one Commission had the disadvantage that it necessitated a very large OC with a subsequent lack of focus. The discussion centred around the possibility that Division VI might be better organised with two or three Commissions that would allow a diverse range of topics to be hosted within each Commission and provide membership for a smaller OC for Division VI.

It was agreed that the new OC for Division VI would consult with the membership with the intention of bring any proposed changes to the next General Assembly.

4. Mailing List

The President had spent a significant amount of time in the past triennium trying to clean up the e-mail addresses of the current membership, significant numbers of whom did not seem to be active in astronomy at present, with many not having published a journal article in the past 10–15 years. The President agreed to update his contacts file once he had been informed by the Secretariat of the details of new members joining the Division at this GA.

5. Website

For several triennia it has been customary that the Division website has resided at the institution of the President with the result that the Division/Commission has several extant websites. It was agreed that significant work was required to update the website and to remove heritage sites from the web. An offer from Dr V Tóth to provide support for this activity was noted with gratitude.

6. Closing remarks

The President would like to thank the officers and members of the Organising Committee for their dedication in supporting the work of Division VI and, in particular, for all their efforts in ranking the large number of proposals for IAU Symposia, Joint Discussions and Special Sessions, particularly those associated with this GA, the results of which were clearly evident in the high-quality meetings currently underway in Rio de Janeiro.

Tom Millar
President of the Division
Queen's University Belfast

Transactions IAU, Volume XXVIIB
Proc. XXVII IAU General Assembly, August 2009
Ian F. Corbett, ed.

doi:10.1017/S1743921310005144

DIVISION VI WORKING GROUPS

Division VI WG	**Astrochemistry**
Division VI WG	**Star Formation**
Division VI WG	**Planetary Nebulae**

1. Working Group Business Meetings

1.1. *Planetary Nebulae*

During the last business meeting of the WG on PN during the GA09, several topics were addressed. The renovation of the WG members was discussed. We agreed that members will be renewed during the PN IAU symposiums, and that the maximum duration of the term will be 10 years. It was decided that the next IAU PN Symp will be held in Puerto de la Cruz, Tenerife, in 2011. It seemed appropriate to hold the symposium in the Canary Islands in recognition of the new 10.4m GTC telescope recently inaugurated on La Palma and its importance to the field of planetary nebulae.

The publication of links of interest in the field of PN on the PNWG webpage (http://www.iac.es/proyecto/PNgroup/) was also discussed.

1.2. **Organising Committee:**

A Manchado (Spain, Chair), M Barlow (UK), R Corradi (Spain), Y-H Chu (USA), S Deguchi (Japan), A Frank (USA), G Jacoby (USA), S Kwok (China), A López (Mexico), W Maciel (Brazil), R Méndez (USA), Q Parker (Australia), D Schoenberner (Germany), L Stanghellini (USA), A Zijlstra (UK)

Transactions IAU, Volume XXVIIB
Proc. XXVII IAU General Assembly, August 2009
Ian F. Corbett, ed.

doi:10.1017/S1743921310005156

DIVISION VII — THE GALACTIC SYSTEM (*SYSTEME GALACTIQUE*)

PRESIDENT	**Ortwin Gerhard**
VICE-PRESIDENT	**Despina Hatzidimitriou**
PAST PRESIDENT	**Patricia A. Whitelock**
BOARD	**Charles J. Lada, Ata Sarajedini, Rosemary F. Wyse, Joseph Lazio**

PARTICIPATING COMMISSIONS

Commission 33	**Structure and Dynamics of the Galactic System**
Commission 37	**Star Clusters and Associations**

DIVISION WORKING GROUPS

The Galactic Center

PROCEEDINGS BUSINESS SESSION, 7 August 2009

1. Introduction

Division VII provides a forum for astronomers studying the Milky Way Galaxy and its constituents. Several meetings directly relevant to his subject were held at the General Assembly in Rio: IAU Symp. 262 *Stellar Populations*, IAU Symp. 265 *Chemical Abundances in the Universe*, IAU Symp. 266 *Star Clusters*, Joint Discussion 5 *Modeling the Milky Way in the Era of Gaia*, and Special Session 5 *The Galactic Plane*. Division VII therefore did not organize a separate science session at Rio, but business meetings were held for both the Division and for Commissions 33 and 37.

2. Business Meeting

The brief Division business meeting was held on 7 August 2009. The president briefly described to those present the developments in the previous months. The suggestions for members of the new Board (see below) had already been approved by the outgoing Board via email as several of its members could not come to Rio. The Working Group on the Galactic Centre will be discontinued in the new triennium. The reports of Division VII, its Commissions 33 and 37, and Working Group over the period 2006-20060 can be found in *Reports on Astronomy, Transactions IAU Volume XXVIIA* (2009, Ed. K. v.d. Hucht). The activities of the Division primarily involve sponsoring IAU Symposia.

3. New Division Board

The established practice of Division VII is for the Board to comprise the Presidents, Vice-Presidents and Past Presidents of Commissions 33 and 37. This process will be continued.

The past practice of the Division has been for the Presidency to alternate between Commissions 33 and 37 from one triennium to the next. Within these terms the new President would be from Commission 37. The new Board is therefore as follows:

- Giovanni Carraro (Italy, Vice-President Commission 37)
- Bruce G. Elmegreen (USA, President Commission 37)

- Ortwin Gerhard (Germany, Past President Commission 33)
- Despina Hatzidimitriou (President, Greece, Past President Commission 37)
- Birgitta Nordström (Denmark, Vice-President Commission 33)
- Rosemary F. Wyse (Vice-President, USA, President Commission 33)

Ortwin Gerhard
President of the Division

Transactions IAU, Volume XXVIIB
Proc. XXVII IAU General Assembly, August 2009
Ian F. Corbett, ed.

doi:10.1017/S1743921310005168

COMMISSION 33

STRUCTURE AND DYNAMICS OF THE GALACTIC SYSTEM

(STRUCTURE ET DYNAMIQUE DU SYSTEME GALACTIQUE)

PRESIDENT **Ortwin Gerhard**
VICE-PRESIDENT **Rosemary F. Wyse**
PAST PRESIDENT **Patricia A. Whitelock**
ORGANIZING COMMITTEE **Yuri N. Efremov, Wyn Evans, Chris Flynn, Jonathan E. Grindlay, Birgitta Nordström, Chi Yuan**

PROCEEDINGS BUSINESS SESSION, 7 August 2009

1. Introduction

Commission 33 held its business session in the afternoon of Friday 7 August 2009. The president briefly described developments during the previous months. A short discussion followed about the activities of the Commission and whether it should play a more active role in view of the many on-going surveys relevant to the subject of the Galaxy.

2. New Commission Organizing Committee

The new organizing committee for the Commission as determined earlier by the present organizing committee was reviewed and confirmed at the meeting. The members of the new organizing committee are:

- President: Rosemary F. Wyse (USA)
- Vice-President: Birgitta Nordström (Denmark)
- Jonathan Bland-Hawthorn (Australia)
- Sofia Feltzing (Sweden)
- Burkhard Fuchs (Germany)
- Takuji Fujimoto (Japan)
- Ortwin Gerhard (Past President, Germany)
- Dante Minniti (Chile)

Ortwin Gerhard
President of the Commission

Transactions IAU, Volume XXVIIB
Proc. XXVII IAU General Assembly, August 2009
Ian F. Corbett, ed.

doi:10.1017/S174392131000517X

COMMISSION 37 — STAR CLUSTERS AND ASSOCIATIONS *(STAR CLUSTERS AND ASSOCIATIONS)*

PRESIDENT	**Bruce Elmegreen**
VICE-PRESIDENT	**Giovanni Carraro**
PAST PRESIDENT	**Despina Hatzidimitriou**
ORGANIZING COMMITTEE	**Richard de Grijs, Dante Minniti, Charles Lada, Gary Da Costa, Monica Tosi, Licai Deng, Young-Wook Lee, Ata Sarajedini**

PROCEEDINGS OF THE BUSINESS SESSION, 11 August 2009

1. Introduction

The business session for Commission 37 was held on 11 August 2009 at the IAU General Assembly in Rio de Janeiro. The meeting was attended by about a dozen members of our Comission, including President Elmegreen, VP Carraro and several committee members. We introduced ourselves and then went through a powerpoint presentation first prepared by outgoing President Hatzidimitriou and revised by incoming President Elmegreen. The contents of the powerpoint presentation are given in this summary.

In what follows, we list past, present and future meetings, publications statistics and important surveys, reviews, and databases about clusters, and then we discuss the procedure for the election of new commission officers.

2. Past Meetings

- Chemical Evolution of Dwarf Galaxies and Stellar Clusters, MPA/ESO/USM/MPE 2008 Joint Astronomy Conference, July 21-25, 2008, Garching (Germany)
- Low-Metallicity Star Formation: From the First Stars to Dwarf Galaxies, IAU Symposium 255, June 16-20, 2008, Rapallo, Liguria (Italy)
- The Galaxy Disk in Cosmological Context, IAU 254 Symposium, June 9-13 2008, Copenhagen (Denmark)
- Nuclear Star Clusters across the Hubble Sequence, February 25-27, 2008, MPI for Astronomy, Heidelberg (Germany)
- Modeling Dense Stellar Systems (MODEST)-8 meeting, December 5-8, 2007, Bonn/Bad Honnef (Germany)
- Young massive star clusters. Initial conditions and environments, September 11-14 2007, Granada (Spain)
- Dynamical Evolution of Dense Stellar Systems, IAU Symposium 246, September 5-9, 2007, Capri (Italy)
- 12 Questions on Star and Massive Star Cluster Formation, ESO workshop, July 3-6, 2007, Garching (Germany)
- Milky Way Halo Conference, 29 May - 2 June 2007, Bonn (Germany)
- Structure formation in the Universe, 27 May - 1 June, 2007, Chamonix (France)
- Galactic & Stellar Dynamics in the Era of High Resolution Surveys, March 16-18, 2007, Strasbourg (France)

• The Dynamics of Star Clusters and Star Cluster Systems, 6-8 November, 2006, Sheffield (UK)
• Modeling Dense Stellar Systems MODEST-7 meeting, IAU GA, 17 - 23 August, 2006, Prague (Chech R.)
• IAU Symposium No. 235, Evolution of Galaxies across the Hubble Time, 14 - 17 August 2006, Prague (Chech R.)
• IAU Symposium No. 237, Triggered Star Formation in a turbulent ISM, 14-18 August, 2006, Prague (Chech R.)
• International School on Galactic and Cosmological N-Body Simulations, July 23 - August 5, 2006, Tonantzintla - Puebla (Mexico).
• Cambridge N-body School, July 30 - August 11, 2006, Cambridge (UK)
• Mass Loss from Stars and the Evolution of Stellar Clusters Workshop, May 29-June 1, 2006, Lunteren (The Netherlands)

3. Meetings at the IAU General Assembly

• Star Clusters: Basic Galactic Building Blocks throughout Time and Space, IAU Symposium 266, 10-14 August 2009, Rio de Janeiro
• Chemical Abundances in the Universe – from stars to planets, IAU Symposium 265, 10-14 August 2009, Rio de Janeiro
• Stellar Populations – Planning for the next decade, IAU Symposium 262, 3-7 August 2009, Rio de Janeiro

4. Upcoming Meetings

• From Stars to Galaxies: Connecting our understanding of star and galaxy formation. *Date:* Wednesday, 7 April 2010 - Saturday, 10 April 2010. *Location:* Gainesville, Florida, USA
• IAU Symposium 270: Computational Star Formation. *Date:* Monday, 31 May 2010 - Friday, 4 June 2010. *Location:* Barcelona, Spain
• EPoS 2010 The Early Phase of Star Formation. *Date:* Monday, 14 June 2010 - Friday, 18 June 2010. *Location:* MPG Conference Center Ringberg Castle, Germany
• The Multi-Wavelength View of Hot, Massive Stars (39th Lige International Astrophysical Colloquium). *Date:* Monday, 5 July 2010 - Friday, 9 July 2010. *Location* Lige, Belgium
• MODEST10, August 29-Sept 3, 2010, Beijing, China
• Great Barriers in High Mass Star Formation. *Date:* Monday, 13 September 2010 - Friday, 17 September 2010. *Location:* Townsville, North Queensland, Australia
• Guillermo Haro Workshop, G. Tenorio-Tagle. *Location:* Puebla Mexico

5. Publications

The topic of star clusters and associations continues to be one of the most widely followed in all of astronomy. It spans the range of interest from stellar properties, to stellar clusters, to star formation and evolution, with considerable overlap in other commissions.

Publications in Refereed Journals in the period from January 2006 to September 2008 tally as follows:
• Globular Clusters: 420 papers (by the Fall of 2008, more than 3100 citations)
• Open Clusters: 270 papers (more than 665 citations)
• Stellar Associations: 30 papers (more than 140 citations)

Some of the **issues** addressed in these publications are:
• the formation and dynamical evolution of star clusters
• stellar evolution and ages
• star clusters as tracers of stellar populations
• studies of specific types of objects within clusters
• nuclear clusters
• extragalactic cluster systems

The authors utilize observations covering an increasing portion of the electromagnetic spectrum, ranging from X-rays to the far-infrared, as well as advanced N-body simulations.

Reviews related to star clusters that appeared in the bibliography from 2006 to September 2008 are:

• A review on the use of globular cluster systems as tracers of galaxy formation and assembly, by Jean P. Brodie and Jay Strader, Annual Reviews of Astronomy and Astrophysics, Vol. 44, Issue 1, pp.193-267 (2006)

• A review on the Evolution of Star Clusters was published by Richard de Grijs and Genevieve Parmentier in the Chinese Journal of Astronomy and Astrophysics, Vol. 7, p 155 (2007)

• Additional reviews regarding a variety of aspects of star cluster research have also appeared in the proceedings of the conferences and meetings mentioned earlier

6. DataBases

Several new cluster **catalogues** have been published:

• M33: Sarajedini, Ata and Mancone, Conor L. 2007, AJ, 134, 447

• ACS survey of Milky Way globulars: Jay Anderson *et al.*, 2008, AJ, 135, 2055

• M31: Sang Chul Kim *et al.* 2007, AJ, 134, 706

Recent **databases** on clusters are:

• Data on Open Clusters in the Milky Way and the Magellanic Clouds can be found in the WEBDA site (http://www.univie.ac.at/webda/), which was originally developed by Jean-Claude Mermilliod from the Laboratory of Astrophysics of the EPFL (Switzerland) and is now maintained and updated by Ernst Paunzen from the Institute of Astronomy of the University of Vienna (Austria).

• Data on Galactic Globular Clusters can be found in the "Catalog of parameters for Milky Way globular clusters" by W.E. Harris (http://www.physics.mcmaster.ca/Globular), as well as in "The Galactic Globular Clusters Database" at Astronomical Observatory of Rome (INAF-OAR: http://venus.mporzio.astro.it/ marco/gc/).

• A Catalogue of Variable Stars in Globular Clusters developed and maintained by Christine Clement can be found in http://www.astro.utoronto.ca/ cclement/.

7. Ballot Procedure for Election of New Officers

An election of new officers was held in March and April 2009 by the outgoing president, vice president, and organizing committee.

Balloting was carried out for the first time. A call for nominations was sent by email. A list of Commission 37 members who were nominated and who accepted the nomination was then put to the vote. The voting took place via email. 82 members of the commission voted. The votes were counted by D. Hatzidimitriou (P) and C. Lada (VP). Not everyone in the Commission could be contacted as 20% of the members had obsolete e-mail addresses.

Several issues came up during the election. The nominating procedure and election were done by email and so the votes were not anonymous. The collation of votes took a lot of housekeeping work. We recommend using a centrally controlled (e.g., IAU Secretariat) user-friendly and secure web site. This would be simpler to manage, ensure anonymity for the voting process, and would be uniform across all commissions.

A number of current organizing committee members were kept in the OC at this time in order to maintain continuity. We expect to replace the longer-term members of the OC in 2012.

Several people who were nominated were not commission members. We will try to get people who are active in stellar cluster research to become members. Perhaps we will prepare a formal letter of invitation by the OC president every three years.

We attempted to maintain a balance in the Organising Committee with respect to theory versus observations, gender and geographical location. This was difficult to achieve by the nomination and voting procedure that was followed. For example, only two women were placed on the ballot. A few others that were nominated either did not accept or were not members of the commission. In any case, none of the women on the ballot were voted in. We should make a concerted effort to nominate more women in the future.

We would have found it useful to have a list of previous members of the Organizing Committee, including previous presidents and vice-presidents. This list could be on the IAU website. Such a list is important because IAU rules do not allow people to serve for more than one term for president and vice-president, or a maximum of two terms for OC members.

We will be attempting to upgrade our Commission web site. We would like to include an updated list of members of the Commission, and a list of former Organizing Committee members, Presidents and Vice-Presidents. We also want updated links to websites of conferences endorsed by the Commission. We should also create a site that is accessible to the Organizing Committee only, to facilitate OC business management (e.g. evaluation of conference proposals). In addition, we want to create a secure site for voting and nominations that all commission members can access.

8. Closing Remarks

Our discussion at the meeting centered on improving communication between members. This includes getting a more complete list of email addresses, and regularly sending out notices of meetings, reviews, databases, and other material of interest to Commission 37 members. We note that the email newsletter, SCYON (edited by H. Baumgardt, E. Paunzen and P. Kroupa), is very successful at spreading important information and research news.

We handed out questionnaires at the Symposium "Star Clusters: Basic Galactic Building Blocks throughout Time and Space, IAU Symposium 266" to get contact information from as many people as possible who were present at the meeting. VP Carraro is collating these and will work toward updating our Commission 37 web page.

Bruce G. Elmegreen
President of the Commission

Transactions IAU, Volume XXVIIB
Proc. XXVII IAU General Assembly, August 2009
Ian F. Corbett, ed.

doi:10.1017/S1743921310005181

DIVISION VIII GALAXIES AND THE UNIVERSE
(LES GALAXIES ET l'UNIVERS)

PRESIDENT **Sadanori Okamura**
VICE-PRESIDENT **Elaine Sadler**
PAST PRESIDENT **Francesco Bertola**
BOARD **Mark Birkinshaw, Françoise Combes, Roger L. Davies, Thanu Padmanabhan, Rachel Webster**

PARTICIPATING COMMISSIONS
Commission 28 **Galaxies**
Commission 47 **Cosmology**

DIVISION WORKING GROUPS
Supernovae

PROCEEDINGS BUSINESS SESSIONS, 12 August 2009

1. Introduction

The business meeting of Division VIII was held on 12 August 2009 14:00-15:00 in room R2.6 together with the business meeting of Commissions 28 and 47.

2. Reports of the Presidents

Division VIII President, Sadanori Okamura, mentioned the conventional rule by which the Organizing Committee (OC) members of Division VIII were elected. The committee consists of the President who is normally promoted from the previous Vice President, Vice President elected by the previous OC members, past President, and five more people, i.e., Presidents and Vice Presidents of the two participating Commissions and a Webmaster.

According to the conventional rule and elections made within the Commissions, the new OC members of Division VIII are the following: Elaine Sadler (President), Françoise Combes (Vice President), Sadanori Okamura (past President), Roger L. Davies (C28 President), Thanu Padmanabhan (C47 President), J. S. Gallagher (C28 Vice President), B. Schmidt (C47 Vice President). There will not be a webmaster any longer and a Secretary will be appointed later by the President.

The President summarized the 2006-2009 triennial report of the Division, which can be found in *Reports on Astronomy, IAU Transactions Volume XXVIIA* (2009, Ed. K. A. van der Hucht), quoting the concluding sentence in the report; "Division VIII, the largest division (with 1544 members), will continue to contribute to our understanding of the properties of galaxies, the formation and evolution of galaxies and large scale structure, and the content and fate of the Universe."

Then, President of Commission 28, Françoise Combes, reported the activity of Commission 28 during 2006-2009 triennium and reported the names of the new OC members. President, Rachel Webster, and Vice President, Thanu Padmanabhan, of Commission 47, could not attend the GA, and Luis Campusano, on behalf of Webster, reported the activity of Commission 47 during 2006-2009 triennium and reported the names of the new OC members.

3. Coordination of Large-Scale Surveys

Mark Birkinshaw, Webmaster, sent a proposal to the President that it would be useful if some kind of coordination is made on large scale surveys because there are currently many preferred survey areas being used, with less overlap in the different wavebands than would be optimal. The President introduced a statement made in the Executive Committee (EC) meeting of IAU officials with Division Presidents on 2 August that an IAU-maintained web page of various large-scale surveys would be useful. He also mentioned that there was a WG on Sky Surveys in Division IX. After some discussion, we decided to make effort for this coordination and handed the job to the incoming President and OC.

4. New IAU Rules

The President explained the new IAU rules, especially those on the election of President, Vice President, and OC members. They became effective at XXVI General Assembly at Prague but unfortunately they were not well disseminated among Presidents until fall 2008. The President also explained that the EC generously allowed us to follow the conventional rule in the election this time and that we should follow the new rule in the next election.

Several points were raised in the folowing discussion.

- The rule says, "Before each General Assembly, the Organizing Committee shall organize an election from among the membership, by electronic or other means suited to its scientific structure, of a new Organizing Committee to take office for the following term. Election procedures should, as far as possible, be similar among the Divisions and require the approval of the Executive Committee."
- We understand that the EC wishes more involvement of the IAU members at large and a more democratic procedure by introducing voting by all members.
- The long-term nature of the commitment required for activity on IAU Commission and Division Boards (normally at least 6 years, and could possibly be as long as 12 years) is somewhat daunting.
- Not all the people are willing to accept the job.
- Our Division has the largest number of members ($\sim$1500; $\sim$1000 in Commission 28, $\sim$650 in Commission 47).
- The OC needs good geographical distribution and gender balance.

We reached the conclusion that blind voting and simple majority rule may not work well. Accordingly, we need a well-designed election system in accord with the EC's intention. This is an important and immediate task to be tackled by the incoming President and OC.

Finally, incoming President, Elaine Sadler, spoke of her hopes and expressed her thanks to the outgoing President and OC members.

Sadanori Okamura
President of the Division

Transactions IAU, Volume XXVIIB
Proc. XXVII IAU General Assembly, August 2009
Ian F. Corbett, ed.

doi:10.1017/S1743921310005193

DIVISION IX OPTICAL & INFRARED TECHNIQUES

PRESIDENT	Andreas Quirrenbach
VICE-PRESIDENT	David R. Silva
PAST PRESIDENT	Rolf-Peter Kudritzki
ORGANIZING COMMITTEE	Michael G. Burton
	Xiangqun Cui
	Ian S. McLean
	Eugene F. Milone
	Jayant Murthy
	Stephen T. Ridgway
	Gražina Tautvaišiene
	Andrei A. Tokovinin
	Guillermo Torres

PARTICIPATING COMMISSIONS

Commission 21	Galactic and Extragalactic Background Radiation
Commission 25	Stellar Photometry & Polarimetry
Commission 30	Radial Velocities
Commission 54	Optical & Infrared Interferometry

DIVISION WORKING GROUPS

Division IX WG	Site Testing Instruments
Division IX WG	Sky Surveys

INTER DIVISION WORKING GROUPS

Division IX-X WG	Encouraging the International Development of Antarctic Astronomy
Inter-Div. IX-X-XI WG	Astronomy from the Moon

PROCEEDINGS BUSINESS SESSION, 14 August 2009

1. Introduction

Division IX provides a forum for astronomers engaged in the planning, development, construction, and calibration of optical and infrared telescopes and instrumentation, as well as observational procedures including data processing. A few years ago, discussions were started about changes in the structure of Division IX, with the aim of bringing it more in line with today's world of large coordinated projects and multi-national observatories. The course of this process, and further steps to be taken in the period from 2009 to 2012, were at the focus of the deliberations at the business meeting of Division IX at the IAU General Assembly in Rio de Janeiro.

2. Re-Structuring of Division IX

The introduction of new directions for Division IX started with the creation of a Working Group on Optical and Infrared Interferometry, which meanwhile has become Commission 54. In preparation for the General Assembly, this commission has informed a large number of astronomers active in optical/infrared interferometry about the commission's activities, and encouraged them to become members of the IAU (if they weren't already) and of Commission 54.

At the General Assembly, the wish of Commission 21 to move from Division III to Division IX, and to change its name to "Galactic and Extragalactic Background Radiation" was approved by the Executive Committee. These changes reflect the current interests and research activities of the majority of the members of the commission, and should help the commission to attract new members.

Several important topics that should fall under the purview of Division IX are currently not covered by the commissions and working groups of the division. It is therefore foreseen to establish new working groups on large telescopes, on medium-size and small telescopes, and on adaptive optics, respectively.

An open question concerns the role of optical and infrared astronomy from space within Division IX. Discussions with the relevant community and with Division XI will be needed to define the appropriate distribution of tasks between the two divisions.

3. Support of the Instrumentation Community by Division IX

Among the important tasks of Division IX are facilitating professional contacts between astronomers interested in optical and infrared instrumentation, and representing the interests of this group within the wider astronomical community. It was noted that the IAU sponsors relatively few symposia with a focus on instrumentation. The largest conferences on astronomical instrumentation are organized on a bi-annual basis by SPIE. These serve many IAU members well, but for others the format is not ideal, and for some the cost of attending may be prohibitively high. Division IX should thus try to prepare suggestions how the IAU could take a stronger role in the organization and sponsorship of instrumentation-related conferences.

Many astronomers engaged in the development of state-of-the-art instrumentation spend a large fraction of their time on the design and construction of instruments, and on managing their teams. Completed instruments are frequently handed over to observatories, giving the original builders only a limited role in their scientific exploitation. As a consequence, the accomplishments of instrument builders are in many cases not reflected in their publication records, which can potentially harm their prospects in hiring and review processes. It was proposed that Division IX should assess this situation, and propose improvements if appropriate.

Andreas Quirrenbach
President of the Division

Transactions IAU, Volume XXVIIB
Proc. XXVII IAU General Assembly, August 2009
Ian F. Corbett, ed.

doi:10.1017/S174392131000520X

COMMISSION 25 — STELLAR PHOTOMETRY AND POLARIMETRY

(STELLAR PHOTOMETRY AND POLARIMETRY)

PRESIDENT	Peter Martinez
VICE-PRESIDENT	Eugene Milone
PAST PRESIDENT	Arlo Landolt
ORGANIZING COMMITTEE	Carme Jordi
	Aleksey Mironov
	Qian Shenbang
	Edward Schmidt
	Christiaan Sterken

PROCEEDINGS BUSINESS SESSIONS, .. August and .. August 2009

1. Introduction

The Business Meeting for Commission 25 was held on the 6th of August 2009. The meeting was chaired by Dr Eugene Milone, Vice President for the 2006-2009 triennium, and incoming President for the 2009-2011 triennium. Dr Milone presented an apology from the President of the Commission, Dr Peter Martinez, who was unable to attend the meeting.

2. Developments in the 2006-2009 triennium

The Commission had presented reports of developments in photometry and polarimetry in the IAU Highlights. This was briefly reviewed. Two particular highlights that were referred to were two important conferences organised by members of the Commission on (i) the the history of photometry, *Photometry: Past and Present,* organised by Eugene Milone, and (ii) the conference *Astronomical Polarimetry 2008 – Science from Small to Large Telescopes,* organised by Pierre Bastien, which took place from 6 to 11 July 2008 in Quebec.

3. Membership

According to the IAU membership database, prior to the start of the 2009 General Assembly, Commission 25 had 230 members from 40 countries. The Commission's membership represented 2.4% of the total IAU membership of 9658.

During the 2006-2009 triennium considerable effort had been devoted to updating the Commission's membership records. This was done by contacting members individually and requesting them to confirm or update their personal details. Many of the members' details were outdated or incorrect. The updated lists will be sent to the IAU Headquarters in Paris.

It was noted that a number of applicants for IAU Membership had listed Commission 25 as one of their preferred Commissions, should they be admitted to the IAU at the 2009 General Assembly. The names of these applicants would be circulated to the Organising Committee for approval as soon as the lists were received from the IAU Headquarters in Paris.

In 2007, the Commission's website was moved from its previous host site at the Vrije Universiteit Brussel to the South African Astronomical Observatory. The URL of the new Commission 25 website is `iau_c25.saao.ac.za`. We thank Dr Christiaan Sterken for having established the Commission's website and for having maintained it for a number of years. The Commission plans to change the C25 website to bring it more in line with the "look and feel" of the official IAU website.

One of the functions of the upgraded C25 website could be to serve as an IAU Photometry and Polarimetry Standards Portal. This matter will be taken up by the Organising Committee of C25 in the coming triennium.

4. Calibration of wide-band filters in use on the HST

The meeting discussed a communication received from Ivan King to make the members of Commission 25 aware of a calibration problem that has so far lacked a solution.

> *"Because of its greater throughput, HST's F606W filter has been used by many observers when the aim has been to go as faint as possible. In the field of star clusters (especially globulars) F606W poses a particular problem of calibration, because its photometric behavior is so bad. Whereas F555W (whose throughput is rather less) is designed to resemble the Johnson V, F606W differs from F555W in ways that are both non- linear (with a sharp bend that cannot be represented by a quadratic term in color) and strong dependence on metallicity.*
>
> *The relationship of F606W to normal photometric systems can be studied by empirical comparisons and by synthetic photometry, and this has been done, to the extent possible, in the photometric standard paper by Sirianni et al. (PASP 117, 1049, 2005), and also in a brief note by myself (King & Anderson, MemSAIt 72, 685, 2001).*
>
> *Unfortunately existing calibration work does not cover the range of metallicity at all well (e.g., nothing at all for the open cluster NGC 6791, with [Fe/H] +0.4), nor does it extend to the very bottom of the main sequence, which has been reached with F606W in several clusters (NGC 6397. M4, Omega Cen, 47 Tuc, NGC 6791). In this faintest range of absolute magnitudes no calibration of F606W exists, because no empirical comparisons have been made, nor do spectrophotometric curves exist that could be used in synthetic photometry."*

In subsequent email discussion on the issue A. T. Young remarked that these non-overlapping passbands are not transformable because the basic information needed for transformations isn't observed. In other words, this system is non-transformable by design.

5. Election of new office bearers for the triennium 2009-2011

The elections for the new officers of Commission 25 took place during May 2009. A total of 221 votes were received from members of the Commission. The results were as follows:

President: Eugene Milone (Canada)
Vice-President: Alistair Walker (Chile - CTIO)
Past President: Peter Martinez (South Africa)
Scientific Organising Committee
Barbara Anthony-Twarog (USA)
Pierre Bastien (Canada)
Jens Knude (Denmark)
Don Kurtz (UK)
John Menzies (South Africa)
Aleksey Mironov (Russian Federation)
Qian Shengbang (China)

6. Closing remarks

As outgoing President of the Commission, I would like to thank the Organising Committee and Members of Commission 25 for all the support, guidance and advice I received during my term of office. In particular, I would like to thank Arlo Landolt and Chris Sterken, who have served the Commission for many, many years. I often consulted them during my term of office and I hope they will still be able to contribute their experience and wisdom to the future leadership of the Commission. Finally, I wish to convey my best wishes to the incoming President, Eugene Milone, the incoming Vice-President, Alistair Walker, and the new SOC. The Commission is in excellent hands for the coming triennium.

Peter Martinez
President of the Commission

Transactions IAU, Volume XXVIIB
Proc. XXVII IAU General Assembly, August 2009
Ian F. Corbett, ed.

doi:10.1017/S1743921310005211

DIVISION IX / COMMISSION 25 / WORKING GROUP INFRARED ASTRONOMY

PRESIDENT **Eugene F. Milone**
VICE-PRESIDENT **Andrew T. Young**
MEMBERS **Roger A. Bell, Michael Bessell, Richard P. Boyl, Martin Cohen, David J.I. Fry, Robert Garrison, Robert Garrison, Ian S. Glass John Graham, Anahi Granada, Lynn Hillenbrand, Robert L. Kurucz, Ian McLean, Matthew Mountain, George Riecke, Rogerio Riffel, Ronald G. Samec, Stephen J. Schiller, Douglas Simons, Michael Skrutskie, C. Russell Stagg, Christiaan L. Sterken, Roger I. Thompson, Alan Tokunaga, Kevin Volk, Robert Wing.**

PROCEEDINGS BUSINESS SESSIONS, 7 August 2009

1. Introduction

The formal origin of the IRWG occured at the Buenos Aires General Assembly, following a Joint Commission meeting at the IAU GA in Baltimore in 1988 that identified the problems with ground-based infrared photometry. The situation is summarized in Milone (1989). In short, the challenges involved how to explain the failure to achieve the milli-magnitude precision expected of infrared photometry and an apparent 3% limit on system transformability. The proposed solution was to redefine the broadband Johnson system, the passbands of which had proven so unsatisfactory that over time effectively different systems proliferated, although bearing the same *JHKLMNQ* designations; the new system needed to be better positioned and centered in the atmospheric windows of the Earth's atmosphere, and the variable water vapour content of the atmosphere needed to be measured in real time to better correct for atmospheric extinction.

The IRWG established criteria for judging the performance of existing infrared passbands and experimented with passband shapes, widths, and placements within the spectral windows of the Earth's atmosphere. The method and coding were initiated and largely carried out by A. T. Young, and, aided by C. R. Stagg, Milone ran the simulations. The full details of the criteria and results of the numerical simulations were presented by Young *et al.* (1994). Subsequent work, described in WG-IR and/or Commission 25 reports, included the use of a newer MODTRAN version (3.7) to check and extend previous work. This part of the program proved so successful in minimizing the effects of water vapour on the source flux transmitted through the passband that the second stage, real-time monitoring of IR extinction, was not pursued, although this procedure remains desirable for unoptimized passbands designed for specific astrophysical purposes.

During the following triennia, the WG concentrated on gathering and presenting evidence of the usefulness of the IRWG infrared passband set. For the near infrared portion of the IRWG set (namely, the *iz*, *iJ*, *iH*, *iK* passbands), field trials were conducted over the years 1999–2003 with an InSb detector in a Dewar mounted on the 1.8-m telescope at the Rothney Astrophysical Observatory of the University of Calgary. The results of those trials and the details of further work were presented in Milone & Young (2005). This paper contained, for the first time, evidence that not only were the IRWG passbands more useful to secure precise transformations than all previous passbands, but that they were also superior in at least one measure of the signal to noise ratio. This evidence was further refined in Milone & Young (2007). As a consequence, the original purpose of the IRWG largely has been achieved, but resistance to the new passband system is still strong, and passbands that somewhat compromise the IRWG recommendations have been advanced in order to provide more throughput, at the cost of precision and standardization. Thus, nonoptimized passbands are still in use at the highest altitude infrared sites. The situation is described (and decried!) in Milone & Young (2007). It would be incorrect to conclude, however, that the community has ignored the work of the IRWG. As noted in previous IRWG reports, there is now a general acceptance of the principles enunciated in Young *et al.* (1994), and incremental movement is occurring. However, the sharp break from the nomenclature of the Johnson passbands has not not been accepted, as the papers below illustrate. Many Infrared astronomers continue to use "$JHKL$" designations even though there is demonstrably no single passband system with those designations. The IRWG suggests that the designations be assigned instead to the atmospheric windows most prominently associated with the original Johnson passbands, as done in Milone & Young (2005). The Mauna Kea near-IR suite of passbands that has been described by Tokunaga & Vacca (2007), is one of the best of the incrementally improved systems, but falls short of the IRWG specifications.

2. IRWG Science and Business Session Meeting

The meeting of the WG in the Rio de Janeiro GA was held during Session 2 on Aug. 7, 2009, and was chaired by Milone. The history of the IRWG was summarized, and an invitation for all photometrists with any interest in the IR to join was issued. The IRWG has had since its inception an open policy of membership, and this invitation was extended at the present GA.

Recent highlights of the work of the IRWG were described. In Milone & Young (2005) and further in Milone & Young (2007), correlations were seen among: our figure of merit, θ, a measure of the distortion of the spectral irradiance of starlight as it descends the Earth's atmosphere; a measure of the Forbes effect (the rapid change in slope of the extinction curve with decreasing airmass); the extinction coefficient between 1 and higher airmasses; and a measure of the signal to noise ratio. Perhaps the most useful contribution in the past triennium was given in Milone & Young (2008), which argues for the suitability of the IRWG passbands to provide millimagnitude precision for variable star infrared photometry, and that this is possible at ANY photometric site irrespective of its elevation. Directed to both professional and amateur astronomical communities, this paper compares extinction coefficients obtained using a sample of old IR filters with those using the IRWG passbands determined from the same night at the RAO. The coefficients for the old passbands are seen to be greater by factors ~ 2 or more, as predicted by the simulations and numerical experiments. The very small Forbes effect seen with the IRWG passbands, permits the use of the Bouguer extinction coefficients to obtain more accurate outside-atmosphere magnitudes than is possible with others, for which the Forbes effect can be very great.

A similar theme was emphasized in a column in the General Assembly's daily newspaper, *Estrela D'alva*, Day 10, p.4. Such a development is now possible thanks to bulk prices (for lots of 9 or greater) for the IRWG passbands by Custom Scientific, Inc. of Phoenix, Arizona, so far the only filter manufacturer who has produced these filters to our specifications. It is to be hoped that photometer manufacturers will offer installation of the IRWG iz, and iH filters, at least, in IR instruments that they sell, in place of the current filters that are not optimum.

3. Other New Developments in IR Astronomy

Among new work carried out during the past triennium is the following:

• Hora, *et al.* (2008) discuss the response functions of the passbands of the Spitzer Infrared Array Camera, and provide corrections, which, when applied, can lead to relative accuracy of $\sim 2\%$;

• Monson & Pierce (2009) discuss the performance of the `BIRCAM` array camera of the Wyoming Red Buttes Observatory, where it is mounted on a 24-in telescope. They equip this instrument with "JHK" filters.

• Sánchez, *et al.* (2007, 2008) discuss the infrared properties of their site, especially the extinction (in the first paper) and determine the sky brightness (in the second) in "JHK_s" passbands for all times of the year at the Calar Alto Observatory in Spain. They find strong variations in sky brightness, with the maximum in the summer, when the temperature (and humidity) is the highest.

• Taylor (2007; 2008a, 2008b) discusses observations made with the Wampler Scanner, including the extinction due to aerosols as well as water vapor.

• van Dokkum, *et al.* (2009) used three overlapping "J" passbands and two "H" passbands to locate the Balmer jump & 400 nm break in galaxies over the z red shift range 1.5 to 3.5 and compare their passbands to the atmospheric transmission curve. The breaking of the conventional passbands into shorter segments, at least in part to improve transmission through the atmosphere, is a welcome step, even if it is not the main purpose of the work.

• Wood-Vasey, *et al.* (2008b), who use "JH" and "Ks" passbands to demonstrate that SN Ia are standard candles. This paper demonstrates that infrared astronomy continues to be a very important tool for distance-scale work.

The wide use of JHK designations for what are clearly not original Johnson passbands highlights the problem that infrared astronomers have been loath to face: There are no standard $JHKL$ passbands, notwithstanding the increasing use of "short" K passbands ("K_s") that cut off the longer-wavelength end of the window, and thus cut down the thermal emission in the passband. The Mauna Kea near-infrared suite, an incremental improvement over older passbands sets, although falling short of the IRWG prescription, typifies the movement of infrared astronomy toward the goals of the WG.

4. Discussion

Several topics were put up for consideration and discussion during the scientific and business session. Among them were:

- Virtual Observatory Contributions;
- Update website content;
- Facilitate manufacture of IRWG filters;
- Extension of standard stars list (for Comm. 25 site);
- Testing of iL, iL', iM, iN, in, iQ passbands;
- Real-time monitoring of H_2O atmospheric content; and
- correlation with extinction effects.

Several of these topics had been discussed recently in Milone & Young (2008). As noted already, IRWG filters are now available at bulk prices from Custom Scientific, Inc., of Phoenix, AZ. The prices may be obtained from that company. Milone expressed willingness to coordinate a bulk purchase, and urged those interested to contact him. The extension of standard stars lists for the Commission 25 website is dependent on the facilities of the former president of Comm. 25, Peter Martinez, who, however, has expressed interest in maintaining the website.

5. Closing remarks

The IRWG has now been in existence for two decades. Although there appeared to be lively interest in the work of the IRWG by those who attended the session, and depite incremental improvements in other IR passbands, it is clear that more work needs to be done to promote the IRWG passbands, especially for their use in variable star light curve acquisition. Astronomers are a conservative lot, and even though there is no standard $JHKLMNQ$ system to which they need to be loyal, infrared astronomers have been particularly reluctant to adopt and try new filters, except to isolate particular spectral features. In most cases they may not want to sacrifice white-light filters (which is what conventional infrared filters have been, effectively) for narrower ones that provide less overall throughput, but are defined by the edges of the atmospheric windows. Such an attitude is followed by those operating in what can be called a "discovery" mode. Photometrists have basically different aims. One legitimate concern is that data taken in other IR passbands may not transform to the IRWG system, with the possible exception of data in the newer Mauna Kea set used at the one site in the world where it is most suitable, and when conditions there are dry. Consequently, it seems that only demonstrations of the superiority of the IRWG set on specific targets will convince many to use these filters.

Therefore, we urge photometrists with a strong interest in precise photometry to give these passbands a try at observatories where photometry is done either with a chopping secondary and LIA system, as at the RAO, or with array cameras.

Eugene F. Milone
Chair of the Working Group

References

van Dokkum, P. G., *et al.* 2009, *PASP*, 121, 2-8

Hora, J. L., *et al.* 2008, *PASP*, 120, 1233-1243

Milone 1989, in: E. F. Milone (ed.), *Infrared Extinction and Standardization*, Proc., Two Sessions of IAU Commissions 25 and 9, Baltimore, MD, USA, 4 August 1988, *Lecture Notes in Physics*, Vol. 341 (Heidelberg: Springer), p. 1

Milone, E. F. & Young, A. T. 2005, *PASP*, 117, 485

Milone, E. F. & Young, A. T. 2007, in: C. Sterken (ed.), *The Future of Photometric, Spectrophotometric, and Polarimetric Standardization*, Proc. Intern. Workhop, Blankenberge, Belgium, 8-11 May 2006, *ASP-CS*, 364, 387

Milone, E. F. & Young, A. T. 2008, *JRASC*, 36, 110

Monson, A. J. & Pierce, M. J. 2009, *PASP*, 121, 728

Sánchez, S. F., *et al.* 2007, *PASP*, 120, 1244

Sánchez, S. F., *et al.* 2008, *PASP*, 120, 1244

Taylor, B. J. 2007, *PASP*, 119, 407

Taylor, B. J. 2008, *PASP*, 120, 602

Taylor, B. J. 2008b, *PASP*, 120, 1183

Tokunaga, A. T. & Vacca, W. D. 2007, in: C. Sterken (ed.), *The Future of Photometric, Spectrophotometric, and Polarimetric Standardization*, Proc. Intern. Workhop, Blankenberge, Belgium, 8–11 May 2006, *ASP-CS*, 364, 409

Wood-Vasey, W. M., *et al.* 2008, *ApJ*, 689, 377

Young, A. T., Milone, E. F., & Stagg, C. R. 1994, *A&AS*, 105, 259

Transactions IAU, Volume XXVIIB
Proc. XXVII IAU General Assembly, August 2009
Ian F. Corbett, ed.

doi:10.1017/S1743921310005223

COMMISSION 30 RADIAL VELOCITIES
(RADIAL VELOCITIES)

PRESIDENT — **Stephane Udry (Switzerland)**
VICE-PRESIDENT — **Willie Torres (USA)**
PAST PRESIDENT — **Birgita Nordström (Denmark)**
ORGANIZING COMMITTEE — **Francis Fekel (USA), Elena Glushkova (Russian Federation), Dimitri Pourbaix (Belgium), Ken Freeman (Australia), Goeff Marcy (USA), Tomaz Zwitter (Slovenia), Catherine Turon (France), Robert Mathieu (USA)**

PROCEEDINGS BUSINESS SESSIONS, August 14, 2009

1. Introduction

The meeting was unfortunately scheduled for the afternoon of the last day of the General Assembly, so attendance was low. Nevertheless, we had some good discussions.

As an introduction, the aims and objectives of the commission were recalled:
– to promote research in the radial velocities of celestial objects
– to disseminate knowledge of radial-velocity research within the astronomical community
– to promote and support international conferences relevant to the eld of radial velocities
– to act as a clearing house for enquiries relating to research in radial velocities
– from time to time to propose resolutions that dene the best practice of conducting radial-velocity research, or which dene technical terms used in conducting this research
– advising the Executive Committee of the IAU on matters relating to radial- velocity research.

The different ways of achieving those goals (communications with members and the Division, triennial report, working group, conferences, election, newsletters, etc) were mentioned, and some comments were made on the corresponding activities of the commission during the last 3 year period.

2. Activities of the commission

2.1. *Members and Organizing Committee*

The first order of business was to accept and welcome the new members of the Organizing Committee, as well as the new president and vice-president. There was no need for an election because the number of nominees was the same as the number of vacant slots. The current composition is the following:
– President: Guillermo Torres (USA)
– Vice-President: Dimitri Pourbaix (Belgium)
– Organizing Committee members:
∘ Continuing till 2012: Robert Mathieu (USA), Geoff Marcy (USA), Catherine Turon (France), Tomaz Zwitter (Slovenia)

∘ Incoming (till 2015): Tsevi Mazeh (Israel), Dante Minniti (Chile), Claire Moutou (France), Francesco Pepe (Switzerland)

– Past President: Stephane Udry (Swizterland)

Several recently nominated new members of C30 were accepted as well, and welcomed into the Commission.

2.2. *Triennial report 2006-2009*

The report was prepared by the Vice President Willie Torres with the help of some members of the commission and the chairmen of the working groups associated with the commission. We can mentioned 3 important topics addressed in the report:

– Exoplanets: The field is driven by a significant push towards higher precision of relative radial velocities of stars allowing for the detection of smaller and smaller mass planets. The lightest one known has a minimum mass of only 1.9 Earth masses. The high-precision radial-velocity surveys are now revealing a large population of super-Earths and Neptune-mass planets. They also contribute to the detection of an increasing number of multi-planet systems. Searches for transiting planets are becoming more and more important, they have greatly improved their efficiency with new candidates being regularly announced. They require a substantial effort in radial-velocity observations to discard false-positive detections and determine the mass of the transiting object to be combined with the radius derived from the transit curve and eventually obtain the mean density of the planet.

– Velocities of stars in general: Large programs of radial velocity observations are leading to the release of several very large data sets (Geneva-Copenhagen survey, Sloan Digital Survey, RAVE, etc). Large surveys in preparation or complementary to future space missions are also still on going as e.g. radial-velocity measurements to define the SIM grid stars or to define a list of standards for GAIA. The period has also seen the characterization of binaries in clusters and galaxies.

– Asteroseismology: with the increase of the efficiency and precision of echelle spectrographs, asterosismic observations of stars are developing giving access to the internal structure of stars.

2.3. *IAU political questions*

Future of Commission 30: Given that a good portion of the membership of Commission 30 (and also of the Organizing Committee) comes from the exoplanet field, which has been driving recent efforts to improve the radial-velocity precision, there was some discussion about the future of Commission 30 after the creation of Commission 53 (Extrasolar Planets). It was felt that there is still a significant purely stellar component in the Commission, and that perhaps it is time to change the title and character of Commission 30 to be more inclusive, e.g., including "Spectroscopy" in the name. It was also felt that we need to be more proactive in recruiting new members that are actively working in these fields (current membership stands at 138 individuals). It was given charge to the new Organizing Committee.

Report from division IX (Optical and Infrared Techniques): There will be some restructuring of the Division, which will move towards the creation of new working groups emphasizing new telescopes and instrumentation, as well as large sky surveys.

3. Reports from the different working groups

As part of the business meeting, we had the reports from the three Working Groups associated with Commission 30:

– Stellar Radial Velocity Bibliography (H. Levato): report on the status of the database.

– Catalogue of Orbital Elements of Spectroscopic Binary Systems (D. Pourbaix): report on the status of the database.

– Radial Velocity Standards (S. Udry): no progress on the list of standards (see hereafter). Presentation of the observational effort to define a list of standard for the GAIA mission (CU6).

Good progress continues to be made on the first two of these subjects with the involved groups regularly updating the corresponding databases. For the *Radial-velocity standards*working group, the situation is a bit less straightforward. Several large projects are measuring radial-velocities at different level of precision for their own needs (exoplanet search, support to space missions, etc). The question was raised about the need for continuation of the effort to establish and maintain a list of radial-velocity standards, given the work that is already being done e.g. by various Doppler teams searching for exoplanets. Furthermore the list is continuously evolving as some of the stars are found to vary when observed over longer periods of time, or when measured at higher precision. The opinion was voiced that up-to-date lists of stars that have "constant" radial velocities down to a certain level can be obtained by contacting those teams directly. It may be that this is enough to satisfy the needs of the radial-velocity community. At this point, most of the participants were still in favor of an official list of standards. The question will however keep the interest of the commission during the next period.

S. Udry
President of the Commission

Transactions IAU, Volume XXVIIB
Proc. XXVII IAU General Assembly, August 2009
Ian F. Corbett, ed.

doi:10.1017/S1743921310005235

COMMISSION 54 — OPTICAL & INFRARED INTERFEROMETRY

(OPTICAL & INFRARED INTERFEROMETRY)

PRESIDENT	**Stephen Ridgway**
VICE-PRESIDENT	**Gerard van Belle**
PAST PRESIDENT	**Guy Perrin**
ORGANIZING COMMITTEE	**Gilles Duvert, Reinhard Genzel, Christopher Haniff, Christian Hummel, Peter Lawson, Peter Tuthill, Farrokh Vakili**

PROCEEDINGS BUSINESS SESSION, 11 August 2009

1. Introduction

Commission 54 held its business meeting on 11 August 2009 at "Botequim" at Rua Visconde de Caravelas 184/186, Humaitá, Botafogo, Rio de Janeiro. Individual members in attendance reported on activities of relevance to C54.

In attendance was Gerard van Belle, European Southern Observatory & incoming C54 vice-president (meeting scribe); Dainis Dravins, Lund Observatory; Andreas Quirrenbach, Universität Heidelberg; Theo ten Brummelaar, CHARA Array, Georgia State University; Brian Mason, United States Naval Observatory & incoming C26 ('Double & Multiple Stars') vice-president; Wes Traub, Jet Propulsion Laboratory; Anahí Granada, Facultad de Ciencias Astronómicas y Geofísicas, UNLP, Argentina; and Pascal Ballester, European Southern Observatory.

2. Proposal for C54-C26 Sponsorship of an IAU Symposium

Brian, along with Gerard, presented a suggestion for jointly sponsored C54-C26 IAU Symposium, tentatively titled (and topiced) as "Interferometry of Binary Stars". The main theme would be the precision astrophysics enabled by astronomical interferometry of binaries, including measurements of masses, radii, and temperatures.

A General Assembly was thought to be the most appropriate venue for such an IAU Symposium for two reasons: first, the desire to continue bringing news of interferometric results and scientific impact to the astronomy community at large; and second, the advantages presented in having the local logistics support provided by a GA. Given the recent (2006, Prague, IAUS240) symposium on binaries, the 2015 GA (Honolulu) was suggested as the 'preferred' venue, but 2012 GA (Bejing) was also noted for consideration.

Other interested bodies would be IAU C30 ("Radial velocity"), C8 ("Astrometry"), and the US Interferometry Commission. Discussion covered widely the particulars of content, challenges of non-GA hosting, and calendar possibilities.

3. CHARA Array

Theo presented some comments on current developments at the CHARA Array (ten Brummelaar *et al.* 2008):

- A proposal for an AO upgrade has been presented to the National Science Foundation's "Academic Research Infrastructure - Recover and Reinvestment" program.

• Worldwide community access to CHARA was being proposed. (After the meeting, this proposal was cleared and details can be found online: http://www.noao.edu/gateway/chara/).

• Beam combiner operations & development continues, with Classic, MIRC, VEGA (Mourard *et al.* 2008), PAVO (Ireland *et al.* 2008), CLIMB, and CHAMP (Berger *et al.* 2008) combiners all working on the sky.

4. OIFITS Standard

Theo noted that one of the main success sorties (thus far) has been the development and implementation of the OIFITS standard (Thureau *et al.* 2006, Young *et al.* 2008). A question was then raised: is the next step the development of a body of common software for use by the community in exploiting this standard? Certain steps forward in this regard have already been made - e.g. NExScI's software tools (http://nexsci.caltech.edu/software/) and the JMMC tools (http://www.jmmc.fr/index.htm) were mentioned as examples. Some questions that popped up:

• Is there a command line tool to throw at an OIFITS file to do a simple uniform disk fit?

• Is there a "C54 Toolbox" - tools for image reconstruction, tools for model fitting?

• How much model fitting is appropriate for C54 to facilitate, versus what is investigator-specific?

Pascal noted the ongoing work at the JMMC on updating calibrator resources, and that OIFITS is now a standard format for AMBER and MIDI pipeline products (Ballester *et al.* 2006); the possibility & challenges of extending the OIFITS standard to the expected data products of VLTI-PRIMA was also discussed. Theo also mentioned that the MIRC & PAVO CHARA Array instrument use OIFITS as their data export format.

5. Very Large Telescope Interferometer (VLTI)

Gerard noted that the Phase-Referenced Imaging and Microarcsecond Astrometry (PRIMA) instrument for the VLTI shipped to the summit and achieved first fringes in September of 2008, with tests at the sub-system level being conducted over the past year. Full system tests should commence by the end of the calendar year (van Belle *et al.* 2008).

Development on the second generation of VLTI instrumentation is beginning with kick-offs for MATISSE (Lopez *et al.* 2008) and GRAVITY (Eisenhauer *et al.* 2009).

6. Keck Interferometer (KI)

Wes noted that the NASA Key Project Science with KI Nulling had achieved nulls to 1% on 30 stars, at about the 100 zodi level. By comparison, Spitzer was sensitive to only the ~1,000 zodi level. These results are being written up by the PIs of the 3 teams conducting the Key Project Science (e.g., Stark *et al.* 2009). Primary limitations on this technique appeared to be differential K-band versus N-band performance as a result of varying amounts of H_2O vapor in the atmosphere above and inside the observatory (Colavita *et al.* 2009).

7. Space Interferometry Mission (SIM)

Wes noted that SIM (Unwin *et al.* 2008) continues to be considered for implementation, notably as part of the ongoing Decadal Survey activities taking place in the US.

8. Large Binocular Telescope (LBT)

Andreas noted that the LBT continues progress towards interferometer operations (Herbst *et al.* 2008). "Generation 1.5" aperture masking instrumentation was suggested as something for the community to consider.

9. Intensity Interferometry (I^2)

Dainis is chairing the I^2 Task Group for the Cherenkov Telescope Array (CTA, http://www.cta-observatory.org/). Intensity interferometry is a possible modest upgrade to the CTA that would make use of the CTA's non-usable 'bright time' when the full moon were in the sky (Dravins & LeBohec 2008). A lively discussion ensued on the more ephemeral details of intensity interferometry.

10. Other Bits and Pieces

Wes noted that IOTA's detector, having had a brief stint at PTI with IONIC, was now moving south to the VLTI to serve as the detector for the proposed PIONIER [sic] 4-way visitor instrument being developed by Laboratoire d'Astrophysique de Grenoble.

The 4×1.8m "outrigger" telescopes, originally intended to augment the Keck Interferometer, are now slated to join the Navy Prototype Optical Interferometer (NPOI) under an agreement between USNO and NASA.

Gerard van Belle
Vice President of the Commission

References

Ballester, P., *et al.* 2006, *Proc. SPIE*, 6270
Berger, D. H., *et al.* 2008, *Proc. SPIE*, 7013
Colavtia, M. M., *et al.* 2009, *PASP*, 121, 1120
Dravins, D. & LeBohec, S. 2008, *Proc. SPIE*, 6986
Eisenhauer, F., *et al.* 2009, Science with the VLT in the ELT Era, 361
Herbst, T. M., Ragazzoni, R., Eckart, A., & Weigelt, G. 2008, *Proc. SPIE*, 7014
Ireland, M. J., *et al.* 2008, *Proc. SPIE*, 7013
Lopez, B., *et al.* 2008, *Proc. SPIE*, 7013
Mourard, D., *et al.* 2008, *Proc. SPIE*, 7013
Stark, C. C., *et al.* 2009, *ApJ*, 703, 1188
ten Brummelaar, T. A., *et al.* 2008, *Proc. SPIE*, 7013
Thureau, N. D., Ireland, M., Monnier, J. D., & Pedretti, E. 2006, *Proc. SPIE*, 6268
Unwin, S. C., *et al.* 2008, *PASP*, 120, 38
van Belle, G. T., *et al.* 2008, *The Messenger*, 134, 6
Young, J. S., Cotton, W. D., Gässler, W., Millan-Gabet, R., Monnier, J. D., Pauls, T. A., & Percheron, I. 2008, *Proc. SPIE*, 7013

Transactions IAU, Volume XXVIIB
Proc. XXVII IAU General Assembly, August 2009
Ian F. Corbett, ed.

doi:10.1017/S1743921310005247

DIVISION IX-X WORKING GROUP

ENCOURAGING THE INTERNATIONAL DEVELOPMENT OF ANTARCTIC ASTRONOMY

PRESIDENT **Michael Burton**

The work of this group is well covered in the report on Special Session 3 given in "Highlights of Astronomy 15", to be published by CUP in 2010.

Transactions IAU, Volume XXVIIB
Proc. XXVII IAU General Assembly, August 2009
Ian F. Corbett, ed.

doi:10.1017/S1743921310005259

DIVISION X RADIO ASTRONOMY
(RADIOASTRONOMIE)

PRESIDENT	**Ren-Dong Nan**
VICE-PRESIDENT	**Russ Taylor**
PAST PRESIDENT	**Luis F. Rodriguez**
BOARD	**Jessica Chapman, Gloria Dubner, Michael Garrett, W. Miller Goss, Jose M. Torrelles, Hisashi Hirabayashi, Chris Carilli, Richard Hills, Prajval Shastri**

PARTICIPATING COMMISSIONS

Commission 40	**Radio Astronomy**

DIVISION WORKING GROUPS

Interference Mitigation
Astrophysically Important Spectral Lines
Global VLBI

INTER DIVISION WORKING GROUPS

Division XII / Division X	**Historic Radio Astronomy**
Division IX / Division X	**Encouraging the International Development of Antarctic Astronomy**
Division IX / Division XI / Division X	**Astronomy from the Moon**

PROCEEDINGS BUSINESS SESSIONS on August 6, 2009 (Rio de Janeiro, Brazil)

1. Introduction

The business meeting of Division X in the IAU 2009GA took place in three sessions during the day of August 6, 2009. The meeting, being well attended, started with the approval for the meeting agenda. Then the triennium reports were made in the first session by the president of Division X, Ren-Dong Nan, and by the chairs of three working groups: "Historic Radio Astronomy WG" by Wayne Orchiston, "Astrophysically Important Lines WG" by Masatoshi Ohishi, and "Global VLBI WG" by Tasso Tzioumis (proxy chair appointed by Steven Tingay). Afterwards, a dozen reports from observatories and worldwide significant projects have been presented in the second session. Business meeting of "Interference Mitigation WG" was located in the third session.

2. Report of DX/COM40

2.1. *Introduction of DX/COM40*

Division X focuses on radio Astronomy, which includes one Commission (COM40), three working groups, and three Inter-Division working groups. Division X counts with 941 individual members till 2009GA from IAU website. And 131 new individual members (including 5 individual members from NCA countries) were approved by this GA. The total member now is 1073 (plus 1 from COM27) at the end of the General Assembly. Twelve Organizing Committee (OC) members including president and vice president were appointed by IAU for 2006-2009; twelve new OC members for 2009-1012 are recommended and have been approved by the EC86.

2.1.1. *Commission*

Some divisions like Division X and Division XI have only one commission, while others have more commissions, with the case of Division IV that has five commissions. The audience was explained that the discussion inside the Organizing Committee as well as broad radio community indicates a preference for keeping the present structure of commission setting, in the sense that Division X will not have more commissions and its Commission 40 will not merge into the other divisions.

2.1.2. *Working Groups*

The activities of working group were summarized and the agenda of business meetings of six working groups was listed in the business meeting of Division X. The ToRs (Terms of Reference) have been renewed by six working groups, and all agree to continue for the next triennium. At present, Division X includes three working groups: "Interference Mitigation WG", "Astrophysically Important Spectral Lines WG" and "Global VLBI WG", plus three Inter-Division working groups: "Historic Radio Astronomy WG", "Astronomy from the Moon WG", and "Encouraging the International Development of Antarctic Astronomy WG".

2.2. *On new OCs of DX/COM40 (2009-2012)*

EC asked Division Presidents to take advantage of email and internet to carry out the nomination of new president, vice-president and members of the Organizing Committee. Division X did it among Organizing Committees, working group chairs and some senior members, which has been agreed by the IAU General Secretary. The president explained the reasons of not involving all members in voting - incomplete email list, local server restrain and invalid voting-tool. As soon as these requirements are met, the new president may involve all members in the future voting process. The new members of the Organizing Committee were nominated and approved, with Russ Taylor as the next president and Jessica Chapman the next vice-president. It was also noted that the five older members of the Organizing Committee completed their three-year-due and were replaced.

Organizing Committee of 2009-2012 consists of Russ Taylor (President, Canada), Jessica Chapman (Vice-President, Australia), Gabriele Giovannini (Italy, new), Justin Jonas (South Africa, new), Joseph Lazio (USA, new), Monica Rubio (Chile, new), Raffaella Morganti (Netherlands, new), Chris Carilli (USA), Hisashi Hirabayashi (Japan), Prajval Shastri (India), Richard Hills (UK) and Ren-Dong Nan (Past President, China).

2.3. *Short review on the activities of 2006-2009*

Short review on the activities in Division X during the past triennium was presented in the business meeting. These included nominating new P, VP and OCs of Division X, proposing candidates to IAU-SNC, proposing candidate speakers for Invited Discourses in the GA, reviewing symposium proposals, coordinating donation and updating Division X webpage, etc. And the triennial report, the Transaction A of Division X (2006-2009) has been outlined at the meeting.

The activities of the Organizing Committee were briefed. It took a short while to explain how to meet the requests for the IAU Symposia to be supported. There was no symposium coordinated by DX/COM40 during this GA, which could be due to large amount of meetings on radio astronomy and its relevant R&Ds, such as ALMA and SKA, incomplete application materials of proposals, and the lack of contacts among division members.

The new OCs are expected to collaborate more closely to promote successful applications for organizing the IAU Symposia in the next triennium.

2.4. *Reports from observatories and projects*

The short reports on the recent developments of the radio observatories and projects have been presented in the 2nd session of the business meeting of Division X, e.g., SKATDP (world, Ken Kellermann), SRT (Italy, Gabriele Giovannini), ASKAP (Australia, Ray Norris), MeerKAT

(South Africa, Adrian Tiplady), FAST (China, Ren-Dong Nan), EVLA (US, Barry Clark), Arecibo ALFA (US, Murray Lewis), eMerlin (UK, Ian Morison), Global VLBI (JIVE/VLBA, Michael Garrett), LOFAR (NL/EU, Michael Garrett), CART (PrepSKA Activities in Canada, R. Taylor).

3. Presentations of the working groups

After the report of DX/COM40 by Ren-Dong Nan in the first session business meeting of Division X, three short presentations of working groups on the work over the last three years have been made respectively, by Wayne Orchiston (Historic Radio Astronomy), Masatoshi Ohishi (Astrophysically Important Lines) and Tasso Tzioumis (Global VLBI). Orchiston emphasized the importance of the preservation of historically significant radio telescopes and associated relics. Ohishi discussed the list of most important spectral lines for the radio bands from 0-275 GHz to 275-1000 GHz. Tasso stressed the cooperation between the ground-based observatories and Space VLBI missions.

4. Business meetings of the working groups

Four working groups (Historic Radio Astronomy WG, Astrophysically Important Spectral Lines WG, Global VLBI WG and Astronomy from the Moon WG) have had their distinct business meetings at the GA. And the business meeting of Interference Mitigation WG was located in the 3rd session of the business meeting of Division X.

As a prestigious radio astronomy prize, the Grote Reber Medal was issued to Dr. Barry Clark, the presentation of which was performed in the business meeting of Historic Radio Astronomy WG.

The Astrophysically Important Spectral Lines WG met on August 7 during the GA, and discussed the proposed updated list of spectral lines between 1,000 and 3,000 GHz. After the debate, the WG has achieved the agreement on the issue.

The business meeting of Global VLBI WG took place on August 7, 2009. Tasso Tzioumis acted as proxy for the new GVWG chair. The ToR of the WG was presented, discussed and refined.

The business meeting of Interference Mitigation WG was Chaired by Tasso Tzioumis. Michael Kesteven presented a technical report on "Current status of RFI mitigation in radio astronomy", which provided a brief summary of the nature and status of different techniques and explored the take-up issue.

5. Closing remarks

The president expressed that the IAU symposium coordinated in Division X should be promoted by the coming OC, and at least more activities on the radio astronomical sciences would be organized in the next General Assembly, held in Beijing in 2012.

Ren-Dong Nan
President of the Division

Transactions IAU, Volume XXVIIB
Proc. XXVII IAU General Assembly, August 2009
Ian F. Corbett, ed.

doi:10.1017/S1743921310005260

DIVISION X WORKING GROUP on RADIO FREQUENCY INTERFERENCE MITIGATION

CHAIR **Tasso Tzioumis (AUS)**
VICE-CHAIR
PAST CHAIR
MEMBERS **Willem Baan (NL), Darrel Emerson USA), Masatoshi Ohishi) Japan), Tomas Gergely) USA), Rick Fisher (USA), John Ponsoby (UK), JAlbert-Jan Boonstra (NL), Ron Ekers (AUS), Wim van Driel (France), Haiyan Zhang (China)**

PROCEEDINGS BUSINESS SESSION, 6 August 2009

1. Introduction

The IAU Working Group on Radio Frequency Interference (RFI) Mitigation was setup in the 2000 IAU GA in Manchester and its mandate was renewed at subsequent IAU GAs in 2003 and 2006. It was noted that that there are important issues related to RFI mitigation that extend beyond the regulatory function of IUCAF, and hence a more extended working group, which may include IUCAF members, was established.

2. Terms of Reference

The IAU Working Group on RFI mitigation was tasked to seek:

1. Technological solutions: inter alia, interference rejection schemes, novel types of modulation manifesting inherently low out-of-band emissions, state-of-the-art RF filter technology and how it may be advanced, antenna null steering, interference recognition, and data editing.

2. Regulatory innovations: new ways of sharing the radio spectrum.

3. Radio-quiet zones: designating remote areas on the Earth's surface where satellite and other broadcasts (emissions) will be restricted in frequency and in time, and where future radio observatories may be located.

4. Institutional innovations: supranational body to examine and test all space vehicles for out-of-band and spurious emissions prior to launch and with power to arrest launch in event of unsatisfactory performance; also, ways the ITU might be constrained by the 1967 Outer Space Treaty and the 1972 Liability Convention and by its obligation to submit itself to higher authority within the UN family of organizations in the event of its interests conflicting with those of other bodies of equal standing in the UN family of organizations.

The terms of reference were revised in 2006 and no further revision was suggested at this meeting.

3. Report on recent progress with technological solutions

3.1. *Workshops*

The last major workshop on RFI mitigation was at the 2004 Penticton SKA meeting (RFI2004), as reported in the previous IAU GA. It was cosponsored by IUCAF and selected contributions were published in Radio Science, 40 (2005). In 2008 a session on "Radio Frequency Interference, Problems and Solutions" was held at the URSI GA (Chicago, August 2008).

An RFI mitigation workshop was planned for 2007 but it did not eventuate. The next workshop (RFI2010) will be held in Gronigen, the Netherlands, on 29-31 March 2010. A first announcement has been circulated but details may change.

3.2. *Reports on RFI Mitigation techniques*

This working group has traditional commissioned reports on RFI mitigation techniques from researchers in this field. These reports have been presented and distributed at each IAU GA. Previous reports by Geoff Bower (USA) in 2003 and Steve Ellingson (USA) in 2006 are available via the Division X website.

A new report by Michael Kesteven (Australia) was commissioned for this IAU GA and it is available in hardcopy and on the IAU website. The subjects covered in this report are given in the following outline:

The Current Status of RFI Mitigation in Radioastronomy

- RFI categories
  - Satellites; Aircraft; Ground-based; Observatory-based
- Proactive Mitigation strategies
  - Regulation
  - Radio Quiet Zones
  - The Observatory environment
- Reactive Mitigation strategies
  - Blanking in time; in frequency; flagging
  - Null steering
  - Adaptive filters
  - Mitigation in the array imaging stage
  - Subspace filtering
  - Generalised spatial filtering
- RFI Mitigation Metrics
- Discussion on Uptake and SKA
- References

The report includes an extensive list of recent references in this field.

4. Regulatory innovations

Regulatory measures are the domain of the International Telecommunications Union (ITU) and a Question on RFI mitigation exists in the ITU-R Study Group 7. Work continues towards the production of a New Report on RFI Mitigation in the ITU-R but political issues and sensitivities need to be addressed.

To facilitate education in the spectrum regulatory environment, IUCAF and other astronomy institutions have been cosponsoring "Summer Schools on Spectrum Management". The previous such summer school was co-sponsored by RadioNet and was conducted in Bologna in 2005. A summer school was in planning for 2008-9 but fell victim to the world financial crisis. The next summer school on spectrum management is planned for early 2010 in Japan.

5. Radio Quiet Zones (RQZ)

Radio Quiet Zones have been used to protect current radio astronomy facilities in many countries. RQZs were also recommended by the OECD Global Science Forum "Task Force on Radio Astronomy and the Radio Spectrum" in 2003. Work in this subject continues in two areas:

(*a*) A Question on RQZ has been raised at the ITU-R and studies are underway towards in ITU-R Report. This issue may become a future item in a World Radio Conference (WRC) but the politics need to be delicately handled.

(*b*) The Square Kilometre Array (SKA), the next generation radio astronomy facility, requires a RQZ. A SKA task group has produced a report on the requirements of a RQZ for the SKA. Both candidate SKA sites in Australia and South Africa have established RQZs to protect current pathfinder instruments (ASKAP; meerKAT) and the SKA in the future. Technical work in this area will continue by the SKA project.

6. Discussion on mitigation uptake and the SKA

The report by M. Kesteven raised some interesting questions that were discussed by the group:

Low level of uptake of RFI mitigation: Despite significant technological developments in RFI mitigation few observatories have implemented on-line, routine RFI mitigation strategies. It appears that traditional simple techniques such as blanking, discarding badly affected data blocks, or rescheduling are the fallback mitigation options.
There appear to be many reasons for this which include the high cost of mitigation and the application of "niche" solutions since there is no universal mitigation technique. It was felt that it is imperative to continue research as RFI is getting worse and mitigation may become critical in the future.

RFI mitigation and the SKA: It is likely that, due to the RQZ on SKA sites and the high computational cost of mitigation techniques, the early SKA may not include RFI mitigation. However, it was strongly felt that the deteriorating RFI environment would necessitate mitigation in the future and research in RFI mitigation for the SKA is essential.

7. Review and discussion on the future of this working group

The meeting reviewed the work of this Working Group, which is primarily reporting on research and other developments in RFI mitigation achieved by other groups. The WG also encourages and cosponsors workshops and reports, even though the WG has few direct resources to allocate.

It was expressed strongly in the meeting that, given that RFI is getting worse and mitigation is becoming more important, the IAU should continue with the working group.

The present chair (Tasso Tzioumis) stood down after 9 years and Willem Baan was appointed the new chair of the working group.

8. Concluding remarks and suggested actions

There was strong support in continuing this working group and it was felt that RFI mitigation will become progressively more important and even critical. The meeting also suggested the following actions for the WG:

- Encourage submission of a suitably modified version of the technical report to the ITU-R for inclusion in its report.
- Help establish a database of references and papers in this field. The references in the technical reports should provide a good start.
- Investigate whether the WG should help organise an IAU Symposium on this subject, perhaps at the next GA.
- Investigate inclusion in the work of this WG of protection of optical/IR dark skies, in consultation and collaboration with IAU Commission 50.

Tasso Tzioumis
Chairman of the Working Group

Transactions IAU, Volume XXVIIB
Proc. XXVII IAU General Assembly, August 2009
Ian F. Corbett, ed.

doi:10.1017/S1743921310005272

DIVISION X WORKING GROUP on HISTORIC RADIO ASTRONOMY

CHAIR	**Wayne Orchiston**
VICE-CHAIR	**Kenneth I. Kellermann**
MEMBERS	**Rodney D. Davies, Suzanne V. Débarbat, Masaki Morimoto, Slava Slysh, Govind Swarup, Hugo van Woerden, Jasper V. Wall, Richard Wielebinski**

PROCEEDINGS BUSINESS SESSIONS, August 2009

1. Introduction

During the Rio General Assembly we held the following meetings of the Working Group: a Business Meeting, a Science Meeting on "The Development of Aperture Synthesis Imaging in Radio Astronomy", and a Science Meeting on "Recent Research".

2. Business Meeting

This meeting was held on Tuesday 4 August, during Session 2. The contents of the WG's Report included in the Division's Triennial Report were summarised. Despite representations from the WG, Stanford University decided to demolish five 60-ft antennas at the Highway 280 site, although the concrete piers with carved names of noted optical and radio astronomers did survive. During the Triennium research on the history of radio astronomy was carried out by more than 40 different astronomers from nine countries, and various papers were published in the Journal of Astronomical History and Heritage. Papers from the Science Meeting on European Radio Astronomy held at the Prague General Assembly were published in a special addition of Astronomische Nachrichten (edited by Richard Wielebinski, Ken Kellermann and Wayne Orchiston). Deaths of the following colleagues during the Triennium were reported: Ron Bracewell, Chris Christiansen, Fred Haddock, Valadimir Kotelnikov, John Kraus, Arkady Kuzmin, Lex Muller, Brian Robinson, Slava Slysh, Henk van der Hulst and Paul Wild.

The Program of Work for the WG during the 2009-2012 Triennium was outlined and discussed, based on the successful program of 2006-2009. A number of new research initiatives were announced.

The following WG Committee for the 2009-2012 Triennium was announced:
Ken Kellermann (Chair - USA), Wayne Orchiston (Vice-Chair - Australia), Rod Davies (UK), James Lequeux (France), Norio Kaifu (Japan), Yuri Ilyasov (Russia), Govind Swarup (India), Hugo van Woerden (The Netherlands), Jasper Wall (Canada) and Richard Wielebinski (Germany)

Under General Business the release of a history of radio astronomy DVD and forthcoming books was announced, and Ken Kellermann reported that NRAO is digitizing early radio astronomy courses from Westfold (Caltech), van der Hulst (Harvard) and Barrett (MIT).

3. Science meeting on "The development of synthesis imaging in radio astronomy"

This meeting was held on Wednesday 5 August during Sessions 3 and 4, and included the presentation of the 2009 Grote Reber Medal to Barry Clark. Six papers (by Barry Clark, Tim Cornwall, Ron Ekers, Bob Frater, Miller Goss, and Radhakristian) were presented during this meeting.

4. Science meeting on "Recent research"

This meeting was also held on Wednesday 5 August, during Session 2. Four papers (by John Dickel, Bruce McAdam, Harry Wendt, Wayne Orchiston, Bruce Slee, Ron Stewart and Harry Wendt) were presented, and four poster papers (by Martin George, Peter Robertson, Wayne Orchiston, Bruce Slee, Ron Stewart and Harry Wendt) were displayed.

Wayne Orchiston
Chairman of the Working Group
Ken Kellermann
Vice-Chairman of the Working Group

Transactions IAU, Volume XXVIIB
Proc. XXVII IAU General Assembly, August 2009
Ian F. Corbett, ed.

doi:10.1017/S1743921310005284

DIVISION XI SPACE & HIGH-ENERGY ASTROPHYSICS

PRESIDENT Günther Hasinger
VICE-PRESIDENT Christine Jones
PAST PRESIDENT Haruyuki Okuda
BOARD João Braga, Noah Brosch, Thijs de Graauw, Leonid Gurvits, George Helou, Ian Howarth, Hideyo Kunieda, Thierry Montmerle, Marco Salvati, Kulinder Pal Singh

INTER-DIVISION WORKING GROUPS

Division IX-XI WG Astronomy from the Moon

No report was received from this Divison or the Working Group.

Transactions IAU, Volume XXVIIB
Proc. XXVII IAU General Assembly, August 2009
Ian F. Corbett, ed.

doi:10.1017/S1743921310005296

DIVISION XII BUSINESS MEETINGS

PRESIDENT	**Malcolm G. Smith**
VICE-PRESIDENT	**Francoise Genova**
PAST PRESIDENT	**Johannes Anderson**
OC MEMBERS	**Steven R. Federman, Alan C. Gilmore**
	Il-Seong Nha, Raymond P. Norris
	Ian E. Robson, Magda G. Stavinschi
	Virginia L. Trimble, Richard J. Wainscoat

PROCEEDINGS OF BUSINESS SESSIONS, 3rd and 4th August, 2009

1. Formal Meetings

Brief meetings were held to confirm the elections of the incoming Division President, Francoise Genova and Vice President, Ray Norris along with the Organizing Committee which will consist of the incoming Presidents of the 7 Commissions (5,6,14,41,46,50 and 55) plus additional nominated members. The incoming Organizing Committee will thus consist of:

Francoise Genova (France), Ray Norris (Australia), Dennis Crabtree (Canada), Olga B. Dluzhnevskaya (Russian Federation), Masatoshi Ohishi (Japan), Rosa M. Ros, (Spain), Clive L.N. Ruggles (UK), Nikolay Samus (Russian Federation), Virginia Trimble (USA), Wim van Driel (France), Glenn Wahlgren (USA), Xiao Chun Sun (People's Republic of China) and Malcolm Smith (Chile, non-voting, ex officio).

2. Informal Meetings

During the two weeks of the General Assembly, many valuable, informal meetings and conversations were held - in the corridors and over restaurant tables at the convention centre - among Division members from a variety of Commissions in order to discuss a range of current topics. These ranged from different ways these members and their Commissions would be co-ordinating their work in following the IAU Strategic Plan, to detailed, last-minute improvements to Resolution B5 to be presented on 13th August at Session 2 of the full General Assembly. Support was obtained (via a lunch-time meeting) from the Astronomers Without Borders group, for co-ordinated, follow-up imaging of astronomically-related, World-Heritage Sites, in support of Commission 41's work with UNESCO. This photography is to be carried out in the context of the IYA, by the TWAN (The World at Night) group. Much more detailed information on these and other matters appears in the various Commission, Working Group and Program Group reports elsewhere in these Transactions. ...

Malcolm G. Smith
Outgoing President of the Division

Transactions IAU, Volume XXVIIB
Proc. XXVII IAU General Assembly, August 2009
Ian F. Corbett, ed.

doi:10.1017/S1743921310005302

COMMISSION 5 — DOCUMENTATION AND ASTRONOMICAL DATA

(DOCUMENTATION AND ASTRONOMICAL DATA)

PRESIDENT	**Ray P. Norris**
VICE-PRESIDENT	**Masatoshi Ohishi**
PAST PRESIDENT	**Franoise Genova**
ORGANIZING COMMITTEE	**Uta Grothkopf, Bob Hanisch, Oleg Malkov, William Pence, Marion Schmitz, Xu Zhou**

PROCEEDINGS OF BUSINESS SESSIONS, 3–7 August 2009

1. Introduction

Commission 5 and its working groups have continued to operate at a high level of activity over the last three years. In an era when the volume of astronomical data generated by next-generation instruments continues to increase dramatically, and data centres and data tools become increasingly central to front-line astronomical research, the activities of Commission 5 are becoming even more significant. However, most of the activities of Commission 5 take place through its working groups. That was reflected in the meetings at the IAU GA, where there was only one short Business Meeting of the Commission as a whole, but several vigorous meetings of the working groups.

2. Commission 5 Business Meeting, Mon 3 August 2009. Chair: Masatoshi Ohishi

2.1. *New Officers of Commission 5*

Nominations were requested from all members of Commission 5 for the new Organising Committee (OC), and all Commission 5 members were then invited to vote for the nominees, as a result of which the following people were appointed:

- Masatoshi Ohishi (Japan): President
- Bob Hanisch (USA): Vice President
- Ray Norris (Australia): Past President
- Heinz Andernach (Mexico)
- Marsha Bishop (USA)
- Elizabeth Griffin (Canada)
- Ajit Kembhavi (India)
- Tara Murphy (Australia)
- Fabio Pasian (Italy)

2.2. *Future Directions of Commission 5*

Although the interpretation of observations has always been central to astronomical research, new frontiers open up when major repositories of digital data are efficiently federated, which place the curation and management of astronomical data in a position of increasing relevance and importance. The activities of Commission 5 and its Working Groups and Task Force are consequently becoming increasingly central to many facets of present and planned research. The meetings held at the recent GA and detailed below reflect both the diversity and the interdependence of those facets.

2.3. *ICSU Data Issues*

The International Council for Science (ICSU) is the umbrella body for scientific unions such as the IAU, and acts on a number of interdisciplinary issues, such as science data, which cut across scientific disciplines. In 2007, ICSU created the SCID (Strategic Committee on Information and Data), which included IAU Commission 5 members, to consider strategic directions for data management across science. Its report, published in June 2008, made a number of recommendations, including merging and restructuring FAGS (Federation of Astronomical and Geophysical data analysis Services) & WDC (World Data Center system) to form a new body, the World Data Services (WDS), which should be better placed to serve the needs of data centres and their users. It also recommended the creation of a new body, the SCCID (Strategic Coordination Committee on Info and Data) to provide ICSU with broad expertise and advice on strategic directions in the area of Scientific Data and Information. Both these recommendations have now been implemented and the IAU is represented by Commission 5 members on each of these new bodies.

2.4. *Possible merging of roles of WG for Astronomical Data (WGAD) and Commission 5*

It was pointed out that there is a great deal of overlap between the WGAD and Commission 5 itself, both in their roles and in their officers, so perhaps the WGAD role should be taken over by Commission 5 OC. Furthermore, while the name of Commission 5 is "Documentation and Astronomical Data", the focus is almost entirely on "data" with very little on "Documentation", so it might be time to rename the Commission.

The suggestion has triggered significant discussion. It was tabled for consideration by the WGAD at its own business meeting, and is to be pursued throughout the triennium.

2.5. *Any Other Business*

It was pointed out that many astronomers still used cgs units in papers, although for several decades almost all students have been taught in the SI system. It is an IAU recommendation that astronomers should use SI units rather than cgs units. The IAU recommendations for SI and non-SI astronomical units can be found on http://www.iau.org/science/publications/proceedings_rules/units/, which is based on the IAU style manual (1989) prepared by G.A. Wilkins (then President of Commission 5).

3. WG – Libraries, Mon 3 August 2009

3.1. *WG Scope and Goals - Bob Hanisch*

The IAU Commission 5 Working Group on Libraries exists to foster communication and understanding between research astronomers and the librarians who support them. Librarians have long been an essential resource to astronomers, helping to track down hard-to-find resources, maintaining access to bibliographic databases, and managing the complex finances associated with journal and book collections. The role of the library and librarian are changing as information becomes increasingly digital in format and increasingly available on the Internet, but the librarian is no less essential than before. We hope that by informing the research community of the librarians' needs, concerns, and aspirations, and providing a forum for hearing the concerns of scientists, that this strong partnership between astronomers and librarians will continue and that libraries and librarians will find strong advocacy regarding their financial needs from the research community they serve.

3.2. *HST bibliographic database and library renovations at STScI – Jill Lagerstrom*

- Library as place: renovating and downsizing
- Metrics: The new HST Bibliography
- see full presentation on http://www.eso.org/sci/libraries/IAU-WGLib/iau09/LagerstromRio.pdf

3.3. *Metrics: Facility identifiers: how they are (or aren't) used – Chris Erdmann*

- Tracking use of ESO data through published papers
- Difficult because authors are very inconsistent in citation of facilities
- Requires major human effort; facilities sometimes identified in text, sometimes in footnotes, facility names can be misspelled
- Common facility list (AAS, Greg Schwarz? based on VO identifiers), incorporated into AASTeX markup
- Authors who do list facilities make mistakes, omissions
- Chris developed tool called FUSE to glean facilities metadata from papers (PDF form)
- Can authors be required to include facilities tagging?
- Arnold noted that dataset identifiers are also not being used often
- FUSE being used by four observatories

Another challenge in metadata management! How does it really work? Uses list of terms, acronyms, as filter followed by human inspection. Can use stop words to eliminate confusing terms like hubble flow. (See http://www.eso.org/sci/libraries/IAU-WGLib/iau09/FacilityIDs_ce_ug.pdf and http://www.eso.org/sci/libraries/IAU-WGLib/iau09/Fuse.pdf for full presentation)

3.4. *Data Curation and Strategies for Survival in the Brave New World of Librarianship - Marsha Bishop*

- Data curation spans small bibliographic records to large data sets
- Metadata: simple? constrained? combination? extent (how many items)?
- Record types: highly structured?
- Updates: who makes updates? under what conditions?
- Data management: who does it? IT staff, library, end-user.
- Preservation: how long to keep data?
- Access: open, closed, or combination
- Exposure: awareness, marketing
- Metadata changes (e.g., facility names can change)
- Data curation requires passion for data retention: librarians.
- Also requires domain expertise! Data scientist.
- Data curation is one small part of libraries responsibilities.
- How can libraries survive current epoch of downsizing, budget cutting/
- Marion Schmidt: what happens to metadata when journals eliminate page numbers? Replace with DOI.

See http://www.eso.org/sci/libraries/IAU-WGLib/iau09/IAU-Pres_DataCura.pdf for full presentation.

3.5. *Links in the Astronomy Data Network - Alberto Accomazzi*

- Types of links include internal, external, computed, curated, contributed
- Useful for attribution, aggregation, preservation, discovery
- Links can be based on graph theory, semantic web, OAI/ORE
- Topic clusters
- Linked Open Data: resource names are URIs, metadata is in RDF format
- Links in astronomy are URLs, static, untyped, and do not use standard vocabularies; not actionable by applications
- How to increase value of links in astronomy? Start with observing proposals and track things forward. OAI/ORE defines how to label the links and relationships in a publication and track the publication through its various states.
- Datanet Data Conservancy
- VAO collaboration: bring semantic awareness to VO applications
- Key technologies are RDF, LOD, OAI/ORE

• Arnold: need to identify the software used for the data processing and analysis.
See http://www.eso.org/sci/libraries/IAU-WGLib/iau09/IAU09.pdf for full presentation.

4. WG – Virtual Observatories, Data Centers and Networks, Mon 3 August 2009

The International Virtual Observatory (IVO) is a truly global endeavour in astronomy. Many projects, each with its own goals, have been set up around the world to develop the IVO. The International Virtual Observatory Alliance (IVOA) links together the VO projects, with the aims of managing communication between VO projects, defining a common roadmap, and managing proposals for the definition and evolution of IVO standards through the IVOA Working Groups.

The IAU WG on Virtual Observatories, Data Centres, and Networks (WG VODC&N) is the standard-bearer of the International Virtual Observatory at IAU, and is the primary point of contact between the IVO and the IAU. Its primary role is to provide an interface between IVOA activities, in particular IVOA standards and recommendations, and other IAU standards, policies, and recommendations. In particular, it promotes VO-related topics (e.g. symposia, GA sessions) which need to be handled by the IAU (Commission 5, Division XII and executive level). It helps facilitate the adoption of VO standards in the broader community, particular in liaison with national and international data centres, and it provides outreach for VO and data management efforts in related fields within the IAU (planetary science, solar astronomy, etc.).

The WG VODC&N brings to the attention of the IVOA Executive any topics it considers to be important for the IVO. It can be consulted by the IVOA Executive on any topic relevant to the international development of the VO. The WG VO consists of members of IVO projects, together with others bringing an external view on the long term vision of the VO, and other stakeholders. It also includes the president of Commission 5, the chair of the IVOA, a representative of the WG FITS (Commission 5), a representative of the WG on Astronomical Data (Commission 5), and a representative of the WG on International Solar Data Access (Division II).

5. WG – Designations, Tue 4 August 2009

5.1. *Purpose*

• The Working Group Designations of IAU Commission 5 clarifies existing astronomical nomenclature and helps astronomers avoid potential problems when designating their sources.

• The WG Designations oversees the IAU Registry for Acronyms for newly discovered astronomical sources of radiation (The Dictionary): http://cdsweb.u-strasbg.fr/cgi-bin/DicForm which is sponsored by the WG and operated by the Centre de Données de Strasbourg (CDS).

• The Clearing House, a subgroup of the WG, screens the submissions for accuracy and conformity to the IAU Recommendations for Nomenclature http://cdsweb.u-strasbg.fr/iau-spec.html

5.2. *Membership*

The WG has 17 members, with one member leaving, and four proposed new members.

5.3. *Acronym registration*

• Over the last three years, 79 proposed acronyms have been received. 61 were accepted, and 9 were revised.

• Lanie Dickel: “The Registry seems to be being used for many of the big surveys at about the amount of submissions per year that we can handle. I think it should continue - a valuable service - and its availability advertised at the IAU.”

• There are still problems with conflicting or ambiguous designations, such as MOST which can either mean “Microvariability and Oscillations of Stars” (a Canadian Satellite launched in 2003) or “Molonglo Observatory Synthesis Telescope” (an Australian telescope built in 1960 – 1980).

5.4. *Changes in designations*

• How have the nomenclature rules changed and how should they be changed in the face of database-oriented astronomy (e.g. case-sensitivity; special character handling; dealing with spaces in names; the name being an index key and not the RA/Dec position, etc.)?

• "Special" characters include Greek letters, hyphens, and spaces
• There is a need for a uniform policy for naming stars and non-stellar objects in other galaxies (and, of course, such objects found in intergalactic space)
• Case-sensitive designations appear in the Literature. E.g. HD 41004B is a star, HD 41004 B b is a planet in Exoplanet Encyclopedia, HD 41004Bb is the name in SIMBAD
• L. Dickel: "... we need to check with the exosolar planet community as they are the ones in the IAU who are designating such objects if I am not mistaken. I think they agreed with the binary/multiple star people to use their kind of designation."
• See: http://ad.usno.navy.mil/wds/wmc/wmc_descrip.txt or
http://www.iau.org/PLANETS_AROUND_OTHER_STARS.247.0.html.
An extract follows: "In order to facilitate international research in the field, and as part of these discussions, the IAU is also developing a system for clear and unambiguous scientific designation of these bodies at all stages during their study, from tentative identification to fully-characterized objects. Such a system must take into account that discoveries are often tentative, later to be confirmed or rejected, possibly by several different methods, and that several planets belonging to the same star may eventually be discovered, again possibly by different means. Thus, considerable care and experience are required in its design."

5.5. *Large Surveys*

• How can we be proactive in communicating with very large survey projects before they start naming things (e.g., Pan-STARRS, LSST)? E.g. "PSO" was submitted by Pan-STARRS, but without a format.
• There needs to be some discussion of how to deal with nomenclature (that is becoming increasingly cumbersome) for the very large surveys either underway now or planned for the immediate future.
• Is it too early to think about GAIA designations?
• Are there other surveys in the pipelines?

5.6. *Journal/Author interactions*

• There needs to be closer interaction with journal editors to ensure consistent adoption of correct nomenclature.
• How do we get individual authors to check the Dictionary before publication?

5.7. *Changing venue for identifying objects*

• Perhaps more controversial: have catalogues grown beyond their usefulness and should we regard images as the basic tools of astronomy? Can we bootstrap our discipline away from the mindset that reduces the cosmos to a collection of elliptical gaussians?
• Coping with and archiving huge piles of data seems to be a continuing problem in astronomy (and elsewhere).

6. WG – FITS Data Format. Tue 4 August 2009

The FITS Working Group is responsible for maintaining and updating the FITS Data Standard Document, which includes specifying standard conventions for time & date, and ensuring that coordinate information attached to data follows the World Coordinate System, and that transformations between systems are well-understood, well-documented, and are specified in the meta-data attached to each FITS file. Data acquired by different systems can then be compared and combined, and data can be transformed from one coordinate system to another without distortion.

The FITS WG has been very active over that past 3 years. The most significant recent development was the release of a new FITS standard, which was formally approved in July 2008 and is now publicly available on the FITS Support Web site at http://fits.gsfc.nasa.gov.

The other main activity of the WG-FITS during that period was to create a registry on the FITS Support Office Web site for documenting the many existing FITS conventions that are in use within the FITS community. As astronomical instrumentation continues to expand in scope and capability, it is inevitable that the FITS standard will continue to evolve.

The success of the FITS WG in managing this process is widely regarded as one of the most effective parts of the IAU. There is a strong need for the FITS WG to continue its role as the international control authority for the FITS data format.

7. TF – Preservation and Digitization of Photographic Plates, Mon 3 August 2009

Many members are involved in studies of long-term variability, which require monitoring over a time-base well exceeding that available through born-digital observations alone. Historic data reproduced accurately from archived plates are thus central to such research. New results can be read in the literature, and are not cited here. Several projects involve *Carte du Ciel* plates; in 2006 the WG-CdC was merged with the PDPP.

Plate Preservation: A North American Astronomical Plate repository has been established at PARI (Pisgah Astronomical Research Institute, North Carolina). The Cambridge (UK) plate store was dismantled and the contents repatriated or disseminated. Requests to create user-friendly access to the RGO plates currently stored in London and to transfer a limited set to the ROE were unsuccessful.

Direct plates: Measurements (accurate to 0.5 μm) of Black Birch, AGK2 and Hamburg Zone astrograph plates with the USNO StarScan plate measuring machine have been completed; the repeatability of StarScan has an error of 0.2 μm. The new data will contribute to the UCAC3 (as reported under Commission 8). Century-old plates from the Sydney Observatory Galactic Survey yield positions and magnitudes that have now been catalogued; early analyses of those data are already contributing new science. A proposal is being formed to provide for a comparison between Gaia results (when available) and CdC measurements along an equatorial belt, to search for changes. A discussion in Paris of the benefits of the unfinished *Carte du Ciel* project heard about the detection of the ISM from CdC plates. At the DAO, part of a 14×14-inch plate from Palomar thought to have imaged an elusive comet was digitized, along with an appropriate control plate. A catalogue of Vatican Observatory Schmidt plates (now digitized), plus 'thumbnail' scans, will be posted on a web-site (yet to be named). In China, the preservation and digitization of the substantial historic collections of direct and objective-prism plates from the National, Qing Dao and Purple Mountain Observatories under the auspices of the Chinese VO project is hampered by lack of resources and a working PDS.

Spectroscopy: The digitizing of selected spectrograms borrowed from various US observatories for telluric ozone research has re-started at the DAO following a major downtime and upgrade of its PDS. A collaboration with the Carnegie Institution to digitize a subset of Mount Wilson spectra for the same purpose is now commencing. The 10-10 PDS from KPNO has been brought in as a first step in setting up an international scanning laboratory. Requests to scan specific spectra continue to arrive. The major objective-prism survey of the Byurakan Astrophysical Observatory has now been digitized, and is in the public domain; early scientific results are already impressive.

8. Meeting of the WG – Astronomical Data, held on Fri 7 August 2009

8.1. *Introduction*

The Working Group on Astronomical Data (WGAD) is the gateway to ICSU CODATA, and enables the astronomical community to comment on, and influence, cross-disciplinary initiatives such as the Science Data Commons, and international legislation which might affect our ability to maintain large public domain databases. WGAD also serves as an electronic forum to discuss data issues, such as the challenge of handling huge data sets resulting from new generation astronomical instruments. For example, a WGAD e-discussion took place in the lead-up to the 2009 IAU GA, and is summarised by Andrew Hopkins below.

WGAD also serves as a forum to debate issues arising from ICSU, such as the Strategic Committee on Information and Data in Science, whose recommendation to restructure FAGS (Federation of Astronomical and Geophysical Services) and WDC (World Data Centres) have now been implemented.

WGAD membership has recently increased from 34 to 58. Other activities over the past three years have included

• Playing an active role in ICSU's "Strategic Committee on Information and Data", resulting in a restructuring of the systems of World Data Centres and Astronomical Data Services, and
• Running a session on "Astronomical Data and the Virtual Observatory" during the CODATA General Assembly in Kiev in October 2008.

In the next three years, WGAD activities will include running a session on "Petabyte databases for next-generation astronomical instruments" at the CODATA GA 2010 in South Africa. The WGAD Chair will continue to represent the IAU at CODATA meetings.

8.2. *WGAD Officers*

Following a nomination and voting process involving all WGAD members, Oleg Malkov has been appointed as Deputy Chair. He is expected to succeed to the Chair when the present holder (Ray Norris) steps down as chair in 2010 (a date chosen to synchronise with CODATA activities).

8.3. *The challenges of petabyte astronomical databases (George Djorgovski)*

George Djorgovski gave an inspirational presentation outlining the challenges and opportunities faced by astronomy as we move into the Petabyte era. He pointed out that time-domain astronomy is possibly the "Killer Application for the Virtual Observatory".

8.4. *The Astronomers Data Manifesto revisited (Ray Norris)*

The Astronomers Data Manifesto was launched in 2006 (see http://www.ivoa.net/cgi-bin/twiki/bin/view/Astrodata/DataManifesto) partly to articulate a future direction for astronomical data, and partly to raise awareness of astronomical data issues. It has already been successful in raising awareness, and had high visibility at the 2006 IAU GA in Prague. It has also been adopted by US National Virtual Observatory as a set of guiding principles. However, awareness has so far largely been confined to the "converted", and data issues still have low visibility amongst the broader astronomical community. Ray Norris argued that Commission 5 should aim to propose it as an IAU Resolution in 2012. However, to achieve that, it needs to gain a broader support base within the IAU. This could be achieved by (a) discussing it with Presidents of other IAU Divisions/Commissions, and (b) by Commission 5 members introducing data-related talks at IAU Symposia and other astronomical meetings.

8.5. *Future Directions of Astronomical Data (Andrew Hopkins)*

A lively e-discussion had been conducted by Andrew Hopkins and WGAD members in the months preceding the IAU GA. The full discussion can be found on http://wgad.pbworks.com/. Key discussion points included:

• The implications of terabyte astronomical databases on the future of the VO and data centres,
• How telescope users might access their data in the future,
• The "Moore's Law" of Sky Surveys,
• How the VO and data mining of surveys could change the way we do our Science,
• How we could improve the flow of data from journals to data centres,
• Managing and preserving legacy data,
• Conquering the digital divide,
• Open access to astronomical archives,
• Using digital object identifiers for astronomical datasets,
• Enabling career recognition for those who manage data, and
• Cross-fertilisation between disciplines.

8.6. *Managing Yesterday's Data: Creating a Resource Worth Having Elizabeth Griffin*

Elizabeth Griffin expressed concern that the Data Manifesto appears to promote the archiving of data that appears in a paper, and argued that the data archive should contain the complete, entire, calibrated dataset and accompanying metadata. What is published in a journal is likely to be only a subset of the original data, and frequently constitutes a subjective selection for a specific interpretation, whereas the original data are an objective record of what was observed. Ray Norris explained that when the Data Manifesto used the word "data", it was intended to refer to published end results such as redshifts, which was what was of interest to most users rather than the original data.

The meeting agreed that the Manifesto's use of the word "data" in that context should be replaced by words such as "published results". It would be amended accordingly.

8.7. *CODATA Report (Ray Norris)*

CODATA is the "Committee on Data for Science and Technology" of the International Council for Science (ICSU). Despite being named a "Committee", it actually functions much like a Scientific Union, with national members, an executive, and a bi-annual General Assembly. However, unlike a union, CODATA transcends all scientific disciplines and so has members from Scientific Unions as well as National members.

The IAU benefits from CODATA in a number of ways. For example, CODATA

- acts in the interest of data for all areas of science (e.g. combating the proposed WIPO legislation),
- explores cross-disciplinary initiatives (e.g. Global Scientific Commons),
- runs a Science Data Conference every two years,
- runs the Data Science Journal,
- is responsible for setting fundamental standards (e.g. speed of light, the charge of the electron),
- may in the future be responsible for other services, such as a registry of registries, ontologies, etc.

CODATA highlights in the last triennium have included:

- New National Members (Both UK and Australia joined CODATA in 2008).
- The CODATA International Conference in Kiev in October 2008.This was attended by about 400 scientists, with the theme of "cultivating an open access environment to scientific data". It included a successful one-day session on "Astronomical Data and the Virtual Observatory".
- a new Strategic Plan.
- ten new Task Groups.
- two new CODATA Initiatives: ("Building European Activities in Public Domain data" and "Collaboration with Global Earth Observation System of Systems").

(see http://www.atnf.csiro.au/people/rnorris/WGAD/ for details).

8.8. *Debate: If WGAD didn't exist, would we invent it? (Initiated by Elizabeth Griffin)*

Elizabeth Griffin argued that it was not clear that we needed both a Commission and a separate WGAD, since the latter's role now overlapped with the goals of Commission 5 and appeared in effect to be doing the job of Commission 5.

Ray Norris argued that:

- Historically, WGAD was set up as a conduit of information between IAU and CODATA, with the WGAD chair being the IAU delegate, and the WGAD being the group of people he/she consults in interactions with CODATA. However, this is not necessarily an argument for continuing in the same way.
- Commission 5 covers "Documentation & Astronomical Data", not just data, and so the interests of WGAD are not congruent with those of Commission 5.
- In recent years, WGAD has also become a valuable forum for e-discussion of data issues. Some of these are triggered by the interactions with CODATA, or by other events, and there has been a wiki discussion in the lead-up to the last two IAU General Assemblies.
- Commission 5 currently has 176 members, and WGAD has 58 members. Only about half the members of WGAD are Commission 5 members, and not all WGAD members are IAU members.
- So if we merged WGAD and Commission 5, we would either need to (a) stop the e-discussions, which would be a loss, (b) include all Commission 5 members in those discussions, which would cause unnecessary "spam" for those Commission 5 members with no interest in data discussions, (c) set up a separate subset of Commission 5 consisting of those people who were interested in participating in e-discussions on data.
- This subset would essentially consist of the current WGAD members, so we would have re-invented WGAD.

Bob Hanisch made the additional comment that dividing up Commission 5 into working groups reduced the size of meetings such as this to below critical mass. There was general agreement with this point.

The meeting then moved on to a more general discussion, including the following key points.

• There was general agreement that the six Commission 5 working groups (WG Astronomical Data, WG Libraries, WG FITS, WG Designations, WG Virtual Observatories, Data Centers & Networks, Task Force on Preservation & Digitization of Photographic Plates) each served a valuable function and should continue.

• There is a perceived problem that the IAU regards working groups as short-term entities, whereas the reality is that many working groups have a valuable (and widely supported) long-term role. Thus, while we have no objection to justifying the existence of each WG every three years, the IAU needs to recognise (in the wording of the bye-laws and working rules) the reality of WG with a long-term role. An action resulted on Division and Commission chairs to raise this with the IAU Executive Committee.

• However, there was also general agreement that splitting up the IAU GA meetings into individual task group meetings was counter-productive, resulting in sub-critical attendance at Commission and WG meetings.

• Poor attendance at Commission 5 and WGAD meetings was perceived to be a problem, although it is acknowledged that most of the real work occurs (mainly by email) between GA, and the low attendance at GA meetings does not indicate low interest in the work of the group as a whole.

• It was therefore agreed that, at the next IAU GA, all the working groups should combine their meetings into a more general Commission 5 meetings, which would include reports from individual WG chairs. However, this needs to be ratified by the majority of Comm5/WG members who were not then present. The Commission 5 President and WG chairs should canvas membership on this issue.

• This does not of course preclude individual specialist WG meetings taking place in those few cases where discussions amongst a few members are appropriate.

• Another problem is that Commission and WG meetings are poorly advertised at the GA, and their agendas are hard to find in the Program. Calling them "Business Meetings" is likely to discourage astronomers who are not interested in "Commission Business" but who may well be interested in excellent general talks such as George Djorgovski's this morning. This resulted in an action on the Commission and Division chairs to raise this matter with the IAU EC.

• There was a discussion on how to raise awareness of data issues amongst the wider astronomical community. IAU GA Sessions on topics such as data management tend to end up preaching to the converted. Instead, we need to ensure that data/VO issues are raised in other symposia and meetings. This resulted in an action on all members to propose talks on astronomical subjects involving data/VO issues at other astronomical meetings which involve data/VO issues.

• A suggestion was made, initially by Elizabeth Griffin, and then widely supported and amplified by other participants, of proposing an IAU Symposium on "Time-Domain Astronomy", which as George Djorgovski pointed out, is possibly the "Killer Application for the VO". Action: the Chair will explore this further with potentially interested people, and (if supported, propose an IAU Symposium in 2011 by 1 Dec 2009.

• Bob Hanisch and Andrew Hopkins also suggested that the IAU should sponsor international astro-informatics or VO workshops. (Later, in the Decadal Plan meeting, Bob suggested to George Miley that this should be part of the IAU Decadal Plan).

Transactions IAU, Volume XXVIIB
Proc. XXVII IAU General Assembly, August 2009
Ian F. Corbett, ed.

doi:10.1017/S1743921310005314

COMMISSION 6 ASTRONOMICAL TELEGRAMS *(TELEGRAMMES ASTRONOMIQUES)*

PRESIDENT A. C. Gilmore
VICE-PRESIDENT N. N. Samus
PAST PRESIDENT K. Aksnes
ORGANIZING COMMITTEE D. W. E. Green, B. G. Marsden, S. Nakano, E. Roemer, J. Ticha, H. Yamaoka

PROCEEDINGS BUSINESS SESSIONS, 10 August 2009

1. Introduction

The President verbally reported that the only scientific matter that he dealt with during the triennium as an appeal over the withholding of a supernova designation from an object observed only in the infra-red with no supporting spectrum.

The new officers and Organising Committee for the Commission had been voted on by email in February 2009 and communicated to the General Secretary soon after, so were taken as read. The new Officers and Organising Committee are as follows: Nikolai N. Samus (President); Hitoshi Yamaoka (Vice President); Alan Gilmore (Past President); Kaare Aksnes; Daniel W. E. Green; Brian G. Marsden; Syuichi Nakano; Timothy Spahr; Jana Ticha; Gareth Williams.

2. Report of the Director

Dan Green referred everyone to his latest Report, included in *IAU Information Bulletin* No. 104 (p. 83; June 2009), distributed at the GA. The following points were noted: Postal circulars will continue into the coming triennium but subscriptions are slowly declining. The CBAT is still needed by the astronomical community as the most trusted source of information on discoveries and for designations of comets and supernovae. This point had been reinforced at a meeting with the VOEvent (Virtual Observatory) organisers, who acknowledged the need for the Bureau to highlight important astronomical discoveries, as automated surveys produce an ever greater flood of data.

Half of the funding of CBAT comes from United States sources. The U.S. National Science Foundation (NSF) is funding half of the Director's salary for two years, due to end in early 2010. [The former Director, Brian Marsden was a U.S. federal employee of the Smithsonian Astrophysical Observatory (SAO), so did not receive a salary from CBAT.] The IAU EC has increased its support of CBAT to around ???? Euros for each of the calendar years 2010, 2011, and 2012.

Previously the CBAT and the Minor Planet Center (MPC) shared a part-time secretary who looked after subscriptions. The MPC will cease charging a subsciption from 2009 October 1. This makes it uneconomical for CBAT to continue to employ the secretary. A proposal to hand subscription administration to a subscription agency is being considered.

The CBAT Director discussed seeking funding for a cometary science center, under which the running of the CBAT might fall, financially. The Director has run most of the day-to-day activities alone (with some help from Director Emeritus Marsden while travelling, and from Assistant Director Williams regarding computer issues). Assistance might come in the form of such a cometary science center, as a large amount of funding for CBAT alone has been difficult to realize. Collaboration with other educational institutions is also being explored, with CBAT as an online teaching tool. Perhaps funding from other national astronomical organisations can be more fully explored, as well.

3. Discussion

Some in the supernova (SN) community are ignoring the CBAT. The big surveys will yield many possible SNe that are unconfirmed. These can get PSN (possible supernova) designations, with the proposal that surveys assign PSN positional designations (good to $0^{s}.01$ in R.A. and to $0''.1$ in Decl.). IAU SN designations can then be given by the CBAT when spectroscopic confirmation is reported, in such cases. Traditionally the CBAT has assigned designations to objects brighter than 20th magnitude close to a host galaxy. Only 1% of designated SNe have proved to be non-SNe.

Following a request from the SN community SNe are now reported only on *Central Bureau Electronic Telegrams (CBETs)*. SNe are reported in the *Circulars* only when line-charges are paid. An unfortunate double-assignment of a SN designation caused problems in the NED database; steps are in place to ensure this does not happen again.

Comet names are issued in the *IAUCs*. The paid CBAT subscription infrastructure will continue after the MPC publications become free.

The new Division XII president (F. Genova) plans to call a meeting of presidents of the Commissions that are in DXII. There is a need to explain to the IAU Executive Committee the importance of supporting the CBAT, Minor Planet Center, and the General Catalogue of Variable Stars team in Moscow as important international centers for the astronomical community that are within the IAU — thus giving the IAU much good visibility, as something important that the IAU does in addition to holding meetings. Many in the astronomical community worldwide see the documenting functions of the IAU as highly important, and if they are not adequately supported, the worldwide community will not see the relevance and uniqueness of the IAU.

4. Closing remarks

Also present at this meeting of Commission 6 besides the outgoing and incoming officers and the CBAT Director were David Tholen, Francoise Genova (incoming Division XII president), and Marion Schmitz.

Alan C. Gilmore
President of the Commission

Transactions IAU, Volume XXVIIB
Proc. XXVII IAU General Assembly, August 2009
Ian F. Corbett, ed.

doi:10.1017/S1743921310005326

COMMISSION 14 ATOMIC AND MOLECULAR DATA (*ATOMIC AND MOLECULAR DATA*)

PRESIDENT **Glenn M. Wahlgren**
VICE-PRESIDENT **Ewine F. van Dishoeck**
PAST PRESIDENT **Steven R. Federman**
ORGANIZING COMMITTEE **Peter Beiersdorfer, Milan Dimitrijevic, Alain Jorrisen, Lyudmila I. Mashonkina, Hampus Nilsson, Farid Salama, Jonathan Tennyson**

PROCEEDINGS BUSINESS SESSIONS, 7 August 2009

Present: P. Bonifacio, E. Biémont, S. Federman, H. Hartman, W. F. Huebner, S. Hubrig, J. Kubat, L. Mashonkina, G. Nave, O. Pintado, N. Piskunov, T. Ryabchikova, G. M. Wahlgren (Chair), G. del Zanna

The meeting was called to order by the Chair, who followed the agenda that had been sent to the membership prior to the meeting. The membership of the Commission stands at approximately 200 members. An exact number will not be available until after learning from the Secretariat on the status of new members. The Chair noted that after the current triennium most of the Organizing Committee (OC) will need to be replaced and that we should not wait until the months before the GA to gauge interest for these positions from the membership. Lively discussions ensued related to the Commission's website, which will be changing to fit a proposed IAU format. The website will become a more enhanced resource for locating web sites for atomic and molecular data. The Chair reported on the extensive participation of the Commission in co-sponsoring meetings at the current GA. Finally, the Chair brought to discussion the desire to implement an evaluation of the Commission's activities with the purpose of determining its future goals and how to achieve them.

Officers: In our commission the Vice President (VP) becomes the President, and a new VP is chosen from among members of the Organizing Committee. The Chair thanked the outgoing President Steven Federman for his service to the commission and to the promotion of laboratory astrophysics. The Chair announced that the new officers are those listed above.

Organizing Committee: Our commission's usual practice is for a member of the Organizing Committee (OC) to serve for two consecutive three year terms, with the past President serving on the OC for three years past their term as President. Officers may serve longer than six years if necessary to complete their service as officers. For the new triennium the entire OC was retained, since they had served for only one term after the reduction in the number of OC members that occurred during the previous triennial period. To keep continuity in the future, the results for a search for two new OC members (Peter Beiersdorfer and Hampus Nilsson) was announced. The Chair remarked with regret on the passing of Sveneric Johansson, who was serving in the position of past President at the time of his passing, and noted his long and valuable service to the astrophysics community. The Chair announced the promotion of Ewine F. van Dishoeck to the position of VP and the new OC as listed above. These changes were ratified by those in attendance.

Working Groups: The Commission's Working Group (WG) structure will be retained for the next triennium and is composed of the WGs Atomic Data, Molecular Data, Collision Processes, and Solids and Their Surfaces. An expanded set of chairpersons for these WGs is being finalized.

Meetings of Interest: The Commission acts to bring together providers and users of atomic and molecular data and to disseminate data. To these goals, a number of meetings serve as forums for these discussions. Meetings of interest to members of the Commission will be posted on its website.

Federman reported on the work of the recently formed working group on Laboratory Astrophysics of the American Astronomical Society. Other continuing forums include the NASA-sponsored Workshop on Laboratory Astrophysics, the yearly EGAS conference, along with ICAMDATA, and DAMOP of the American Physical Society. The most recent ASOS meeting occurred in 2007 in Lund, Sweden, and the next in the series is scheduled to take place in Berkeley, CA, USA in 2010. It became clear from the meeting attendees that there are a number of regional and international meetings addressing aspects of laboratory astrophysics that provide opportunities for the exchange of ideas and information among data providers and users.

General Assembly Commission 14 Science Meeting: In an effort to bring together providers and users of atomic and molecular data, the commission sponsored a science session immediately following the business meeting. Willing participants were each given a few minutes to present the work performed at their facility in providing fundamental atomic and molecular data or to present a science case for needed data.

Among the facilities providing data, presentations were made by E. Biémont (MONS; atomic lifetime measurements at the Lund Laser Centre, the creation of the DREAM and DESIRE atomic line databases for heavy elements, relativisitc calculations), S. Federman (Univ. Toledo; beam-foil atomic lifetimes at ultraviolet wavelengths, synchrotron measurements of molecules), H. Hartman (Lund Univ.; atomic lifetime measurements in the 1 ns to 1 μs range, forbidden line transition probabilities, planning for VUV absorption line measurements), G. Nave (NIST; Fourier transform spectroscopy, spectroscopy using EBIT, databases of atomic data), and G. del Zanna (Univ. Cambridge; calculation of atomic data through the Iron Project, CHIANTI, and VAMDC. VAMDC (Virtual Atomic and Molecular Data Centre, www.vamdc.eu) is a recently formed EU-funded international project to provide the infrastructure and system integration of atomic and molecular data uses in astrophysics, atmospheric physics, fusion, environmental sciences and industrial applications. G. M. Wahlgren (CUA/NASA-GSFC) mentioned his collaborative efforts with Lund Observatory and NIST to provide hyperfine structure constants for elements where line structure affects abundance analyses.

Data users presented needs that go beyond transition wavelengths and oscillator strengths. L. Mashonkina discussed data needs for non-LTE spectrum modeling, including radiative rate coefficients and photo-ionization cross sections and collision rates for electron impact excitation and ionization. For the latter, no such data are available for the rare earth elements and one must rely on a hydrogenic approximation. P. Bonifacio stressed the need for atomic line broadening parameters. W. F. Huebner discussed comet chemistry analysis, in particular for hydrocarbons and the data needs for studies of dissociation and ionization processes.

Glenn M. Wahlgren and Ewine F. van Dishoeck
President of the Commission and Vice-President 2009-2012, respectively

Transactions IAU, Volume XXVIIB
Proc. XXVII IAU General Assembly, August 2009
Ian F. Corbett, ed.

doi:10.1017/S1743921310005338

COMMISSION 41 — HISTORY OF ASTRONOMY
(HISTORY OF ASTRONOMY)

PRESIDENT — **Nha Il-Seong**
VICE-PRESIDENT — **Clive Ruggles**
PAST PRESIDENT — **Alexander Gurshtein**
ORGANIZING COMMITTEE — **David DeVorkin, Teije de Jong, Rajesh Kochhar, Tsuko Nakamura, Wayne Orchiston, Antonio A. P. Videira, Brian Warner**

PROCEEDINGS, BUSINESS SESSIONS

1. General Matters

The Business Meeting of Commission 41 was held in Sessions 2 on 5 August 2009, with Acting President Clive Ruggles (UK) in the Chair. He called for a moment of silence for those members who had passed away in the last triennium, including Prof. Xi Zezong (b. 1927, d. 2008 Dec 27) and Prof. Chen Meidong (b. 1936, d. 2008 Dec 30).

2. President's Report

Clive Ruggles began by noting that, as Vice President, he had taken over the role of Acting President in November 2008 at the request of the President, Nha Il-Seong, following an eye operation. He also noted that Rajesh Kochhar had taken over the role of Secretary from David DeVorkin in August 2007, following the resignation of the latter owing to pressures of work.

He noted the Commission's high level of activity in the International Year of Astronomy, and in particular the following three co-sponsored meetings: 1 "Astronomy education between past and future", Special Session 4 at the Rio GA, 2009 Aug 6-10, jointly with IAU C46 (Co-Chairs: Rajesh Kochhar, Jean Pierre de Greve and Ed Guinan). 2 "Accelerating the rate of astronomical discovery", Special Session 5 at the Rio GA, 2009 Aug 11-14, jointly with IAU C5 (Co-Chairs: Ray Norris and Clive Ruggles). 3 "Astronomy and its instruments before and after Galileo", a symposium organized by the Historical Instruments WG, and supported by INAF and the Astronomical Observatory of Padova, in Venice, 2009 Sep 28 - Oct 3 (Co-Chairs: Luisa Pigatto and Clive Ruggles).

He proceeded to note the signing of the formal Memorandum of Understanding between the IAU and UNESCO on Astronomy and World Heritage, and the subsequent formation of the IAU Working Group on Astronomy and World Heritage, chaired by Clive Ruggles with vice-chair Gudrun Wolfschmidt. The activities of the WG promise to make a real difference in helping to promote astronomical heritage sites on to the World Heritage List.

He noted with pleasure the effective completion, with the publication in the February 2009 issue of Journal for the History of Astronomy of the second part of Robert W. Smith's article on the development of extragalactic astronomy 1885-1965, of the General History of Astronomy. This hugely ambitious project was proposed by C41 back in 1970, at a time when the history of astronomy was still relatively undeveloped internationally as an academic discipline.

The aim was to provide a major synthesis of the field on a quite unprecedented scale. While not achieving final publication in the coherent form initially envisaged, the books and articles making up the GHA nonetheless comprise an extraordinarily comprehensive reference collection stretching from pre-literate astronomy to the modern era. Further information and a complete list of the books and articles making up the GHA are available on the Commission website http://www.historyofastronomy.org/.

He reported that no Newsletters have been produced in the last triennium, but that they had to a large extent been superseded by the new website (http://www.historyofastronomy.org/), which was completely restructured and overhauled in 2008 thanks to a small grant from the IUHPS/DHST.

He reported that Hilmar Duerbeck (Germany) had taken over from Steve Dick (USA) in 2008 November, in order to lead the WG's preparations for the 2012 Transit. He then proposed a formal vote of thanks to Steve Dick for all his work as WG Chair, culminating in IAU Colloquium 196 at the time of the 2004 Transit.

For information on other C41/ICHA activities during the triennium, the Acting President referred members to the existing triennial report, mentioning in particular:
o the "Astrum 2009" exhibition being organised in October 2009 in Rome by C41 members in collaboration with the Vatican Observatory;
o the activities of the thriving history of astronomy research group within the "Systmes de Rfrence Temps-Espace" Department within the Paris Observatory; and
o the 6th International Conference on Oriental Astronomy held in July 2008 at James Cook University, Australia.

The Acting President concluded by noting the "Oxford IX" International Symposium on Archaeoastronomy, to be held in January 2011 in Lima, Peru, and reported that the OC had already pledged the Commission's support for this important cross-disciplinary event. The meeting endorsed this support, and discussion followed about putting the meeting forward as a possible IAU Symposium. It was felt that any support from the IAU should be matched by support by international academic bodies of equal standing in some of the other fields contributing to the conference, such as the IUHPS.

The President's Report was unanimously accepted.

3. Secretary's Report

Rajesh Kochhar reported on the conduct of the election for the new OC. This took place in 2009 April, in accordance with the IAU's requirements that were strictly enforced this year. While some difficulties arose because of the discrepancies between the Commission's own membership list and the IAU's list, the election was nonetheless fair and democratic. He recorded the Commission's very grateful thanks to Michael Hoskin for acting as scrutineer.

The Secretary's Report was unanimously accepted.

4. Membership

The Acting President reported on the meeting for outgoing Commission Presidents, at which there had been a discussion on procedures for admitting existing IAU members into new Commissions.

He also noted that discrepancies continue to exist between the Commission's own membership list and the IAU's list, and undertook to submit a list of names to the IAU Secretariat of IAU members admitted to the Commission at the last two GA Business Meetings but whose names still do not appear on the IAU list.

The meeting approved the admission of the following IAU members to C41:
Jean-Louis Bougeret (France)
Chang Heon-Young (Korea)
Kim Yongchul [Yong-Cheol] (Korea)
Harry Nussbaumer (Switzerland)
Jos Osrio (Portugal)

5. Admission of Non-IAU Members to ICHA

The subject then turned to the Inter-Union Commission on the History of Astronomy (ICHA), of which all C41 members are members ipso facto. The Acting President noted that, despite the clear new procedures for the admission of non-IAU members to the ICHA that had been adopted at Prague, there had been no applicants since the first 6 months of the triennium. He suggested that new ways must be sought by the incoming OC of publicising the Commission admission procedures to professional historians. In order that the broader interests of the ICHA are adequately represented on the C41 OC, it was agreed that moves should be move to identify a suitable person to take responsibility for IUHPS/DHST liaison and to co-opt them on to the OC.

6. Working Groups

The meeting noted the IAU EC's new directive that WG Chairs should not serve for more than 6 years, and approved the continuation of each of C41's five WGs.

7. Closing Business

The meeting noted the new membership of the new OC:
President Clive Ruggles (UK)
Vice-President Rajesh Kochar (India)
Immediate Past President Nha Il-Seong (Korea)
Organizing Committee Juan A. Belmonte Avils (Spain), Brenda Corbin (USA), Teije de Jong (Netherlands), Ray Norris (Australia), Luisa Pigatto (Italy), Mitsuru Soma (Japan), Chris Sterken (Belgium), Sun Xiaochun (China)

The Acting President then handed over the meeting to himself as new President. In this capacity he then closed the meeting.

8. Report on C41 Science Meeting, Sessions 3 and 4, 5 August 2009

by I. S. Glass, W. Orchiston & Mitsuru Sma

The meeting generated considerable general interest and was well-attended.

The first session was chaired by Ian S Glass (SAAO). Though J.H. Seiradakis's lecture extended well into the tea break, such was the interest that the entire audience stayed to the end.

14.00-14.30 Flavia Pedroza Lima (Fundao Planetrio da Cidade do Rio de Janeiro, Instituto de Geocincias - UNICAMP, Rio de Janeiro, Brazil; flaviapedroza@gmail.com) & Silvia F. de M. Figueira (Instituto de Geocincias - UNICAMP, Rio de Janeiro, Brazil): "Ethnoastronomy in Brazil" 14.30-15.00 Oscar T. Matsuura (Rua Itacema, 199 apto. 43, So Paulo, Brazil; otmats@terra.com.br): "Pioneering Astronomical Activities in the New World"
15.00-15.30 John H. Seiradakis (Department of Physics, Aristotle University of Thessaloniki, Thessaloniki, Greece; jhs@astro.auth.gr): "The Antikythera Mechanism: An Ancient Greek Astronomical Computer"

The second session was chaired by Mitsuru Sma (NAOJ).

16.00-16.30 Ian Glass (South African Astronomical Observatory, South Africa; isg@saao.ac.za): "Alpha and Proxima Centauri: The Quest for the Nearest Star"
16.30-17.00 Erik Hg (Niels Bohr Institute/Copenhagen University, Copenhagen, Denmark; erik.hoeg@get2net.dk): "Seventy-five Years of Photoelectric Astrometry: From Experiments in 1925 to the Tycho-2 Catalogue"
17.00-17.30 Mitsuru Sma & Kiyotaka Tanaikawa (National Astronomical Observatory of Japan, Tokyo, Japan; Mitsuru.Soma@nao.ac.jp; tanikawa@exodus.mtk.nao.ac.jp): "The Investigation of the Earth's Rotation From Ancient Occultation Records"

9. POSTER PAPERS

Clifford Cunningham, Brian Marsden & Wayne Orchiston (Centre for Astronomy, James Cook University, Townsville, Australia; Clifford.Cunningham@jcu.edu.au; Brian. Marsden@jcu.edu. au; Wayne.Orchiston@jcu.edu.au): "How the First Dwarf Planet Became the Asteroid Ceres"

Hong-Jin Yang (Korean Astronomy and Space Science Institute, Daejeon, Korea; hjyang@kasi. re.kr), Changbom Park (Korea Institute for Advanced Study, Seoul, Korea; cbp@kias.re.kr) and Myeong-Gu Park (Kyungpook National University, Daegu, Korea; mgp@knu.ac.kr): "Astronomical Aspects of Korean Dolmens"

Jefferson Sauter, Irakli Simonia, Richard Stephenson & Wayne Orchiston (Centre for Astronomy, James Cook University, Townsville, Australia; Jefferson.Sauter@jcu. edu.au; Irakli. Simonia@jcu.edu.au; Richard.Stephenson@jcu.edu.au; Wayne. Orchiston@jcu.edu.au): "The Legendary 4th Century AD Georgian Solar Eclipse: Fact or Fantasy?"

John Pearson, Wayne Orchiston & Kim Malville (Centre for Astronomy, James Cook University, Townsville, Australia, John.Pearson1@jcu.edu.au; Wayne.Orchiston @jcu.eu.au; kimmalville@hotmail.com): "Forty Years of Solar Eclipse Science at the Lick Observatory and Paradigm Shifts in Solar Physics"

Stella Cottam, Wayne Orchiston & Richard Stephenson (Centre for Astronomy, James Cook University, Townsville, Australia; Stella.Cottam@jcu.edu.au; Wayne. Orchiston@ jcu.edu.au; Richard.Stephenson@jcu.edu.au): "The 7 August 1869 Total Solar Eclipse and the Popularisation of Astronomy in the USA"

Tomoko Fujiwara (Center for Research and Advancement in Higher Education, Kyushu University, Japan; tomokof@rche.kyushu-u.ac.jp): "Detection of Unknown Objects in Tycho Brahe's Astronomiae Instauratae Progymnasmata"

Clive Ruggles
Vice-President and Acting President of the Commission

Transactions IAU, Volume XXVIIB
Proc. XXVII IAU General Assembly, August 2009
Ian F. Corbett, ed.

doi:10.1017/S174392131000534X

COMMISSION 41 WORKING GROUP on ASTRONOMY AND WORLD HERITAGE

CHAIR **Clive Ruggles**
VICE-CHAIR **Gudrun Wolfschmidt**
MEMBERS **Ennio Badolati, Alan Batten, Juan Belmonte, Ragbir Bhathal, Peter Brosche, Suzanne Dbarbat, David DeVorkin, Hilmar W. Duerbeck, Priscilla Epifania, Roger Ferlet, Jos Funes, Ian S. Glass, Elizabeth Griffin, Alexander Gurshtein, John Hearnshaw, George Helou, Bambang Hidayat, Thomas Hockey, Jarita Holbrook, Manuela Incerti, S. O. Kepler, Rajesh Kochhar, Edwin C. Krupp, Kurt Locher, Penka Maglova-Stoeva, Areg Mickaelian, Bjorn R. Pettersen, Mara Cristina Pineda de Caras, Gennadiy Pinigin, Luciana Pompeia, Zhanna Pozhalova, Shi Yun-li, Irakli Simonia, Francoise Le Guet Tully, Richard Wainscoat**

PROCEEDINGS BUSINESS SESSIONS August 2009

What follows is a short report on the Business Meeting of the Astronomy and World Heritage Working Group held on Thursday August 6, 2009. This was the first formal Business

Meeting of the Working Group since its formation following the signing of the Memorandum of Understanding between the IAU and UNESCO on Astronomy and World Heritage in October 2008.

A fuller account of the activities of the Working Group may be found in the "Astronomy and World Heritage" by Clive Ruggles that appears in Chapter I of these Transactions.

1. Officers and Members

The chair Clive Ruggles, the vice-chair Gudrun Wolfschmidt and the secretary Tom Hockey were confirmed in office. The existing membership was confirmed, which includes some non-IAU members. This is necessary in a WG whose remit is inherently cross-disciplinary. It was agreed to invite all Thematic Study authors to join the WG, including those who are not IAU members. The meeting noted that there is a good balance with regard to gender and to geographical distribution.

The following were confirmed as new WG members: Dieter Engels (Germany), Tomoko Fujiwara (Japan), Brian Mason (USA), Nha Il-Seong (Korea), Jos Osrio (Portugal), Malcolm Smith (Chile), and Yang Hong-Jin (Korea).

2. Terms of Reference

The WG's Terms of Reference were confirmed, as follows:

1. To work on behalf of the IAU to help ensure that cultural properties and artefacts significant in the development of astronomy, together with the intangible heritage of astronomy, are duly studied, protected and maintained, both for the greater benefit of humankind and to the potential benefit of future historical research.

(a) To fulfil, on behalf of the IAU, its commitments under its Memorandum of Understanding with UNESCO to promote Astronomy and World Heritage and to explore and implement UNESCO's Astronomy and World Heritage Initiative.

(b) To liaise with other international and national bodies concerned with astronomical history and heritage, in so far as their interests and activities impinge on these aims, to help achieve these aims.

2. To work, in conjunction with IAU C41 (History of Astronomy), IAU C46 (Education) and other Commissions and Working Groups within the IAU as appropriate, to enhance public interest, understanding, and support in the field of astronomical heritage.

Notes

(a) The range of properties and objects in question includes ancient sites and monuments with demonstrable links to the sky (such as Stonehenge), instruments of all ages, archives, and historical observatories.

(b) This initiative aims to ensure the recognition, promotion and preservation of achievements in science through the nomination to the World Heritage List of properties whose outstanding significance to humankind derives in significant part from their connection with astronomy.

3. Progress to Date

The main work item for the WG so far has been the Thematic Study on the Heritage Sites of Astronomy and Archaeoastronomy in the context of the UNESCO World Heritage Convention, which is being prepared in collaboration with ICOMOS,the Advisory Body to UNESCO that assesses WHL applications relating to cultural heritage. This will be edited by Clive Ruggles together with Michel Cotte of ICOMOS. The editors have completed, in draft form, an overview that will constitute the opening chapter. The 14 themes for the Thematic Study have been defined, namely Earlier Prehistory; Pre-Columbian and indigenous America; Later Prehistoric Europe; Later prehistoric and indigenous Africa, Asia, Australasia and Oceania; Ancient and medieval Far East; India; Mesopotamia and the Middle East; Ancient Egypt; The Classical World; Arabic and Islamic Astronomy; Medieval astronomy in Europe; Astronomy from the Renaissance to the mid-twentieth century; Contemporary astronomy and astrophysics; and Space astronomy. The authors have identified and nine draft chapters have been received. The intention is to complete the Thematic Study by January 2010. The three sessions following the Business Meeting were devoted to a detailed discussion of the Thematic Study.

A second key activity is to create a public database of properties with a relationship to astronomy that will serve as a forum for information gathering and exchange. The database will be produced by transferring and then expanding UNESCO's Astronomy and World Heritage "Timeframe", which is already available on the WHC website, but which is restricted to sites that are already on the World Heritage List.

It was noted that the September 2009 issue of UNESCO's quarterly magazine World Heritage Review will be devoted to astronomical and science heritage, with the leading article written by Clive Ruggles.

The Working Group has developed strong links with the Starlight Initiative and was represented at its meeting in Fuerteventura in March 2009. A significant outcome of this meeting was the establishment of a "Dark Skies Advisory Group" whose remit is to support IUCN activities relating to dark skies (IUCN being the Advisory Body to UNESCO that assesses WHL applications relating to natural heritage). The WG is represented on this Advisory Group by its Chair.

4. Future meetings

It was noted that a number of meetings will take place during 2009 as part of the cycle of activities planned by the UNESCO-IAU Astronomy and World Heritage Initiative within the framework of the IYA Global Cornerstone Project on Astronomy and World Heritage. In particular, immediately following the GA there is an international conference dedicated to the theme of "Astronomy and World Heritage: Across Time and Continents" to be held at Kazan in the Russian Federation. All of these meetings will provide an opportunity for WG members to obtain further input for and feedback on the Thematic Study and the WG's other activities.

Clive Ruggles

Chair of the Working Group

Transactions IAU, Volume XXVIIB
Proc. XXVII IAU General Assembly, August 2009
Ian F. Corbett, ed.

doi:10.1017/S1743921310005351

COMMISSION 46 ASTRONOMY EDUCATION AND DEVELOPMENT

PRESIDENT	**Magda D. Stavinschi**
VICE-PRESIDENT	**Rosa M. Ros**
PAST PRESIDENT	**Jay M. Pasachoff**
ORGANIZING COMMITTEE	**Johannes Andersen**
	Susana Deustua
	Jean-Pierre de Greve
	Edward F. Guinan
	Hans J. Haubold
	John B. Earnshaw
	Barrie W. Jones
	Rajesh K. Kochhar
	Kam-Ching Leung
	Laurence A. Marschall
	John R. Percy
	Silvia Torres-Peimbert

PROCEEDINGS BUSINESS SESSIONS, 5 and 13 August 2009,

1. Business

Commission 46 had held two sessions of business meetings during the General Assembly in Rio de Janeiro. Both of them were chaired by Magda Stavinschi.

The first meeting took place on August 5th from 14:00 to 17:30.
The programme of this meeting was as follows:

Magda Stavinschi: General report of the IAU C46 (2006-2009), 20 min
John Hearnshaw: Report for the PG for the World-wide development of Astronomy for 2006-2009, 20 min
Jean Pierre de Greve: ISYA: Istanbul 2008, and future prospects, 20 min
Kam-Ching Leung: Exchange Astronomers Program, 10 min
Barrie Jones: Newsletters & National liaisons, 20 min
Rosa M. Ros: Courses organized in Ecuador & Chile in cooperation with UNESCO, 15 min
Ed Guinan: The IAU Teaching Astronomy for Development (TAD)Program - Recent Developments and Opportunities, 15 min
Paulo S. Bretones: The Latin-American Journal of Astronomy Education (RELEA), 10min
William van Altena: A new book of astrometry, an important tool for education, 10 min
Julio Blanco Zarate, Astronomy education in Uruguay, 5 min

The meeting ended with a general discussion of 25 min.

The report of the President summarized the standing of Commission 46 from 2006 to 2009. The website of the Commission was the first topic considered. The situation was not satisfactory. After several efforts, and with two site addresses http://www.iau.org/EDUCATION.8.0.html and http://www.iau46.obspm.fr really the Commission does not have an active website which is essential for its activities.

The chairs of the PG active groups presented the following reports:

PG International Schools for Young Astronomers (ISYA), chaired by Jean-Pierre de Greve, organized the International School in 2007 - Malaysia, 2008 - Turkey and 2009 - Trinidad & Tobago.
PG Teaching Astronomy for Development (TAD), chaired by Edward Guinan & Laurence Marschall, visited Vietnam, Mongolia, Morocco, Nicaragua, Kenya, South Africa, Trinidad & Tobago, Philippines, D.P.R. Korea.
PG for the Worldwide Development of Astronomy (WWDA), chaired by John Hearnshaw, visited 19 nations which currently have individual IAU members but these countries are not members of IAU
PG for Newsletters & Webpage, chaired by Barrie Jones and vice-chaired by Chantal Balkovski, published the Newsletter twice a year
PG on Public Education at the times of the Solar Eclipses & Transit Phenomena (PE-TSETP), chaired by Jay Pasachoff, offers timely advice to countries that will experience a Solar Eclipse and transits of the Sun by Venus and Mercury (http://www.eclipses.info).
PG Network for Astronomy School Education (NASE), chaired by Rosa M. Ros, which will begin their activities in 2010 had organized a pilot course in Ecuador and Peru in cooperation with UNESCO in 2009.
There are two Special PG. PG for Collaborative Programs, chaired by Hans Haubold, in order to carry out activities with UNESCO, COSPAR, UN, ICSU, and other international organizations. And PG for Exchange of Astronomers (EA), chaired by John Percy and vice-chaired by Kam-Ching Leung, with detailed rules and application procedures in http://physics.open.ac.uk/IAU46/travel.html

The list of Events supported by Commission 46 is:

"Joint European and National Astronomy Meeting 2007: Our non-stable Universe". Yerevan, Armenia
JENAM SpS5 Astronomy Education in Europe
20-25 August, 2007

IAU GA, Rio de Janeiro
Young Astronomers Events
3-14 August 2009

IAU GA, Rio de Janeiro
SPS4 Astronomy Education between Past and Future
6, 7, 10 August, 2009

IAU GA, Rio de Janeiro
SPS5 Accelerating the Rate of Astronomical Discovery
11 - 14 August, 2009

IAU JOINT SYMPOSIA, Venice
Astronomy & its Instruments - Before and After Galileo
28 September - 3 October, 2009

IAU - IMU Symposium, Madrid
"Mathematics and Astronomy, a joint long journey"
November 23 - 27, 2009

IAU - EAAE Course, Madrid
"Adventure in Teaching Astronomy"
November 27 - December 1, 2009

After the talks a short discussion of halt an hour took place. The main topic was the website situation.

The second meeting took place on August 13th from 9:00 to 13:30. The program of this meeting was as follows:
Magda Stavinschi: Appointing the new OC, 20 min

Jay Pasachoff: Report on the PG on Public Understanding at the Times of Eclipses, including the results of the 22 July 2009 TSE in China. 20min
Kevin Govender: Activities and Future of the Developing Astronomy Global Cornerstone, 20 min
John Hearnshaw: Future plans of PGWWDA for 2009-12, 10 min
- Rosa M. Ros: "Network for Astronomy School Education" NASE in the future, 20min

The new Organizing Committee was appointed:

President: Rosa M. Ros (Spain)
Vice-President: John Hearnshaw (New Zealand)
Retiring President: Magda Stavinschi (Romania)

SOC (Chairs of PG):
Jean-Pierre de Greve, Ed Guinan, Hans Haubold, Barrie Jones, Larry Marschall, Jay Pasachoff and Beatriz Garca

After the talks a general discussion on the "Questions for participants at NASL LUNCH" took place. The website situation had been discussed again. The offer from IAU EC of support for the website's expenses was well accepted.

Some changes regarding PGs were considered. In particular the PG for Exchanges of Books, Journals and Materials was not considered necessary, at present. This facility can be included in the TAD PG. The PGs on Newsletter and National Liaisons, both of them chaired by Barrie Jones, had been merged into a single PG. The PG on Exchange of Astronomers had been considered necessary but it seams convenient to introduce some changes in order to promote and increase its visibility. Finally the decision was to include their tasks in the new Office generated by the Decanal Plan.

In the general debate the situation of several Cornerstone Projects of IYA2009 after this year had been considered. In particular Galileo Teacher Training Project, Universe Awareness and Developing Astronomy Globally were interested in being included in the Commission 46. The decision was not taken in order to wait for the end of the International Year and the general situation for all the Cornerstone Projects.

Magda G. Stavinschi *President of the Commission* Rosa M. Ros *Incoming President of the Commission*

Transactions IAU, Volume XXVIIB
Proc. XXVII IAU General Assembly, August 2009
Ian F. Corbett, ed.

doi:10.1017/S1743921310005363

COMMISSION 50 — PROTECTION OF EXISTING AND POTENTIAL OBSERVING SITES

PRESIDENT — **Wim van Driel**
VICE-PRESIDENT — **Richard Green**
PAST PRESIDENT — **Richard Wainscoat**
ORGANIZING COMMITTEE — **Elizabeth Alvarez del Castillo, Carlo Blanco, Dave Crawford, Margarita Metaxa, Masatoshi Ohishi, Woody Sullivan, Tasso Tzioumis.**

PROCEEDINGS BUSINESS SESSION, 7 August 2009

The past President of Division XII, Malcolm Smith, took the chair of the business meeting of C50, in the absence of the past President of C50, Richard Wainscoat. He described the procedure of the election in 2006 of the then-time vice-president of C50, Richard Wainscoat, as President of C50, after the untimely deatch of the then-time C50 President, Hugo Schwarz. He also described the procedure for electing the incoming OC members and officers.

The outcome of the elections is as follows:
President: Wim van Driel (France)
Vice-president: Richard Green (USA)
Past President: Richard Wainscoat (USA)
Board: Elizabeth Alvarez del Castillo (USA), Carlo Blanco (Italy),Dave Crawford (USA), Margarita Metaxa (Greece), Masatoshi Ohishi (Japan), Woody Sullivan (USA), Tasso Tzioumis (Australia).

President of C50's Working Group on Controlling Light Pollution: Richard Wainscoat (USA).

The meeting reviewed Resolution B5 in Defence of the night sky and the right to starlight, which was proposed by the IAU GA EC WG IYA2009 Cornerstone Project Dark Skies Awareness and the members of the Starlight Declaration, and supported by C50.

A report was given on the coming threat of white lights, an account was given of the work done for the International Commission on Illumination (CIE) by C50 members, followed by a status report on the "Starlight" reserves initiative and on the current status of the IYA Dark-Skies Awareness Cornerstone. The meeting was finished by a session on the protection of radio astronomy observing sites, with a tutorial on the subject and a report of future issues in radio spectrum management.

Wim van Driel
President of the Commission

Transactions IAU, Volume XXVIIB
Proc. XXVII IAU General Assembly, August 2009
Ian F. Corbett, ed.

doi:10.1017/S1743921310005375

COMMISSION 55 — COMMUNICATING ASTRONOMY WITH THE PUBLIC

PARTAGER L'ASTRONOMIE

PRESIDENT	**Ian E. Robson**
VICE-PRESIDENT	**Dennis R. Crabtree**
SECRETARY	**Lars Lindberg Christensen**
ORGANIZING COMMITTEE	**Oscar Alvarez Pomares, Augusto Damineli Neto, Richard T. Fienberg, Anne Green, Ajit K. Kembhavi, Birgitta Nordström, Kazuhiro Sekiguchi, Patricia Ann Whitelock, Jin Zhu**

COMMISSION 55 WORKING GROUPS

Div. XII / Commission 55 WG	**Communicating Astronomy Journal**
Div. XII / Commission 55 WG	**New Ways of Communicating Astronomy with the Public**
Div. XII / Commission 55 WG	**CAP Conferences**
Div. XII / Commission 55 WG	**Best Practices**
Div. XII / Commission 55 WG	**VAMP**
Div. XII / Commission 55 TF	**Washington Charter**

TRIENNIAL REPORT 2006-2009

1. Introduction

Since its formation at the XXVI General Assembly in Prague in 2006, amazing progress has been made by Commission 55, all due to the work of the key activists and enthusiasts. The web-page for the Commission contains a wealth of information and is one of the key foundations and tools for the Commission. The web address is
http://www.communicatingastronomy.org

The focus over the past three years has, not surprisingly, been on the International Year of Astronomy 2009 (IYA2009). This has included membership on the IAU Executive Working Group, which in itself has been no mean task. As a result, the ability to make as much progress as we had intended across the entire portfolio of C55 has been limited. Nevertheless, I am confident that the effort expounded on IYA2009 has been extremely successful so far and furthermore, this has brought about a very useful secondary focus, especially in the areas of New Media and the VAMP work (see below) and in other ways that were not envisaged in the Working Groups - such as the Galileoscope and FETTU to name but two.

2. The Washington Charter - led by Dennis Crabtree

Progress has been limited due to effort availability but the current picture is that it has been endorsed by: 19 professional astronomical societies or agencies and 12 universities, labs, facilities and other organisations.

3. VAMP - Virtual Astronomy Multimedia Project - led by Robert Hurt

Truly spectacular progress has been made on this project; there has been enormous progress on agreement on metadata and tools, culminating in the launch of 'Portal to the Universe' (http://www.portaltotheuniverse.org/) at the NAM/JENAM meeting in the UK in April 2009 and sponsored by the IAU and ESO. Portal to the Universe is a global, one-stop clearinghouse for online astronomy content, with news, blogs, video and audio podcasts, images, videos targeting the complete range from laypeople, press, educators, to scientists. During its first five months of operation it has aggregated 3,000 press releases; 1,800+ podcast episodes; 16,000+ blog posts and received more than 300,000 visitors. VAMP is so far an amazing success story and the web address for the work is at: http://virtualastronomy.org/

4. Best Practices - led by Lars Lindberg Christensen

This project has also been impacted by the IYA2009 activities but, importantly, some of the aims and best practices have been directly incorporated into the IYA activities through instructions and guidelines for organisers along with the all-important and often neglected aspect of evaluation.

5. Communicating Astronomy Journal - led by Pedro Russo

The CAP Journal has been another spectacular success story. From the initial meetings and discussions with the Astronomy Educational Review amongst others, the CAP Journal was launched in October 2007. It has a formal organisational structure and editorial board and provides a refereeing process for submitted papers. Six editions have been published to date, containing over 60 articles. There are 1,500 hard-copy subscribers and 2,700 on-line subscribers. Thanks are given to ESO who have supported the editorial and shipping costs (see later). The journal is incredibly well produced and the covers themselves are collectors' items of spectacular imagery. The web address for the journal is at http://www.capjournal.org.

6. New Ways of Communicating Astronomy with the Public - led by Michael West

This work has also taken off in a big way, albeit through a life of its own rather than through the Group. This is one of the amazing spin-offs that has transpired and it has resulted through the efforts of a small number of devoted 'geeks/gurus'. As well as blogging and podcasts, Twitter has recently appeared on the scene alongside Facebook and Second Life. These are all activities that are blossoming and illustrate one of the key features for the future (see below). In other areas, key examples of new ways of communicating are 'citizen science' projects like 'Galaxy Zoo', which has demonstrated that this is a vibrant way forward for the future, especially for the younger generation of producers and communicators.

7. Communicating Astronomy with the Public Conference - led by Ian Robson

There has been one CAP conference since Prague, this was CAP2007 in October 2007 in Athens. The conference was co-organized by the National Observatory of Athens and the Eugenides Planetarium and was a tremendous success, with over 200 participants. As was to be expected, much of the focus was on looking forward to IYA2009. One of the interesting features is that the CAP conferences continues to attract a significant audience from C46 (Education), even though we strive to focus on specific outreach activities. The Proceedings were published in book format as well as on-line.

8. Next Three Years

The Structure of Commission 55 for the coming three years will be as follows:

PRESIDENT	Dennis R. Crabtree
VICE-PRESIDENT	Lars Lindberg Christensen
SECRETARY	Pedro Russo
ORGANIZING COMMITTEE	Oscar Alvarez Pomares, Augusto Damineli Neto, Richard T. Fienberg, Anne Green, Ajit K. Kembhavi, Anja Anderson, Ian Robson, Kazuhiro Sekiguchi, Patricia Ann Whitelock, Jin Zhu

The Business Meeting was very well attended; this was especially notable seeing as there was an unfortunate clash for the outreach/education contingent. The proposed Working Groups for the next 3 years was the topic of much discussion during the business meeting, with a range of views being expressed and some clarification being required for the IAU regarding eligibility of Working Group Chairs. The Washington Charter, CAP Journal and CAP Conferences remain unchanged, while Robert Hurt takes over as Chair of the VAMP group. The Best Practices Group has been would up, with the activity being subsumed into other Working Groups. The most discussion surrounded the 'New Ways of Communicating Astronomy' with the Public Group, which had been proposed to have been wound up and the work subsumed into VAMP. During the debate it was clear that a better way forward was for the Group to continue with a change of name to "New Media", and to cater for the likes of Facebook, Twitter and what might follow. It was agreed that Pamela Gay would be an ideal Chair for this Group and Lars Lindberg Christensen Co-Chair.

Further discussion centred around the 'citizen science' projects and the work of the amateurs. It was felt that the projects such as 'Galaxy Zoo' readily fitted into the remit of 'New Media' but that the role of the amateurs was far from clear. Opening up astronomy centres, clubs, participating in communicating with the public was a clear and essential part of Commission 55's remit, and the roles of the amateurs in this area should be supported. However, the work of the other distinct band of 'professional amateurs' (dedicated novae, supernovae and Solar System observers) was very different and fitted far better in other Commissions or Divisions within the IAU. It was not believed that there was a single solution for the amateur community within the IAU.

Looking forward to the next three years, planning is already well underway for the CAP2010 conference. This will be held in Cape Town, from March 15th-19th 2010. The first announcement has been released and further details for the conference can be found at: http://www.communicatingastronomy.org. The conference will focus on the legacy from IYA2009 and the ongoing plans for the future. There is also an associated workshop for communicating astronomy in the developing world. One of the questions that I raised was because CAP2010 was out of synch (it should have been in 2009 but we had agreed everyone would be too busy and 2010 would give the opportunity for a look-back) whether the next CAP should be in late 2011 or in 2012. From the discussion there were a number of good reasons proposed that suggests that late 2011 would be the preferred option.

It is also clear that we need to 'up' our membership; everyone should update their IAU membership list and we (Dennis Crabtree) should petition that C55 should be exempt from the '3 Commission rule'. A note will be included in the next IAU electronic newsletter requesting IAU members to join C55. We also need to make the 'next big push' on the Washington Charter following IYA2009. We need to garner more keen supporters who can influence their 'local' agencies etc. This should be one of the main thrusts and targets of the coming three years.

Portal to the Universe is an enormous success, but can it become a one-stop image repository or should we be looking to offload this or to look for other services that currently exist? Is it further expandable and if so by how much?

The CAP Journal is now in need of direct sponsorship for the printing costs and a number of suggestions were made as to possible sources as the cost is rather modest given the quality of the product. Perhaps sponsorship as in the IYA2009 brochures might be a model to follow.

Finally, there remains the very difficult problem of what do we/can we retain both nationally and globally, post IYA2009. In this I refer to all the fabulous work that has gone into IYA2009 web-pages. Maintaining web-pages requires a resource and unless one has access to a ready supply of volunteers, or has enough staff that this activity can be 'hidden', the resource means cash. This will be a serious problem for many countries and how to retain the legacy activity, often manifest through web-pages, will be a serious question facing many involved in the outreach activities.

In conclusion, let us say how satisfying the last three years have been. Commission 55 is up, running and fully committed. It has more than delivered on its targets and the work that has gone into IYA2009 activity throughout the world (and outside of C55) has been tremendous. The Commission is now poised to move forward and take the legacy of IYA2009 into the future through the Washington Charter, New Media etc to ensure that Outreach Activity is a respected and rewarded activity for all those who practice it.

Ian E. Robson
president of the Commission
Dennis R. Crabtree
vice-president of the Commission
Lars Lindberg Christensen
secretary of the Commission

Transactions IAU, Volume XXVIIB
Proc. XXVII IAU General Assembly, August 2009
Ian F. Corbett, ed.

doi:10.1017/S1743921310005387

CHAPTER VII
XXVII GENERAL ASSEMBLY

1. IAU Statutes Rio de Janeiro, Brazil, 4 August 2009

I OBJECTIVE

1. The International Astronomical Union (hereinafter referred to as the Union) is an international non-governmental organization. Its objective is to promote the science of astronomy in all its aspects.

II DOMICILE AND INTERNATIONAL RELATIONS

2. The legal domicile of the Union is Paris, France.

3. The Union adheres to, and co-operates with the body of international scientific organizations through: the International Council for Science (ICSU). It supports and applies the policies on the Freedom, Responsibility, and Ethics in the Conduct of Science defined by ICSU.

III COMPOSITION OF THE UNION

4. The Union is composed of:

4.a. National Members (adhering organizations)

4.b.Individual Members (adhering persons)

IV NATIONAL MEMBERS

5. An organization representing a national professional astronomical community, desiring to promote its participation in international astronomy and supporting the objective of the Union, may adhere to the Union as a National Member. Exceptionally, a National Member may represent the community in the territory of more than one nation, provided that no part of that community is represented by another National Member.

6. An organization desiring to join the Union as a National Member while developing professional astronomy in the community it represents may do so:

6.a. on an interim basis, on the same conditions as above, for a period of up to nine years. After that time, it must apply to become a National Member on a permanent basis, or its membership in the Union will terminate.

6.b. on a prospective basis for a period of up to six years if its community has less than six Individual Members. After that time it must apply to become a National Member on either an interim or permanent basis or its membership in the Union will terminate.

7. A National Member is admitted to the Union on a permanent, interim, or prospective basis by the General Assembly. It may resign from the Union by so informing the General Secretary, in writing.

8. A National Member may be either:

8.a. the organization by which scientists of the corresponding nation or territory adhere to ICSU or:

8.b. an appropriate National Society or Committee for Astronomy, or:

8.c. an appropriate institution of higher learning.

9. The adherence of a National Member is automatically suspended if its annual contributions, as defined in Articles 23c and 23e below have not been paid for five years; it resumes, upon the approval of the Executive Committee, when the arrears in contributions have been paid in full. After five years of suspension of a National Member, the Executive Committee may recommend to the General Assembly to terminate the Membership.

10. A National Member is admitted to the Union in one of the categories specified in the Bye-Laws.

V. INDIVIDUAL MEMBERS

11. A professional scientist who is active in some branch of astronomy may be admitted to the Union by the Executive Committee as an Individual Member. An Individual Member may resign from the Union by so informing the General Secretary, in writing.

VI. GOVERNANCE

12. The governing bodies of the Union are:

12.a. The General Assembly;
12.b. The Executive Committee; and
12.c. The Officers.

VII. GENERAL ASSEMBLY

13. The General Assembly consists of the National Members and of Individual Members. The General Assembly determines the overall policy of the Union.

13.a. The General Assembly approves the Statutes of the Union, including any changes therein.

13.b. The General Assembly approves Bye-Laws specifying the Rules of Procedure to be used in applying the Statutes.

13.c. The General Assembly elects an Executive Committee to implement its decisions and to direct the affairs of the Union between successive ordinary meetings of the General Assembly. The Executive Committee reports to the General Assembly.

13.d. The General Assembly appoints a Finance Committee, consisting of one representative of each National Member having the right to vote on budgetary matters according to Article 14.a. , to advise it on the approval of the budget and accounts of the Union. The General Assembly also appoints a Finance Sub-Committee to advise the Executive Committee on its behalf on budgetary matters between General Assemblies.

13.e. The General Assembly appoints a Special Nominating Committee to prepare a suitable slate of candidates for election to the incoming Executive Committee.

13.f. The General Assembly appoints a Nominating Committee to advise the Executive Committee on the admission of Individual Members.

14. Voting at the General Assembly on issues of a primarily scientific nature, as determined by the Executive Committee, is by Individual Members. Voting on all other matters is by National Member. Each National Member authorises a representative to vote on its behalf.

14.a. On questions involving the budget of the Union, the number of votes for each National Member is one greater than the number of its category, referred to in article 10. National Members with prospective or interim status, or which have not paid their dues for years preceding that of the General Assembly, may not participate in the voting.

14.b. On questions concerning the administration of the Union, but not involving its budget, each National Member has one vote, under the same condition of payment of dues as in Article 14.a.

14.c. National Members may vote by correspondence on questions concerning the agenda for the General Assembly.

14.d. A vote is valid only if at least two thirds of the National Members having the right to vote by virtue of article 14.a. participate in it by either casting a vote or signalling an abstention. An abstention is not considered a vote cast.

15. The decisions of the General Assembly are taken by an absolute majority of the votes cast. However, a decision to change the Statutes requires the approval of at least

two thirds of all National Members having the right to vote by virtue of article 14.a. Where there is an equal division of votes, the President determines the issue.

16. Changes in the Statutes or Bye-Laws can only be considered by the General Assembly if a specific proposal has been duly submitted to the National Members and placed on the Agenda of the General Assembly by the procedure and deadlines specified in the Bye-Laws.

VIII. EXECUTIVE COMMITTEE

17. The Executive Committee consists of the President of the Union, the President-Elect, six Vice-Presidents, the General Secretary, and the Assistant General Secretary, elected by the General Assembly on the proposal of the Special Nominating Committee.

IX. OFFICERS

18. The Officers of the Union are the President, the General Secretary, the President-Elect, and the Assistant General Secretary. The Officers decide short-term policy issues within the general policies of the Union as decided by the General Assembly and interpreted by the Executive Committee.

X. SCIENTIFIC DIVISIONS

19. As an effective means to promote progress in the main areas of astronomy, the scientific work of the Union is structured through its Scientific Divisions. Each Division covers a broad, well-defined area of astronomical science, or deals with international matters of an interdisciplinary nature. As far as practicable, Divisions should include comparable fractions of the Individual Members of the Union.

20. Divisions are created or terminated by the General Assembly on the recommendation of the Executive Committee. The activities of a Division are organized by an Organizing Committee chaired by a Division President. The Division President and a Vice-President are elected by the General Assembly on the proposal of the Executive Committee, and are ex officio members of the Organizing Committee.

XI. SCIENTIFIC COMMISSIONS

21. Within Divisions, the scientific activities in well-defined disciplines within the subject matter of the Division may be organized through scientific Commissions. In special cases, a Commission may cover a subject common to two or more Divisions and then becomes a Commission of all these Divisions.

22. Commissions are created or terminated by the Executive Committee upon the recommendation of the Organizing Committee(s) of the Division(s) desiring to create or terminate them. The activities of a Commission are organized by an Organizing Committee chaired by a Commission President. The Commission President and a Vice-President

are appointed by the Organizing Committee(s) of the corresponding Division(s) upon the proposal of the Organizing Committee of the Commission.

XII. BUDGET AND DUES

23. For each ordinary General Assembly the Executive Committee prepares a budget proposal covering the period to the next ordinary General Assembly, together with the accounts of the Union for the preceding period. It submits these, with the advice of the Finance Sub-Committee, to the Finance Committee for consideration before their submission to the vote of the General Assembly.

23.a. The Finance Committee examines the accounts of the Union from the point of view of responsible expenditure within the intent of the previous General Assembly, as interpreted by the Executive Committee. It also considers whether the proposed budget is adequate to implement the policy of the General Assembly. It submits reports on these matters to the General Assembly before its decisions concerning the approval of the accounts and of the budget.

23.b. The amount of the unit of contribution is decided by the General Assembly as part of the budget approval process.

23.c. Each National Member pays annually a number of units of contribution corresponding to its category. The number of units of contribution for each category shall be specified in the Bye-Laws.

23.d. A vote on matters under article 23 is valid only if at least two thirds of the National Members having the right to vote by virtue of article 14.a. cast a vote. In all cases an abstention is not a vote, but a declaration that the Member declines to vote.

23.e. National Members having interim status pay annually one half unit of contribution.

23.f. National Members having prospective status pay no contribution.

23.g. The payment of contributions is the responsibility of the National Members. The liability of each National Members in respect of the Union is limited to the amount of contributions due through the current year.

XIII. EMERGENCY POWERS

24. If, through events outside the control of the Union, circumstances arise in which it is impracticable to comply fully with the provisions of the Statutes and Bye-Laws of the Union, the Executive Committee and Officers, in the order specified below, shall take such actions as they deem necessary for the continued operation of the Union. Such action shall be reported to all National Members as soon as this becomes practicable, until an ordinary or extraordinary General Assembly can be convened. The following is the order of authority: The Executive Committee in meeting or by correspondence; the

President of the Union; the General Secretary; or failing the practicability or availability of any of the above, one of the Vice-Presidents.

XIV. DISSOLUTION OF THE UNION

25. A decision to dissolve the Union is only valid if taken by the General Assembly with the approval of three quarters of the National Members having the right to vote by virtue of article 14.a. Such a decision shall specify a procedure for settling any debts and disposing of any assets of the Union.

XV. FINAL CLAUSES

26. These Statutes enter into force on 4 August 2009.

27. The present Statutes are published in French and English versions. For legal purposes, the French version is authoritative.

2. Statuts de l'UAI, Rio de Janeiro, Brazil, 4 août 2009

I OBJECTIF

1. L'Union Internationale Astronomique (mentionnée ci-après, "l'Union") est une organisation non gouvernementale internationale. Son objectif est de promouvoir la science de l'astronomie sous tous ses aspects.

II DOMICILIATION ET RELATIONS INTERNATIONALES

2. Le domicile légal de l'Union est situé Paris, en France.
3. L'Union adhre à, et coopre avec, l'ensemble des organisations scientifiques internationales à travers le Conseil International pour la Science (ICSU). Elle soutient et applique les directives sur la Liberté, la Responsabilité, et l'Ethique dans la Conduite des sciences définies par le ICSU.

III COMPOSITION DE L'UNION

4. L'Union se compose d':
   4.a. Adhérents Nationaux (organisations)
   4.b. Adhérents Individuels (personnes physiques)

IV MEMBRES NATIONAUX

5. Une organisation représentant une communauté astronomique professionnelle nationale, désireuse de développer sa participation sur la scène de l'astronomie internationale et soutenant l'objectif de l'Union, peut adhérer à l'Union en qualité d'Adhèrent National. A titre exceptionnel, un Adhérent National peut représenter la communauté dans le territoire de plus d'une nation, à condition qu'aucune partie de cette communauté ne soit représentée par une autre Adhérent National.

6. Une organisation désireuse de rejoindre l'Union en qualité d'Adhérent National tout en développant l'astronomie professionnelle dans la communauté qu'elle représente peut le faire:

6.a. De manière temporaire, selon les conditions précitées, pour une période maximale de neuf ans. Passé ce délai, elle doit demander à devenir Adhérent National de manière permanente, ou son adhésion à l'Union sera résiliée.

6.b. De manière prospective, pour une période maximale de six ans si sa communauté compte moins de six Adhérents Individuels. Passé ce délai, elle devra demander à devenir Adhérent National de manière temporaire ou permanente ou bien son adhésion sera résiliée.

7. Un Adhérent National est admis dans l'Union de manière permanente, temporaire ou prospective par l'Assemblée Générale. Elle peut se retirer de l'Union en informant de son retrait le secrétaire Général, par écrit.

8. Un Adhérent National peut tre :

8.a. L'organisation par laquelle les scientifiques de la nation correspondante ou du territoire correspondant adhérent à la ICSU ou :

8.b. Une Société ou Comité National(e) Compétent(e) d'Astronomie, ou:

8.c. Un établissement compétent d'enseignement supérieur.

9. L'adhésion d'un Adhérent National est automatiquement suspendue si ses cotisations annuelles, telles que définies aux Articles 23c et 23e ci-dessous n'ont pas été payées pendant cinq ans ; elle sera rétablie, sur approbation du Comité de Direction, lorsque les arriérés relatifs à ses cotisations auront été payés en totalité. Après cinq ans de suspension d'un Adhérent National, le Comité de Direction peut recommander à l'Assemblée Générale de résilier l'Adhésion.

10. Un Adhérent National est admis dans l'Union à travers l'une des catégories spécifiées dans les Règlements.

V. ADHERENTS INDIVIDUELS

11. Un scientifique professionnel qui est actif dans un domaine de l'astronomie peut tre admis à l'Union par le Comité de Direction en qualité d'Adhérent Individuel. Un Adhérent Individuel peut se retirer de l'Union en informant de son retrait le Secrétaire Général, par écrit.

VI. DIRECTION

12. La Direction de l'Union est :

12.a. L'Assemblée Générale;
12.b. Le Comité de Direction; et
12.c. Les Membres de la Direction.

VII. ASSEMBLEE GENERALE

13. L'Assemblée Générale se compose d'Adhérents Nationaux et d'Adhérents Individuels. L'Assemblée Générale détermine la ligne directive générale de l'Union.

13.a. L'Assemblée Générale ratifie les Statuts de l'Union, y compris tous changements apportés à ces Statuts.

13.b. L'Assemblée Générale ratifie les Règlements spécifiant les Règles de Procédure devant tre suivies lors de l'application des Statuts.

13.c. L'Assemblée Générale élit un Comité de Direction afin de mettre en oeuvre ses décisions et de diriger les affaires de l'Union entre les réunions ordinaires successives de l'Assemblée Générale. Le Comité de Direction transmet ses comptes-rendus à l'Assemblée Générale.

13.d. L'Assemblée Générale nomme un Comité de Finance, composé d'un représentant pour chaque Adhérent National ayant droit de vote sur des questions budgétaires selon 14.a., afin de le conseiller sur l'approbation du budget et des comptes de l'Union. L'Assemblée Générale nomme aussi un Sous-comité de Finance pour conseiller le Comité de Direction en son nom sur les questions budgétaires dans l'intervalle des Assemblées Générales.

13.e. L'Assemblée Générale nomme un Comité Spécial de Nomination afin de préparer un éventail de candidats appropriés pour l'élection du prochain Comité de Direction.

13.f. L'Assemblée Générale nomme un Comité de Nomination pour conseiller le Comité de Direction sur l'admission des Adhérents individuels.

14. Les votes à l'Assemblée Générale qui portent sur des questions de nature essentiellement scientifique, telles que déterminées par le Comité de Direction, se font par les Adhérents Individuels. Les Votes qui portent sur toutes les autres questions se font par les Adhérents Nationaux. Chaque Adhérent National autorise un représentant à voter en son nom.

14.a. Concernant les questions impliquant le budget de l'Union, le nombre de voix pour chaque Adhérent National est supérieur au nombre de sa catégorie, comme indiqué à l'article 10. Les Adhérents Nationaux dont le statut est temporaire ou prospectif, ou qui n'ont pas payé leurs cotisations pendant des années avant celle de l'Assemblée Générale n'auront pas le droit de participer au vote.

14.b. Concernant les questions relatives à l'administration de l'Union, mais n'impliquant pas le budget, chaque Adhérent National aura une voix, selon les mmes conditions relatives au paiement des cotisations citées en l'Article 14.a.

14.c. Les Adhérents Nationaux peuvent voter par correspondance sur des questions relatives à l'ordre du jour de l'Assemblée Générale.

14.d. Un vote n'est valide que si au moins deux tiers des Adhérents Nationaux ayant le droit de voter en vertu de l'article 14.a. y participent soit en votant soit en signalant une abstention. Une abstention n'est pas considérée comme l'expression d'un vote.

15. Les décisions de l'Assemblée Générale sont prises à la majorité absolue des voix. Cependant, la décision de modifier les Statuts exige l'approbation d'au moins deux tiers de tous les Adhérents Nationaux ayant droit de voter en vertu de l'article 14.a . Lorsqu'il y a égalité des voix, il revient au Président de statuer sur l'issue du vote.

16. Les changements apportés aux Statuts ou aux Règlements ne peuvent tre examinés par l'Assemblée Générale que si une proposition spécifique a été dment soumise aux Adhérents Nationaux et mentionnée à l'ordre du jour de l'Assemblée Générale par la procédure et dans des délais spécifiés dans les Règlements .

VIII. COMITE DE DIRECTION

17. Le Comité de Direction se compose du Président de l'Union, du Président désigné, de six Vice-présidents, du Secrétaire Général, et du Secrétaire Général Adjoint, élus par l'Assemblée Générale sur proposition du Comité Spécial de Nomination.

IX. MEMBRES DE LA DIRECTION

18. Les Membres de l'Union sont le Président, le Secrétaire Général, le Président désigné et le Secrétaire Général Adjoint. Les Membres de la Direction décident de questions de directives sur le court terme dans le cadre des directives générales de l'Union telles que décidées par l'Assemblée Générale et interprétées par le Comité de Direction.

X. SECTIONS SCIENTIFIQUES

19. En tant que moyen efficace d'inciter aux progrès dans les principaux domaines de l'astronomie, le travail scientifique de l'Union est structuré à travers ses Divisions Scientifiques. Chaque Division recouvre un large domaine bien défini de la science astronomique, ou traite de sujets internationaux de nature interdisciplinaire. Autant que possible, les Sections se répartiront en nombre comparable les Membres Individuels de l'Union.
20. Les sections sont créés ou annulées par l'Assemblée Générale sur recommandation du Comité de Direction. Les activités d'une Section sont organisées par un Comité Organisationnel présidé par le Président de Section. Le Président de Section et le Vice-Président sont élus par l'Assemblée Générale sur proposition du Comité de Direction, et sont des membres d'office du Comité Organisationnel.

XI. COMMISSIONS SCIENTIFIQUES

21. A l'intérieur des Sections, les activités scientifiques réparties en disciplines bien définies à l'intérieur de la matière de la Section peuvent tre organisées à travers des Commissions Scientifiques. Dans certains cas spécifiques, une Commission peut recouvrir une matière commune à une ou deux Sections et devient alors la Commission de l'ensemble de ces Sections.

22. Les Commissions sont créées ou annulées par le Comité de Direction sur recommandation du ou des Comité(s) de Direction de la ou les Section(s) désirant les créer

ou les annuler. Les activités d'une Commission sont organisées par un Comité Organisationnel présidé par le Président d'une Commission. Le Président de la Commission et un Vice-président sont nommés par le ou les Comité(s) Organisationnel(s) de la ou des Section(s) correspondantes sur proposition du Comité Organisationnel de la Commission.

XII. BUDGET ET COTISATIONS

23. Pour chaque Assemblée Générale, le Comité de Direction prépare une proposition de budget couvrant la période de la prochaine Assemblée Générale, accompagnée des comptes de l'Union pour la période précédente. Il les soumet, avec le conseil du Souscomité de Finance, au Comité de Finance en vue de leur examen avant leur soumission au vote de l'Assemblée Générale.

23.a. Le Comité de Finance examine les comptes de l'Union par rapport aux dépenses raisonnables prévues par l'Assemblée Générale précédente, selon l'interprétation qu'en fait le Comité de Direction. Il se penche aussi sur l'adéquation ou non du budget proposé en vue de la mise en uvre de la ligne directive de l'Assemblée Générale. Il soumet des rapports sur ces sujets à l'Assemblée Générale avant ses décisions concernant l'approbation des comptes et du budget.

23.b. Le montant des unités de cotisations est décidé par l'Assemblée Générale comme entrant dans le processus d'approbation du budget.

23.c. Chaque Adhérent National verse annuellement un nombre d'unités de cotisations correspondant à sa catégorie. Le nombre d'unités de cotisation pour chaque catégorie sera spécifié dans les Règlements.

23.d. Un vote sur des questions en vertu de l'article 23 n'est valable que si les deux tiers des Adhérents Nationaux ayant le droit de voter en vertu de l'article 14.a. votent. Dans tous les cas, une abstention ne constitue par un vote, mais une déclaration selon laquelle que le Membre décide de ne pas voter.

23.e. Les adhérents ayant un statut intérimaire payent annuellement une demie unité de cotisation.

23.f. Les Adhérents Nationaux ayant un statut prospectif ne payent aucune cotisation.

23.g. Le paiement des contributions est de la responsabilité des Adhérents Nationaux. La responsabilité de chaque Adhérent National relative à l'Union est limitée au montant des cotisations dues au cours de l'année.

XIII. PLEINS POUVOIRS

24. Si, en cas d'événements dépassant le contrôle de l'Union, des circonstances font qu'il est impossible de respecter pleinement les dispositions des Statuts et les Règlements de l'Union, le Comité de Direction et les Membres de la Direction, selon l'ordre spécifié ci-dessous, agiront comme il leur semblera nécessaire de le faire pour la continuité du fonctionnement de l'Union. De telles actions seront rapportées aux Adhérents Nationaux dès que cela sera possible, jusqu'à ce qu'une Assemblée Générale Ordinaire ou extraordinaire soit convoquée. L'ordre hiérarchique est le suivant: Le Comité de Direction en réunion ou par correspondance; le Président de l'Union; le Secrétaire Général; ou à défaut de possibilité ou de disponibilité de ce qui précède, un des Vice-présidents.

XIV. DISSOLUTION DE L'UNION

25. La décision de dissoudre l'Union n'est valable que si prise par l'Assemblée Générale avec l'approbation des trois quarts des Adhérents Nationaux ayant le droit de voter en vertu de l'article 14.a. Une telle décision spécifiera une procédure visant à régler toutes dettes et disposant des actifs de l'Union.

XV. DERNIERES CLAUSES

26. Ces Statuts entrent en vigueur le 4 aot 2009.

27. Les présents statuts sont publiés en français et en anglais. Pour toute fin juridique, la version française fera autorité.

3. IAU Bye-Laws Rio de Janeiro, Brazil, 4 August 2009

I MEMBERSHIP

1. An application for admission to the Union as a National Member shall be submitted to the General Secretary by the proposing organization at least eight months before the next ordinary General Assembly.

2. The Executive Committee shall examine the application and resolve any outstanding issues concerning the nature of the proposed National Member and the category of membership (VII.25). Subsequently, the Executive Committee shall forward the application to the General Assembly for decision, with its recommendation as to its approval or rejection.

3. The Executive Committee shall examine any proposal by a National Member to change its category of adherence to a more appropriate level. If the Executive Committee is unable to approve the request, either party may refer the matter to the next General Assembly.

4. Individual Members are admitted by the Executive Committee upon the nomination of a National Member or the President of a Division. The Executive Committee shall publish the criteria and procedures for membership, and shall consult the Nominating Committee before approving applications for admissions as Individual Members.

II. GENERAL ASSEMBLY

5. The ordinary General Assembly meets, as a rule, once every three years. Unless determined by the previous General Assembly, the place and date of the ordinary General Assembly shall be fixed by the Executive Committee and be communicated to the National Members at least one year in advance.

6. The President may summon an extraordinary General Assembly with the consent of the Executive Committee, and must do so at the request of at least one third of the National Members. The date, place, and agenda of business of an extraordinary General Assembly must be communicated to all National Members at least two months before the first day of the Assembly.

7. Matters to be decided upon by the General Assembly shall be submitted for consideration by those concerned as follows, counting from the first day of the General Assembly:

7.a. A motion to amend the Statutes or Bye-Laws may be submitted by a National Member or by the Executive Committee. Any such motion shall be submitted to the General Secretary at least nine months in advance and be forwarded, with the recommendation of the Executive Committee as to its adoption or rejection, to the National Members at least six months in advance.

7.b. The General Secretary shall distribute the budget prepared by the Executive Committee to the National Members at least eight months in advance. Any motion to modify this budget, or any other matters pertaining to it, shall be submitted to the General Secretary at least six months in advance. Any such motion shall be submitted, with the advice of the Executive Committee as to its adoption or rejection, to the National Members at least four months in advance.

7.c. Any motion or proposal concerning the administration of the Union, and not affecting the budget, by a National Member, or by the Organizing Committee of a Scientific Division of the Union, shall be placed on the Agenda of the General Assembly, provided it is submitted to the General Secretary, in specific terms, at least six months in advance.

7.d. Any motion of a scientific character submitted by a National Member, a Scientific Division of the Union, or by an ICSU Scientific Committee or Program on which the Union is formally represented, shall be placed on the Agenda of the General Assembly, provided it is submitted to the General Secretary, in specific terms, at least six months in advance.

7.e. The complete agenda, including all such motions or proposals, shall be prepared by the Executive Committee and submitted to the National Members at least four months in advance.

8. The President may invite representatives of other organizations, scientists in related fields, and young astronomers to participate in the General Assembly. Subject to the agreement of the Executive Committee, the President may authorise the General Secretary to invite representatives of other organizations, and the National Members or other appropriate IAU bodies to invite scientists in related fields and young astronomers.

III. SPECIAL NOMINATING COMMITTEE

9. The Special Nominating Committee consists of the President and past President of the Union, a member proposed by the retiring Executive Committee, and four members selected by the Nominating Committee from among twelve candidates proposed by Presidents of Divisions, with due regard to an appropriate distribution over the major branches of astronomy.

9.a. Except for the President and immediate past President, present and former members of the Executive Committee shall not serve on the Special Nominating Committee. No two members of the Special Nominating Committee shall belong to the same nation or National Member.

9.b. The General Secretary and the Assistant General Secretary participate in the work of the Special Nominating Committee in an advisory capacity, and the President-Elect may participate as an observer.

10. The Special Nominating Committee is appointed by the General Assembly, to which it reports directly. It assumes its duties immediately after the end of the General Assembly and remains in office until the end of the ordinary General Assembly next

following that of its appointment, and it may fill any vacancy occurring among its members.

IV. OFFICERS AND EXECUTIVE COMMITTEE

11.

11.a. The President of the Union remains in office until the end of the ordinary General Assembly next following that of election. The President-Elect succeeds the President at that moment.

11.b. The General Secretary and the Assistant General Secretary remain in office until the end of the ordinary General Assembly next following that of their election. Normally the Assistant General Secretary succeeds the General Secretary, but both officers may be re-elected for another term.

11.c. The Vice-Presidents remain in office until the end of the ordinary General Assembly following that of their election. They may be immediately re-elected once to the same office.

11.d. The elections take place at the last session of the General Assembly, the names of the candidates proposed having been announced at a previous session.

12. The Executive Committee may fill any vacancy occurring among its members. Any person so appointed remains in office until the end of the next ordinary General Assembly.

13. The past President and General Secretary become advisors to the Executive Committee until the end of the next ordinary General Assembly. They participate in the work of the Executive Committee and attend its meetings without voting rights.

14. The Executive Committee shall formulate Working Rules to clarify the application of the Statutes and Bye-Laws. Such Working Rules shall include the criteria and procedures by which the Executive Committee will review applications for Individual Membership; standard Terms of Reference for the Scientific Commissions of the Union; rules for the administration of the Unions financial affairs by the General Secretary; and procedures by which the Executive Committee may conduct business by electronic or other means of correspondence. The Working Rules shall be published electronically and in the Transactions of the Union.

15. The Executive Committee appoints the Union's official representatives to other scientific organizations.

16. The Officers and members of the Executive Committee cannot be held individually or personally liable for any legal claims or charges that might be brought against the Union.

V. SCIENTIFIC DIVISIONS

17. The Divisions of the Union shall pursue the scientific objects of the Union within their respective fields of astronomy. Activities by which they do so include the

encouragement and organization of collective investigations, and the discussion of questions relating to international agreements, cooperation, or standardization.

They shall report to each General Assembly on the work they have accomplished and such new initiatives as they are undertaking.

18. Each Scientific Division shall consist of:

18.a. an Organizing Committee, normally of 6-12 persons, including the Division President and Vice-President, and a Division Secretary appointed by the Organizing Committee from among its members.

18.b. members of the Union appointed by the Organizing Committee in recognition of their special experience and interests. The Committee is responsible for conducting the business of the Division.

19. Normally, the Division President is succeeded by the Vice-President at the end of the General Assembly following their election, but both may be re-elected for a second term. Before each General Assembly, the Organizing Committee shall organize an election from among the membership, by electronic or other means suited to its scientific structure, of a new Organizing Committee to take office for the following term. Election procedures should, as far as possible, be similar among the Divisions and require the approval of the Executive Committee.

20. Each Scientific Division may structure its scientific activities by creating a number of Commissions. In order to monitor and further the progress of its field of astronomy, the Division shall consider, before each General Assembly, whether its Commission structure serves its purpose in an optimum manner. It shall subsequently present its proposals for the creation, continuation or discontinuation of Commissions to the Executive Committee for approval.

21. With the approval of the Executive Committee, a Division may appoint Working Groups to study well-defined scientific issues and report to the Division. Unless specifically re-appointed by the same procedure, such Working Groups cease to exist at the next following General Assembly.

VI. SCIENTIFIC COMMISSIONS

22. A Scientific Commission shall consist of:

22.a. A President and an Organizing Committee consisting of 4-8 persons elected by the Commission membership, subject to the approval of the Organizing Committee of the Division;

22.b. Members of the Union, appointed by the Organizing Committee, in recognition of their special experience and interests, subject to confirmation by the Organizing Committee of the Division.

23.A Commission is initially created for a period of six years. The parent Division may recommend its continuation for additional periods of three years at a time, if sufficient justification for its continued activity is presented to the Division and the Executive Committee. The activities of a Commission is governed by Terms of Reference, which are based on a standard model published by the Executive Committee and are approved by the Division.

24. With the approval of the Division, a Commission may appoint Working Groups to study well-defined scientific issues and report to the Commission. Unless specifically re-appointed by the same procedure, such Working Groups cease to exist at the next following General Assembly.

VII. ADMINISTRATION AND FINANCES

25. Each National Member pays annually to the Union a number of units of contribution corresponding to its category as specified below; National Members with interim status pay annually one half unit of contribution, and those with prospective status pay no dues.

Categories as defined in article 10 of the Statutes:

I	II	III	IV	V	VI	VII	VIII	IX	X	XI	XII
1	2	4	6	10	14	20	27	35	45	60	80

Number of units of contribution

26. The income of the Union is to be devoted to its objects, including:

26.a. the promotion of scientific initiatives requiring international co-operation;

26.b. the promotion of the education and development of astronomy world-wide;

26.c. the costs of the publications and administration of the Union.

27. Funds derived from donations are reserved for use in accordance with the instructions of the donor(s). Such donations and associated conditions require the approval of the Executive Committee.

28. The General Secretary is the legal representative of the Union. The General Secretary is responsible to the Executive Committee for not incurring expenditure in excess of the amount specified in the budget as approved by the General Assembly.

29. The General Secretary shall consult with the Finance Sub-Committee (cf. Statutes §13.d) in preparing the accounts and budget proposals of the Union, and on any other matters of major importance for the financial health of the Union. The comments and advice of the Finance Sub-Committee shall be made available to the Officers and Executive Committee as specified in the Working Rules.

30. An Administrative office, under the direction of the General Secretary, conducts the correspondence, administers the funds, and preserves the archives of the Union.

31. The Union has copyright to all materials printed in its publications, unless otherwise arranged.

VIII. FINAL CLAUSES

32. These Bye-Laws enter into force on 4 August 2009.

33. The present Bye-Laws are published in French and English versions. For legal purposes, the French version is authoritative.

Note that the French version of the Bye-Laws is avaiable on the IAU web site:
<http://www.iau.org/static/administration/byelaws2009_french.pdf >

4. IAU Working Rules August 2009

INTRODUCTION AND RATIONALE

The Statutes of the International Astronomical Union (IAU) define the goals and organizational structure of the Union, while the Bye-Laws specify the main tasks of the various bodies of the Union in implementing the provisions of the Statutes. The Working Rules are designed to assist the membership and governing bodies of the Union in carrying out these tasks in an appropriate and effective manner. Each of the sections below is preceded by an introduction outlining the goals to be accomplished by the procedures specified in the succeeding paragraphs. The Executive Committee updates the Working Rules as necessary to reflect current procedures and to optimize the services of the IAU to its membership.

I. NON-DISCRIMINATION

The International Astronomical Union (IAU) follows the regulations of the International Council for Science (ICSU) and concurs with the actions undertaken by their Standing Committee on Freedom in the Conduct of Science on non-discrimination and universality of science (cf. §22 below).

II. NATIONAL MEMBERSHIP

The aim of the rules for applications for National Membership is to ensure that the proposed National Member adequately represents an astronomical community not already represented by another Member, and that such membership will be of maximum benefit for the community concerned (cf. Statutes §IV).

1. Applications for National Membership should therefore clearly describe the following essential conditions:

1.a. the precise definition of the astronomical community to be represented by the proposed Member;

1.b. the present state and expected development of that astronomical community;

1.c. the manner in which the proposed National Member represents this community;

1.d. whether the application is for membership on a permanent, interim, or prospective basis; and

1.e. the category in which the prospective National Member wishes to be classified (cf. Bye-Laws Article 25).

1.f. the process by which the National Membership annual dues will be paid promptly and in full.

2. Applications for National Membership shall be submitted to the General Secretary, who will forward them to the Executive Committee for review as provided in the Statutes.

III. INDIVIDUAL MEMBERSHIP

Professional scientists whose research is directly relevant to some branch of astronomy are eligible for election as Individual Members of the Union (cf. Statutes Article V). Individual Members are normally admitted by the Executive Committee on the proposal of a National Member. However, Presidents of Divisions may also propose individuals for membership in cases when the normal procedure is not applicable or practicable (cf. Bye-Laws Article 4). The present rules are intended to ensure that all applications for membership are processed on a uniform basis, and that all members are fully integrated in and contributing to the activities of the Union.

3. The term "Professional Scientist" shall normally designate a person with a doctoral degree (Ph.D.) or equivalent experience in astronomy or a related science, and whose professional activities have a substantial component of work related to astronomy.

4. National Members and Division Presidents may propose Individual Members who fall outside the category of professional scientist but who have made major contributions to the science of astronomy, e.g., through education or research related to astronomy. Such proposals should be accompanied by a detailed motivation for what should be seen as exceptions to the rule.

5. Eight months before an ordinary General Assembly, National Members and Presidents of Divisions will be invited to propose new Individual Members; these proposals should reach the General Secretary no later than five months before the General Assembly. Late proposals will normally not be taken into consideration. Proposals from Presidents of Divisions will be communicated by the IAU 3 months before the General Assembly to the relevant National Members, if any, who may add the person(s) in question to their own list of proposals.

6. National Members shall promptly inform in writing the General Secretary of the death of any Individual Member represented by them. National Members are also urged to propose the deletion of Individual Members who are no longer active in astronomy by including a written agreement of the member concerned. Such proposals should be submitted to the General Secretary at the same time as proposals for new Individual Members.

7. Proposals for membership shall include the full name, date of birth, and nationality of the candidate, postal and electronic addresses, the University, year, and subject of the M.Sc./Ph.D. or equivalent degree, current affiliation and occupation, the proposing National Member or Division, the Division(s) and/or Commission(s) which the candidate wishes to join, and any further detail that might be relevant.

8. The Nominating Committee shall normally comprise all National Representatives present at the General Assembly. It shall examine all proposals for individual membership and advise the Executive Committee on their approval or rejection. To assist in this, the Nominating Committee shall appoint a standing Nominating Sub-Committee

which remains in office until the next General Assembly to deal with membership issues. This Sub-Committee shall normally comprise not more than 10 members drawn from the representatives of the National Members such that there is a spread of representation of all categories of national membership. The Nominating Sub-Committee shall appoint its own Chair.

9. The General Secretary shall submit all proposals for Individual Membership the to the Nominating Committee via the Nominating Sub-Committee for review, coconsolidated into two lists:

9.a one containing all proposals by National Members; and

]9.b one containing all proposals by Presidents of Divisions, in accordance with Bye-Law Article 4.

10. The Nominating Sub-Committee shall examine all proposals for individual membership and advise the Nominating Committee on their approval or rejection. The Nominating Committee shall then examine the recommendations of the Nominating Sub-Committee and advise the Executive Committee on the proposals for individual membership.

11. In exceptional cases, the Executive Committee may, on the proposal of a Division, admit an Individual Member between General Assemblies. Such proposals shall be prepared as described above (cf. Article 2) and submitted with a justification of the request to bypass the normal procedure. The Executive Committee shall consult the Nominating Committee or relevant National Member before approving such exceptions to the normal procedure.

12. The General Secretary shall maintain updated lists of all National and Individual Members, and shall make these available to the membership in electronic form. The procedures for dissemination of these lists shall be set by the General Secretary in such a way that the membership directory be properly protected against unintended or inappropriate use.

IV. RESOLUTIONS OF THE UNION

Traditionally, the decisions and recommendations of the Union on scientific and organizational matters of general and significant importance are expressed in the Resolutions of the Union. In order for such Resolutions to carry appropriate weight in the international community, they should address astronomical matters of significant impact on the international society, or matters of international policy of significant importance for the international astronomical community as a whole.

Resolutions should be adopted by the Union only after thorough preparation by the relevant bodies of the Union. The proposed resolution text should be essentially complete before the beginning of the General Assembly, to allow Individual and National Members time to study them before discussion and debate by the General Assembly. The following procedures have been designed to accomplish this.

13. Proposals for Resolutions to be adopted by the Union may be submitted by a National Member, by the Executive Committee, a Division, a Commission or a Working Group. They should address specific issues of the nature described above, define the objectives to be achieved, and describe the action(s) to be taken by the Officers, Executive Committee, or Divisions to achieve these objectives.

14. Resolutions proposed for vote by the Union fall in three categories as set out in Article 14 of the Union's Statutes:

14.a. Resolutions with implications for the budget of the Union (Statute 14a); or
14.b. Resolutions affecting the administration of the Union but without financial implications (Statute 14b).
14.c Resolutions of a primarily scientific nature (Statute 14).

Proposals for Resolutions should be submitted on standard forms appropriate for each type, which are available from the IAU Secretariat. They may be submitted in either English or French and will be discussed and voted upon in the original language. Upon submission each proposed Resolution is posted on the Union web site. When the approved Resolutions are published, a translation to the other language will be provided.

15. Resolutions with implications for the budget of the Union must be submitted to the General Secretary at least nine months before the General Assembly in order to be taken into account in the budget for the impending triennium. All other Resolutions must be submitted to the General Secretary three months before the beginning of the General Assembly. The Executive Committee may decide to accept late proposals in exceptional circumstances.

16. Before being submitted to the vote of the General Assembly, proposed Resolutions will be examined by the Executive Committee, Division Presidents, and by a Resolutions Committee, which is nominated by the Executive Committee. The Resolutions Committee consists of at least three members of the Union, one of whom should be a member of the Executive Committee, and one of whom should be a continuing member from the previous triennium. It is appointed by the General Assembly during its final session and remains in office until the end of the following General Assembly.

17. The Resolutions Committee will examine the content, wording, and implications of all proposed Resolutions promptly after their submission. In particular, it will address the following points:

i. suitability of the subject for an IAU Resolution;
ii. correct and unambiguous wording;
iii. consistency with previous IAU Resolutions.

The Resolutions Committee may refer a Resolution back to the proposers for revision or withdrawal if it perceives significant problems with the text, but can neither withdraw nor modify its substance on its own initiative. The Resolutions Committee advises the Executive Committee whether the subject of a proposed Resolution is primarily a matter of policy or primarily scientific. The Resolutions Committee will also notify the Executive Committee of any perceived problems with the substance of a proposed Resolution.

18. The Executive Committee will examine the substance and implications of all proposed Resolutions. Proposed Resolutions shall be published in the General Assembly Newspaper before the final session. The Resolutions Committee will present the proposals during a plenary session of the General Assembly with its own recommendations, and those of the Executive Committee, if any, for their approval or rejection. A representative of the body proposing the Resolution will be given the opportunity to defend the Resolution in front of the General Assembly, after which a general discussion shall take place.

19. Resolutions with implications for the budget of the Union are voted upon by the National Members during the final plenary session of the General Assembly. Other resolutions may be voted upon by the National Members or by Individual Members as appropriate according to the Statutes of the Union by correspondence after the General Assembly. The Union will facilitate electronic discussion of all Resolutions on the Union website in advance of a vote either at the General Assembly or by correspondence.

V. EXTERNAL RELATIONS

Contacts with other international scientific organizations, national and international public bodies, the media, and the public are increasing in extent and importance. In order to maintain coherent overall policies in matters of international significance, clear delegation of authority is required. Part of this is accomplished by having the Union's representatives in other scientific organization appointed by the Executive Committee (cf. Bye-Laws Article 15). Supplementary rules are given in the following.

20. Representatives of the Union in other scientific organizations are appointed by the Executive Committee upon consultation with the Division(s) in the field(s) concerned.

21. In other international organizations, e.g. in the United Nations Organization, the Union is normally represented by the General Secretary or Assistant General Secretary, as decided by the Executive Committee.

22. The Union strongly supports the policies of the International Council for Science (ICSU) as regards the freedom and universality of science. Participants in IAU sponsored activities who feel that they may have been subjected to discrimination are urged, first, to seek clarification of the origin of the incident, which may have been due to misunderstandings or to the cultural differences encountered in an international environment. Should these attempts not prove successful, contact should be made with the General Secretary who will take steps to resolve the issue.

23. Public statements that are attributed to the Union as a whole can be made only by the President, the General Secretary, or the Executive Committee. The General Secretary may, in consultation with the relevant Division, appoint Individual Members of the Union with special expertise in questions that attract the attention of media and the general public as IAU spokespersons on specific matters.

VI. FINANCIAL MATTERS

The great majority of the Union's financial resources are provided by the National Members, as laid out in the Statutes Article XII and Bye-Laws Article VII. The purpose of the procedures described below is twofold: (i) to provide the best possible advice and guidance to the General Secretary and Executive Committee in planning and managing the Unions financial affairs, and (ii) to provide National Members with a mechanism for continuing input to and oversight over these affairs between and in preparation for the General Assemblies. The procedures adopted to accomplish this are as follows:

24. At the end of each of its final sessions the General Assembly, at the proposal of the Finance Committee, appoints a Finance Sub-Committee of 5-6 members, including a Chair. The Finance Sub-Committee remains in office until the end of the next General Assembly (cf. Statutes Article 13.d.) and cooperates with the National Members, Finance Committee, Executive Committee and General Secretary in the following manner:

24.a. After the end of each year the General Secretary will call for a legal audit of the accounts by a properly licensed, external auditor. The auditor will make a report addressed to the General Assembly. The General Secretary provides the Finance Sub-Committee with the auditor's report and summary reports covering the financial performance of the Union as compared to the approved budget, together with an analysis of any significant departures, and information on any Executive Committee approvals of budget changes. Upon receipt of the above reports from the General Secretary, the Finance Sub-Committee examines the accounts of the Union in the light of the corresponding budget and any relevant later decisions by the Executive Committee. It reports its findings and recommendations to the Executive Committee at its next meeting. The Finance Sub-Committee may at any time, at the request of the Executive Committee or the General Secretary, or on its own initiative, advise the General Secretary and/or the Executive Committee on any aspect of the Union's financial affairs.

24.b. Early in the year preceding that of a General Assembly, the General Secretary shall submit a preliminary draft of the budget for the next triennium to the Finance Sub-Committee for review. The draft budget, updated as appropriate following the comments and advice of the Finance Sub-Committee, is submitted to the Executive Committee for approval at the EC meeting in the year preceding that of a General Assembly, together with the report of the Finance Sub-Committee. The final budget proposal as approved by the Executive Committee is subsequently submitted to the National Members with a statement of the views of the Finance Sub-Committee on the proposal.

24.c. Before the first session of a General Assembly, the Finance Sub-Committee shall submit a report, including the auditor's reports, to the Executive Committee and the Finance Committee on its findings and recommendations concerning the development of the Union's finances over the preceding triennium. The Finance Sub-Committee shall also prepare a slate of candidates for the composition of the Finance Sub-Committee in the next triennium, preferably providing a balance between new and continuing members.

24.d. The report of the Finance Sub-Committee, together with the audited detailed accounts and the earlier comments on the proposed budget for the next triennium, will form a suitable basis for the discussions of the Finance Committee leading to its recommendations to the General Assembly concerning the approval of the accounts for the previous triennium and the budget for the next triennium, as well as the new Finance Sub-Committee to serve during that period.

25. The General Secretary is responsible for managing the Union's financial affairs according to the approved budget (cf. Bye-Laws Article 28).

25.a. In response to changing circumstances, the Executive Committee may approve such specific changes to the annual budgets as are consistent with the intentions of the General Assembly when the budget was approved.

25.b. Unless authorized by the Executive Committee, the General Secretary shall not approve expenses exceeding the approved budget by more than 10% of any corresponding major budget line or 2% of the total budget in a given year, whichever is larger. This restriction does not apply in cases when external funding has been provided for a specific purpose, e.g. travel grants to a General Assembly.

25.c. Unless specifically identified in the approved budget, contractual commitments in excess of 50,000 or with performance terms in excess of 3 years require the additional approval of the Union President.

26. The National Representatives, in approving the accounts for the preceding triennium, discharge the General Secretary and the Executive Committee of liability for the period in question.

VII. RULES OF PROCEDURE FOR THE EXECUTIVE COMMITTEE

The Executive Committee must respond quickly to events and thus it needs to be able to have discussions and take decisions on a relatively short timescale and without meeting in person. The following rules, as required by Bye-Law 14, are designed to facilitate EC action in a flexible manner, while giving such decisions the same legal status as those taken at actual physical meetings.

27. The Executive Committee should meet in person at least once per year. In years of a General Assembly it should meet in conjunction with and at the venue of the General Assembly. In other years, the Executive Committee decides on the date and venue of its regular meeting. The meetings of the Executive Committee are chaired by the President or, if the President is unavailable, by the President Elect or by one of the Vice-Presidents chosen by the Executive Committee to serve in this capacity.

28. The date and venue of the next regular meeting of the Executive Committee shall be communicated at least six months in advance to all its members and the Advisors, and to all Presidents of Divisions. Any of these persons may then propose items for inclusion in the Draft Agenda of the meeting before the date posted on the IAU Deadlines page.

29. Outgoing and incoming Presidents of Divisions are invited to attend all non-confidential sessions of the outgoing and incoming Executive Committee, respectively, in the year of a General Assembly. The President will invite Presidents of Divisions to attend the meetings of the Executive Committee in the years preceding a General Assembly. Division Presidents attend these sessions with speaking right, but do not participate in any voting.

30. The Executive Committee may take official decisions if at least half of its members participate in the discussion and vote on an issue. Decisions are taken by a simple majority of the votes cast. In case of an equal division of votes, the Chair's vote decides the issue. Members who are unable to attend may, by written or electronic correspondence with the President before the meeting, authorize another member to vote on her/his behalf or submit valid votes on specific issues.

31. If events arise that require action from the Executive Committee between its regular meetings, the Committee may meet by teleconference or by such electronic or other means of correspondence as it may decide. In such cases, the Officers shall submit a clear description of the issue at hand, with a deadline for reactions. If the Officers propose a specific decision on the issue, the decision shall be considered as approved unless a majority of members vote against it by the specified deadline. In case of a delay in communication, or if the available information is considered insufficient for a decision, the deadline shall be extended or the decision deferred until a later meeting at the request of at least two members of the Executive Committee.

32. The Officers of the IAU should, as a rule, meet once a year at the IAU Secretariat in order to discuss all matters of importance to the Union. The other members of the Executive Committee and the Division Presidents shall be invited to submit items for discussion at the Officers' Meetings and shall receive brief minutes of these Meetings.

33. Should any member of the Executive Committee have a conflict of interest on a matter before the Executive Committee that might compromise their ability to act in the best interests of the Union, they shall declare their conflict of interest. The remaining members of the Executive Committee determine the appropriate level of participation in such issues for members with a potential conflict of interest.

VIII. SCIENTIFIC MEETINGS AND PUBLICATIONS

Meetings and their proceedings remain a major part of the activities of the Union. The purpose of scientific meetings is to provide a forum for the development and dissemination of new ideas, and the proceedings are a written record of what transpired.

34. The General Secretary shall publish in the Transactions and on the IAU web site rules for scientific meetings organized or sponsored by the Union.

35. The proceedings of the General Assemblies and other scientific meetings organized or sponsored by the Union shall, as a rule, be published. To ensure prompt publication of Proceedings of IAU Symposia and Colloquia, the Assistant General Secretary is authorized to oversee the production of the material for the Proceedings. The Union shall publish an Information Bulletin at regular intervals to keep Members informed of current and future events in the Union. The Union shall also publish a more informal, periodic Newsletter which it distributes electronically to its members. The Executive Committee decides on the scope, format, and production policies for such publications, with due regard to the need for prompt publication of new scientific results and to the financial implications for the Union. At the present time, publications are in printed and in electronic form.

36. Divisions, Commissions, and Working Groups shall, with the approval of the Executive Committee, be encouraged to issue Newsletters or similar publications addressing issues within the scope of their activity.

IX. TERMS OF REFERENCE FOR DIVISIONS

The Divisions are the scientific backbone of the IAU. They have a main responsibility for monitoring the scientific and international development of astronomy within their subject areas, and for ensuring that the IAU will address the most significant issues of the time with maximum foresight, enterprising spirit, and scientific judgment. To fulfill this role IAU Divisions should maintain a balance between innovation and continuity. The following standard Terms of Reference have been drafted to facilitate that process, within the rules laid down in the Statutes Article X and the Bye-Laws Article V.

37. As specified in Bye-Law 18, the scientific affairs of the Division are conducted by an Organizing Committee of up to 12 members of the Division, headed by the Division President, Vice-President, and Secretary. Thus, all significant decisions of the Division require the approval of the Organizing Committee, and the President and Vice-President are responsible for organizing the work of the Committee so that its members are consulted in a timely manner. Contact information for the members of the Organizing Committee shall be maintained at the Division web site.

38. Individual Members of the Union are admitted to membership in a Division by its Organizing Committee (cf. Bye-Laws Article 18). Individual Members active within the field of activity of the Division and interested in contributing to its development should contact the Division Secretary, who will consult the Organizing Committee on the admission of the candidates.

38.a. The Division Secretary shall maintain a list of Division members for ready consultation by the community, including their Commission memberships if any. Updates to the list shall be provided to the IAU Secretariat on a running basis.

38.b. Members may resign from a Division by so informing the Division Secretary.

39. The effectiveness of the Division relies strongly on the scientific stature and dedication of its President and Vice-President to the mission of the Division. The Executive Committee, in proposing new Division Presidents and Vice-Presidents for election by the General Assembly, will rely heavily on the recommendations of the Organizing Committee of the Division. In order to prepare a strong slate of candidates for these positions, and for the succession on the Organizing Committee itself, the following procedures apply:

39.a. Candidates are proposed and selected from the membership of the Division on the basis of their qualifications, experience, and stature in the fields covered by the Division. In addition, the Organizing Committees should have proper gender balance and broad geographical representation.

39.b. At least six months before a General Assembly, the Organizing Committee submits to the membership of the Division a list of candidates for President, Vice-President (for which there should be at least two persons willing to serve), Secretary, and the Organizing Committee for the next triennium. The Organizing Committee requests nominations from the membership in preparing this list, and then conducts a vote among Division members for the above offices, the results of which are reported to the General Secretary at least three months before the General Assembly. The Vice-President is normally nominated to succeed the President. The outgoing President participates in the deliberations of the new Organizing Committee in an advisory capacity.

39.c. If more names are proposed than there are positions to be filled on the new Organizing Committee, the outgoing Organizing Committee devises the procedure by which the requisite number of candidates is elected by the membership. The resulting list is communicated to the General Secretary at least two months before the General Assembly. The General Secretary may allow any outstanding issues to be resolved at the business meeting of the Division during the General Assembly. If for any reason the Organizing Committee has not been able to arrange for the election of new officers and an Organizing Committee by two months before the GA, the EC will nominate a VP and Organizing Committee at its first General Assembly meeting.

39.d. A member of the Organizing Committee normally serves a maximum of two terms, unless elected Vice-President of the Division in her/his second term. Presidents may serve for only one term.

39.e. The Organizing Committee decides on the procedures for designating the Division Secretary, who maintains the web site, records of the business and membership of the Division, and other rules for conducting its business by physical meetings or by correspondence.

40. A key responsibility of the Organizing Committee is to maintain an internal organization of Commissions and Working Groups in the Division which is conducive to the fulfillment of its mission. The Organizing Committee shall take the following steps to accomplish this task in a timely and effective manner:

40.a. Within the first year after a General Assembly - with the business meeting of the Commission at the General Assembly itself as a natural starting point - the Organizing Committee shall discuss with its Commissions, and within the Organizing Committee itself, if changes in its Commission and Working Group structure may enable it to accomplish its mission better in the future. As a rule, Working Groups should be created (following the rules in Bye-Law 21 and Bye-Law 23) for new activities that are either of a known, finite duration or are exploratory in nature. If experience, possibly from an existing Working Group, indicates that a major section of the Division's activities require a coordinating body for a longer period (a decade or more), the creation of a new Commission may be in order.

40.b. Whenever the Organizing Committee is satisfied that the creation of a new Working Group or Commission is well motivated, it may take immediate action as specified in Bye-Law 21 or Bye-Law 23. In any case, the Organizing Committee submits its complete proposal for the continuation, discontinuation, or merger of its Commissions and Working Groups to the General Secretary at least three months before the next General Assembly.

40.c. The President and Organizing Committee maintain frequent contacts with the other IAU Divisions to ensure that any newly emerging or interdisciplinary matters are addressed appropriately and effectively.

X. TERMS OF REFERENCE FOR COMMISSIONS

The role of the Commissions is to organize the work of the Union in specialized subsets of the fields of their parent Division(s), when the corresponding activity is judged to be of considerable significance over times of a decade or more. Thus, new Commissions may be created when fields emerge that are clearly in sustained long-term development and where the Union may play a significant role in promoting this development at the international level. Similarly, Commissions may be discontinued when their work can be accomplished effectively by the parent Division. In keeping with the many-sided activities of the Union, Commissions may have purely scientific as well as more organizational and/or interdisciplinary fields. They will normally belong and report to one of the IAU Divisions, but may be common to two or more Divisions. The following rules apply if a Division has more than one Commission.

41. The activities of a Commission are directed by an Organizing Committee of 4-8 members of the Commission, headed by a Commission President and Vice-President (cf. Bye-Laws Article 22). A member of the Organizing Committee normally serves a maximum of two terms, unless elected Vice-President of the Commission in her/his second term. Presidents may serve for only one term. All members of the Organizing Committee are expected to be active in this task, and are to be consulted on all significant actions of the Commission. The Organizing Committee appoints a Commission Secretary who maintains the records of the membership and activities of the Commission in co-operation with the Division Secretary and the IAU Secretariat. Contact information for the members of the Organizing Committee shall be maintained at the Commission web site.

42. Individual Members of the Union, who are active in the field of the Commission and wish to contribute to its progress, are admitted as members of the Commission by the Organizing Committee. Interested Members should contact the Commission Secretary, who will bring the request before the Organizing Committee for decision. Members may resign from the Commission by notifying the Commission Secretary. Before each General Assembly, the Organizing Committee may also decide to terminate the Commission membership of persons who have not been active in the work of the Commission; the individuals concerned shall be informed of such planned action before it is put into effect. The Commission Secretary will report all changes in the Commission membership to the Division Secretary and the IAU Secretariat.

43. At least six months before a General Assembly, the Organizing Committee submits to the membership of the Commission a list of candidates for President, Vice-President (for which there should be the names of two persons willing to serve), the Organizing Committee, and heads of Program Groups for the next triennium. The Organizing Committee requests nominations from the membership in preparing this list, and then conducts a vote among the members for the above offices, the results of which are reported to the General Secretary at least three months before the General Assembly. The Vice-President is normally nominated to succeed the President. The outgoing President participates in the deliberations of the new Organizing Committee in an advisory

capacity. If more names are proposed than available elective positions, the outgoing Organizing Committee devises the procedure by which the requisite number of candidates is elected by the membership. The resulting list is submitted to the Organizing Committee of the parent Division(s) for approval before the end of the General Assembly. Members of the Organizing Committee normally serve a maximum of two terms, unless elected Vice-President of the Commission. Presidents may serve for only one term.

44. At least six months before each General Assembly, the Organizing Committee shall submit to the parent Division(s) a report on its activities during the past triennium, with its recommendation as to whether the Commission should be continued for another three years, or merged with one or more other Commissions, or discontinued. If a continuation is proposed, a plan for the activities of the next triennium should be presented, including those of any Working Groups which the Commission proposes to maintain during that period.

45. The Organizing Committee decides its own rules for the conduct of its business by physical meetings or (electronic) correspondence. Such rules require approval by the Organizing Committee of the parent Division(s).

Transactions IAU, Volume XXVIIB
Proc. XXVII IAU General Assembly, August 2009
Ian F. Corbett, ed.

doi:10.1017/S1743921310005399

CHAPTER VIII

RULES AND GUIDELINES FOR IAU SCIENTIFIC MEETINGS

Updated post General Assembly Rio de Janeiro, Brazil, August 2009

1. INTRODUCTION

The program of IAU scientific meetings is one of the most important means by which the IAU pursues its goal of promoting astronomy through international collaboration. A large fraction of the Union's budget is devoted to the support of these IAU scientific meetings. The IAU Executive Committee (EC) places great emphasis on maintaining high scientific standards, coverage of a balanced spectrum of topics, and an appropriately broad and international flavour for the program of IAU meetings. In that respect, the ICSU rules on non-discrimination in the access of qualified scientists from all parts of the world to any IAU meeting apply. The ICSU rules on non-discrimination are described in the document "Freedom, Responsibility and Universality of Science", available on <http://www.icsu.org/Gestion/img/ICSU_DOC_DOWNLOAD/2205_DD_FILE _Freedom_Responsibility_Universality_of_Science_booklet.pdf.>

The number of meetings that the IAU supports financially is restricted to nine IAU Symposia per year (19,000 EUR each); one Regional IAU Meeting (19,000 EUR) per year in the years between General Assemblies; and two co-Sponsored Meetings per year (up to 5,000 EUR each). Accordingly, not all meeting proposals worthy of support can be awarded IAU sponsorship.
The IAU Colloquium Series (usually three per year) was terminated after IAU Colloquium No. 200 (October 2005), to the benefit of the IAU Symposium Series (now nine per year).

Regular contact with the organizers during the preparation and conduct of IAU scientific meetings is maintained by the IAU Assistant General Secretary (AGS) and the President of the Coordinating Division (see below).

2. IAU SYMPOSIA / IAU GA SYMPOSIA

The IAU Symposium Series is the scientific flagship of the IAU. Symposia are organized on suitably broad, yet well-defined, scientific themes of considerable general interest, and normally last 5 days. IAU Symposia are intended to significantly advance the field by seeking answers to current key questions and/or clarify emerging concepts in invited reviews, invited papers, contributed papers and poster papers. Therefore their programs should consist of reviews and previews and should provide ample time for discussion.

Proposals for IAU Symposia in a certain calendar year, backed by a coordinating IAU Division and endorsed by a reasonable number of supporting IAU Divisions, IAU Commissions, and/or IAU Working Groups, have to be submitted to the IAU Proposal Web Server, before 1 December of the year two years before the intended Symposium.

The scientific merit of each IAU Symposium proposal will be evaluated by the twelve IAU Division Presidents (DPs), taking into consideration comments and advice received from the Organizing Committees of their IAU Divisions, and from IAU Commissions and/or IAU Working Groups. The DPs will communicate their recommendation for selection to the IAU Executive Committee (EC). The EC will decide on and announce the final selection of the nine IAU Symposia to be held in a certain year, in Spring of the year before. The decision on the selection, including conditions to be fulfilled before final approval, will be communicated to the proposers by the Assistant General Secretary (AGS) in a letter of award, accompanied by an official form listing the essential facts of the meeting as approved by the EC. Any revision of the details recorded on this form will require prior approval by the AGS.

The Assistant General Secretary (AGS), in consultation with the organizers of the individual Symposia, decides on the distribution of the financial support to each Symposium.

IAU Symposium Proceedings are published in the IAU Proceedings' Series by the IAU Publisher, Cambridge University Press (CUP). In the year of an IAU General Assembly, six of the nine IAU Symposia of that year will be scheduled as GA Symposia within the scientific program of the GA and held at the GA venue. A GA Symposium normally lasts 3.5 days. For GA Symposia, the GA Local Organizing Committee (GA NOC/LOC) will handle the local organization. The General Secretary (GS), in consultation with the organizers of the individual GA Symposia, coordinates the financial support to be allocated to each of the GA Symposia.

In the year of a GA, the three IAU Symposia not associated with that GA should not be scheduled within three months before or after the dates of that GA. The Executive Committee may, in exceptional circumstances, decide to support a Symposium within 3 months of GA. All IAU Symposium proposals have to be submitted to the IAU Proposal Web Server.

2.1. SELECTION CRITERIA FOR IAU SPONSORSHIP OF SYMPOSIA

The following guidelines for obtaining IAU sponsorship should be observed by prospective proposers:

(a) An IAU Symposium should have a well-defined and scientifically relevant theme, should be scheduled at a propitious time for significant progress in the field, and should be of interest to young researchers as well as senior experts.

(b) While the IAU embraces all fields in astronomy, a proposed IAU Symposium program should maintain a broad and balanced scope and cover the main active fields at appropriate intervals. Accordingly, even scientifically strong proposals in the same or largely overlapping fields can only be approved at some intervals. While some themes have developed series of IAU Symposia with intervals of 3-5 years, approval for those is not automatically guaranteed, since each proposal will be judged on its own scientific merits.

(c) Scientific programs of proposed IAU Symposia should be well balanced, to be demonstrated by the proposed draft program and the proposed draft list of key speakers.

(d) Given the international nature of the Union, IAU Symposia are by definition internationally oriented. This requires a well-balanced geographical and gender distribution of both the proposed Scientific Organizing Committee (SOC) and the proposed key speakers. Normally, substantially less than half of the SOC membership and of the key speakers should come from a single country. As a matter of course, the SOC membership should reflect in a balanced way the current activity in the field.

(e) Presentation of scientific results at an IAU Symposium is by invitation of the SOC chairperson. Suitably qualified scientists working in the field may seek invitations. It is the policy of the IAU to promote the full participation of astronomers worldwide in its Symposium program.

(f) It is essential that no restriction based on gender, race, colour, nationality, and religious or political affiliation be imposed on the full participation of all bona fide scientists in any aspect of the organization and conduct of IAU Symposia, either by its organizers or by the authorities of the host country. Approval of a proposal for an IAU meeting requires explicit guarantees that this principle will be respected. The statement that the ICSU rules on non-discrimination in open access to the meeting will be strictly observed must be explicitly confirmed before a proposal will receive final approval by the EC. A summary of the measures taken to ensure this should be given; the signatures of both the SOC and LOC chairpersons are required.

(g) In association with IAU meetings, educational activities may be organized, like International Schools for Young Astronomers (ISYAs) and Teachers' Workshops. By taking advantage of the presence of many expert national and foreign scientists, 1-day or 2-day events may be organized for the benefit of university and high-school educators in the country hosting the meeting. In the past, such initiatives have generally been well received and successful. While the scientific quality of the proposed Symposium will remain the primary selection criterion for IAU sponsorship, a good parallel educational program will certainly add to the overall merit of a proposal.

2.2. **PROPOSAL PREPARATION FOR IAU SYMPOSIA**

2.2.1. *GENERALITIES*

Normally, the initiative to propose a scientific meeting for IAU sponsorship originates from a group of scientists in a certain field. In collaboration with colleagues worldwide, they prepare a draft scientific program and nominations for the members of a candidate Scientific Organizing Committee, who will be responsible for the scientific aspects of the meeting from its inception to its conclusion. Responsibility for the preparation and timely submission of the final proposal rests with the chairperson of the candidate SOC.

Prospective meeting organizers should contact the AGS well in advance of their intended proposal submission, by sending a Letter of Intent (LoI, section 2.2.2).

An electronic application form and procedures have been designed so as to ensure that the information necessary for the evaluation of the proposals by the IAU Division Presidents and the IAU EC will be complete and in a uniform format, allowing objective comparison between proposals as far as possible. Therefore proposals with all entries properly answered have to be submitted electronically to the IAU Proposal Web Server.

2.2.2. *LETTER OF INTENT FOR AN IAU SYMPOSIUM TO BE PROPOSED*

Before submitting a proposal for an IAU Symposium, proposers must submit a Letter of Intent (LoI) via the IAU Proposal Web Server where it will be sent to the AGS, with a copy to the president of the desired Coordinating IAU Division associated with the scientific field of the meeting. The LoI format is given on the web page and should state:

(a) the title of the intended IAU Symposium;
(b) the full name(s) of the proposed SOC chairperson(s);
(c) the desired Coordinating IAU Division for the intended IAU Symposium;
(d) the Symposium venue and the preferred Symposium dates; and
(e) a short list of topics to be addressed at the Symposium (about 10 topics).

The deadline for the submission of Letters of Intent is 15 September. Letters of Intent received will be posted and updated on the IAU web site, informing prospective proposers of other existing plans for IAU Symposium proposals. This is in order to avoid unnecessary competition between proposals and to stimulate possible collaborations between otherwise competing groups.

2.2.3. *TOPIC AND TITLE OF PROPOSED IAU SYMPOSIUM*

The title of a Symposium should state the topic of the meeting as concisely and succinctly as possible. Long and detailed titles do not catch the eye, and are cumbersome for the announcement of a Symposium as well as on the cover of its subsequent Proceedings. Symposium titles should be no longer than 10 words (or 70 characters including spaces) in total. The AGS may request the proposers to modify their title to meet this requirement.

Any change of title or date of an IAU Symposium, after it has been accepted by the EC, requires the prior approval of the AGS.

2.2.4. *COORDINATING IAU DIVISION; SUPPORTING IAU DIVISIONS, IAU COMMISSIONS AND/OR IAU WORKING GROUPS*

An IAU Symposium can be proposed by individual members of the IAU, by an IAU Working Group, or by an IAU Commission. An IAU Division should accept the coordinating responsibilities for an IAU Symposium proposal as Coordinating Division.

When supporting Divisions, Commissions, and/or Working Groups are listed in the proposal, a report of the communication between the proposers and the above should be submitted together with the proposal.

Proposals must be submitted electronically to the IAU Proposal Web Server before the posted deadline. Normally, this deadline will be 1 December, two years before the year of the proposed Symposium.

2.2.5. *SCIENTIFIC ORGANIZING COMMITTEE*

The composition of the proposed SOC is a key element in assessing the scientific value of a proposal. The SOC of a Symposium has the overall responsibility for its scientific standards and should make sure to cover the principal topics of the field to be covered.

The SOC should normally not be larger than sixteen persons and should represent an optimum scientific, gender, and geographic distribution. Normally, an institution should not be represented on the SOC by more than one person. It is customary, but not required, that SOC members are members of the IAU. The composition of the SOC should reflect in a positive way the intent of the ICSU Statement on Freedom in the Conduct of Science.

SOC membership is subject to approval by the IAU EC, as part of the approval process. Any change of SOC membership, after a Symposium has been accepted by the EC, requires the prior approval of the IAU AGS.

The SOC is responsible for the scientific, gender, and geographical balance of the meeting in five main aspects:

- before the Symposium:

(a) in the definition of the scientific program of the Symposium, including the choice and distribution of topics for individual sessions, and the selection of invited reviews, invited papers, contributed papers and poster papers;

(b) in the choice of key speakers for invited reviews, and of session chairs;

(c) in providing in the proposal a list of about 10 preliminary scientific program topics, for announcement in the IAU Information Bulletin;

- after approval of the Symposium:

(d) in providing a list of individuals qualifying for IAU Travel Grants, with amounts recommended (for criteria, see below). That list must be submitted for approval to the AGS at least 5 months before the start of the Symposium;

- after the Symposium:

(e) within 1 month after a Symposium, the SOC chair person must send to the AGS the Post Meeting Report of the Symposium.

2.2.6. *POST MEETING REPORT*

The Symposium Post Meeting Report must contain:

(i) a copy of the final scientific program, listing invited review speakers and session chairs;

(ii) a list of participants, including their distribution on gender (this does not apply to GA Symposia)

(iii) a list of recipients of IAU grants, stating amount, country, and gender; (this does not apply to GA Symposia)

(iv) receipts signed by the recipients of IAU Grants; (this does not apply to GA Symposia)

(v) a report to the IAU EC summarizing the scientific highlights of the meeting (1-2 pages).

The Post Meeting Report form is available here <link >.
That site links also to compilations of previous years' IAU Post Meeting Reports.
The Symposium Post Meeting Report must be send to the AGS within one month after the Symposium.

2.2.7. *LOCAL ORGANIZING COMMITTEE*

The Symposium LOC, to be identified in the proposal, is responsible for all aspect of the local arrangements associated with the Symposium. Those tasks include booking and preparation of meeting rooms, provisions for modern audio-visual facilities, for coffee and tea breaks, arranging for necessary transportation for meeting participants, for ensuring that accommodation within reasonable price levels is available, for providing assistance to meeting participants with their bookings, and for providing or offering advice on access to affordable, reliable child care. In addition, the LOC should prepare and schedule social events as appropriate.
IAU Symposia hosted by an IAU General Assembly do not have their own LOCs.

2.2.8. *EDITORS OF THE PROCEEDINGS*

It speaks for itself that the success of a Symposium and its Proceedings depends in the first place on arranging for the best possible scientific program and on selecting the best possible speakers, keeping in mind the need for a balanced distribution of gender and geographical origin. It is of paramount importance that the Proceedings of an IAU Symposium will be published timely, i.e., **within 6 months after the Symposium**, as a valuable record of the Symposium for future reference. Arrangements for Authors and Editors for the publication of Proceedings of IAU Symposia are summarized in the <README >files in the directories below:
<...static/scientific_meetings/authors_2009/>
<..../static/scientific_meetings/editors_2009/>

Arrangements for Authors and Editors for the publication of Proceedings of IAU GA Symposia are summarized in the <README >files in the directories below:
<..../static/scientific_meetings/authors_ga2009/>
<..../static/scientific_meetings/editors_ga2009/ >

Full names and addresses of the proposed Editors must be given in the proposal. One of the proposed Editors should be marked as Chief Editor, with prime responsibility for contact with the IAU AGS and with the IAU Publisher, Cambridge University Press.

The contract between the IAU and CUP stipulates that the Proceedings of an IAU Symposium will be published within 6 months after that Symposium. Since CUP needs three months for its processing and publishing of a complete Proceedings' volume, **Editors must complete their editing task within the first 3 months after their Symposium** . This requires that all Authors have to deliver their completed manuscripts to the Chief Editor before or during the Symposium. Authors may be allowed to submit a revised version of their manuscript to the Chief Editor within four weeks after the Symposium.

Thus, Editors are committed to submit the final complete manuscript of the Proceedings of their IAU Symposium to CUP within 3 months after their IAU Symposium. Editors should realize that editing an IAU Symposium Proceedings volume can be a full time job for three months. Any change of Editors, after a Symposium has been accepted by the EC, requires the prior approval of the IAU AGS.

2.2.8.1 IAU EDITORIAL BOARD

In order to ensure the quality of the IAU Symposium Proceedings Series, to strengthen the working relation among the IAU Proceedings' Editors and the IAU AGS, and to provide a platform for communication, the EC has established the IAU Editorial Board (EB) for the Series.
The EB serves as a communication and support platform for the Chief Editors of all IAU Symposium Proceedings, where they can exchange experience and ask for advice, whenever necessary, in their efforts to ensure that all papers published in their Proceedings are of the highest quality and that their Proceedings are published on time, i.e., within 6 months after their Symposium.
Members of the IAU EB for the Proceedings of IAU Symposia held in a certain year are:

(a) all Chief Editors of the Proceedings of IAU Symposia of that year (the working members);

(b) the IAU AGS (chair); and

(c) the IAU GS plus three or four members appointed by the EC for a period of at least three years (advisory members).

The constitution of the EB of a certain year will be listed in all nine IAU Symposium Proceedings volumes of that year. EB members of a certain year will receive copies of all nine IAU Symposium Proceedings volumes of that year.

2.2.9. *REGISTRATION FEE*

2.2.9.1 REGISTRATION FEE FOR IAU SYMPOSIA NOT ASSOCIATED WITH AN IAU GENERAL ASSEMBLY

A determined effort must be made to keep the Symposium registration fee affordable to all. Such effort should include the use of low-cost meeting facilities and finding local sponsorship. Proposers should carefully specify what services the registration fee will cover. The current upper limit of the registration fee for IAU Symposia is US$ 325.- including the price of the Proceedings (2009 participant's price: 48.- GBP, 2010 49.- GBP).

2.2.9.2 REGISTRATION FEE FOR IAU SYMPOSIA HOSTED BY AN IAU GENERAL ASSEMBLY

For IAU Symposia held as part of an IAU GA, participants are required to pay the full registration fee for that GA.

2.2.10. *VENUE AND ACCOMMODATION*

The proposed venue should be reasonably accessible and affordable. The venue should have modern audio-visual facilities, and ensure that all poster presentations will be on display during the whole duration of the Symposium, preferably in the tea/coffee break areas.

In order to enable interested and qualified colleagues from all countries around the world to attend a Symposium, affordable accommodation should be available. It is recognized that some resorts offer conference room, board and lodging together in one location, which is most favourable for all scientific interactions during a Symposium. In case such a resort is expensive, efforts should be made to secure additional financial sponsoring, in order to keep the participants' costs affordable. Access to nearby, affordable child care is important, and the need for support for such services should be taken into account in the distribution of the IAU grant to attendees of the meeting.

2.2.11. SUBMITTING THE PROPOSAL

When all above requirements are observed, a completed proposal for an IAU Symposium should be submitted electronically to the IAU Proposal Web Server, before the posted deadline: December 1 of the year two years before the intended Symposium.

2.3. *IAU GRANTS FOR IAU SYMPOSIA*

2.3.1. *IAU GRANTS FOR IAU SYMPOSIA NOT ASSOCIATED WITH AN IAU GENERAL ASSEMBLY*

IAU Grants are intended to cover in part expenses associated with attendance of participants at the Symposium. Symposium organizers will receive IAU Grant funds to a maximum of 19,000 EUR per Symposium.

Participants of IAU Symposia may apply for an IAU Grant, using the form available on <..../static/meetings/GrantSymposiumRIM.pdf >.

A proposal for the distribution of IAU Grants to individual participants will be drafted by the SOC chairperson and sent to the AGS for approval. The IAU wishes to support qualified scientists to whom only limited means of support are available, e.g., colleagues from economically less privileged countries and young scientists. An IAU Grant should be the seed money in ensuring the participation of a selected beneficiary, including participants with young children, rather than adding comfort for colleagues whose attendance is already assured. In addition, a reasonable gender and geographical distribution is expected in the IAU Grant distribution proposal. Normally, no more than 1/3 of the IAU Grant funds for a Symposium should be allocated to a single country.

Within these general guidelines, it is left to the judgment of the SOC how to formulate its proposal for IAU Grant distribution, maintaining the overall scientific standard of the conference as the primary criterion. The recommendation for the distribution of the IAU Grants shall be sent by the SOC chairperson to the AGS, specifying for each person: name, nationality, full mailing and e-mail address, amount of proposed grant (in EUR), any amount reserved for child care, and title and nature of contribution (invited review, invited paper, contributed paper, poster paper). The recommendation of the SOC should reach the AGS no later than 5 months before the Symposium. This deadline is necessary in order to ensure timely notification to grant beneficiaries and completion of visa formalities.

After approval by the AGS of the IAU Grant distribution proposal of the SOC, individual IAU Grant notification letters will be mailed by the AGS to the recipients, with a copy to the SOC chair and LOC chair. Any disagreement between the SOC and the AGS on the award of grants shall be referred to the IAU General Secretary, whose decision shall be final.

Normally, the LOC will open a bank account in the name of the Symposium (or use a bank account of its institute or university) to which the IAU Secretariat will transfer the allocated IAU Grant funds. Individual IAU Grants will be paid by the LOC chairperson to recipients, upon their arrival and registration at the Symposium, against a signature of receipt.

2.3.2. *IAU GRANTS FOR IAU SYMPOSIA HOSTED BY AN IAU GENERAL ASSEMBLY*

In case of IAU Symposia hosted by a GA, IAU GA Grants are intended to cover in part expenses associated with the participant's attendance of the entire GA. Requests for financial support for attending a GA must use the electronic Application Form for an IAU Grant available at <...../grants_prizes/iau_grants/ga_events/ >.

Full completion of the Application Form is mandatory, including submission of an Abstract if relevant. The deadline for receiving these applications will be such, that enough time is left to the IAU Secretariat to prepare relevant summaries of applications, to be sent to the SOCs of the different scientific meetings for ranking, before a final selection is made by the IAU GS. Successful applicants will receive their allocated grant upon arrival and registration at the GA.

2.4. *WEB SITE FOR AN IAU SYMPOSIUM*

As soon as a successful applicant has been informed by the AGS of the approval of her/his proposed IAU Symposium, the SOC and LOC are kindly requested to create a website for that Symposium, containing, inter alia, all those parts of the information given above which are essential to know for the participants of that IAU meeting. The URL of the Symposium website should be communicated to the IAU AGS as soon as available.

2.5. *IAU GENERAL ASSEMBLY JOINT DISCUSSIONS AND SPECIAL SESSIONS*

A Joint Discussion (JD) held at an IAU General Assembly addresses scientific themes of interest to two or more IAU Commissions. A JD normally lasts 0.5 - 1.5 days during the GA.

A Special Session (SpS) held at IAU General Assemblies addresses important topics that concern the IAU, and is more specialized than a JD. A SpS may focus on timely issues ranging from recent scientific events to educational activities. A SpS normally lasts 0.5 - 3.0 days during the GA. JDs and SpSs may be proposed by individual IAU members through two or more IAU Commissions.

Proposals for IAU GA JDs and SpSs, backed by a coordinating IAU Division and endorsed by a reasonable number of IAU Divisions, IAU Commissions and IAU Working Groups, have to be submitted to the IAU Proposal Web Server:
<..../science/meetings/proposals/lop/ >, before 1 December of the year two years before the General Assembly.

The scientific merit of each IAU GA JD and SpS proposal will be evaluated by the twelve IAU Division Presidents (DPs), taking into consideration comments and advice received from the Organizing Committees of their IAU Divisions, and from IAU Commissions and/or IAU Working Groups. The DPs will communicate their recommendation for selection to the IAU Executive Committee (EC). The EC will decide on and announce the final selection of the nine IAU Symposia to be held in a certain year, in Spring of the year before. The decision on the selection, including conditions to be fulfilled before final approval, will be communicated to the proposers by the Assistant General Secretary (AGS) in a letter of award, accompanied by an official form listing the essential facts of the meeting as approved by the EC. Any revision of the details recorded on this form will require prior approval by the AGS.

The IAU General Secretary (GS), in consultation with the organizers of the individual JDs and SpSs, will coordinate the financial support to be allocated to each of the GA JDs and SpS. Participants in JDs and SpS may apply for IAU GA Grants. (The application form for a GA Travel Grant is available here <link >)

The local organization of all JDs and SpSs is in the hands of the GA Local Organizing Committee (GA NOC/LOC).

The proceedings of JDs and SpSs will be published in the IAU series Highlights of Astronomy by the IAU Publisher, Cambridge University Press, and edited by the IAU General Secretary (GS).

All IAU GA JD and SpS proposals have to be submitted to the IAU Proposal Web Server

2.6. *SELECTION CRITERIA FOR IAU SPONSORSHIP OF A GA JOINT DISCUSSION OR A GA SPECIAL SESSION*

The following guidelines for obtaining IAU sponsorship should be observed by prospective proposers:

(a) Each JD and SpS should have a well-defined and scientifically relevant theme, should be scheduled at a propitious time for significant progress in the field, and should be of interest to young researchers as well as senior experts.

(b) The proposed scientific program of a JD or SpS should be well balanced, to be demonstrated by the proposed draft program and the proposed draft list of key speakers.

(c) Given the international nature of the IAU, JDs and SpSs are by definition internationally oriented. This requires a well-balanced geographical and gender distribution of both the proposed Scientific Organizing Committee (SOC) and the proposed key speakers. Normally, substantially less than half of the SOC membership and of the key speakers should come from a single country. As a matter of course, the SOC membership should reflect in a balanced way the current activity in the field.

2.7. *PROPOSAL PREPARATION FOR A GA JOINT DISCUSSION OR A SPECIAL SESSION*

2.7.1. *GENERALITIES*

Normally, the initiative to propose an IAU GA JD or SpS originates from a group of scientists in a certain field. In collaboration with colleagues worldwide, they should prepare a draft scientific program and nominations for the members of a candidate Scientific Organizing Committee (SOC), which will be responsible for the scientific aspects of the meeting from its inception (the proposal) to its conclusion (the proceedings). Responsibility for the preparation and timely submission of the final proposal rests with the chairperson of the candidate SOC. Prospective meeting organizers should inform the IAU well in advance of their intended proposal submission by submitting a Letter of

Intent (LoI, section 3.2.2) before 15 September via the IAU Proposal Web Server to the AGS.

An electronic application form and procedures have been designed to ensure that the information necessary for the evaluation of the proposals by the IAU Division Presidents and the IAU EC will be complete and in a uniform format, allowing objective comparison between proposals as far as possible. Proposals, with all entries properly completed, have to be submitted electronically to the IAU Proposal Web Server.

2.7.2. *LETTER OF INTENT FOR A JOINT DISCUSSION OR A SPECIAL SESSION*

Before submitting a proposal for a Joint Discussion or Special Session, proposers must submit a Letter of Intent (LoI) via the IAU Proposal Web Server to the AGS, also copied to the President of the desired Coordinating IAU Division associated with the scientific field of the meeting. The LoI should state:

(a) the title of the intended JD or SpS;
(b) the full name(s) of the proposed SOC chairperson(s);
(c) the desired Coordinating IAU Division for the intended JD or SpS; and
(d) a short list of topics to be addressed at the intended JD or SpS (about 10 topics).

A list of received Letters of Intent will be posted and updated on the IAU website, informing prospective proposers of other existing plans for intended IAU GA Symposium, JD and SpS proposals. This, in order to avoid unnecessary competition between proposals and to stimulate possibly collaboration between otherwise competing groups.

2.8. *TOPIC AND TITLE OF A JOINT DISCUSSION OR A SPECIAL SESSION*

The title of a Joint Discussion or Special Session should state the topic of the meeting as concisely and succinctly as possible. JD/SpS titles should be no longer than 10 words (or 70 characters including spaces) in total.

Any change of title of a JD or SpS, after it has been accepted by the EC, requires the prior approval of the AGS.

2.8.1. *COORDINATING IAU DIVISION; SUPPORTING IAU DIVISIONS, IAU COMMISSIONS, AND/OR IAU WORKING GROUPS*

Joint Discussions and Special Sessions can be proposed by individual members of the IAU, by an IAU Working Group, or by an IAU Commission. An IAU Division should accept the coordinating responsibilities for a JD/SpS proposal as Coordinating Division.

When supporting Divisions, Commissions, and/or Working Groups are listed in the proposal, a report of the communication between the proposers and the above should be submitted together with the proposal.

Proposals must be submitted electronically to the IAU Proposal Web Server before the posted deadline. Normally, this deadline will be 1 December two years before the year of the General Assembly.

2.8.2. *SCIENTIFIC ORGANIZING COMMITTEE*

The composition of the proposed JD/SpS SOC is a key element in assessing the scientific value of a proposal. The SOC of a JD or SpS has the overall responsibility for its scientific standards and should make sure that the principal topics of the field of the JD/SpS will be covered. The SOC should normally not be larger than sixteen persons and should represent an optimum scientific, geographic and gender distribution. Normally, an institution should not be represented on the SOC by more than one person. It is customary, but not required, that SOC members are members of the IAU. The composition of the SOC should reflect in a positive way the intent of the ICSU Statement on Freedom in the Conduct of Science.

SOC membership is subject to approval by the IAU EC, as part of the approval process. Any change of SOC membership, after a JD/SpS has been accepted by the EC, requires the prior approval of the IAU AGS.

The JD/SpS SOC is responsible for the scientific, gender, and geographical balance of the meeting in three main aspects:

(a) in the definition of the scientific program, including the choice and distribution of topics for individual sessions, and the selection of invited reviews, invited papers, contributed papers, and poster papers;

(b) in the choice of key speakers for invited reviews and of session chairs; and

(c) in providing in the proposal a list of about 10 preliminary scientific program topics, for announcement in the IAU Information Bulletin.

2.8.3. EDITORS OF PROCEEDINGS

It speaks for itself that the success of a Joint Discussion or Special Session and its proceedings depends in the first place on arranging for the best possible scientific program and on selecting the best possible speakers.

The proceedings will be published in the IAU series Highlights of Astronomy by Cambridge University Press, with the GS as the Editor-in-Chief. It is of paramount importance that the proceedings of all JDs and SpS will be published timely, i.e., within 6 months after the GA , as a valuable record of the General Assembly for future reference.

Arrangements for JD/SpS Authors and JD/SpS Editors for the publication of the JD proceedings are summarized in the <README >files on the links below:
<http://www.iau.org/static/scientific_meetings/authors_proceedings/ >
<http://www.iau.org /static/scientific_meetings/editors_proceedings/ >

Full names and addresses of the proposed JD/SpS Editors must be given in the proposal. In the proposal form, one of the proposed JD/SpS Editors should be marked as Chief Editor, with prime responsibility for contact with the IAU GS, who will act as Editor-in-Chief for the Highlights of Astronomy volume.

In the contract between the IAU and CUP it is stipulated that the Highlight of Astronomy will be published within 6 months after the GA . Since CUP needs three months for its processing and publishing of a complete proceedings' manuscript, the Editors must complete their editing task within the first 3 months after the GA . This requires that all JD/SpS Authors have to deliver their completed manuscripts to the JD Editor before or during the JD/SpS. Authors may be allowed to submit a revised version of their manuscript to their JD/SpS Editor within four weeks after the GA.

Thus, JD/SpS Editors are committed to submit the final complete manuscript of the proceedings of their JD to the IAU GS within three months after the IAU GA.

Any change of Editors, after a JD or SpS has been accepted, requires the prior approval of the IAU AGS.

2.8.4. REGISTRATION FEE FOR THE GA

To attend a Joint Discussion or Special Session held as part of an IAU GA, participants are required to pay the full registration fee for that GA.

2.8.5. SUBMITTING THE PROPOSAL

When all above requirements are observed, completed proposals for IAU JDs and SpS should be submitted electronically to the IAU Proposal Web Server, before 1 December of the year two years before the General Assembly.

2.9. *TRAVEL GRANTS FOR GA JOINT DISCUSSIONS AND SPECIAL SESSIONS*

IAU GA Travel Grants are intended to cover in part expenses associated with the participant's attendance during the entire GA, including support for child care if needed. Requests for financial support for attending a GA should complete the electronic Application Form for an IAU Travel Grant available at:

<http://www.iau.org /grants_prizes/iau_grants/ga_events/ >

Full completion of the Application Form is mandatory, including submission of an Abstract if relevant. The deadline for receiving these applications will be such, that enough time is left to the IAU Secretariat to prepare relevant summaries of applications, to be sent to the SOCs of the different scientific meetings for ranking, before a final selection is made by the IAU GS. Successful applicants will receive their allocated grant upon arrival and registration at the GA.

2.10. *WEBSITE FOR A GA JOINT DISCUSSION OR SPECIAL SESSION*

As soon as a successful applicant has been informed by the AGS of the approval of her/his proposed IAU JD or SpS, the SOC is kindly requested to create a website for that IAU meeting, containing, inter alia, those parts of the above information which are essential for the participants of that IAU JD or SpS. The URL of the website should be communicated to the IAU AGS as soon as available.

3. REGIONAL IAU MEETINGS (RIMs)

The IAU sponsors two series of Regional IAU Meetings (RIMs): a series of triennial meetings in the Asian-Pacific region (APRIM, since 1978) , and a series of triennial meetings in the Latin-American region (LARIM, since 1978). A past series of twelve European Regional IAU Meetings (1974-1990) has effectively been succeeded by the series of Joint European and National Astronomy Meetings (JENAM), under the auspices of the European Astronomical Society.

APRIMs and LARIMs are held by invitation of a national astronomical society in, respectively, the Asian-Pacific region and the Latin-American region, in years between GAs. Their purpose, in addition to the discussion of specific scientific topics, is to promote contacts between scientists in the regions concerned, especially young astronomers, including those with young children. Therefore, a much wider range of scientific topics, a larger SOC, and a larger total attendance are expected for RIMS than for IAU Symposia. The Proceedings of RIMs are usually published by a regional publisher or in a regional astronomical publication series.

4. EDUCATIONAL ASPECTS OF IAU SCIENTIFIC MEETINGS

At IAU meetings, International Schools for Young Astronomers (ISYAs), Teachers' Workshops, or similar educational activities may be organized adjacent to the scientific meeting. By taking advantage of the presence of distinguished national and foreign scientists, one- or two-day events may be organized for the benefit of university and high-school astronomy educators in the country hosting the meeting. In the past, such initiatives have been generally very successful and well received by their audiences. Stimulating and improving the teaching of science, and of astronomy in particular, is becoming increasingly urgent, and parallel educational activities of the type described above, in connection with IAU scientific meetings, are encouraged. While the quality of the proposed scientific program will remain the primary selection criterion for IAU sponsorship, a good parallel education program will certainly add to the overall merit of a proposal.

5. CO-SPONSORING OF MEETINGS

The IAU may decide to co-sponsor meetings that are organized by other scientific unions. Main organizational and financial responsibility for such meetings rests with the main sponsoring union. The IAU expects to be represented in the Scientific Organizing Committee concerned and to be consulted about publication of the proceedings and other major issues. The IAU may make a financial contribution to that meeting.

Transactions IAU, Volume XXVIIB
Proc. XXVII IAU General Assembly, August 2009
Ian F. Corbett, ed.

doi:10.1017/S1743921310005405

CHAPTER IX

DIVISIONS, COMMISSIONS AND WORKING GROUPS

Proposed Structure and Composition of IAU Divisions, Commissions and Working Groups 2009 - 2012
Updated post General Assembly Rio de Janeiro, Brazil, August 2009 but subject to constant updating on the IAU web site

Discontinued Commissions and Working Groups

Div. I/WG - Second Realization of the International Celestial Reference Frame
Div. I/Comm. 8/ WG - Densification of the Optical Reference Frame
Div. II/WG - Solar and Interplanetary Nomenclature
Inter- Div. IV-V-IX /WG - Standard Stars
Div. VII /WG - Galactic Center
Div.IX/ - Detectors
Div.XII/Comm.46/ PG - Exchange of Astronomers
Div.XII/Comm.46/ PG - National Liaisons on Astronomy Education
title changed to National Liaisons on Astronomy Education and Newsletter
Div.XII/Comm.46/ PG - Exchange of Books and Journals
Div.XII/Comm.46/PG - Newsletter
PG suspended - included in PG National Liaisons on Astronomy Education and Newsletter
EC -WG - Publishing

New Commissions and Working Groups

Div.II/WG - Communicating Heliophysics
Div II/WG - Comparative Solar Minima
Div. XII/Com. 46/WG - Network for Astronomy Education NASE
Div.XII/Com.55/WG - New Media
Div. XII/ WG - Johannes Kepler
Div. XII/Com.46/ - PG for Collaborative Programs

Division I Fundamental Astronomy

URL: <http://astro.cas.cz/iaudiv1 >

President
Dennis D. McCarthy
US Naval Observatory
3450 Massachusetts Ave NW
US Washington DC 20392-5420
United States

Phone: +1 202 762 1837
Fax: +1 202 762 1563
<dennis.mccarthy@usno.navy.mil >
<http://www.usno.navy.mil/>

Vice-President
Sergei A. Klioner
Technical University Dresden
Lohrmann Observatory
Mommsenstr 13
DE 01062 Dresden
Germany
Phone: +49 351 4633 2821
Fax: +49 351 4633 7019
<Sergei.Klioner@tu-dresden.de >
<http://astro.geo.tu-dresden.de/>

Organizing Committee

Dafydd Wyn Evans (United Kingdom), Catherine Y. Hohenkerk (United Kingdom), Mizuhiko Hosokawa (Japan), Cheng-Li Huang (China Nanjing), George H. Kaplan (United States), Zoran Kneževi (Serbia, Republic of), Richard N. Manchester (Australia), Alessandro Morbidelli (France), Gérard Petit (France), Harald Schuh (Austria), Michael H. Soffel (Germany), Jan Vondrák (Czech Republic), Norbert Zacharias (United States).

PARTICIPATING COMMISSIONS AND COMMISSION WORKING GROUPS

Division I Commission 4 Ephemerides

P: Kaplan, George (United States), <gkaplan@usno.navy.mil>
VP: Hohenkerk, Catherine (United Kingdom), <Catherine.Hohenkerk@UKHO.gov.uk>
OC: Arlot, Jean-Eudes (France), Bangert, John (United States), Bell, Steven (United Kingdom), Folkner, William (United States), Lara, Martin (Spain), Pitjeva, Elena (Russian Federation), Urban, Sean (United States), Vondrák, Jan (Czech Republic), URL: <http://ssd.jpl.nasa.gov/iau-comm4/ >

Division I Commission 7 Celestial Mechanics & Dynamical Astronomy

P: Kneževi, Zoran (Serbia, Republic of), <zoran@aob.rs>VP: Morbidelli, Alessandro (France), <morby@obs-nice.fr>OC: Athanassoula, Evangelia (France), Laskar, Jacques (France), Malhotra, Renu (United States), Mikkola, Seppo (Finland), Peale, Stanton (United States), Roig, Fernando (Brazil).
URL: <http://copernico.dm.unipi.it/comm7/ >

Division I Commission 8 Astrometry

P: Evans, Dafydd (United Kingdom), <dwe@ast.cam.ac.uk>VP: Zacharias, Norbert (United States), <nz@usno.navy.mil>OC: Andrei, Alexandre (Brazil), Brown, Anthony (Netherlands), Gouda, Naoteru (Japan), Kumkova, Irina (Russian Federation), Popescu, Petre (Romania), Souchay, Jean (France), Unwin, Stephen (United States), Zhu, Zi (China Nanjing).
URL: <http://www.ast.cam.ac.uk/iau_comm8/ >

Division I Commission 19 Rotation of the Earth

P: Schuh, Harald (Austria), <harald.schuh@tuwien.ac.at>
VP: Huang, Cheng-Li (China Nanjing), <clhuang@shao.ac.cn>
S: Seitz, Florian (Germany), <seitz@bv.tum.de>
OC: Bizouard, Christian (France), Chao, Benjamin (China Taipei), Gross, Richard (United States), Kosek, Wieslaw (Poland), Malkin, Zinovy (Russian Federation), Richter, Bernd (Germany), Salstein, David (United States), Titov, Oleg (Australia).

URL: <http://www.iau-comm19.org/ >

Division I Commission 31 Time
P: Manchester, Richard (Australia), <Dick.Manchester@csiro.au>
VP: Hosokawa, Mizuhiko (Japan), <hosokawa@crl.go.jp>
OC: Arias, Elisa (France), Gang, Zhang (China Nanjing), Tuckey, Philip (France), Zharov, Vladimir (Russian Federation).
URL: <http://www.astro.oma.be/IAU/COM31/ >

Division I Commission 52 Relativity in Fundamental Astronomy
P: Petit, Gérard (France), <gpetit@bipm.org>
VP: Soffel, Michael (Germany), <soffel@rcs.urz.tu-dresden.de>
OC: Brumberg, Victor (Russian Federation), Capitaine, Nicole (France), Fienga, Agnès (France), Guinot, Bernard (France), Huang, Cheng (China Nanjing), Klioner, Sergei (Germany), Mignard, François (France), Seidelmann, P. (United States), Wallace, Patrick (United Kingdom).
URL: <http://astro.geo.tu-dresden.de/RIFA/ >

Division I Commission 8 WG Densification of the Optical Reference Frame
Chair: Zacharias, Norbert (United States), <nz@usno.navy.mil>

Division I Commission 19 WG High-frequency & Sudden Variations in Earth Orientation Chair: Rothacher, Markus (Switzerland), <markus.rothacher@ethz.ch>

Division I Commission WG Relativity in Fundamental Astronomy
Division I Fundamental Astronomy Working Groups
Division I WG Second Realization of International Celestial Reference Frame
Chair: Ma, Chopo (United States), <cma@gemini.gsfc.nasa.gov>
URL: <http://rorf.usno.navy.mil/ICRF2/ >

Division I WG Numerical Standards in Fundamental Astronomy Chair: Luzum, Brian (United States), <brian.luzum@usno.navy.mil>
URL: <http://maia.usno.navy.mil/NSFA.html >

Division I WG Astrometry by Small Ground-Based Telescopes
Chair: Thuillot, William (France), <thuillot@imcce.fr>
URL: <http://www.imcce.fr/hosted_sites/iau_wgnps/astrom.html >

Division I Fundamental Astronomy Inter-Division Working Groups

Inter-Div. III-I WG Cartographic Coordinates & Rotational Elements
Chair: Archinal, Brent (United States), <barchinal@usgs.gov>
URL: <http://astrogeology.usgs.gov/Projects/WGCCRE/ >

Inter-Div. III-I WG Natural Satellites
Chair: Arlot, Jean-Eudes (France), <Jean-Eudes.Arlot@obspm.fr>
URL: <http://www.imcce.fr/host/iau_wgnps/iauwg.html >

Division II Sun & Heliosphere

URL: <http://www.iac.es/proyecto/iau_divii/IAU-DivII/main/index.php >

President
Valentin Martnez Pillet
Instituto de Astrofsica de Canarias
C/ Va Lctea s/n
ES 38200 La Laguna, Tenerife
Spain
Phone: +34 922 605 237
Fax: +34 922 605 210
<vmp@iac.es >
<http://www.iac.es/>

Vice-President
James A. Klimchuk
NASA/GSFC, Goddard Space Flight Center
Solar Physics Lab
Code 671
US Greenbelt MD 20771, United States
Phone: +1 301 286 9060
Fax: +1 301 286 7194
<James.A.Klimchuk@nasa.gov >
<http://www.nasa.gov/centers/goddard/home/index.html>

Organizing Committee

Gianna Cauzzi (Italy), Natchimuthuk Gopalswamy (United States), Alexander Kosovichev (United States), Ingrid Mann (Belgium), Karel J. Schrijver (United States), Lidia van Driel-Gesztelyi (France)

Secretary

Lidia van Driel-Gesztelyi
MSSL, University College London, U.K.
Paris Observatory, LESIA, 92195 Meudon, France
Konkoly Observatory, Budapest, Hungary
Phone: +33 1 45 07 79 00
Fax: +33 1 45 07 79 75
Email: <Lidia.vanDriel@obspm.fr >
Organization website: <http://www.mssl.ucl.ac.uk/>

PARTICIPATING COMMISSIONS AND COMMISSION WORKING GROUPS

Division II Commission 10 Solar Activity

P: van Driel-Gesztelyi, Lidia (France), <Lidia.vanDriel@obspm.fr>
VP: Schrijver, Karel (United States), <schrijver@lmsal.com>
OC: Charbonneau, Paul (Canada), Fletcher, Lyndsay (United Kingdom), Hasan, S. Sirajul (India), Hudson, Hugh (United States), Kusano, Kanya (Japan), Mandrini, Cristina (Argentina), Peter, Hardi (Germany), Vrnak, Bojan (Croatia, the Republic of), Yan, Yihua (China Nanjing)
URL: <http://www.mssl.ucl.ac.uk/iau_c10/index.html >

Division II Commission 12 Solar Radiation & Structure

P: Kosovichev, Alexander (United States), <akosovichev@solar.stanford.edu>
VP: Cauzzi, Gianna (Italy), <gcauzzi@arcetri.astro.it>
OC: Asplund, Martin (Germany), Bogdan, Thomas (United States), Brandenburg, Axel (Sweden), Cauzzi, Gianna (Italy), Christensen-Dalsgaard, Jrgen (Denmark), Cram, Lawrence (Australia), Gan, Weiqun (China Nanjing), Gizon, Laurent (Germany), Heinzel, Petr (Czech Republic), Kuznetsov, Vladimir (Russian Federation), Rovira, Marta (Argentina), Shchukina, Nataliya (Ukraine), Venkatakrishnan, P. (India), Warren Jr, Wayne (United States)
URL: <http://www.iac.es/proyecto/iau_divii/IAU-Com12/main/index.php >

Division II Commission 49 Interplanetary Plasma & Heliosphere

P: Gopalswamy, Natchimuthuk (United States), <Nat.Gopalswamy@nasa.gov>
VP: Mann, Ingrid (Belgium), <imann@uni-muenster.de>
OC: Briand, Carine (France), Lallement, Rosine (France), Lario, David (United States), Manoharan, P. (India), Shibata, Kazunari (Japan), Webb, David (United States)
URL: <http://www.lesia.obspm.fr/iau_49/ >

Division II Sun & Heliosphere Working Groups

Division II WG Communicating Heliophysics
Chair: Briand, Carine (France), <carine.briand@obspm.fr>

Division II WG WG - Comparative Solar Minima
Chair: Gibson, Sarah (United States), <sgibson@ucar.edu>

Division II WG Solar Eclipses
Chair: Pasachoff, Jay (United States), <jay.m.pasachoff@williams.edu>
URL: <http://www.williams.edu/Astronomy/IAU_eclipses/ >

Division II WG International Solar Data Access
Chair: Bentley, Robert (United Kingdom), <rdb@mssl.ucl.ac.uk>
URL: <http://www.mssl.ucl.ac.uk/grid/iau/DivII_WG_IntDataAccess.html >

Division II WG International Collaboration on Space Weather
Chair: Webb, David (United States), <David.Webb@hanscom.af.mil>

Division III Planetary Systems Sciences

URL: <http://www.ss.astro.umd.edu/IAU/div3/ >

President
Karen J. Meech
University of Hawaii Honolulu
Inst of Astronomy
2680 Woodlawn Dr
US Honolulu HI 96822
United States
Phone: +1 808 956 6828
Fax: +1 808 956 9580
<meech@ifa.hawaii.edu >
<http://www.ifa.hawaii.edu/>

Vice-President
Giovanni B. Valsecchi
INAF
IASF Sezione di Roma
Via Fosso del Cavaliere 100
Tor Vergata
IT 00133 Roma, Italy
Phone: +39 06 499 34446
Fax: +39 06 206 60188
<giovanni@iasf-roma.inaf.it >
<http://www.iasf-roma.inaf.it./>

Organizing Committee

Dominique Bockelee-Morvan (France), Alan Paul Boss (United States), Alberto Cellino (Italy), Guy Joseph Consolmagno (Vatican City State), Julio Angel Fernández (Uruguay), William M. Irvine (United States), Daniela Lazzaro (Brazil), Patrick Michel (France), Keith Stephen Noll (United States), Rita M. Schulz (Netherlands), Jun-ichi Watanabe (Japan), Makoto Yoshikawa (Japan), Jin Zhu (China Nanjing)

PARTICIPATING COMMISSIONS AND COMMISSION WORKING GROUPS

Division III Commission 15 WG Task Force on Cometary Magnitudes
Chair: Tancredi, Gonzalo (Uruguay), <gonzalo@fisica.edu.uy>
URL: <http://www.lowtem.hokudai.ac.jp/iau-c15-wg/index.html >

Division III Commission 15 Physical Studies of Comets & Minor Planets
P: Cellino, Alberto (Italy), <cellino@to.astro.it>
VP: Bockelee-Morvan, Dominique (France), <dominique.bockelee@obspm.fr>
OC: Bockelee-Morvan, Dominique (France), Davidsson, Bjrn (Sweden), Dotto, Elisabetta (Italy), Fitzsimmons, Alan (United Kingdom), Jenniskens, Petrus (United States), Lupishko, Dmitrij (Ukraine), Mothé-Diniz, Thais (Brazil), Tancredi, Gonzalo (Uruguay), Wooden, Diane (United States)
URL: <http://iau15.space.swri.edu >

Division III Commission 16 Physical Study of Planets & Satellites
P: McGrath, Melissa (United States), <melissa.a.mcgrath@nasa.gov>
VP: Lemmon, Mark (United States), <lemmon@tamu.edu>
OC: KIM, SANG JOON (Korea, Rep of), Ksanfomality, Leonid (Russian Federation), Lara, Luisa (Spain), Morrison, David (United States), Tejfel, Viktor (Kazakhstan), Yanamandra-Fisher, Padma (United States). URL: <http://www.iaa.es/IAUComm16 >

Division III Comm. 20 Positions & Motions of Minor Planets, Comets & Satellites
P: Yoshikawa, Makoto (Japan), <makoto@isas.jaxa.jp>
URL: <http://www.astro.uu.se/IAU/c20/ >

Division III Commission 22 Meteors, Meteorites & Interplanetary Dust
P: Watanabe, Jun-ichi (Japan), <jun.watanabe@nao.ac.jp>
VP: Jenniskens, Petrus (United States), <pjenniskens@mail.arc.nasa.gov>
OC: Borovicka, Jir (Czech Republic), Campbell-Brown, Margaret (Canada), Consolmagno, Guy (Vatican City State), Jenniskens, Petrus (United States), Jopek, Tadeusz (Poland), Vaubaillon, Jérémie (France), Williams, Iwan (United Kingdom)

URL: <http://meteor.asu.cas.cz/IAU >

Division III Commission 51 Bio-Astronomy
P: Irvine, William (United States), <irvine@astro.umass.edu>
VP: Ehrenfreund, Pascale (Netherlands), <p.ehrenfreund@chem.leidenuniv.nl>
OC: Cosmovici, Cristiano (Italy), Kwok, Sun (China Nanjing), Levasseur-Regourd, Anny-Chantal (France), Morrison, David (United States), Udry, Stephane (Switzerland)
URL: <http://www.dtm.ciw.edu/boss/c51index.html >

Division III Commission 53 Extrasolar Planets (WGESP)
P: Boss, Alan (United States), <boss@dtm.ciw.edu>
VP: Lecavelier des Etangs, Alain (France), <lecaveli@iap.fr>
OC: Bodenheimer, Peter (United States), Cameron, Andrew (United Kingdom), Kokubo, Eiichiro (Japan), Mardling, Rosemary (Australia), Minniti, Dante (Chile), Queloz, Didier (Switzerland)
URL: <http://www.ciw.edu/boss/IAU/div3/wgesp/ >

Division III Commission 15 WG Physical Studies of Comets
Chair: Boice, Daniel (United States), <dboice@swri.edu>
URL: <http://atlas.sr.unh.edu/IAU_Comm15/ >

Division III Commission 15 WG Physical Studies of Minor Planets
Chair: Gil-Hutton, Ricardo (Argentina), <rgilhutton@casleo.gov.ar>
URL: <http://atlas.sr.unh.edu/IAU_Comm15/ >

Division III Commission 20 WG Motions of Comets
Chair: Fernández, Julio (Uruguay), <julio@fisica.edu.uy>
URL: <http://www.astro.uu.se/IAU/c20/wgcomet.html >

Division III Commission 20 WG Distant Objects
Chair: Marsden, Brian (United States), <bmarsden@cfa.harvard.edu>
URL: <http://www.astro.uu.se/IAU/c20/ >

Division III Commission 22 WG Professional-Amateur Cooperation in Meteors
Chair: Ryabova, Galina (Russian Federation), <ryabova@niipmm.tsu.ru>

Division III Planetary Systems Sciences Working Groups

Division III WG Small Bodies Nomenclature (SBN)
Chair: Tichá, Jana (Czech Republic), <jticha@klet.cz>
URL: <http://www.ss.astro.umd.edu/IAU/csbn/ >

Division III WG Planetary System Nomenclature (WGPSN)
Chair: Schulz, Rita (Netherlands), <rschulz@rssd.esa.int>
URL: <http://planetarynames.wr.usgs.gov/append2.html >

Division III Planetary Systems Sciences Inter-Division Working Groups

Inter-Div. III-I WG Cartographic Coordinates & Rotational Elements
Chair: Archinal, Brent (United States), <barchinal@usgs.gov>
URL: <http://astrogeology.usgs.gov/Projects/WGCCRE/ >

Inter-Div. III-I WG Natural Satellites
Chair: Arlot, Jean-Eudes (France), <Jean-Eudes.Arlot@obspm.fr>
URL: <http://www.imcce.fr/host/iau_wgnps/iauwg.html >

Division IV Stars

URL: <http://clavius.as.arizona.edu/iaudiv4/ >

President
Christopher Corbally
University of Arizona
Steward Observ-Vatican Observ Research Grp
US Tucson AZ 85721
United States

Phone: +1 520 621 3225
Fax: +1 520 621 1532
<corbally@as.arizona.edu >
<http://vaticanobservatory.org/>

Vice-President
Francesca D'Antona
INAF
Osservatorio Astronomico di Roma
Via Frascati 33
Monte Porzio Catone
IT 00040 Roma, Italy
Phone: +39 06 942 86447
Fax: +39 06 9447 243
<dantona@mporzio.astro.it >
<http://www.oa-roma.inaf.it/>

Organizing Committee

Martin Asplund (Germany), Corinne Charbonnel (Switzerland), Jose A. Durantez Docobo (Spain), Richard O. Gray (United States), Nikolai E. Piskunov (Sweden)

PARTICIPATING COMMISSIONS AND COMMISSION WORKING GROUPS

Division IV Commission 26 Double & Multiple Stars

P: Docobo, Jose A. (Spain), <joseangel.docobo@usc.es>
VP: Mason, Brian (United States), <bdm@usno.navy.mil>
OC: Arenou, Frédéric (France), Balega, Yurij (Russian Federation), Oswalt, Terry (United States), Pourbaix, Dimitri (Belgium), Scardia, Marco (Italy), Scarfe, Colin (Canada), Tamazian, Vakhtang (Spain)
URL: <http://ad.usno.navy.mil/wds/dsl.html >

Division IV Commission 29 Stellar Spectra

P: Piskunov, Nikolai (Sweden), <piskunov@astro.uu.se>
VP: Cunha, Katia (Brazil), <katia@on.br>
OC: Aoki, Wako (Japan), Asplund, Martin (Germany), Bohlender, David (Canada), Carpenter, Kenneth (United States), Melendez, Jorge (Portugal), Rossi, Silvia (Brazil), Smith, Verne (United States), Soderblom, David (United States), Wahlgren, Glenn (United States)
URL: <http://www.iiap.res.in/personnel/parthasarathy/IAUcom29.html >

Division IV Commission 35 Stellar Constitution

P: Charbonnel, Corinne (Switzerland), <Corinne.Charbonnel@unige.ch>
VP: Limongi, Marco (Italy), <marco@oa-roma.inaf.it>
OC: Fontaine, Gilles (Canada), Isern, Jorge (Spain), Lattanzio, John (Australia), Leitherer, Claus (United States), van Loon, Jacco (United Kingdom), Weiss, Achim (Germany), Yungel'son, Lev (Russian Federation)
URL: <http://iau-c35.stsci.edu/ >

Division IV Commission 36 Theory of Stellar Atmospheres

P: Asplund, Martin (Germany), <asplund@mpa-garching.mpg.de>
VP: Puls, Joachim (Germany), <uh101aw@usm.uni-muenchen.de>
OC: Allende Prieto, Carlos (United States), Ayres, Thomas (United States), Berdyugina, Svetlana (Finland), Gustafsson, Bengt (Sweden), Hubeny, Ivan (United States), Ludwig, Hans (France), Mashonkina, Lyudmila (Russian Federation), Randich, Sofia (Italy)
URL: <http://www.galax.obspm.fr/IAU36/ >

Division IV Commission 45 Stellar Classification
P: Gray, Richard (United States), <grayro@appstate.edu>
VP: Nordström, Birgitta (Denmark), <birgitta@astro.ku.dk>
OC: Burgasser, Adam (United States), Eyer, Laurent (United States), Gupta, Ranjan (India), Hanson, Margaret (United States), Irwin, Michael (United Kingdom), Soubiran, Caroline (France)
URL: <http://www.phys.appstate.edu/com45 >

Division IV Commission 26 WG Binary & Multiple System Nomenclature
Chair: Hartkopf, William (United States), <wih@usno.navy.mil>

Division IV Stars Working Groups

Division IV WG Massive Stars
Chair: Owocki, Stanley (United States), <owocki@bartol.udel.edu>
URL: <http://www.astroscu.unam.mx/massive_stars/index.php >

Division IV WG Abundances in Red Giants
Chair: Lattanzio, John (Australia), <john.lattanzio@sci.monash.edu.au>
URL: <http://www.maths.monash.edu.au/~johnl/wgarg// >

Division IV Stars Inter-Division Working Groups
Inter-Div. IV-V WG Active B Stars
Chair: Peters, Geraldine (United States), <gjpeters@mucen.usc.edu>
URL: <http://www.astro.virginia.edu/~dam3ma/benews/iauwg_abs.html >

Inter-Div. IV-V WG Ap & Related Stars
Chair: Mathys, Gautier (Chile), <gmathys@eso.org>
URL: <http://ams.astro.univie.ac.at/apn/varia/IAUgroup.html >

Division V Variable Stars

URL: <http://www.konkoly.hu/IAUDV/ >

President
Steven D. Kawaler
Iowa State University
Dept. of Physics and Astronomy
A323 Zaffarano Hall
US Ames IA 50011-3160
United States

Phone: +1 515 294 9728
Fax: +1 515 294 6027
<sdk@iastate.edu >
<http://www.iastate.edu/>

Vice-President
Ignasi Ribas
Inst de Ciencies de l'Espai
Fac de Ciencies Campus UAB
Torre C5-parell
2a planta
ES 08193 Bellaterra, Barcelona
Spain
Phone: +34 93 581 4371
Fax: +34 93 581 4363
<iribas@ieec.uab.es >

Organizing Committee

Michel Breger (Austria), Edward F. Guinan (United States), Gerald Handler (Austria), Slavek M. Rucinski (Canada)

Participating Commissions and Commission Working Groups

Division V Commission 27 Variable Stars

P: Handler, Gerald (Austria), <gerald.handler@univie.ac.at>
VP: Pollard, Karen (New Zealand), <karen.pollard@canterbury.ac.nz>
OC: Bedding, Timothy (Australia), Cateln, Mrcio (Chile), Cunha, Margarida (Portugal), Eyer, Laurent (United States), Jeffery, Christopher (United Kingdom), Kepler, S. (Brazil), Kolenberg, Katrien (Austria), Mkrtichian, David (Korea, Rep of), Olah, Katalin (Hungary)
URL: <http://www.konkoly.hu/IAUC27/ >

Division V Commission 42 Close Binary Stars

small P: Ribas, Ignasi (Spain), <iribas@ieec.uab.es>
VP: Richards, Mercedes (United States), <mrichards@astro.psu.edu>
OC: Bradstreet, David (United States), Harmanec, Petr (Czech Republic), Kaluzny, Janusz (Poland), Mikolajewska, Joanna (Poland), Munari, Ulisse (Italy), Niarchos, Panayiotis (Greece), Olah, Katalin (Hungary), Pribulla, Theodor (Slovakia), Scarfe, Colin (Canada), Torres, Guillermo (United States)
URL: <http://www.konkoly.hu/IAUC42/index.html >

Division V Variable Stars Working Groups

Division V WG Spectroscopic Data Archiving

Chair: Griffin, R. (Canada), <Elizabeth.Griffin@nrc-cnrc.gc.ca>
URL: <http://www.konkoly.hu/SVO/ >

Division V Variable Stars Inter-Division Working Groups

Inter-Div. IV-V WG Active B Stars

Chair: Peters, Geraldine (United States), <gjpeters@mucen.usc.edu>
URL: <http://www.astro.virginia.edu/d̃am3ma/benews/iauwg_abs.html >

Inter-Div. IV-V WG Ap & Related Stars

Chair: Mathys, Gautier (Chile), <gmathys@eso.org>
URL: <http://ams.astro.univie.ac.at/apn/varia/IAUgroup.html >

Division VI Interstellar Matter

URL: <http://www.div6.qub.ac.uk >

President
You-Hua Chu
University of Illinois Urbana
Astronomy Dept
103 Astron Bldg
1002 W Green St
US Urbana IL 61801
United States
Phone: +1 217 333 5535
Fax: +1 217 244 7638
<yhchu@illinois.edu >
<http://www.astro.uiuc.edu>

Vice-President
Sun Kwok
University of Hong Kong
Faculty of Science
Chong Yuet Ming Physics Bldg
Pokfulam Rd
HK Hong Kong
Hong Kong, China Nanjing
Phone: +852 2859 2682
Fax: +852 2858 4620
<sunkwok@hku.hk >

Organizing Committee

Dieter Breitschwerdt (Germany), Michael G. Burton (Australia), Sylvie Cabrit (France), Paola Caselli (Italy), Elisabete M. de Gouveia Dal Pino (Brazil), Neal J. Evans (United States), Thomas Henning (Germany), Mika J. Juvela (Finland), Bon-Chul Koo (Korea, Rep of), Michal Rozyczka (Poland), Laszlo V. Tth (Hungary), Masato Tsuboi (Japan), Ji Yang (China Nanjing)

PARTICIPATING COMMISSIONS AND COMMISSION WORKING GROUPS

Division VI Commission 34 Interstellar Matter

P: Millar, Thomas (United Kingdom), <Tom.Millar@qub.ac.uk>
VP: Chu, You-Hua (United States), <yhchu@illinois.edu>
OC: Breitschwerdt, Dieter (Germany), Burton, Michael (Australia), Cabrit, Sylvie (France), Caselli, Paola (Italy), de Gouveia Dal Pino, Elisabete (Brazil), Dyson, John (United Kingdom), Ferland, Gary (United States), Juvela, Mika (Finland), Koo, Bon-Chul (Korea, Rep of), Kwok, Sun (China Nanjing), Lizano, Susana (Mexico), Rozyczka, Michal (Poland), Tóth, Laszlo (Hungary), Tsuboi, Masato (Japan), Yang, Ji (China Nanjing)
URL: <http://www1.ast.leeds.ac.uk/IAU34/IAU34news.html >

Division VI Interstellar Matter Working Groups
Division VI WG Star Formation

Chair: Palla, Francesco (Italy), <palla@arcetri.astro.it>

Division VI WG Astrochemistry

Chair: van Dishoeck, Ewine (Netherlands), <ewine@strw.leidenuniv.nl>
URL: <http://www.strw.leidenuniv.nl/ĩau34/ >

Division VI WG Planetary Nebulae

Chair: Manchado, Arturo (Spain), <amt@iac.es>
URL: <http://www.iac.es/project/PNgroup/wg/ >

Division VII Galactic System

URL: <http://www.mpe.mpg.de/IAU_DivVII/ >

President
Despina Hatzidimitriou
University of Crete
Dept Physics
P.O. Box 2208
GR 710 03 Heraklion
Greece
Phone: +30 2 810 394 212
Fax: +30 2 810 394 201
Email: <dh@physics.uoc.gr >
Organization website:
<http://www.uoc.gr/>

Vice-President
Rosemary F. Wyse
Johns Hopkins University
Physics-Astronomy Dept CAS
Charles and 34th Street
US Baltimore MD 21218-2686
United States
Phone: +1 410 516 5392
Fax: +1 410 516 5096
Email: <wyse@pha.jhu.edu >
Organization website:
<http://www.jhu.edu/>

Organizing Committee

Giovanni Carraro (Italy), Bruce Gordon Elmegreen (United States), Birgitta Nordström (Denmark)

PARTICIPATING COMMISSIONS AND COMMISSION WORKING GROUPS

Division VII Commission 33 Structure & Dynamics of the Galactic System

P: Wyse, Rosemary (United States), <wyse@pha.jhu.edu>
VP: Nordstr¨om, Birgitta (Denmark), <birgitta@astro.ku.dk>
OC: Bland-Hawthorn, Jonathan (Australia), Feltzing, Sofia (Sweden), Fuchs, Burkhard (Germany), Minniti, Dante (Chile)
URL: <http://www.mpe.mpg.de/IAU_Comm33/ >

Division VII Commission 37 Star Clusters & Associations

P: Elmegreen, Bruce (United States), <bge@watson.ibm.com>
VP: Carraro, Giovanni (Italy), <giovanni.carraro@unipd.it>
OC: Da Costa, Gary (Australia), de Grijs, Richard (United Kingdom), Deng, LiCai (China Nanjing), Lada, Charles (United States), Lee, Young (Korea, Rep of), Minniti, Dante (Chile), Sarajedini, Ata (United States), Tosi, Monica (Italy)
URL: <http://www.sc.eso.org/g̃carraro/IAUComm37.html >

Division VII Galactic System Working Groups

Division VII WG Galactic Center

Chair: Lazio, Joseph (United States), <Lazio@nrl.navy.mil>

Division VIII Galaxies & the Universe

URL: <http://www.star.bris.ac.uk/iau/ >

President
Elaine M. Sadler
University of Sydney
School of Physics A28
AU Sydney NSW 2006
Australia
Phone: +61 2 9351 2622
Fax: +61 2 9351 7726
<ems@physics.usyd.edu.au >
<http://www.physics.usyd.edu.au/>

Vice-President
Françoise Combes
Observatoire de Paris
LERMA, Bat. A
61 Av de l'Observatoire
FR 75014 Paris, France
Phone: +33 1 40 51 20 77
Fax: +33 1 40 51 20 02
<francoise.combes@obspm.fr >
<http://lerma.obspm.fr/>

Organizing Committee

Roger L. Davies (United Kingdom), John S. Gallagher III (United States), Thanu Padmanabhan (India)

PARTICIPATING COMMISSIONS AND COMMISSION WORKING GROUPS

Division VIII Commission 28 Galaxies

P: Davies, Roger (United Kingdom), <rld@astro.ox.ac.uk>
VP: Gallagher III, John (United States), <jsg@astro.wisc.edu>
OC: Courteau, Stephane (Canada), Dekel, Avishai (Israel), Franx, Marijn (Netherlands), Jog, Chanda (India), Nakai, Naomasa (Japan), Rubio, Monica (Chile), Tacconi, Linda (Germany), Terlevich, Elena (United Kingdom)
URL: <http://aramis.obspm.fr/IAU28/index.php >

Division VIII Commission 47 Cosmology

P: Padmanabhan, Thanu (India), <paddy@iucaa.ernet.in>
VP: Schmidt, Brian (Australia), <brian@mso.anu.edu.au>
OC: Bunker, Andrew (United Kingdom), Ciardi, Benedetta (Germany), Jing, Yipeng (China Nanjing), Koekemoer, Anton (United States), Lahav, Ofer (United Kingdom), Lefevre, Olivier (France), Scott, Douglas (Canada)
URL: <http://www.exp-astro.phys.ethz.ch/IAU47/ >

Division VIII Galaxies & the Universe Working Groups

Division VIII WG Supernovae

Chair: Hillebrandt, Wolfgang (Germany), <wfh@mpa-garching.mpg.de>
URL: <http://www.star.bris.ac.uk/iau/news/sn_wg_ga_26.pdf >

Division IX Optical & Infrared Techniques

URL: <http://www.ifa.hawaii.edu/users/kud/iau/DivIX.htm >

President
Andreas Quirrenbach
Universitt Heidelberg
Landessternwarte
Koenigstuhl 12
DE 69117 Heidelberg Germany
Phone: +49 6221 54 1792
Fax: +49 6221 54 1702
<A.Quirrenbach@lsw.uni-heidelberg.de >
<http://www.lsw.uni-heidelberg.de/>

Vice-President
David Richard Silva
National Optical Astronomy Observatory
950 North Cherry Avenue
Tucson AZ 85719
United States
Phone: +1 520 318 8281
Fax: +1 520 318 8170
<dsilva@noao.edu >
<http://www.noao.edu>

Organizing Committee

Michael G. Burton (Australia), Xiangqun Cui (China Nanjing), Ian S. McLean (United States), Eugene F. Milone (Canada), Jayant Murthy (India), Stephen T. Ridgway (United States), Gražina Tautvaišiene (Lithuania), Andrei A. Tokovinin (Chile), Guillermo Torres (United States)

Participating Commissions and Commission Working Groups

Division IX Commission 21 Galactic and Extragalactic Background Radiation

P: Witt, Adolf (United States), <awitt@dusty.astro.utoledo.edu>
VP: Murthy, Jayant (India), <jmurthy@yahoo.com>
OC: Baggaley, William (New Zealand), Dwek, Eli (United States), Gustafson, Bo (United States), Levasseur-Regourd, Anny-Chantal (France), Mann, Ingrid (Belgium), Mattila, Kalevi (Finland), Watanabe, Jun-ichi (Japan)
URL: <http://www.astro.ufl.edu/g̃ustaf/IAUCom21/IAU_Com_21.html >

Division IX Commission 25 Stellar Photometry and Polarimetry

P: Milone, Eugene (Canada), <milone@ucalgary.ca>
VP: Walker, Alistair (Chile), <awalker@ctio.noao.edu>
OC: Anthony-Twarog, Barbara (United States), Bastien, Pierre (Canada), Knude, Jens (Denmark), Kurtz, Donald (United Kingdom), Menzies, John (South Africa), Mironov, Aleksey (Russian Federation), Qian, Shengbang (China Nanjing)
URL: <http://iau_c25.saao.ac.za/ >

Division IX Commission 30 Radial Velocities

P: Torres, Guillermo (United States), <torres@cfa.harvard.edu>
VP: Pourbaix, Dimitri (Belgium), <pourbaix@astro.ulb.ac.be>
OC: Marcy, Geoffrey (United States), Mathieu, Robert (United States), Mazeh, Tsevi (Israel), Minniti, Dante (Chile), Moutou, Claire (France)Pepe, Francesco (Switzerland), Turon, Catherine (France), Zwitter, Tomaž (Slovenia, the Republic of)
URL: <http://www.ctio.noao.edu/science/iauc30/iauc30.html >

Division IX Commission 54 Optical & Infrared Interferometry

P: Ridgway, Stephen (United States), <ridgway@noao.edu>
VP: van Belle, Gerard (Germany), <gvanbell@eso.org>
OC: Duvert, Gilles (France), Genzel, Reinhard (Germany), Haniff, Christopher (United Kingdom), Hummel, Christian (Germany), Lawson, Peter (United States), Monnier, John (United States), Tuthill, Peter (Australia), Vakili, Farrokh (France)
URL: <http://olbin.jpl.nasa.gov/iau/ >

Division IX Commission 25 WG Infrared Astronomy

Chair: Milone, Eugene (Canada), <milone@ucalgary.ca>
URL: <http://www.ucalgary.ca/m̃ilone/IRWG/ >

Division IX Commission 30 WG Radial-Velocity Standard Stars
Chair: Udry, Stephane (Switzerland), <stephane.udry@obs.unige.ch>
URL: <http://obswww.unige.ch/ũdry/std/std.html >

Division IX Commission 30 WG Stellar Radial Velocity Bibliography
Chair: Levato, Orlando (Argentina), <hlevato@icate-conicet.gob.ar>
URL: <http://www.casleo.gov.ar/catalogo/catalogo.htm >

Division IX Commission 30 WG Catolog of Orbital Elements of Spectroscopic Binary Systems
Chair: Pourbaix, Dimitri (Belgium), <pourbaix@astro.ulb.ac.be>
URL: <http://sb9.astro.ulb.ac.be/ >

Division IX Optical & Infrared Techniques Working Groups
Division IX WG Adaptive Optics

Division IX WG Large Telescope Projects

Division IX WG Site Testing Instruments
Chair: Tokovinin, Andrei (Chile), <atokovinin@ctio.noao.edu>
URL: <http://www.ctio.noao.edu/science/iauSite/ >

Division IX WG Small Telescope Projects

Division IX WG Sky Surveys
Chair: Parker, Quentin (Australia), <qap@physics.mq.edu.au>
URL: <http://www.skysurveys.org >

Division IX Optical & Infrared Techniques Inter-Division Working Groups

Inter-Div. IX-X-XI WG Astronomy from the Moon
Chair: Falcke, Heino (Germany), <hfalcke@mpifr-bonn.mpg.de>
URL: <http://www.cfa.harvard.edu/moon/ >

Inter-Div. IX-X WG Encouraging the International Development of Antarctic Astronomy
Chair: Burton, Michael (Australia), <M.Burton@unsw.edu.au>
URL: <http://www.phys.unsw.edu.au/jacara/iau >

Division X Radio Astronomy

URL: <http://www.bao.ac.cn/IAU_COM40/ >

President
Russell A. Taylor
Dept Physics-Astronomy
2500 University Dr NW
CA Calgary AB T2N 1N4
Canada
Phone: +1 403 220 5385
Fax: +1 403 289 3331
<russ@ras.ucalgary.ca >
<http://www.ucalgary.ca/phas/>

Vice-President
Jessica Mary Chapman
CSIRO/ATNF
PO Box 76
AU Epping NSW 1710
Australia
Phone: +61 2 9372 4196
Fax: +61 2 9372 4310
<Jessica.Chapman@csiro.au >
<http://www.atnf.csiro.au/>

Organizing Committee

Christopher L. Carilli (United States), Gabriele Giovannini (Italy), Richard E. Hills (United Kingdom), Hisashi Hirabayashi (Japan), Justin L. Jonas (South Africa), Joseph Lazio (United States), Raffaella Morganti (Netherlands), Ren-Dong Nan (China Nanjing), Monica Rubio (Chile), Prajval Shastri (India)

PARTICIPATING COMMISSIONS AND COMMISSION WORKING GROUPS

Division X Commission 40 Radio Astronomy

P: Nan, Ren-Dong (China Nanjing), <nrd@bao.ac.cn>
VP: Taylor, Russell (Canada), <russ@ras.ucalgary.ca>
OC: Carilli, Christopher (United States), Chapman, Jessica (Australia), Dubner, Gloria (Argentina), Garrett, Michael (Netherlands), Goss, W. Miller (United States), Hills, Richard (United Kingdom), Hirabayashi, Hisashi (Japan), Rodriguez, Luis (Mexico), Shastri, Prajval (India), Torrelles, Jose (Spain)
URL: <http://www.bao.ac.cn/IAU_COM40/ >

Division X Radio Astronomy Working Groups
Division X WG Global VLBI

Chair: Tingay, Steven (Australia), <s.tingay@curtin.edu.au>
URL: <http://www.bao.ac.cn/IAU_COM40/WG/WgVLBI.html >

Division X WG Interference Mitigation

Chair: Baan, Willem (Netherlands), <baan@astron.nl>
URL: <http://www.bao.ac.cn/IAU_COM40/WG/WgRF.html >

Division X WG Astrophysically Important Spectral Lines

Chair: Ohishi, Masatoshi (Japan), <masatoshi.ohishi@nao.ac.jp>
URL: <http://www.bao.ac.cn/IAU_COM40/WG/WgSL.html >

Division X Radio Astronomy Inter-Division Working Groups
Inter-Div. IX-X-XI WG Astronomy from the Moon

Chair: Falcke, Heino (Germany), <hfalcke@mpifr-bonn.mpg.de>
URL: <http://www.cfa.harvard.edu/moon/ >

Inter-Div. IX-X WG Encouraging the International Development of Antarctic Astronomy

Chair: Burton, Michael (Australia), <M.Burton@unsw.edu.au>
URL: <http://www.phys.unsw.edu.au/jacara/iau >

Inter-Div. X-XII WG Historic Radio Astronomy
Chair: Kellermann, Kenneth (United States), <kkellerm@nrao.edu>
URL: <http://www.bao.ac.cn/IAU_COM40/WG/WgHistRA.html >

Division XI Space & High Energy Astrophysics

URL: <http://www.mpe.mpg.de/IAU_DivXI/ >

President
Christine Jones
Harvard Smithsonian CfA
High Energy Astrophysics Division
MS 3, 60 Garden Str
US Cambridge MA 02138-1516
United States
Phone: +1 617 495 7137
Fax: +1 617 495 7356
<cjf@cfa.harvard.edu >
<http://cfa-www.harvard.edu/hco/>

Vice-President
Noah Brosch
Tel Aviv University
Dept Phys-Astronomy/Wise Observatory
Ramat Aviv
PO Box 39040
IL Tel Aviv 69978, Israel
Phone: +972 3 640 7413
Fax: +972 3 640 8179
<noah@wise.tau.ac.il >
<http://wise-obs.tau.ac.il/>

Organizing Committee

Matthew G. Baring (United States), Martin Adrian Barstow (United Kingdom), Joo Braga (Brazil), Evgenij M. Churazov (Russian Federation), Jean Eilek (United States), Hideyo Kunieda (Japan), Jayant Murthy (India), Isabella Pagano (Italy), Hernan Quintana (Chile), Marco Salvati (Italy), Kulinder Pal Singh (India), Diana Mary Worrall (United Kingdom)

PARTICIPATING COMMISSIONS AND COMMISSION WORKING GROUPS

Division XI Commission 44 Space & High Energy Astrophysics

P: Hasinger, Gnther (Germany), <guenther.hasinger@ipp.mpg.de>
VP: Jones, Christine (United States), <cjf@cfa.harvard.edu>
OC: Braga, Joo (Brazil), Brosch, Noah (Israel), de Graauw, Thijs (Chile), Gurvits, Leonid (Netherlands), Helou, George (United States), Howarth, Ian (United Kingdom), Kunieda, Hideyo (Japan), Montmerle, Thierry (France), Okuda, Haruyuki (Japan), Salvati, Marco (Italy), Singh, Kulinder (India)
URL: <http://www.mpe.mpg.de/IAU_DivXI/ >

Division XI Space & High Energy Astrophysics Working Groups

Division XI WG Particle Astrophysics

Chair: Schlickeiser, Reinhard (Germany), <office@tp4.ruhr-uni-bochum.de>
URL: <http://iau.physik.rub.de/ >

Division XII Union-Wide Activities

URL: <http://www.iaudivisionxii.org/ >

President
Françoise Genova
Observatoire Astronomique Strasbourg
11 r de l'Université
FR 67000 Strasbourg
France
Phone: +33 3 68 85 24 76
Fax: +33 3 68 85 24 32
<genova@astro.u-strasbg.fr >
<http://astro.u-strasbg.fr/observatoire/>

Vice-President
Raymond P. Norris
CSIRO/ATNF
PO Box 76
AU Epping NSW 1710
Australia
Phone: +61 2 9372 4416
Fax: +61 2 9372 4310
<ray.norris@csiro.au >
<http://www.atnf.csiro.au/>

Organizing Committee

Dennis Crabtree (Canada), Olga B. Dluzhnevskaya (Russian Federation), Masatoshi Ohishi (Japan), Rosa M. Ros (Spain), Clive L.N. Ruggles (United Kingdom), Nikolaj N. Samus' (Russian Federation), Xiaochun Sun (China Nanjing), Virginia L. Trimble (United States), Wim van Driel (France), Glenn Michael Wahlgren (United States)

PARTICIPATING COMMISSIONS AND COMMISSION WORKING GROUPS

Division XII Commission 55 WG Virtual Astronomy Multimedia project

Division XII Commission 55 WG New Media

Division XII Commission 55 WG Communicating Astronomy Journal
Chair: Russo, Pedro (Germany), <prusso@eso.org>

Division XII Commission 55 WG Washington Charter
Chair: Crabtree, Dennis (Canada), <Dennis.Crabtree@nrc-cnrc.gc.ca>

Division XII Commission 55 WG CAP Conferences
Chair: Robson, Ian (United Kingdom), <eir@roe.ac.uk>

Division XII Commission 5 Documentation & Astronomical Data
P: Ohishi, Masatoshi (Japan), <masatoshi.ohishi@nao.ac.jp>
VP: Hanisch, Robert (United States), <hanisch@stsci.edu>
OC: Andernach, Heinz (Mexico), Bishop, Marsha (United States), Griffin, R. (Canada), Kembhavi, Ajit (India), Murphy, Tara (Australia), Pasian, Fabio (Italy)
URL: <http://www.atnf.csiro.au/people/rnorris/IAUC5/ >

Division XII Commission 6 Astronomical Telegrams
P: Samus, Nikolai (Russian Federation), <samus@sai.msu.ru>
VP: Yamaoka, Hitoshi (Japan), <yamaoka@phys.kyushu-u.ac.jp>
OC: Aksnes, Kaare (Norway), Green, Daniel (United States), Marsden, Brian (United States), Nakano, Syuichi (Japan), Spahr, Timothy (United States), Tich, Jana (Czech Republic), Williams, Gareth (United States)
URL: <http://cfa-www.harvard.edu/iau/Commission6.html >

Division XII Commission 14 Atomic & Molecular Data
P: Wahlgren, Glenn (United States), <wahlgren@milkyway.gsfc.nasa.gov>
VP: van Dishoeck, Ewine (Netherlands), <ewine@strw.leidenuniv.nl>
OC: Beiersdorfer, Peter (United States), Dimitrijevic, Milan (Serbia, Republic of), Jorissen, Alain (Belgium), Mashonkina, Lyudmila (Russian Federation), Nilsson, Hampus (Sweden), Salama, Farid (United States), Tennyson, Jonathan (United Kingdom)
URL: <http://iacs.cua.edu/IAUC14 >

Division XII Commission 41 History of Astronomy
P: Ruggles, Clive (United Kingdom), <rug@leicester.ac.uk>
VP: Kochhar, Rajesh (India), <rkochhar2000@yahoo.com>
S: Belmonte Avilés, Juan Antonio (Spain), <jba@ll.iac.es>
OC: Corbin, Brenda (United States), de Jong, Teije (Netherlands), Norris, Raymond (Australia), Pigatto, Luisa (Italy), Sma, Mitsuru (Japan), Sterken, Christiaan (Belgium), Sun, Xiaochun (China Nanjing)
URL: <http://www.historyofastronomy.org >

Division XII Commission 46 Astronomy Education & Development
P: Ros, Rosa (Spain), <ros@mat.upc.es>
VP: Hearnshaw, John (New Zealand), <john.hearnshaw@canterbury.ac.nz>
OC: Balkowski-Mauger, Chantal (France), de Greve, Jean-Pierre (Belgium), Deustua, Susana (United States), Guinan, Edward (United States), Haubold, Hans (Austria), Jones, Barrie (United Kingdom), Kochhar, Rajesh (India), Malasan, Hakim (Indonesia), Marschall, Laurence (United States), Miley, George (Netherlands), Pasachoff, Jay (United States), Torres-Peimbert, Silvia (Mexico)
URL: <http://iau46.obspm.fr/ >

Division XII Commission 50 Protection of Existing & Potential Observatory Sites
P: van Driel, Wim (France), <wim.vandriel@obspm.fr>
VP: Green, Richard (United States), <rgreen@as.arizona.edu>
OC: Alvarez del Castillo, Elizabeth (United States), Blanco, Carlo (Italy), Crawford, David (United States), Metaxa, Margarita (Greece), Ohishi, Masatoshi (Japan), Sullivan, III, Woodruff (United States), Tzioumis, Anastasios (Australia)
URL: <http://www.ctio.noao.edu/iau50/ >

Division XII Commission 55 Communicating Astronomy with the Public
P: Crabtree, Dennis (Canada), <Dennis.Crabtree@nrc-cnrc.gc.ca>
VP: Christensen, Lars (Germany), <lars@eso.org>
OC: Alvarez Pomares, Oscar (Cuba), Damineli Neto, Augusto (Brazil), Fienberg, Richard (United States), Green, Anne (Australia), Kembhavi, Ajit (India), Russo, Pedro (Germany), Sekiguchi, Kazuhiro (Japan), Whitelock, Patricia (South Africa), Zhu, Jin (China Nanjing)
URL: <http://www.communicatingastronomy.org >

Division XII Commission 5 WG Astronomical Data
Chair: Norris, Raymond (Australia), <ray.norris@csiro.au>
URL: <http://www.atnf.csiro.au/people/rnorris/WGAD/ >

Division XII Commission 5 WG Libraries

Division XII Commission 5 WG FITS
Chair: Pence, William (United States), <William.D.Pence@nasa.gov>
URL: <http://fits.gsfc.nasa.gov/iaufwg/ >

Division XII Commission 5 WG Nomenclature
Chair: Schmitz, Marion (United States), <zb4ms@ipac.caltech.edu>
URL: <http://vizier.u-strasbg.fr/Dic/iau-spec.htx >

Division XII Commission 5 WG Virtual Observatories, Data Centers & Networks
Chair: Hanisch, Robert (United States), <hanisch@stsci.edu>
URL: <http://cdsweb.u-strasbg.fr/IAU/wgvo.html >

Division XII Commission 5 WG Task Force on Preservation & Digitization of Photographic Plates
Chair: Griffin, R. (Canada), <Elizabeth.Griffin@nrc-cnrc.gc.ca>

Division XII Commission 14 WG Atomic Data
URL: <http://iacs.cua.edu/IAUC14/ >

Division XII Commission 14 WG Collision Processes
Chair: Dimitrijevic, Milan (Serbia, Republic of), <mdimitrijevic@aob.bg.ac.yu>
URL: <http://iacs.cua.edu/IAUC14/ >

Division XII Commission 14 WG Molecular Data
URL: <http://iacs.cua.edu/IAUC14/ >

Division XII Commission 14 WG Solids & Their Surfaces
Chair: Vidali , Gianfranco (United States), <gvidali@syr.edu>
URL: <http://iacs.cua.edu/IAUC14/ >

Division XII Commission 41 WG Archives
Chair: Chinnici, Ileana (Italy), <chinnici@astropa.unipa.it>

Division XII Commission 41 WG Historic Radio Astronomy
Chair: Orchiston, Wayne (Australia), <Wayne.Orchiston@jcu.edu.au>

Division XII Commission 41 WG Historical Instruments
Chair: Schechner, Sara (United States)
URL: <http://www.oapd.inaf.it/museo/PagineInglesi/History%20of%20astronomy/HI_WG/hi_wgroup.htm >

Division XII Commission 41 WG Transits of Venus
Chair: Duerbeck, Hilmar (Germany), <hilmar@uni-muenster.de>

Division XII Commission 41 WG Astronomy and World Heritage
Chair: Ruggles, Clive (United Kingdom), <rug@leicester.ac.uk>
URL: <http://www.astronomicalheritage.org/ >

Division XII Commission 50 WG Controlling Light Pollution
Chair: Wainscoat, Richard (United States), <rjw@ifa.hawaii.edu>
URL: <http://www.ctio.noao.edu/light_pollution/iau50/ >

Division XII Union-Wide Activities Working Groups
Division XII WG Johannes Kepler
Chair: Mahoney, Terence (Spain), <tjm@iac.es>

Division XII Union-Wide Activities Inter-Division Working Groups
Inter-Div. X-XII WG Historic Radio Astronomy
Chair: Kellermann, Kenneth (United States), <kkellerm@nrao.edu>
URL: <http://www.bao.ac.cn/IAU_COM40/WG/WgHistRA.html >

Executive Committee Working Groups

Future Large Scale Facilities
This WG is currently being reformed.

Women in Astronomy
Chairs: Sarah T. Maddison (Australia) & Francesca Primas (Germany)
Email: smaddison@swin.edu.au & fprimas@eso.org

OC: Sarah Maddison, Francesca Primas, Conny Aerts (Belgium), Geoffery Clayton (USA), Franoise Combes (France), Gloria Dubner (Argentina), Luigina Feretti (Italy), Anne Green (Australia), Elizabeth Griffin (Canada), Yanchun Liang (China), Yuko Motizuki (Japan), Birgitta Nordström (Denmark).
<http://astronomy.swin.edu.au/IAU-WIAWG/members/OC.html >

International Year of Astronomy 2009
Chair: Catherine Cesarsky (France)
Members: Christensen, Lars (ESO), Corbett, Ian (France/UK), Crabtree, Dennis (Canada), Deustua, Susana (USA), Govender, Kevindran (South Africa), Hemenway, Mary (USA), Kaifu, Norio (Japan), Robson, Ian (UK).
<http://www.astronomy2009.org/ >

This WG will be disbanded in the course of 2010.

IAU General Assemblies
Chair: Daniela Lazarro (Brazil)
Members: Jan Palouš (Czech Republic), GA 2012 (t.b.d), GA2015 (t.b.d).

Advisory Committee on Hazards of Near-Earth Objects
Chair: David Morrison (USA)

Transactions IAU, Volume XXVIIB
Proc. XXVII IAU General Assembly, August 2009
Ian F. Corbett, ed.

doi:10.1017/S1743921310005417

CHAPTER X

NATIONAL MEMBERSHIP

Year(s) of Adherence	National Member	Category	Individual Members
1927	**ARGENTINA** CONICET Rivadavia 1917 AR Buenos Aires 1033 Argentina	II	134
1935, 1994	**ARMENIA** Armenian National Academy of Sciences Marshal Baghramian av 24 AM 375019 Yerevan Armenia	I	24
1939	**AUSTRALIA** Australian Academy of Sciences Ian Potter House Gordon St. G.P.O. Box 783 AU Canberra ACT 2601 Australia	IV	262
1955	**AUSTRIA** Österreichische Akademie der Wissenschaften Dr-I Seipel Platz 2 AT 1010 Vienna Austria	I	49
1920	**BELGIUM** Koninkl. Vlaamse Acad. voor Wetensch. en Kunsten Académie Royale des Sciences des Lettres et des Beaux-Arts Palais des Acadmies Rue Ducale 1 BE 1000 Bruxelles Belgium	IV	117

1998	**BOLIVIA** Academia Nacional de Ciencias de Bolivia Av 16 de Julio No. 1732 BO La Paz Bolivia	Interim	0
1961	**BRAZIL** Conselho Nacional de Desenvolvimento Científico e Tecnológico CNPq Av W3 Norte Quadra 507 B Caixa Postal 11-1142 BR Brasilia, DF 70740 Brazil	II	172
1957	BULGARIA Bulgarian Academy of Sciences 1 15 Noemvri Str BG Sofia 1040 Bulgaria	I	57
1920	**CANADA** National Research Council of Canada International Relations Office Project Liaison NRC-ISCU Desk Ms. Starlene Buchanan Room 1038M 100 Sussex Drive CA Ottawa ON K1A 0R6 Canada	V	245
1947	CHILE National Astronomical Observatory Camino El Observatorio 1515 CL Santiago Chile	II	90
1935	CHINA NANJING Chinese Academy of Sciences Purple Mountain Observatory 2 West Beijing Rd CN Nanjing 210008 China PR	VI	405
1959	CHINA TAIPEI Academia Sinica Institute of Astronomy & Astrophysics 128 Academia Rd Sec. 2 Nankang TW Taipei 115 Taiwan	II	51

2009	COSTA RICA University of Costa Rica University of Costa Rica CR San Jose Costa Rica	Interim	0
1935, 1994	CROATIA, THE REPUBLIC OF Croatian Academy of Sciences & Arts Hrvatsko Astronomsko Drutvo Kaciceva 26 HR 41000 Zagreb Hrvatska Croatia, the Republic of	I	16
2001	CUBA Academia de Ciencias de Cuba Capitolio Nacional CP 10240 CU 12400 La Habana Cuba	Interim	7
1922, 1993	CZECH REPUBLIC Academy of Sciences of Czech Republic Narodni 3 CZ 117 20 Praha 1 Czech Republic	III	92
1922	DENMARK Kongelige Danske Videnskabernes Selskab H C Andersens Bvd 35 DK 1553 Copenhagen V Denmark	III	63
1925	EGYPT ASRT 101 Kasr El-Eini Str EG Cairo 11516 Egypt	III	56
1935, 1992	ESTONIA Estonian Academy of Sciences Kohtu 6 EE 10130 Tallinn Estonia	I	23
1948	FINLAND The Finnish Academy of Science and Letters Suomalainen Tiedeakatemia Mariankatu 5 FI 00170 Helsinki Finland	II	67

1920	FRANCE Comit Franais des Unions Scientifiques Internati Acadmie des Sciences COFUSI 23 Quai Conti FR 75006 Paris France	VII	699
1951	GERMANY Rat Deutscher Sternwarten Gojenbergsweg 112 DE 21029 Hamburg Germany	VII	533
1920	GREECE General Secretariat for Research-Technology Hellenic Ministry of Development 14-18 Messogeion Ave GR 115 27 Athens Greece	III	108
2009	HONDURAS Universidad Nacional Autonoma de Honduras Observatrio Astronomico Centroamericano de Suyapa Ciudad Universitaria HN Tegucigalpa M.D.C. Honduras	Interim	2
1947	HUNGARY Hungarian Academy of Sciences Nador Street 7 HU 1051 Budapest Hungary	II	48
1988	ICELAND Director Office of Financial Affairs Ministry of Education, Sciences and Culture Solvholsgata 4 Dunhaga 3 IS 150 Reykjavik Iceland	I	4
1964	INDIA Indian National Science Academy Bahadur Shah Zafar Marg IN New Delhi 110 002 India	IV	222
1964, 1979	INDONESIA Indonesian Institute of Sciences Gedung Widya Graha Jl. Jenderal Gatot Subroto 10 ID Jakarta 12710 Indonesia	I	16

1969	IRAN, ISLAMIC REP OF University of Tehran Enghlab Ave IR Tehran Iran, Islamic Rep of	I	23
1947	IRELAND The Royal Irish Academy 19 Dawson Street IE 2 Dublin Ireland	I	44
1954	ISRAEL Israel Academy of Sciences & Humanities Albert Einstein Square PO Box 4040 IL Jerusalem 91040 Israel	III	75
1920	ITALY Istituto Nazionale di AstroFisica INAF Via del Parco Mellini 84 IT 00136 Roma Italy	VII	568
1920	JAPAN Science Council of Japan 7-22-34 Roppongi Minato-ku JP Tokyo 106-8555 Japan	VII	598
1973	KOREA, REP OF Korean Astronomical Society 61 Hwa-am Dong Yusung Ku KR Taejon 305-348 Korea, Rep of	II	109
1996	LATVIA University of Latvia Raina blvd 19 LV 1586 Riga Latvia	I	17
2006	LEBANON National Council for Scientific Research PO Box 11-8281 Riad El Solh 1107 LB 2260 Beirout Lebanon	Interim	4

1935, 1993	LITHUANIA Lithuanian Academy of Sciences MTP-1 Gedimino Prospekt 3 LT 2600 Vilnius Lithuania	I	15
1988	MALAYSIA Kementerian Sains Teknologi dan Inovasi Aras 5, Block 2, Menara PjH, Presint 2 MY Putrajaya 62100 Malaysia	I	9
1921	MEXICO Universidad Nacional Autonoma de Mexico UNAM Instituto de Astronoma Circuito exterior s/n Ciudad Universitaria MX Mexico DF 04510 Mexico	III	111
2006	MONGOLIA Mongolian Academy of Sciences International Relations Department Sukhbaatar Sq.3 A.Amar Street-1 MN 210620a Ulaanbaatar 11 Mongolia	Interim	0
1988, 2001	MOROCCO Cadi Ayyad University Presidence Prince Moulay Abdellah Avenue PO Box 511 MA Marrakech Morocco	Interim	9
1922	NETHERLANDS Koninklijke Nederlandse Akadamie van Wetenschappen Kloveniersburgwal 29 Box 19121 NL 1000 GC Amsterdam Netherlands	V	208
1964	NEW ZEALAND Royal Society of New Zealand PO Box 598 NZ Wellington 6140 New Zealand	II	28
2003	NIGERIA Nigerian Academy of Science University of Lagos PMB 1004 Univ of Lagos Post Office Akoka-Yaba NG Lagos Nigeria	I	10

1922	NORWAY Det Norske Videnskaps-Akademi Drammensveien 78 NO 0271 Oslo 2 Norway	II	33
2009	PANAMA Universidad de Panama Departamento de Fisica Facultad de Ciencias Naturales PA Ciudad de Panama Panama	Interim	0
1988	PERU Consejo Nacional de Ciencia y Tecnologia Apartado Postal 1984 PE Lima 100 Peru	Interim	3
2000	PHILIPPINES c/o Chief, AGSSB Science Garden Complex Agham Rd, Diliman PH Quezon City 1101 Philippines	Interim	4
1922	POLAND Polish Academy of Sciences Palac Kultury i Nauki Plac Defilad 1 PL 00-901 Warsaw Poland	III	149
1924	PORTUGAL Secção Portuguesa das Uniões Internacionais SPUIAGG Rua da Artilharia Um 107 PT 1099-052 Lisboa Portugal	II	43
1922	ROMANIA Romanian Academy of Sciences Calea Victoriei 125 RO 71102 Bucharest Romania	I	36
1935 1992	RUSSIAN FEDERATION Russian Academy of Sciences Foreign Relations Dept Leninskij Prosp 14 RU 117901 Moscow Russian Federation	V	369

1988	SAUDI ARABIA KACST Box 6086 SA Riyadh 11442 Saudi Arabia	I	12
1935, 2003	SERBIA, REPUBLIC OF Astronomical Observatory Society of Astronomers of Serbia Volgina 7 RS 11160 Belgrade Serbia, Republic of	I	34
1922, 1993	SLOVAKIA Slovak Academy of Sciences Scientific Secretary Stefanikova 49 SK 814 38 Bratislava Slovakia	I	38
1938	SOUTH AFRICA National Research Foundation P.O. Box 2600 ZA 0001 Pretoria South Africa	III	71
1922	SPAIN Comision Nacional de Astronomia Ministerio de Fomento Inst Geografico General Ibanez de Ibero 3 ES 28003 Madrid Spain	IV	303
1925	SWEDEN Royal Swedish Academy of Sciences The Foreign Secretary Box 50005 SE 104 05 Stockholm Sweden	III	111
1923	SWITZERLAND Swiss Academy of Sciences Schwarztorstrasse 9 CH 3007 Bern Switzerland	III	76
1935, 1993	TAJIKISTAN Tajik Academy of Sciences Institute of Astronomy Bukhoro Str 22 TJ 734042 Dushanbe Tajikistan	I	6

2006	THAILAND National Astronomical Research Inst Physics Bldg (Temporary Office) Chiang Mai University TH Chiang Mai 50200 Thailand	I	14
1961	TURKEY Turkish Astronomical Society Sabanci Üniversitesi Karaköy Iletişim Merkezi Bankalar Cad. 2 Karaköy TR 34420 Istanbul Turkey	I	40
1935, 1993	UKRAINE National Academy of Science of Ukraine 54 Volodymyrska St UA Kyiv 01601 Ukraine	III	188
1920	UNITED KINGDOM The Royal Astronomical Society Burlington House Piccadilly GB London W1J OBQ United Kingdom	VII	525
1920	UNITED STATES Board on International Scientific Organizations The National Academies c/o Dr. Kathie Bailey Mathae 500 Fifth St NW US Washington DC 20001 United States	X	2578
1932	VATICAN CITY STATE Governatorato Citta del Vaticano VA 00120 Citta del Vaticano Vatican City State	I	2
1953	VENEZUELA Centre de Investigaciones de Astronoma- CIDA Apdo Postal 264 VE Mérida 5101 A Venezuela	I	19
2009	VIET NAM Hanoi National University of Education Vietnam Astronomy Society 136 Xuan Thuy Street Cau Giay District VN Hanoi Viet Nam	Interim	4

Transactions IAU, Volume XXVIIB
Proc. XXVII IAU General Assembly, August 2009
Ian F. Corbett, ed.

doi:10.1017/S1743921310005429

CHAPTER XI

INDIVIDUAL MEMBERSHIP

Argentina

Abadi, Mario
Aguero, Estela
Ahumada, Javier
Ahumada, Andrea
Alonso, Maria
Alonso, Maria
Aquilano, Roberto
Arias, Maria
Arnal, Edmundo
Azcarate, Diana
Bagala, Liria
Bajaja, Esteban
Barba, Rodolfo
Bassino, Lilia
Baume, Gustavo
Beaugé, Christian
Benaglia, Paula
Benvenuto, Omar
Bosch, Guillermo
Brandi, Elisande
Branham, Richard
Brunini, Adrian
Buccino, Andrea
Calderón, Jesús
Cappa de Nicolau, Cristina
Carpintero, Daniel
Carranza, Gustavo
Castagnino, Mario
Castelletti, Gabriela
Cellone, Sergio
Cichowolski, Silvina
Cidale, Lydia
Cincotta, Pablo
Cionco, Rodolfo
Claria, Juan
Colomb, Fernando
Combi, Jorge
Cora, Sofia
Crsico, Alejandro
Costa, Andrea
Cruzado, Alicia
Dasso, Sergio
De Biasi, Maria
Di Sisto, Romina
Diaz, Ruben
Domnguez, Mariano
Donzelli, Carlos
Dubner, Gloria
Feinstein, Carlos
Feinstein, Alejandro
Fernández, Silvia
Fernández, Laura
Fernández Lajús, Eduardo
Ferrer, Osvaldo
Filloy, Emilio
Forte, Juan
Gamen, Roberto
Gangui, Alejandro
Garca, Beatriz
Garcia, Lambas
Garcia, Lia
Giacani, Elsa
Gil-Hutton, Ricardo
Giordano, Claudia
Giorgi, Edgard
Goldes, Guillermo
Gómez, Daniel
Gomez, Mercedes
González, Jorge
Grosso, Monica
Günthardt, Guillermo
Hernandez, Carlos
Hol, Pedro
Iannini, Gualberto
Lapasset, Emilio
Levato, Orlando
López, Garcia
López, José
López, Carlos
López Fuentes, Marcelo
López Garcia, Francisco
Luna, Homero
Machado, Marcos
Malaroda, Stella
Mallamaci, Claudio
Mandrini, Cristina
Manrique, Walter
Marabini, Rodolfo
Marraco, Hugo
Martin, Maria
Mauas, Pablo
Melita, Mario
Merlo, David
Milone, Luis
Morras, Ricardo
Hermán, Hernan
Muzzio, Juan
Navone, Hugo
Nunez, Josue
Olano, Carlos
Orellana, Mariana
Orellana, Rosa
Orsatti, Ana
Panei, Jorge
Paron, Sergio
Pedrosa, Susana
Pellizza, Leonardo
Perdomo, Raúl
Perez, Maria
Piacentini, Ruben
Piatti, Andrés
Pintado, Olga
Plastino, Angel
Pöppel, Wolfgang
Rabolli, Monica
Reynoso, Estela
Ringuelet, Adela
Romero, Gustavo
Rovero, Adrin
Rovira, Marta
Saffe, Carlos
Sahade, Jorge
Sistero, Roberto
Solivella, Gladys
Tignalli, Horacio
Tissera, Patricia
Valotto, Carlos
Vazquez, Ruben
Vega, E.
Vergne, Mara
Vieytes, Mariela
Mónica, Monica
Vucetich, Héctor
Wachlin, Felipe

Armenia

Abrahamian, Hamlet
Gigoyan, Kamo
Gurzadyan, Vahagn
Gyulbudaghian, Armen
Hambaryan, Valeri
Harutyunian, Haik
Hovhannessian, Rafik
Kalloglian, Arsen
Kandalyan, Rafik
Khachikian, Edward
Magakian, Tigran
Mahtessian, Abraham
Malumian, Vigen
Melikian, Norair
Mickaelian, Areg
Nikoghossian, Arthur
Parsamyan, Elma
Petrosian, Artaches
Pikichian, Hovhannes
Sanamian, V.
Sedrakian, David
Shakhbazian, Romelia
Shakhbazyan, Yurij
Yengibarian, Norair

Australia

Argast, Dominik
Argo, Megan
Ashley, Michael
Bailes, Matthew
Bailey, Jeremy
Ball, Lewis
Barnes, David
Bedding, Timothy
Bessell, Michael
Bhat, Ramesh
Bicknell, Geoffrey
Biggs, James
Bignall, Hayley
Birch, Peter
Blair, David
Bland-Hawthorn, Jonathan
Boyle, Brian
Braun, Robert
Bray, Robert
Briggs, Franklin
Brooks, Kate
Brown, Michael
Bryant, Julia
Burman, Ronald
Burton, Michael
Butcher, Harvey
Cairns, Iver
Calabretta, Mark
Cally, Paul
Campbell-Wilson, Duncan
Cane, Hilary
Cannon, Russell
Carter, Bradley
Caswell, James
Chapman, Jessica
Chapman, Jacqueline
Clay, Roger
Colless, Matthew
Costa, Marco
Couch, Warrick
Cram, Lawrence
Cramer, Neil
Crocker, Roland
Croom, Scott
Green, James
Greenhill, John
Hall, Peter
Harvey-Smith, Lisa
Hau, George
Haynes, Raymond
Haynes, Roslynn
Hobbs, George
Hollow, Robert
Hopkins, Andrew
Horiuchi, Shinji
Horton, Anthony
Hotan, Aidan
Hughes, Stephen
Humble, John
Hunstead, Richard
Hurley, Jarrod
Hyland, Harry
Ireland, Michael
Jackson, Carole
Jauncey, David
Jerjen, Helmut
Johnston, Helen
Jones, Paul
Jones, Heath
Kalnajs, Agris
Karakas, Amanda
Keay, Colin
Kedziora-Chudozer, Lucyna
Keller, Stefan
Kennedy, Hans
Kesteven, Michael
Kilborn, Virginia
Killeen, Neil
Kiss, Laszlo
Kobayashi, Chiaki
Koribalski, Brbel
Kuncic, Zdenka
Lambeck, Kurt
Landt, Hermine
Lattanzio, John
Lawrence, Jon
Lawson, Warrick
Lewis, Geraint
Norris, John
Norris, Raymond
O'Byrne, John
O'Sullivan, John
O'Toole, Simon
Onken, Christopher
Orchiston, Wayne
Page, Arthur
Pandey, Birendra
Parker, Quentin
Peterson, Bruce
Phillips, Christopher
Pimbblet, Kevin
Pongracic, Helen
Power, Chris
Pracy, Michael
Prentice, Andrew
Proctor, Robert
Protheroe, Raymond
Rees, David
Reynolds, John
Reynolds, Cormac
Robertson, James
Robinson, Garry
Russell, Kenneth
Ryder, Stuart
Sackett, Penny
Sadler, Elaine
Sault, Robert
Savage, Ann
Schinckel, Antony
Schmidt, Brian
Sharma, Dharma
Sharp, Robert
Sheridan, K.
Shobbrook, Robert
Shortridge, Keith
Slee, O.
Smith, Craig
Sood, Ravi
Sparrow, James
Stathakis, Raylee
Staveley-Smith, Lister
Stello, Dennis

Croton, Darren
Cunningham, Maria
Curran, Stephen
Da Costa, Gary
Davis, John
Davis, Tamara
Dawson, Bruce
De Blok, Erwin
Dickey, John
Dodson, Richard
Donea, Alina
Dopita, Michael
Drinkwater, Michael
Duldig, Marc
Durrant, Christopher
Edwards, Paul
Ekers, Ronald
Elford, William
Ellingsen, Simon
Ellis, Graeme
Ellis, Simon
Erickson, William
Evans, Robert
Feain, Ilana
Ferrario, Lilia
Fiedler, Russell
Filipović, Miroslav
Fluke, Christopher
Forbes, Duncan
Forbes, J.
Francis, Paul
Frater, Robert
Freeman, Kenneth
Gaensler, Bryan
Gaensler, Bryan
Galloway, David
Gascoigne, S.
George, Martin
Germany, Lisa
Gingold, Robert
Glazebrook, Karl
Godfrey, Peter
Graham, Alister
Green, Anne
Li, Bo
Liffman, Kurt
Lineweaver, Charles
Lomb, Nicholas
Lopez-Sanchez, Angel
Lovell, James
Luck, John
Lugaro, Maria
Luo, Qinghuan
Mackie, Glen
Maddison, Sarah
Madsen, Gregory
Malin, David
Manchester, Richard
Mardling, Rosemary
Marsden, Stephen
Mathewson, Donald
McAdam, Bruce
McClure-Griffiths, Naomi
McConnell, David
McCulloch, Peter
McGee, Richard
McGregor, Peter
McKinnon, David
McLean, Donald
McNaught, Robert
McSaveney, Jennifer
Melatos, Andrew
Melrose, Donald
Meyer, Martin
Middelberg, Enno
Mills, Bernard
Mills, Franklin
Milne, Douglas
Monaghan, Joséph
Morgan, Peter
Mould, Jeremy
Muller, Erik
Murdoch, Hugh
Murphy, Tara
Murray, James
Nelson, Graham
Newell, Edward
Nikoloff, Ivan
Stewart, Ronald
Stibbs, Douglas
Storey, John
Storey, Michelle
Sutherland, Ralph
Tango, William
Taylor, Kenneth
Tingay, Steven
Tinney, Christopher
Tisserand , Patrick
Titov, Oleg
Trampedach, Regner
Tuohy, Ian
Turtle, A.
Tuthill, Peter
Tzioumis, Anastasios
Urquhart, James
van der Borght, Rene
Vaughan, Alan
Voronkov, Maxim
Wagner, Alexander
Walker, Mark
Walsh, Andrew
Wardle, Mark
Waterworth, Michael
Watson, Frederick
Watson, Robert
Webb, John
Webster, Rachel
Wellington, Kelvin
Wen, Linqing
Westmeier, Tobias
White, Graeme
Whiteoak, John
Whiting, Matthew
Wickramasinghe, D.
Wilson, Peter
Wood, Peter
Wright, Alan
Wyithe, Stuart
Yong, David
Zealey, William

Austria

Argast, Dominik
Argo, Megan
Ashley, Michael
Bailes, Matthew
Bailey, Jeremy
Ball, Lewis
Barnes, David
Bedding, Timothy
Bessell, Michael
Bhat, Ramesh
Bicknell, Geoffrey
Auner, Gerhard
Handler, Gerald
Hanslmeier, Arnold
Hartl, Herbert
Haubold, Hans
Haupt, Hermann
Hensler, Gerhard
Hron, Josef
Jackson, Paul
Kapferer, Wolfgang
Kerschbaum, Franz
Koeberl, Christian
Kolenberg, Katrien
Pfleiderer, Jorg
Pilat-Lohinger, Elke
Posch, Thomas
Pötzi, Werner
Rakos, Karl
Recchi, Simone
Rucker, Helmut
Schindler, Sabine
Schnell, Anneliese
Schober, Hans
Schuh, Harald
Stift, Martin

Barden, Marco
Böhm, Asmus
Breger, Michel
Dorfi, Ernst
Dvorak, Rudolf
Firneis, Maria
Firneis, Friedrich
Goebel, Ernst
Kömle, Norbert
Lebzelter, Thomas
Leubner, Manfred
Lichtenegger, Herbert
Maitzen, Hans
Nittmann, Johann
Paladino, Rosita
Paunzen, Ernst
Temmer, Manuela
Temporin, Sonia
Theis, Christian
Veronig, Astrid
Weber, Robert
Weinberger, Ronald
Weiss, Werner
Zeilinger, Werner

Belgium

Acke, Bram
Aerts, Conny
Alvarez, Rodrigo
Andries, Jesse
Arnould, Marcel
Arpigny, Claude
Baes, Maarten
Berghmans, David
Biémont, Emile
Blommaert, Joris
Blomme, Ronny
Borkowski, Virginie
Braes, Lucien
Briquet, Maryline
Bruyninx, Carine
Burger, Marijke
Callebaut, Dirk
Carrier, Fabien
Chamel, Nicolas
Claeskens, Jean-François
Clette, Frederic
Cuypers, Jan
David, Marc
De Cat, Peter
De Cuyper, Jean-Pierre
De Grève, Jean-Pierre
de Groof, Anik
de Keyser, Johan
de Loore, Camiel
de Ridder, Joris
de Rijcke, Sven
de Rop, Yves
Decin, Leen
Defraigne, Pascale
Dehant, Véronique
Dejaiffe, Rene
Dejonghe, Herwig
Denis, Carlo
Denoyelle, Jozef
Dommanget, Jean
Eenens, Philippe
Elst, Eric
Fremat, Yves
Füzfa, Andre
Gabriel, Maurice
Gerard, Jean-Claude
Goossens, Marcel
Goriely, Stephane
Gosset, Eric
Grevesse, Nicolas
Groenewegen, Martin
Hensberge, Herman
Heynderickx, Daniel
Hochedez, Jean-François
Houziaux, Leo
Hutsemekers, Damien
Huygen, Eric
Jamar, Claude
Jehin, Emmanuel
Jorissen, Alain
Karatekin, Özgür
Keppens, Rony
Lampens, Patricia
Lapenta, Giovanni
Lemaire, Joséph
Lemaître, Anne
Magain, Pierre
Manfroid, Jean
Mann, Ingrid
Nakos, Theodoros
Noels, Arlette
Østensen, Roy
Palmeri, Patrick
Paquet, Paul
Pauwels, Thierry
Perdang, Jean
Pireaux, Sophie
Poedts, Stefaan
Pourbaix, Dimitri
Quinet, Pascal
Rauw, Gregor
Rayet, Marc
Renson, P.
Reyniers, Maarten
Robe, Henri
Roosbeek, Fabian
Roth, Michel
Royer, Pierre
Runacres, Mark
Sauval, A.
Scuflaire, Richard
Siess, Lionel
Simon, Paul
Siopis, Christos
Smeyers, Paul
Sterken, Christiaan
Surdej, Jean
Swings, Jean-Pierre
Tassoul, Jean-Louis
Thoul, Anne
van de Steene, Griet
van der Linden, Ronald
van Dessel, Edwin
van Eck, Sophie
van Hoolst, Tim
van Rensbergen, Walter
van Santvoort, Jacques
van Winckel, Hans
Vandenbussche, Bart
Verheest, Frank
Verhoelst, Tijl
Verschueren, Werner
Voitenko, Yuriy
Vreux, Jean
Waelkens, Christoffel
Yseboodt, Marie
Zander, Rodolphe

Brazil

Abraham, Zulema
Alcaniz, Jailson
Aldrovandi, Ruben
Alencar, Silvia
Andrei, Alexandre
Angeli, Claudia
do Nascimento Jr., José
Dottori, Horacio
Drake, Natalia
Ducati, Jorge
Emilio, Marcelo
Falceta-Goncalves, Diego
Opher, Reuven
Ortiz, Roberto
Oscare Giacaglia, Giorgio
Pastoriza, Miriani
Pavani, Daniela
Penna, Jucira

Arany-Prado, Lilia
Assafin, Marcelo
Baptista, Raymundo
Barbuy, Beatriz
Barroso Jr., Jair
Batalha, Celso
Benevides Soares, Paulo
Berman, Marcelo
Bica, Eduardo
Boechat-Roberty, Heloisa
Bonatto, Charles
Braga, João
Bruch, Albert
Canalle, João
Capelato, Hugo
Caproni, Anderson
Carruba, Valerio
Carvalho, Joel
Carvano, Jorge
Chan, Roberto
Chian, Abraham
Cid Fernandes Junior, Roberto
Corradi, Wagner
Correia, Emilia
Costa, Joaquim
Couto da Silva, Telma
Cuisinier, François
Cunha, Katia
Cypriano, Eduardo
D'Amico, Flavio
da Costa, José
da Costa, Luiz
da Rocha, Vieira
da Silva, Licio
da Silveira, Enio
Daflon, Simone
Dal Ri Barbosa, Cassio
Damineli Neto, Augusto
Dantas, Christine
de Aguiar, Odylio
de Almeida, Amaury
de Araujo, Francisco
de Carvalho, Reinaldo
de Gouveia Dal Pino, Elisabete
de la Reza, Ramiro
de Lima, José
de Medeiros, José
de Souza, Ronaldo
De Souza Pellegrini, Paulo
del Peloso, Eduardo
Dias da Costa, Roberto
Diaz, Marcos
Faundez-Abans, Max
Fernandes, Francisco
Ferraz Mello, Sylvio
Foryta, Dietmar
Franco, Gabriel Armando
Freitas Mourão, Ronaldo
Friaca, Amancio
Giménez de Castro, Carlos
Giuliatti Winter, Silvia
Gomes, Rodney
Gonçalves, Denise
Gonzales, Walter
Gregorio-Hetem, Jane
Gruenwald, Ruth
Hetem Jr., Annibal
Idiart, Thais
Jablonski, Francisco
Jafelice, Luiz
Janot Pacheco, Eduardo
Jatenco-Pereira, Vera
Jayanthi, Udaya
Kaufmann, Pierre
Kepler, S.
Lanfranchi, Gustavo
Lazzaro, Daniela
Leão, João Rodrigo
Leister, Nelson
Lépine, Jacques
Lima, Botti
Lima Neto, Gastao
Lopes, Dalton
Lorenz-Martins, Silvia
Machado, Maria
Maciel, Walter
Magalhaes, Antonio
Maia, Marcio
Marques, Dos
Marranghello, Guilherme
Martin, Inácio
Martin, Vera
Matsuura, Oscar
Medina Tanco, Gustavo
Meliani, Mara
Mendes, Luiz
Mendes, Da
Mendes de Oliveira, Cláudia
Miranda, Oswaldo
Mothé-Diniz, Thais
Neves de Araujo, José
Novello, Mario
Oliveira, Alexandre
Pereira, Claudio
Piazza, Liliana
Picazzio, Enos
Plana, Henri
Pompeia, Luciana
Poppe, Paulo
Porto de Mello, Gustavo
Prires Martins, Lucimara
Quast, Germano
Raulin, Jean-Pierre
Ribeiro, Marcelo
Ribeiro, Andre Luis
Rocha-Pinto, Hélio
Rodrigues, Claudia
Rodrigues de Oliveira Filho, Irapuan
Roig, Fernando
Rosa, Reinaldo
Rossi, Silvia
Santiago, Basilio
Santos, Nilton
Santos Jr., João
Saraiva, Maria de Fatima
Sawant, Hanumant
Scalise Jr., Eugenio
Schuch, Nelson
Silva, Adriana
Soares, Domingos
Sodré, Laerte
Steiner, João
Storchi-Bergmann, Thaisa
Tateyama, Claudio
Teixeira, Ramachrisna
Telles, Eduardo
Tello Bohorquez, Camilo
Torres, Carlos Alberto
Tsuchida, Masayoshi
Vaz, Luiz Paulo
Veiga, Carlos
Videira, Antonio
Viegas, Sueli
Vieira, Martins
Vieira Neto, Ernesto
Vilas-Boas, José
Vilhena, Rodolpho
Villas da Rocha, Jaime
Villela Neto, Thyrso
Voelzke, Marcos
Willmer, Christopher
Winter, Othon
Yokoyama, Tadashi

Bulgaria

Antonova, Antoaneta
Antov, Alexandar
Bachev, Rumen
Bojurova, Eva
Konstantinova-Antova, Renada
Kovachev, Bogomil
Kraicheva, Zdravka
Kunchev, Peter
Russev, Ruscho
Russeva, Tatjana
Semkov, Evgeni
Shkodrov, Vladimir

Bonev, Tanyu
Chapanov, Yavor
Dechev, Momchil
Duchlev, Peter
Filipov, Latchezar
Georgiev, Leonid
Georgiev, Tsvetan
Golev, Valery
Iliev, Ilian
Ivanov, Georgi
Ivanova, Violeta
Kirilova, Daniela
Kjurkchieva, Diana
Kolev, Dimitar
Komitov, Boris
Kurtev, Radostin
Marchev, Dragomir
Markov, Haralambi
Markova, Nevjana
Mihov, Boyko
Nedialkov, Petko
Nikolov, Andrej
Nikolov, Nikola
Ovcharov, Evgeni
Panov, Kiril
Petrov, Georgi
Petrov, Nikola
Peykov, Zvezdelin
Popov, Vasil
Radeva, Veselka
Stanishev, Vallery
Stateva, Ivanka
Stavrev, Konstantin
Stoev, Alexey
Strigatchev, Anton
Tomov, Nikolai
Tsvetkov, Milcho
Tsvetkov, Tsvetan
Tsvetkova, Katja
Umlenski, Vasil
Valcheva, Antoniya
Veltchev, Todor
Yankulova, Ivanka
Zamanov, Radoslav
Zhekov, Svetozar

Canada

Abraham, Roberto
Aikman, G.
Auman, Jason
Avery, Lorne
Babul, Arif
Bailin, Jeremy
Balogh, Michael
Barmby, Pauline
Bartel, Norbert
Bastien, Pierre
Basu, Dipak
Basu, Shantanu
Batten, Alan
Beaudet, Gilles
Beaulieu, Sylvie
Bell, Morley
Bennett, Philip
Bergeron, Pierre
Bishop, Roy
Bochonko, D.
Bohlender, David
Bolton, Charles
Bond, John
Borra, Ermanno
Brassard, Pierre
Bridges, Terry
Brooks, Randall
Broten, Norman
Brown, Peter
Brown, Jo-Anne
Browning, Matthew
Burke, J.
Caldwell, John
Cami, Jan
Campbell-Brown, Margaret
Cannon, Wayne
Carignan, Claude
Carlberg, Raymond
Charbonneau, Paul
Clark, Thomas
Clark, Thomas
Forbes, Douglas
Fort, David
Foster, Tyler
Gaizauskas, Victor
Gallagher, Sarah
Galt, John
Garrison, Robert
Ghizaru, Mihai
Gower, Ann
Gray, David
Gregory, Philip
Griffin, R.
Guenther, David
Gulliver, Austin
Hall, Patrick
Halliday, Ian
Hanes, David
Harris, Gretchen
Harris, William
Hartwick, F.
Hawkes, Robert
Henriksen, Richard
Hesser, James
Hickson, Paul
Higgs, Lloyd
Holmgren, David
Houde, Martin
Hube, Douglas
Hudson, Michael
Hutchings, John
Innanen, Kimmo
Irwin, Alan
Irwin, Judith
Israel, Werner
Jarrell, Richard
Jayawardhana, Ray
Johnstone, Douglas
Joncas, Gilles
Jones, James
Jones, Carol
Kaspi, Victoria
Navarro, Julio
Naylor, David
Nemec, James
Nicholls, Ralph
Page, Don
Pathria, Raj
Peeters, Els
Pen, Ue-Li
Percy, John
Pineault, Serge
Plume, René
Pritchet, Christopher
Pryce, Maurice
Purton, Christopher
Racine, René
Rice, John
Richardson, Eric
Richer, Harvey
Robb, Russell
Robert, Carmelle
Roberts, Scott
Roger, Robert
Rogers, Christopher
Rosvick, Joanne
Routledge, David
Rucinski, Slavek
Rutledge, Robert
Safi-Harb, Samar
Sarty, Gordon
Scarfe, Colin
Schade, David
Scott, Douglas
Scrimger, J.
Seaquist, Ernest
Sher, David
Short, Christopher
Sigut, Aaron
Sills, Alison
Smith, Nigel
Smylie, Douglas
Sreenivasan, S.

Clarke, Thomas
Clarke, David
Claude, Stéphane
Clement, Maurice
Climenhaga, John
Clutton-Brock, Martin
Connors, Martin
Côté, Stéphanie
Côté, Patrick
Couchman, Hugh
Courteau, Stéphane
Coutts-Clement, Christine
Crabtree, Dennis
Crampton, David
Davidge, Timothy
Dawson, Peter
de Robertis, Michael
Demers, Serge
Deupree, Robert
Dewdney, Peter
Di Francesco, James
Dobbs, Matt
Dougherty, Sean
Douglas, R.
Doyon, Rene
Drissen, Laurent
Duley, Walter
Duncan, Martin
Dupuis, Jean
Durand, Daniel
Dutil, Yvan
Dyer, Charles
English, Jayanne
Fahlman, Gregory
Feldman, Paul
Feldman, Paul
Fernie, J.
Ferrarese, Laura
Fich, Michel
Fletcher, J.
Fontaine, Gilles
Kavelaars, JJ.
Kenny, Harold
Koehler, James
Kouba, Jan
Kronberg, Philipp
Lake, Kayll
Lamontagne, Robert
Landecker, Thomas
Landstreet, John
Laurin, Denis
Leahy, Denis
Legg, Thomas
Lester, John
Levine, Robyn
Locke, Jack
Lowe, Robert
MacLeod, John
Marlborough, J.
Martel, Hugo
Martin, Peter
Matthews, Jaymie
Matthews, Brenda
McCall, Marshall
McCutcheon, William
McIntosh, Bruce
Menon, T.
Merriam, James
Michaud, Georges
Milone, Eugene
Mitalas, Romas
Mitchell, George
Mochnacki, Stephan
Moffat, Anthony
Moffat, John
Monin, Dmitry
Moorhead, James
Morbey, Christopher
Morgan, Lawrence
Moriarty-Schieven, Gerald
Morton, Donald
Nadeau, Daniel
St-Louis, Nicole
Stagg, Christopher
Stairs, Ingrid
Steinbring, Eric
Stetson, Peter
Stil, Jeroen
Sutherland, Peter
Talon, Suzanne
Tapping, Kenneth
Tatum, Jeremy
Taylor, Russell
Taylor, James
Tikhomolov, Evgeniy
Turner, David
Vallee, Jacques
van den Bergh, Sidney
van Kerkwijk, Marten
VandenBerg, Don
Venn, Kimberly
Veran, Jean-Pierre
Vishniac, Ethan
Vorobyov, Eduard
Wade, Gregg
Walker, Gordon
Webb, Tracy
Webster, Alan
Wehlau, Amelia
Welch, Gary
Welch, Douglas
Wesemael, François
Wesson, Paul
Widrow, Larry
Wiegert, Paul
Willis, Anthony
Wilson, William
Wilson, Christine
Woodsworth, Andrew
Wu, Yanqin
Yang, Stephenson
Yee, Howard

Chile

Abbott, Tim
Alcaino, Gonzalo
Alloin, Danielle
Alvarez, Hector
Aparici, Juan
Barrientos, Luis
Bensby, Thomas
Boehnhardt, Hermann
Borissova, Jordanka
Bronfman, Leonardo
Campusano, Luis
Carrasco, Guillermo
Catelán, Márcio
Celis, Leopoldo
Costa, Edgardo
Curé, Michel
Infante, Leopoldo
Ivanov, Dmitrij
Jordan, Andres
Kaufer, Andreas
Knee, Lewis
Krzeminski, Wojciech
Kunkel, William
Liller, William
Lindgren, Harri
Lo Curto, Gaspare
Lopez, Sebastian
Loyola, Patricio
Mathys, Gautier
Mauersberger, Rainer
Maury, Alain
May, J.
Nyman, Lars-Aake
Peck, Alison
Pedreros, Mario
Phillips, Mark
Pompei, Emanuela
Quintana, Hernan
Rawlings, Mark
Richtler, Tom
Rivinius, Thomas
Rojo, Patricio
Roth, Miguel
Rubio, Monica
Ruiz, Maria
Saviane, Ivo
Sawada, Tsuyoshi
Schmidtobreick, Linda

De Buizer, James
de Graauw, Thijs
de Gregorio-Monsalvo, Itziar
de Propris, Roberto
Dumke, Michael
Fouque, Pascal
Fox, Andrew
Garay, Guido
Geisler, Douglas
Gieren, Wolfgang
Gutierrez-Moreno, A.
Hamuy, Mario
Hardy, Eduardo
Hubrig, Swetlana
Maza, José
Melnick, Jorge
Melo, Claudio
Méndez Bussard, René
Mennickent, Ronnald
Meza, Andres
Michalska, Gabriela
Minniti, Dante
Moerchen, Margaret
Morrell, Nidia
Motta, Véronica
Naef, Dominique
Noel, Fernando
Nürnberger, Dieter
Smette, Alain
Smith, Malcolm
Smith, Robert
Sterzik, Michael
Stritzinger, Maximilian
Suntzeff, Nicholas
Szeifert, Thomas
Tokovinin, Andrei
Torres, Carlos
Vogt, Nikolaus
Walker, Alistair
West, Michael
Wilson, Thomas
Yegorova, Irina

China Nanjing

Ai, Guoxiang
Bao, Shudong
Bi, Shao
Bian, Yulin
Cao, Huilai
Cao, Li
Cao, Xinwu
Cao, Zhen
Chang, Jin
Chang, Ruixiag
Chao, Zhang
Chen, DaMing
Chen, Daohan
Chen, Dong
Chen, Guoming
Chen, Jiansheng
Chen, Li
Chen, Li
Chen, Meidong
Chen, Peng Fei
Chen, Xuefei
Chen, Xuelei
Chen, Yafeng
Chen, Yang
Chen, Yang
Chen, Yongjun
Chen, Yuqin
Chen, Zhiyuan
Cheng, Fuzhen
Chu, Yaoquan
Chun, Sun
Cui, Chenzhou
Cui, Shizhu
Cui, Wen-Yuan
Cui, Xiangqun
Cui, Zhenhua
Dai, Zigao
Deng, LiCai
Deng, YuanYong
Deng, Zugan
Di, Xiaohua
Ding, Mingde
Li, Yong
Li, Yuanjie
Li, Zhengxing
Li, Zhi
Li, Zhigang
Li, Zhiping
Li, Zhongyuan
Lian, Luo
Liang, Shiguang
Liang, Yanchun
Liao, Dechun
Liao, Xinhao
Lin, Qing
Lin, Weipeng
Lin, Xuan-bin
Lin, Yi-qing
Liu, Caipin
Liu, Ciyuan
Liu, Fukun
Liu, Lin
Liu, Qingzhong
Liu, Xiang
Liu, Xinping
Liu, Yongzhen
Liu, Yu
Liu, Zhong
Lou, Yu-Qing
Lu, BenKui
Lu, Chunlin
Lu, Fangjun
Lu, Jufu
Lu, Tan
Lu, Xiaochun
Lu, Ye
Luo, Ali
Luo, Dingchang
Luo, Shaoguang
Luo, Xianhan
Ma, Jingyuan
Ma, Jun
Ma, Lihua
Ma, Wenzhang
Wu, De Jin
Wu, Dong
Wu, Guichen
Wu, Haitao
Wu, Jianghua
Wu, Lianda
Wu, Shaoping
Wu, Shouxian
Wu, Wentao
Wu, Xiangping
Wu, Xinji
Wu, Xue-bing
Wu, Xuejun
Wu, Yuefang
Xi, Zezong
Xia, Xiao-Yang
Xia, Yifei
Xiang, Shouping
Xiao, Naiyuan
Xie, Xianchun
Xiong, Da Run
Xiong, Jianning
Xiong, Yaoheng
Xu, Aoao
Xu, Chongming
Xu, Dawei
Xu, Jiayan
Xu, Jin
Xu, Jun
Xu, Renxin
Xu, Weibiao
Xu, Zhi
Xue, Li
Xue, Suijian
Yan, Jun
Yan, Yihua
Yang, Changgen
Yang, Dehua
Yang, Fumin
Yang, Ji
Yang, Jing
Yang, Lei

Dong, Xiao-Bo
Dong, Xiaojun
Du, Lan
Esamdin, Ali
Esimbek, Jarken
Fan, Junhui
Fan, Yu
Fan, Zuhui
Fang, Cheng
Fang, Liu
Feng, Long Long
Fong, Chugang
Fu, Jianning
Fu, Jian-Ning
Fu, Yanning
Gan, Weiqun
Gang, Zhang
Gao, Buxi
Gao, Yu
Gao, Yuping
Geng, Lihong
Gong, Xuefei
Gu, Qiusheng
Gu, Sheng-hong
Gu, Wei-Min
Gu, Xuedong
Guo, Hongfang
Guo, Jianheng
Han, JinLin
Han, Tianqi
Han, Yanben
Han, Zhanwen
Hao, Jinxin
He, Jinhua
He, Miao-fu
He, XiangTao
Hong, Sun
Hong, Wu
Hong, Xiaoyu
Hong, Zhang
Hou, Jinliang
Hu, Fuxing
Hu, Hongbo
Hu, Tiezhu
Hu, Wenrui
Hu, Xiaogong
Hu, Yonghui
Hu, Zhong Wen
Hua, Yu
Huang, Cheng
Huang, Cheng-Li
Huang, Guangli
Huang, He
Huang, Jiehao
Huang, Keliang
Huang, Runqian
Huang, Tianyi
Huang, YongFeng
Ji, Haisheng
Ji, Jianghui
Ma, Xingyuan
Ma, Yuehua
Ma, YuQian
Mao, Rui-Qing
Mao, Weijun
Min, Wang
Min, Yuan
Ming, Bai
Nan, Ren-Dong
Ning, Zongjun
Pan, Liande
Pei, Chunchuan
Peng, Bo
Peng, Qingyu
Peng, Qiuhe
Ping, Jinsong
Qi, Guanrong
Qian, Shengbang
Qiao, Guojun
Qiao, Rongchuan
Qiming, Wang
Qin, Bo
Qin, Yi-Ping
Qin, Zhihai
Qiu, Yaohui
Qiu, Yulei
Qu, Qinyue
Qu, Zhong Quan
Rong, Jianxiang
Rui, Qi
Shan, Hongguang
Shao, Zhengyi
Shen, Kaixian
Shen, Zhiqiang
Sheng, Wan
Shi, Huli
Shi, Huoming
Shi, Jianrong
Shi, Shengcai
Shu, Chenggang
Song, Jinan
Song, Liming
Song, Qian
Su, Cheng-yue
Su, Ding-qiang
Sun, Fuping
Sun, Jin
Sun, Xiaochun
Sun, Yisui
Tan, Huisong
Tang, Yuhua
Tang, Zheng
Tao, Jin-he
Tao, Jun
Tian, Wenwu
Wang, Ding-Xiong
Wang, Dongguang
Wang, Feilu
Wang, Gang
Wang, Guo-min
Yang, Shijie
Yang, Shimo
Yang, Tinggao
Yang, Xiaohu
Yang, Xuhai
Yang, Zhigen
Yang, Zhiliang
Yao, Qijun
Yao, Yongqiang
Yao, Zhi
Ye, Shuhua
Ye, Wenwei
Yi, Meiliang
Yi, Zhaohua
Yong, Zhou
You, Junhan
Yu, Dai
Yu, Nanhua
Yu, Wang
Yu, Wenfei
Yu, Zhiyao
Yuan, Feng
Yuan, Xiangyan
Yuan, Ye-fei
Zeng, Qin
Zhang, Baozhou
Zhang, Bo
Zhang, Fenghui
Zhang, Haiyan
Zhang, Haotong
Zhang, Hongbo
Zhang, Huawei
Zhang, Jialu
Zhang, Jian
Zhang, JiangShui
Zhang, Jiaxiang
Zhang, Jingyi
Zhang, Jun
Zhang, Li
Zhang, Mei
Zhang, Qiang
Zhang, Shuang Nan
Zhang, Tong-Jie
Zhang, Wei
Zhang, Wei
Zhang, Weiqun
Zhang, Xiaobin
Zhang, Xiaoxiang
Zhang, Xiuzhong
Zhang, Xizhen
Zhang, Yang
Zhang, Yanxia
Zhang, You-Hong
Zhang, Zhongping
Zhao, Changyin
Zhao, Donghai
Zhao, Gang
Zhao, Haibin
Zhao, Juan
Zhao, Jun Liang

Jiang, Aimin
Jiang, Biwei
Jiang, Chongguo
Jiang, Dongrong
Jiang, Ing-Guey
Jiang, Xiaojun
Jiang, Xiaoyuan
Jiang, Yun
Jiang, Zhaoji
Jin, Biaoren
Jin, WenJing
Jin, Zhenyu
Jing, Hairong
Jing, Yipeng
Kong, Xu
Lei, Chengming
Li, Guangyu
Li, Guoping
Li, Hui
Li, Ji
Li, Jinling
Li, Jinzeng
Li, Kejun
Li, Li
Li, Qi
Li, Qibin
Li, Tipei
Li, Wei
Li, Xiangdong
Li, Xiaohui
Li, Xiaoqing
Li, Xinnan
Li, Yan
Wang, Hongchi
Wang, Hong-Guang
Wang, Huaning
Wang, Huiyuan
Wang, Jiaji
Wang, Jiancheng
Wang, Jian-Min
Wang, Jingxiu
Wang, Jingyu
Wang, Jun-Jie
Wang, Junxian
Wang, Kemin
Wang, Min
Wang, Na
Wang, Rongbin
Wang, Shen
Wang, Shouguan
Wang, Shui
Wang, Shujuan
Wang, Tinggui
Wang, Xiao-bin
Wang, Xunhao
Wang, Yanan
Wang, Yiping
Wang, Yong
Wang, Zhengming
Wang, Zhenru
Wei, Daming
Wei, Jianyan
Wei, Liu
Wei, Wenren
Wenlei, Shan
Wu, Bin
Zhao, Yongheng
Zhao, You
Zheng, Weimin
Zheng, Xiaoping
Zheng, Xinwu
Zheng, Xuetang
Zheng, Yong
Zhong, Gu
Zhong, Min
Zhou, Daoqi
Zhou, Guiping
Zhou, Hongnan
Zhou, Hongyan
Zhou, Jianfeng
Zhou, Jianjun
Zhou, Ji-Lin
Zhou, Xu
Zhou, Yonghong
Zhou, Youyuan
Zhu, Jin
Zhu, LiChun
Zhu, Liying
Zhu, Nenghong
Zhu, Wenbai
Zhu, Wenyao
Zhu, Xingfeng
Zhu, Yaozhong
Zhu, Yongtian
Zhu, Zheng
Zhu, Zhenxi
Zhu, Zi
Zou, Zhenlong
Zuo, Yingxi

China - Hong Kong

Chan, Kwing
Nakashima, Jun-ichi
Cheng, Kwongsang
Kwok, Sun

China Taipei

Cai, Michael
Chang, Hsiang-Kuang
Chao, Benjamin
Chen, Alfred
Chen, An-Le
Chen, Huei-Ru
Chen, Lin-Wen
Chen, Pisin
Chen, Wen Ping
Chin, Yi-nan
Chiueh, Tzihong
Chou, Chih-Kang
Chou, Dean-Yi
Chou, Yi
Fu, Hsieh-Hai
Fu-Shong, Kuo
Gir, Be
Hasegawa, Tatsuhiko
Hirano, Naomi
Hsiang-Kuang, Tseng
Hsu, Rue-Ron
Huang, Yi-Long
Huang, Yinn-Nien
Hwang, Chorng-Yuan
Hwang, Woei-yann
Ip, Wing-Huen
King, Sun-Kun
Ko, Chung-Ming
Kuan, Yi-Jehng
Lai, Shih-Ping
Lee, Thyphoon
Lee, Wo-Lung
Lim, Jeremy
Ling, Chih-Bing
Liou, Guo Chin
Liu, Sheng-Yuan
Matsushita, Satoki
Nee, Tsu-Wei
Ng, Kin-Wang
Ngeow, Chow Choong
Shen, Chun-Shan
Shinsuke , Abe
Shu, Frank
Sun, Wei-Hsin
Ting, Yeou-Tswen
Tsai, Chang-Hsien
Tsao, Mo
Tsay, Wean-Shun
Wu, Hsin-Heng
Wu, Jiun-Huei
Yuan, Kuo-Chuan

Croatia, Republic of

Andreic, Zeljko
Bozic, Hrvoje
Brajša, Roman
Dadic, Zarko
Dominis Prester, Dijana
Jurdana-Šepić, Rajka
Kotnik-Karuza, Dubravka
Maričić, Darije
Martinis, Mladen
Pavlovski, Kresimir
Roa, Dragan
Ruždjak, Vladimir
Solarić, Nikola
Spoljaric, Drago
Vršnak, Bojan
Vujnovic, Vladis

Cuba

Alvarez Pomares, Oscar
Boytel, Jorge
Cárdenas, Rolando
Doval, Jorge M.
Garcia, Eduardo
Quirós, Israel
Taboada, Ramon

Czech Republic

Ambroz, Pavel
Bárta, Miroslav
Bičák, Jiří
Borovicka, Jiří
Bouška, Jiří
Brož, Miroslav
Bumba, Vaclav
Bursa, Milan
Bursa, Michal
Ceplecha, Zdenek
Dale, James
Dovciak, Michal
Durech, Josef
Ehlerová, Soňa
Fárník, František
Fischer, Stanislav
Grygar, Jiří
Hadrava, Petr
Harmanec, Petr
Heinzel, Petr
Hejna, Ladislav
Heyrovský, David
Horáček, Jiří
Horák, Jiří
Horsky, Jan
Hudec, Rene
Jungwiert, Bruno
Jurčák, Jan
Karas, Vladimir
Karlický, Marian
Kašparová, Jana
Kawka, Adela
Kleczek, Josip
Klokocnik, Jaroslav
Klvana, Miroslav
Korčáková, Daniela
Kostelecký, Jan
Koten, Pavel
Kotrč, Pavel
Koubsky, Pavel
Kraus, Michaela
Krtička, Jiří
Kubát, Jiří
Kulhánek, Petr
Lala, Petr
Marková, Eva
Mayer, Pavel
Meszaros, Attila
Meszárosová, Hana
Mikulášek, Zdeněk
Moravec, Zdeněk
Nickeler, Dieter
Novotný, Jan
Dušan Odstrcil, Dusan
Ouhrabka, Miroslav
Palouš, Jan
Pecina, Petr
Perek, Luboš
Pešek, Ivan
Podolský, Jiří
Polechová, Pavla
Pravec, Petr
Prouza, Michael
Ron, Cyril
Ružička, Adam
Ruzickova-Topolova, B.
Schwartz, Pavol
Sehnal, Ladislav
Semerák, Oldrich
Sidlichovsky, Milos
Sima, Zdislav
Šimek, Miloš
Simon, Vojtech
Skoda, Petr
Sobotka, Michal
Šolc, Martin
Spurny, Pavel
Štefl, Vladimir
Štefl, Stanislav
Stuchlík, Zdeněk
Šubr, Ladislav
Švestka, Jiří
Tichá, Jana
Tlamicha, Antonin
Vandas, Marek
Vennes, Stéphane
Vokrouhlicky, David
Vondrák, Jan
Votruba, Viktor
Vykutilová, Marie
Wolf, Marek
Wünsch, Richard

Denmark

Andersen, Johannes
Andersen, Anja
Arentoft, Torben
Baerentzen, Jørn
Borysow, Aleksandra
Brandt, Søren
Bruntt, Hans
Budiz-Jøgensen, Carl
Haugbølle, Troels
Helmer, Leif
Helt, Bodil
Hjorth, Jens
Høg, Erik
Hornstrup, Allan
Jensen, Brian
Johansen, Karen
Nissen, Poul
Nordlund, Aake
Nordström, Birgitta
Norgaard-Nielsen, Hans
Novikov, Igor
Olsen, Erik
Olsen, Hans
Pedersen, Holger

Chenevez, Jérôme
Christensen, Per
Christensen-Dalsgaard, Jørgen
Clausen, Jens
Dorch, Sren Bertil
Field, David
Frandsen, Soeren
Fynbo, Johan
Galsgaard, Klaus
Gammelgaard, Peter
Grove, Lisbeth
Grundahl, Frank
Hannestad, Steen
Jørgensen, Uffe
Jørgensen, Henning
Kjaergaard, Per
Kjeldsen, Hans
Knude, Jens
Kristensen, Leif
Linden-Vørnle, Michael
Lund, Niels
Madsen, Jes
Malesani, Daniele
Milvang-Jensen, Bo
Moesgaard, Kristian
Naselsky, Pavel
Pedersen, Kristian
Petersen, J.
Pethick, Christopher
Rasmussen, Ib
Sommer-Larsen, Jesper
Tauris, Thomas
Thejll, Peter
Thomsen, Bjarne
Toft, Sune
Ulfbeck, Ole
Vedel, Henrik
Watson, Darach
Westergaard, Niels

Egypt

Abd El Hamid, Rabab
Abdelkawi, M.
Abou'el-ella, Mohamed
Abulazm, Mohamed
Ahmed, Abdel-aziz
Ahmed, Imam
Ahmed, Mostafa
Aiad, A.
Amer, Morsi
Awad, Mervat
Awadalla, Nabil
Beheary, Mohamed
El Basuny, Ahmed
El Nawaway, Mohamed
El Raey, Mohamed
El Shahawy, Mohamad
El-Saftawy, Magdy
El-Sharawy, Mohamed
Gaber, Ali
Galal, A.
Gamaleldin, Abdulla
Ghobros, Roshdy
Hady, Ahmed
Hamdy, M.
Hamid, S.
Hanna, Magdy
Hassan, Inal
Helali, Yhya
Ismail, Hamed
Ismail, Mohamed
Issa, Issa
Kahil, Magd
Kamal, Fouad
Kamel, Osman
Khalil, Khalil
Khalil, Nisreen
Mahmoud, Farouk
Marie, Mohamed
Mikhail, Joséph
Morcos, Abd El Fady
Nawar, Samir
Osman, Anas
Rassem, Mohamed
Saad, Abdel-naby
Saad, Nadia
Saad, Somaya
Selim, Hadia
Shalabiea, Osama
Shaltout, Mosalam
Sharaf, Mohamed
Soliman, Mohamed
Tadross, Ashraf
Tawadros, Maher
Wanas, Mamdouh
Yousef, Shahinaz
Youssef, Nahed

Estonia

Annuk, Kalju
Einasto, Maret
Einasto, Jaan
Gramann, Mirt
Haud, Urmas
Kipper, Tonu
Kolka, Indrek
Leedjärv, Laurits
Malyuto, Valeri
Nugis, Tiit
Pelt, Jaan
Pustõnski, Vladislav-Veniamin
Saar, Enn
Sapar, Arved
Sapar, Lili
Suhhonenko, Ivan
Tago, Erik
Tamm, Antti
Tenjes, Peeter
Traat, Peeter
Veismann, Uno
Vennik, Jaan
Viik, Tõnu

Finland

Aittola, Marko
Babkovskaia, Natalia
Berdyugin, Andrei
Berdyugina, Svetlana
Donner, Karl
Dümmler, Rudolf
Flynn, Chris
Haikala, Lauri
Kultima, Johannes
Lähteenmäki, Anne
Lainela, Markku
Laurikainen, Eija
Lehtinen, Kimmo
Lehto, Harry
Lumme, Kari
Markkanen, Tapio
Rautiainen, Pertti
Reunanen, Juha
Riehokainen, Aleksandr
Riihimaa, Jorma
Roos, Matts
Salo, Heikki
Sillanpaa, Aimo
Tähtinen, Leena

Hakala, Pasi
Hannikainen, Diana
Hänninen, Jyrki
Harju, Jorma
Heinamaki, Pekka
Huovelin, Juhani
Jetsu, Lauri
Juvela, Mika
Kaasalainen, Mikko
Käpylä, Petri
Karttunen, Hannu
Katajainen, Seppo
Kocharov, Leon
Korpi, Maarit
Kotilainen, Jari
Mattila, Kalevi
Mikkola, Seppo
Muinonen, Karri
Nevalainen, Jukka
Niemi, Aimo
Nilsson, Kari
Nurmi, Pasi
Oja, Heikki
Piirola, Vilppu
Piironen, Jukka
Pohjolainen, Silja
Portinari, Laura
Poutanen, Juri
Rahunen, Timo
Raitala, Jouko
Takalo, Leo
Teerikorpi, Veli
Terasranta, Harri
Tiuri, Martti
Tornikoski, Merja
Tuominen, Ilkka
Usoskin, Ilya
Valtaoja, Esko
Valtonen, Mauri
Vilhu, Osmi
Virtanen, Jenni
Wiik, Kaj
Zheng, Jia-Qing

France

Abgrall, Hervé
Aboudarham, Jean
Acker, Agnes
Adami, Christophe
Aime, Claude
Alard, Christophe
Alecian, Evelyne
Alecian, Georges
Alejandra, Recio-Blanco
Alimi, Jean-Michel
Allard, Nicole
Allard, France
Allegre, Claude
Aly, Jean-Jacques
Amram, Philippe
Andrillat, Yvette
Arduini-Malinovsky, Monique
Arenou, Frédéric
Arias, Elisa
Arlot, Jean-Eudes
Arnaud, Jean-Paul
Arnaud, Monique
Artru, Marie-Christine
Artzner, Guy
Asseo, Estelle
Assus, Pierre
Athanassoula, Evangelia
Aubier, Monique
Audouze, Jean
Aurière, Michel
Aussel, Herve
Auvergne, Michel
Azzopardi, Marc
Babusiaux, Carine
Baglin, Annie
Balanca, Christian
Balkowski-Mauger, Chantal
Balland, Christophe
Ballereau, Dominique
Balmino, Georges
Baluteau, Jean-Paul
Barban, Caroline
Dumont, Anne-Marie
Dumont, René
Dumont, Simone
Duriez, Luc
Durouchoux, Philippe
Durret, Florence
Dutrey, Anne
Duval, Marie-France
Duvert, Gilles
Egret, Daniel
Eidelsberg, Michèle
Emsellem, Eric
Enard, Daniel
Encrenaz, Pierre
Encrenaz, Thérèse
Erard, Stéphane
Falgarone, Edith
Famaey, Benoit
Faurobert-Scholl, Marianne
Feautrier, Nicole
Felenbok, Paul
Ferlet, Roger
Ferrando, Philippe
Ferrari, Cecile
Ferrari, Marc
Ferrière, Katia
Fienga, Agnès
Fillion, Jean-Hugues
Floquet, Michèle
Florsch, Alphonse
Folini, Doris
Fornasier, Sonia
Fort, Bernard
Forveille, Thierry
Fossat, Eric
Foy, Renaud
Fraix-Burnet, Didier
François, Patrick
Freire Ferrero, Rubens
Fresneau, Alain
Friedjung, Michael
Frisch, Helène
Mercier, Claude
Metris, Gilles
Meyer, Jean-Paul
Mianes, Pierre
Michard, Raymond
Michel, Patrick
Michel, Eric
Michel, Laurent
Mignard, Franois
Millet, Jean
Milliard, Bruno
Minier, Vincent
Mirabel, Igor
Mochkovitch, Robert
Moncuquet, Michel
Monier, Richard
Monin, Jean-Louis
Monnet, Guy
Montmerle, Thierry
Montmessin, Franck
Morbidelli, Alessandro
Moreau, Olivier
Moreels, Guy
Morel, Pierre-Jacques
Mosser, Benoît
Motch, Christian
Mouchet, Martine
Mouradian, Zadig
Mourard, Denis
Muller, Richard
Muratorio, Gerard
Nadal, Robert
Namouni, Fathi
Neiner, Coralie
Nguyen-Quang, Rieu
Noens, Jacques-Clair
Nollez, Gerard
Nottale, Laurent
Oblak, Edouard
Ochsenbein, François
Omnes, Roland
Omont, Alain

Baret, Bruny
Barge, Pierre
Barlier, François
Barret, Didier
Barriot, Jean-Pierre
Barucci, Maria
Basa, Stéphane
Baudry, Alain
Beaulieu, Jean-Philippe
Beckmann, Volker
Bely-Dubau, Françoise
Ben-Jaffel, Lofti
Benaydoun, Jean-Jacques
Bendjoya, Philippe
Benest, Daniel
Bensammar, Slimane
Bergeat, Jacques
Bergeron, Jacqueline
Berruyer-Desirotte, Nicole
Bertaux, Jean-Loup
Berthier, Jerôme
Berthomieu, Gabrielle
Bertout, Claude
Beuzit, Jean-Luc
Bézard, Bruno
Bienayme, Olivier
Bijaoui, Albert
Billebaud, Françoise
Binetruy, Pierre
Biraud, Franois
Birlan, Mirel
Biver, Nicolas
Bizouard, Christian
Blamont, Jacques-Emile
Blanchard, Alain
Blazit, Alain
Bocchia, Romeo
Bockelee-Morvan, Dominique
Boehm, Torsten
Boer, Michel
Boily, Christian
Bois, Eric
Boisse, Patrick
Boissier, Samuel
Boisson, Catherine
Bommier, Veronique
Bonazzola, Silvano
Bonifacio, Piercarlo
Bonnarel, François
Bonneau, Daniel
Bonnet, Roger
Bonnet-Bidaud, Jean-Marc
Bontemps, Sylvain
Borde, Suzanne
Borde, Pascal
Borgnino, Julien
Bosma, Albert
Bottinelli, Lucette
Boucher, Claude
Frisch, Uriel
Froeschlé, Christiane
Froeschlé, Claude
Froeschlé, Michel
Fulchignoni, Marcello
Gabriel, Alan
Gambis, Daniel
Garcia, Rafael
Gargaud, Muriel
Garnier, Robert
Gautier, Daniel
Gay, Jean
Gendre, Bruce
Genova, Françoise
Georgelin, Yvon
Gerard, Eric
Gerbaldi, Michèle
Gerin, Maryvonne
Gillet, Denis
Giraud, Edmond
Goldbach, Claudine
Goldwurm, Andrea
Gomez, Ana
Gonczi, Georges
Gontier, Anne-Marie
González, Jean-François
Gordon, Charlotte
Goret, Philippe
Götz, Diego
Gouguenheim, Lucienne
Gounelle, Matthieu
Goupil, Marie-José
Gouttebroze, Pierre
Granveaud, Michel
Grec, Gérard
Grenier, Isabelle
Greve, Albert
Grewing, Michael
Grosso, Nicolas
Gry, Cécile
Guelin, Michel
Guibert, Jean
Guiderdoni, Bruno
Guilloteau, Stéphane
Guinot, Bernard
Hadamcik, Edith
Halbwachs, Jean-Louis
Hameury, Jean-Marie
Hammer, François
Harvey, Christopher
Hayli, Abraham
Haywood, Misha
Hebrard, Guillaume
Heck, André
Henon, Michel
Henoux, Jean-Claude
Herpin, Fabrice
Hestroffer, Daniel
Heudier, Jean-Louis
Pagani, Laurent
Paletou, Frédéric
Pariat, Etienne
Parisot, Jean-Paul
Paturel, Georges
Paul, Jacques
Paumard, Thibaut
Pecker, Jean-Claude
Pedoussaut, André
Pellas, Paul
Pelletier, Guy
Pello, Roser
Pequignot, Daniel
Perault, Michel
Péroux, Céline
Perrier-Bellet, Christian
Perrin, Jean-Marie
Perrin, Guy
Petit, Jean-Marc
Petit, Gérard
Petit, Pascal
Petitjean, Patrick
Petrini, Daniel
Philippe, Salome
Picat, Jean-Pierre
Pick, Monique
Pierre, Marguerite
Pineau des Forets, Guillaume
Plez, Bertrand
Pollas, Christian
Poquerusse, Michel
Porquet, Delphine
Poulle, Emmanuel
Poyet, Jean-Pierre
Prantzos, Nikos
Prevot-Burnichon, Marie-Louise
Prieur, Jean-Louis
Proisy, Paul
Proust, Dominique
Provost, Janine
Prugniel, Philippe
Puget, Jean-Loup
Puy, Denis
Quemerais, Eric
Querci, François
Querci, Monique
Rabbia, Yves
Raoult, Antoinette
Rapaport, Michel
Rayrole, Jean
Reboul, Henri
Reeves, Hubert
Reinisch, Gilbert
Renard, Jean-Baptiste
Requieme, Yves
Reylé, Céline
Riazuelo, Alain
Ricort, Gilbert
Rieutord, Michel

Bouchet, Franois
Bouchet, Patrice
Bougeard, Mireille
Bougeret, Jean-Louis
Bouigue, Roger
Boulanger, François
Boulesteix, Jacques
Bouvier, Jerôme
Boyer, René
Brahic, André
Braine, Jonathan
Breysacher, Jacques
Briand, Carine
Briot, Danielle
Brouillet, Nathalie
Brun, Allan
Brunet, Jean-Pierre
Bryant, John
Buat, Véronique
Buchlin, Eric
Buecher, Alain
Burgarella, Denis
Burkhart, Claude
Cabrit, Sylvie
Calame, Odile
Cambresy, Laurent
Capitaine, Nicole
Caplan, James
Casandjian, Jean-Marc
Casoli, Fabienne
Casse, Michel
Castets, Alain
Catala, Claude
Cayatte, Veronique
Cayrel, Roger
Cayrel de Strobel, Giusa
Cazenave, Anny
Cecconi, Baptiste
Celnikier, Ludwik
Cesarsky, Catherine
Chabrier, Gilles
Chadid, Merieme
Chalabaev, Almas
Chamaraux, Pierre
Chambe, Gilbert
Chapront, Jean
Chapront-Touze, Michelle
Charlot, Patrick
Charlot, Stéphane
Charpinet, Stéphane
Chassefiere, Eric
Chaty, Sylvain
Chauvineau, Bertrand
Chelli, Alain
Chevrel, Serge
Chopinet, Marguerite
Clairemidi, Jacques
Colas, François
Colin, Jacques
Heydari-Malayeri, Mohammad
Heyvaerts, Jean
Hill, Vanessa
Hill, Adam
Hoang, Binh
Hua, Chon Trung
Hubert-Delplace, Anne-Marie
Hui bon Hoa, Alain
Humphries, Colin
Imbert, Maurice
Irigoyen, Maylis
Isabelle, Tallon-Bosc
Israel, Guy
Jacq, Thierry
Jasniewicz, Gerard
Jedamzik, Karsten
Joly, François
Joly, Monique
Jorda, Laurent
Josselin, Eric
Joubert, Martine
Jung, Jean
Kahane, Claudine
Kandel, Robert
Katz, David
Klein, Karl
Kneib, Jean-Paul
Koch-Miramond, Lydie
Koechlin, Laurent
Koutchmy, Serge
Kovalevsky, Jean
Krikorian, Ralph
Kunth, Daniel
Kupka, Friedrich
Labeyrie, Antoine
Labeyrie, Jacques
Lachieze-Rey, Marc
Lacour, Sylvestre
Lafon, Jean-Pierre
Lagache, Guilaine
Lagage, Pierre-Olivier
Lagrange, Anne-Marie
Lallement, Rosine
Laloum, Maurice
Lamy, Philippe
Lançon, Ariane
Lannes, Andre
Laques, Pierre
Laskar, Jacques
Lasota, Jean-Pierre
Latour, Jean
Launay, Jean-Michel
Launay, Françoise
Laurent, Claudine
Laval, Annie
Lazareff, Bernard
Le Bertre, Thibaut
Le Borgne, Jean-François
Le Bourlot, Jacques
Robillot, Jean-Maurice
Robin, Annie
Robley, Robert
Rocca-Volmerange, Brigitte
Roques, Sylvie
Roques, Françoise
Rostas, François
Rothenflug, Robert
Rouan, Daniel
Roudier, Thierry
Roueff, Evelyne
Rousseau, Jeanine
Rousseau, Jean-Michel
Rousselot, Philippe
Rousset, Gérard
Royer, Frédéric
Rozelot, Jean-Pierre
Sadat, Rachida
Sahal-Brechot, Sylvie
Saissac, Joséph
Salez, Morvan
Samain, Denys
Sami, Dib
Sanchez, Norma
Sareyan, Jean-Pierre
Sauty, Christophe
Sbordone, Luca
Schatzman, Evry
Scheidecker, Jean-Paul
Schmieder, Brigitte
Schneider, Jean
Scholl, Hans
Schuller, Peter
Schumacher, Gerard
Segonds, Alain-Philippe
Semel, Meir
Semelin, Benoit
Sibille, François
Sicardy, Bruno
Signore, Monique
Simon, Jean-Louis
Simon, Guy
Simonneau, Eduardo
Sirousse, Zia
Sivan, Jean-Pierre
Slezak, Eric
Sol, Helene
Soubiran, Caroline
Soucail, Genevive
Souchay, Jean
Souriau, Jean-Marie
Spallicci di Filottrano, Alessandro
Spielfiedel, Annie
Spite, François
Spite, Monique
Stasinska, Grazyna
Stee, Philippe
Stehlé, Chantal
Steinberg, Jean-Louis

Collin, Suzy
Colom, Pierre
Colombi, Stéphane
Combes, Michel
Combes, Françoise
Comte, Georges
Connes, Pierre
Connes, Janine
Corbett, Ian
Cornille, Marguerite
Coudé du Foresto, Vincent
Coupinot, Gérard
Courtès, Georges
Courtin, Régis
Coustenis, Athena
Couteau, Paul
Couturier, Pierre
Cox, Pierre
Crézé, Michel
Crifo, Françoise
Crovisier, Jacques
Cruvellier, Paul
Cruzalèbes, Pierre
Cuby, Jean-Gabriel
Cuny, Yvette
d'Hendecourt, Louis
Daigne, Gérard
Daigne, Frederic
Daniel, Jean-Yves
Dauphole, Bertrand
Davoust, Emmanuel
de Bergh, Catherine
de Kertanguy, Amaury
de la Noë, Jérôme
de Lapparent-Gurriet, Valérie
de Laverny, Patrick
de Viron, Olivier
Débarbat, Suzanne
Decourchelle, Anne
Deharveng, Jean-Michel
Deharveng, Lise
Delaboudiniere, Jean-Pierre
Delannoy, Jean
Deleflie, Florent
Delmas, Christian
Delsanti, Audrey
Démoulin, Pascal
Denisse, Jean-François
Dennefeld, Michel
Descamps, Pascal
Désert, François-Xavier
Désesquelles, Jean
Despois, Didier
Divan, Lucienne
Dohlen, Kjetil
Dolez, Noel
Dollfus, Audouin
Donas, José
Donati, Jean-François
Le Contel, Jean-Michel
Le Coroller, Hervé
Le Floch, André
Le Guet Tully, Françoise
Le Squeren, Anne-Marie
Leach, Sydney
Lebre, Agnes
Lebreton, Yveline
Lecavelier des Etangs, Alain
Lefebvre, Michel
LeFevre, Jean
Le Févre, Olivier
Lega, Elena
Leger, Alain
Lehnert, Matthew
Lelievre, Gérard
Lellouch, Emmanuel
Lemaire, Philippe
Lemaire, Jean-Louis
Lemaître, Gérard
Lena, Pierre
Leorat, Jacques
Lequeux, James
Lerner, Michel-Pierre
Leroy, Jean-Louis
Leroy, Bernard
Lesteven, Soizick
Lestrade, Jean-François
Levasseur-Regourd, Anny-Chantal
Lignières, François
Losco, Lucette
Loucif, Mohammed
Louise, Raymond
Loulergue, Michelle
Loup, Cécile
Louys, Mireille
Lucas, Robert
Ludwig, Hans
Luminet, Jean-Pierre
Madden, Suzanne
Magnan, Christian
Maillard, Jean-Pierre
Malbet, Fabien
Malherbe, Jean-Marie
Mamon, Gary
Mangeney, André
Marcelin, Michel
Marchal, Christian
Marco, Olivier
Maret, Sébastien
Martic, Milena
Martin, Jean-Michel
Martin, François
Martins, Fabrice
Masnou, Françoise
Masnou, Jean-Louis
Mathez, Guy
Mathias, Philippe
Mathis, Stéphane
Stellmacher, Götz
Stellmacher, Irène
Corbel, Stéphane
Sygnet, Jean-François
Tagger, Michel
Tallon, Michel
Talon, Raoul
Tanga, Paolo
Tarrab, Irene
Tchang-Brillet, Lydia
Terquem, Caroline
Terzan, Agop
Theureau, Gilles
Thevenin, Frederic
Thiry, Yves
Thomas, Claudine
Thuillot, William
Thum, Clemens
Tobin, William
Tran-Minh, Françoise
TranMinh, Nguyet
Trottet, Gerard
Truong, Bach
Tuckey, Philip
Tully, John
Turck-Chièze, Sylvaine
Turon, Catherine
Vakili, Farrokh
Valls-Gabaud, David
Valtier, Jean-Claude
van Driel, Wim
van Driel-Gesztelyi, Lidia
van't Veer-Menneret, Claude
van't-Veer, Frans
Vapillon, Loic
Vaubaillon, Jérémie
Vauclair, Gérard
Vauclair, Sylvie
Vauglin, Isabelle
Verdet, Jean-Pierre
Vergez, Madeleine
Vernin, Jean
Vernotte, François
Véron, Philippe
Véron, Marie-Paule
Vial, Jean-Claude
Viala, Yves
Viallefond, François
Vidal, Jean-Louis
Vidal-Madjar, Alfred
Vienne, Alain
Vigroux, Laurent
Vilkki, Erkki
Vilmer, Nicole
Vinet, Jean-Yves
Viton, Maurice
Vollmer, Bernd
Volonte, Sergio
Vuillemin, André

Doressoundiram, Alain
Dourneau, Gérard
Downes, Dennis
Drossart, Pierre
Dubau, Jacques
Dubois, Pascal
Dubois, Marc
Dubout, Renée
Dubus, Guillaume
Duc, Pierre-Alain
Ducourant, Christine
Dufay, Maurice
Dulieu, François
Dulk, George
Maucherat, Jean
Maurice, Eric
Maurogordato, Sophie
Mauron, Nicolas
Mavrides, Stamatia
Mazure, Alain
McCarroll, Ronald
Megessier, Claude
Mein, Pierre
Mein, Nicole
Mékarnia, Djamel
Mellier, Yannick
Meneguzzi, Maurice
Merat, Parviz
Walch, Jean-Jacques
Wenger, Marc
Widemann, Thomas
Wilson, Brian
Wlerick, Gérard
Wozniak, Hervé
Yamasaki, Tatsuya
Zahn, Jean-Paul
Zamkotsian, Frederic
Zarka, Philippe
Zavagno, Annie
Zeippen, Claude
Zorec, Juan
Zylka, Robert

Germany

Abraham, Peter
Albrecht, Rudolf
Altenhoff, Wilhelm
Andersen, Michael
Anzer, Ulrich
Appenzeller, Immo
Arnaboldi, Magda
Arp, Halton
Arshakian, Tigran
Aschenbach, Bernd
Asplund, Martin
Aurass, Henry
Axford, W.
Aznar Cuadrado, Regina
Baade, Dietrich
Baade, Robert
Baars, Jacob
Bahner, Klaus
Baier, Frank
Bailer-Jones, Coryn
Ballester, Pascal
Balthasar, Horst
Banday, Anthony
Barrow, Colin
Bartelmann, Matthias
Barwig, Heinz
Baschek, Bodo
Bastian, Ulrich
Baumgardt, Holger
Beck, Rainer
Becker, Werner
Bender, Ralf
Benvenuti, Piero
Berkefeld, Thomas
Berkhuijsen, Elly
Bernstein, Hans-Heinrich
Beuermann, Klaus
Bien, Reinhold
Biermann, Peter
Birkle, Kurt
Bleyer, Ulrich
Böhringer, Hans
Herrmann, Dieter
Hessman, Frederic
Hildebrandt, Joachim
Hilker, Michael
Hillebrandt, Wolfgang
Hippelein, Hans
Hirth, Wolfgang
Hofmann, Wilfried
Holmberg, Johan
Homeier, Derek
Hopp, Ulrich
Horedt, Georg
Huchtmeier, Walter
Hünsch, Matthias
Huettemeister, Susanne
Hugentobler, Urs
Hummel, Christian
Hummel, Wolfgang
Ilyin, Ilya
Isserstedt, Joerg
Jahnke, Knud
Jahreiss, Hartmut
Janka, Hans
Jockers, Klaus
Johansson, Peter
Jordan, Stefan
Junkes, Norbert
Just, Andreas
Kähler, Helmuth
Käufl, Hans Ulrich
Kahabka, Peter
Kalberla, Peter
Kallenbach, Reinald
Kanbach, Gottfried
Kauffmann, Guinevere
Kaufmann, Jens
Kegel, Wilhelm
Keller, Hans-Ulrich
Keller, Horst
Kelz, Andreas
Kendziorra, Eckhard
Kentischer, Thomas
Richichi, Andrea
Richter, Gotthard
Richter, Bernd
Richter, Philipp
Ripken, Hartmut
Ritter, Hans
Rix, Hans-Walter
Rödiger, Elke
Roemer, Max
Röpke, Friedrich
Roeser, Hermann-Josef
Roeser, Siegfried
Röser, Hans-Peter
Rohlfs, Kristen
Romaniello, Martino
Ros Ibarra, Eduardo
Rosa, Michael
Rosswog, Stephan
Roth, Markus
Roy, Alan
Ruder, Hanns
Rüdiger, Günther
Rupprecht, Gero
Russo, Pedro
Saglia, Roberto
Sakelliou, Irini
Sams, Bruce
Savolainen, Tuomas
Schaefer, Gerhard
Schaffner-Bielich, Jurgen
Schaifers, Karl
Schilbach, Elena
Schilke, Peter
Schindler, Karl
Schinnerer, Eva
Schleicher, Helmold
Schlemmer, Stephan
Schlichenmaier, Rolf
Schlickeiser, Reinhard
Schlosser, Wolfhard
Schlüter, A.
Schmadel, Lutz

Börner, Gerhard
Börngen, Freimut
Boffin, Henri
Boller, Thomas
Bomans, Dominik
Bonaccini, Domenico
Boschan, Peter
Bothmer, Volker
Brandt, Peter
Brauninger, Heinrich
Breitschwerdt, Dieter
Brinkmann, Wolfgang
Britzen, Silke
Brosche, Peter
Brüggen, Marcus
Bruls, Jo
Büchner, Joerg
Bues, Irmela
Burkert, Andreas
Burwitz, Vadim
Butler, Keith
Calamida, Annalisa
Camenzind, Max
Cameron, Robert
Carlson, Arthur
Carsenty, Uri
Cattaneo, Andrea
Cesarsky, Diego
Chini, Rolf
Christensen, Lars
Christensen, Lise
Christlieb, Norbert
Ciardi, Benedetta
Cullum, Martin
Curdt, Werner
d'Odorico, Sandro
Da Rocha, Cristiano
Dachs, Joachim
Dall, Thomas
de Boer, Klaas
de Zeeuw, Pieter
Degenhardt, Detlev
Deinzer, W.
Deiss, Bruno
Delplancke, Francoise
Dennerl, Konrad
Dettmar, Ralf-Jürgen
Deubner, Franz-Ludwig
Dick, Wolfgang
Diehl, Roland
Diercksen, Geerd
Dobler, Wolfgang
Dobrzycka, Danuta
Dobrzycki, Adam
Dorschner, Johann
Drechsel, Horst
Dreizler, Stefan
Duerbeck, Hilmar
Duschl, Wolfgang
Kerber, Florian
Khanna, Ramon
King, Lindsay
Kippenhahn, Rudolf
Kirk, John
Kissler-Patig, Markus
Kitsionas, Spyridon
Klahr, Hubert
Klare, Gerhard
Klein, Ulrich
Klessen, Ralf
Kley, Wilhelm
Kliem, Bernhard
Klinkhamer, Frans
Klioner, Sergei
Klose, Sylvio
Kneer, Franz
Knudsen, Kirsten
Koehler, Peter
Köhler, Rainer
Koester, Detlev
Kohoutek, Lubos
Kollatschny, Wolfram
Korhonen, Heidi
Kraus, Alexander
Krause, Marita
Krautter, Joachim
Kreykenbohm, Ingo
Kreysa, Ernst
Krichbaum, Thomas
Krivov, Alexander
Krivova, Natalie
Kroll, Peter
Kroupa, Pavel
Krüger, Harald
Krügel, Endrik
Kühne, Christoph
Kürster, Martin
Kundt, Wolfgang
Kuntschner, Harald
Lal, Dharam
Lehman, Holger
Leibundgut, Bruno
Leinert, Christoph
Lemke, Dietrich
Lemke, Michael
Lenhardt, Helmut
Lenzen, Rainer
Lesch, Harald
Li, Li-Xin
Liebscher, Dierck-E
Lobanov, Andrei
Lombardi, Marco
Lüst, Reimar
Luks, Thomas
Lutz, Dieter
Madsen, Claus
Mainieri, Vincenzo
Mandel, Holger
Schmid-Burgk, J.
Schmidt, H.
Schmidt, Robert
Schmidt, Wolfgang
Schmidt, Wolfram
Schmidt-Kaler, Theodor
Schmitt, Jürgen
Schmitt, Dieter
Schneider, Peter
Schöller, Markus
Schoenberner, Detlef
Schoenfelder, Volker
Scholz, Ralf-Dieter
Scholz, M.
Schramm, Thomas
Schröder, Rolf
Schubart, Joachim
Schuessler, Manfred
Schuh, Sonja
Schulz, R.
Schutz, Bernard
Schwan, Heiner
Schwartz, Rolf
Schwekendiek, Peter
Schwenn, Rainer
Schwope, Axel
Scorza de Appl, Cecilia
Sedlmayer, Erwin
Seggewiss, Wilhelm
Seifert, Walter
Seitz, Stella
Seitz, Florian
Sengbusch, Kurt
Shaver, Peter
Sherwood, William
Shetty, Rahul
Siebenmorgen, Ralf
Sieber, Wolfgang
Sigwarth, Michael
Simon, Klaus
Soffel, Michael
Solanki, Sami
Solf, Josef
Sollazzo, Claudio
Soltau, Dirk
Springel, Volker
Springer, Tim
Spruit, Henk
Spurzem, Rainer
Stahl, Otmar
Staubert, Rüdiger
Staude, Juergen
Staude, Hans
Stecklum, Bringfried
Steffen, Matthias
Steiner, Oskar
Steinert, Klaus
Steinle, Helmut
Steinmetz, Matthias

Eisenhauer, Frank
Eislöffel, Jochen
Elias II, Nicholas
Engels, Dieter
Ensslin, Torsten
Fahr, Hans
Falcke, Heino
Feitzinger, Johannes
Fendt, Christian
Fichtner, Horst
Fiebig, Dirk
Fosbury, Robert
Fraenz, Markus
Freudling, Wolfram
Fricke, Klaus
Fried, Josef
Fritze, Klaus
Fuchs, Burkhard
Fürst, Ernst
Gail, Hans-Peter
Gandorfer, Achim
Gavignaud, Isabelle
Geffert, Michael
Gehren, Thomas
Genzel, Reinhard
Geppert, Ulrich
Gerhard, Ortwin
Geyer, Edward
Giacconi, Riccardo
Gilmozzi, Roberto
Gizon, Laurent
Glatzel, Wolfgang
Glindemann, Andreas
Glover, Simon
Gottlöber, Stefan
Gouliermis, Dimitrios
Graham, David
Grahl, Bernd
Grebel, Eva
Gredel, Roland
Groote, Detlef
Grosbol, Preben
Grossmann-Doerth, U.
Groten, Erwin
Grothkopf, Uta
Grün, Eberhard
Gënther, Eike
Guertler, Joachin
Guesten, Rolf
Gussmann, Ernst
Gutcke, Dietrich
Haefner, Reinhold
Hänel, Andreus
Haerendel, Gerhard
Hagen, Hans-J'urgen
Hahn, Gerhard
Hainut, Olivier
Hamann, Wolf-Rainer
Hammer, Reiner
Mann, Gottfried
Mannheim, Karl
Marsch, Eckart
Matas, Vladimir
Matthews, Owen
Mattig, W.
Mebold, Ulrich
Meeus, Gwendolyn
Mehlert, Dörte
Meinig, Manfred
Meisenheimer, Klaus
Meister, Claudia
Menten, Karl
Meusinger, Helmut
Meyer, Friedrich
Meyer-Hofmeister, Eva
Mezger, Peter
Moehler, Sabine
Moehlmann, Diedrich
Möllenhoff, Claus
Møller, Palle
Moorwood, Alan
Mücket, Jan
Müller, Volker
Müller, Ewald
Müller, Thomas
Müller, Andreas
Mundt, Reinhard
Nagel, Thorsten
Neckel, Heinz
Neckel, Th.
Neeser, Mark
Nesis, Anastasios
Neuhaeuser, Ralph
Neukum, Gerhard
Nothnagel, Axel
Notni, Peter
Ocvirk, Pierre
Oestreicher, Roland
Oetken, L.
Oleak, H.
Ossendrijver, Mathieu
Ott, Heinz-Albert
Padovani, Paolo
Parmentier, Geneviève
Patat, Ferdinando
Patnaik, Alok
Patzer, A. Beate C.
Pauldrach, Adalbert
Peter, Hardi
Petr-Gotzens, Monika
Pflug, Klaus
Philipp, Sabine
Pinkau, Klaus
Pitz, Eckhart
Popescu, Cristina
Porcas, Richard
Preuss, Eugen
Preuss, Oliver
Stix, Michael
Stoecker, Horst
Stoehr, Felix
Storm, Jesper
Strassmeier, Klaus
Strong, Andrew
Stumpff, Peter
Stutzi, Jürgen
Szostak, Roland
Tacconi, Linda
Tacconi-Garman, Lowell
Tanaka, Yasuo
Tarenghi, Massimo
Teixeira, Paula
Teriaca, Luca
Thielheim, Klaus
Thomas, Hans
Traulsen, Iris
Trefftz, Eleonore
Treumann, Rudolf
Trümper, Joachim
Tscharnuter, Werner
Tueg, Helmut
Tuffs, Richard
Ulmschneider, Peter
Ulrich, Marie-Helene
van Belle, Gerard
van den Ancker, Mario
Völk, Heinrich
Voges, Wolfgang
Voigt, Hans
Volkmer, Reiner
von Borzeszkowski, H.
von der Lühe, Oskar
Wagner, Stefan
Walter, Hans
Wambsganss, Joachim
Warmels, Rein
Wedemeyer-Böhm, Sven
Wehrse, Rainer
Weidemann, Volker
Weigelt, Gerd
Weilbacher, Peter
Weis, Kerstin
Weiss, Achim
Weiss, Axel
Werner, Klaus
West, Richard
White, Simon
Wiedemann, Günter
Wiegelmann, Thomas
Wiehr, Eberhard
Wielebinski, Richard
Wielen, Roland
Wilms, Jörn
Wilson, P.
Winnewisser, Gisbert
Wisotzki, Lutz
Wittkowski, Markus

Hanuschik, Reinhard
Harris, Alan
Hartogh, Paul
Hasinger, Günther
Hauschildt, Peter
Hayama, Kazuhiro
Hazlehurst, John
Heber, Ulrich
Hefele, Herbert
Hegmann, Michael
Heidt, Jochen
Hempel, Marc
Hempelmann, Alexander
Henkel, Christian
Henning, Thomas
Héraudeau, Philippe
Herbstmeier, Uwe
Hering, Roland
Primas, Francesca
Przybilla, Norbert
Puls, Joachim
Quinn, Peter
Quirrenbach, Andreas
Raedler, K.
Rauch, Thomas
Rauer, Heike
Reffert, Sabine
Reich, Wolfgang
Reif, Klaus
Reimers, Dieter
Reiners, Ansgar
Reinsch, Klaus
Reiprich, Thomas
Rejkuba, Marina
Rendtel, Juergen
Renzini, Alvio
Wittmann, Axel
Witzel, Arno
Woehl, Hubertus
Wolf, Bernhard
Wolf, Rainer
Wolfschmidt, Gudrun
Wucknitz, Olaf
Wunner, Guenter
Wyrowski, Friedrich
Yamada, Shoichi
Zensus, J-Anton
Zickgraf, Franz
Ziegler, Bodo
Ziegler, Harald
Zimmermann, Helmut
Zinnecker, Hans
Zwaan, Martin

Greece

Alissandrakis, Costas
Anastasiadis, Anastasios
Antonacopoulos, Gregory
Antonopoulou, Evgenia
Arabelos, Dimitrios
Asteriadis, Georgios
Avgoloupis, Stavros
Banos, George
Bellas-Velidis, Ioannis
Bonanos, Alceste
Boumis, Panayotis
Bozis, George
Caranicolas, Nicholas
Caroubalos, Constantinos
Charmandaris, Vassilis
Chatzichristou, Eleni
Christopoulou, Panagiota
Chryssovergis, Michael
Contadakis, Michael
Contopoulos, George
Contopoulos, Ioannis
Daglis, Ioannis
Danezis, Emmanuel
Dapergolas, Anastasios
Dara, Helen
Deliyannis, Jean
Dialetis, Dimitris
Dionysiou, Demetrios
Efthymiopoulos, Christos
Evangelidis, E.
Georgantopoulos, Ioannis
Geroyannis, Vassilis
Gontikakis, Constantin
Goudas, Constantine
Goudis, Christos
Hadjidemetriou, John
Hantzios, Panayiotis
Harsoula, Mirella
Hatzidimitriou, Despina
Hilaris, Alexander
Hiotelis, Nicolaos
Isliker, Heinz
Kalvouridis, Tilemachos
Katsiyannis, Athanassios
Kazantzis, Panayotis
Kokkotas, Konstantinos
Kontizas, Evangelos
Kontizas, Mary
Korakitis, Romylos
Krimigis, Stamatios
Kylafis, Nikolaos
Laskarides, Paul
Liritzis, Ioannis
Malandraki, Olga
Manimanis, Vassilios
Mastichiadis, Apostolos
Mavraganis, Anastasios
Mavridis, Lyssimachos
Mavromichalaki, Helen
Merzanides, Constantinos
Metaxa, Margarita
Moussas, Xenophon
Niarchos, Panayiotis
Nicolaidis, Efthymios
Nindos, Alexander
Papadakis, Iossif
Papaelias, Philip
Papathanasoglou, Dimitrios
Papayannopoulos, Theodoros
Patsis, Panos
Persides, Sotirios
Pinotsis, Antonis
Plionis, Manolis
Preka-Papadema, Panagiota
Prokakis, Theodore
Rovithis, Peter
Rovithis-Livaniou, Helen
Sarris, Eleftherios
Sarris, Emmanuel
Seimenis, John
Seiradakis, John
Sergis, Nick
Sinachopoulos, Dimitris
Skokos, Charalambos
Spyrou, Nicolaos
Stathopoulou, Maria
Svolopoulos, Sotirios
Terzides, Charalambos
Theodossiou, Efstratios
Tritakis, Basil
Tsamparlis, Michael
Tsikoudi, Vassiliki
Tsinganos, Kanaris
Tsiropoula, Georgia
Tziotziou, Konstantinos
Vardavas, Ilias
Varvoglis, Harry
Veis, George
Ventura, Joséph
Vlachos, Demetrius
Vlahakis, Nektarios
Vlahos, Loukas
Xilouris, Emmanouel
Xiradaki, Evangelia
Zachariadis, Theodosios
Zachilas, Loukas
Zafiropoulos, Basil
Zikides, Michael

Honduras

Pineda de Caras, Maria
Ponce, Gustavo

Hungary

Almar, Ivan
Bagoly, Zsolt
Balazs, Lajos
Balázs, Bela
Baranyi, Tuende
Barcza, Szabolcs
Barlai, Katalin
Benkoe, Jozsef
Érdi, Bálint
Fejes, Istvan
Forgács-Dajka, Emese
Frey, Sandor
Gerlei, Otto
Grandpierre, Attila
Györi, Lajos
Hegedues, Tibor
Horváth, András
Ill, Marton
Illes, Almar
Jankovics, Istvan
Jurcsik, Johanna
Kalman, Bela
Kanyo, Sandor
Kelemen, János
Koevari, Zsolt
Kollath, Zoltan
Kovacs, Agnes
Kovacs, Geza
Kun, Maria
Lovas, Miklos
András Ludmány, Andras
Olah, Katalin
Paparo, Margit
Patkos, Laszlo
Petrovay, Kristof
Szabados, Laszlo
Szabó, Robert
Szatmary, Karoly
Szécsényi-Nagy, Gábor
Szego, Karoly
Szeidl, Béla
Tóth, Lászlo
Tóth, Imre
Vargha, Magda
Veres, Ferenc
József Vinkó, Jozsef
Wynn-Williams, C.
Zsoldos, Endre

Iceland

Björnsson, Gunnlaugur
Saemundsson, Thorsteinn
Gudmundsson, Einar
Jakobsson, Pall

India

Acharya, Bannanje
Agrawal, P.
Ahmad, Farooq
Alladin, Saleh
Ambastha, Ashok
Anandaram, Mandayam
Ananthakrishnan, Subramanian
Ansari, S.M.
Antia, H.
Anupama, G.
Apparao, K.
Ashok, N.
Babu, G.
Bacham, Eswar
Bagare, S.
Bagla, Jasjeet
Balasubramanian, V.
Balasubramanyam, Ramesh
Baliyan, Kiran
Ballabh, Goswami
Bandyopadhyay, A.
Banerjee, Dipankar
Banerji, Sriranjan
Banhatti, Dilip
Basu, Baidyanath
Bhandari, Rajendra
Bhandari, N.
Bharadwaj, Somnath
Gupta, Sunil
Gupta, Yashwant
Gupta, Surendra
Hallan, Prem
Hasan, S. Sirajul
Iyengar, Srinivasan
Iyer, B.
Jain, Rajmal
Jog, Chanda
Joshi, Umesh
Kantharia, Nimisha
Kapoor, Ramesh
Kasturirangan, K.
Kaul, Chaman
Kembhavi, Ajit
Khare, Pushpa
Kilambi, G.
Kochhar, Rajesh
Konar, Sushan
Krishna, Swamy
Krishna, Gopal
Krishnan, Thiruvenkata
Kulkarni, Prabhakar
Kulkarni, Vasant
Lal, Devendra
Majumdar, Subhabrata
Mallik, D.
Manchanda, R.
Rao, M.
Rao, Pasagada
Rastogi, Shantanu
Rautela, B.
Raveendran, A.
Ravindranath, Swara
Ray, Alak
Rengarajan, Thinniam
Sagar, Ram
Saha, Swapan
Sahni, Varun
Saikia, Dhruba
Sanwal, Basant
Sapre, Ashok
Sarma, N.
Sastri, Hanumath
Sastry, Ch.
Sastry, Shankara
Saxena, P.
Saxena, A.
Sen, Asoke
Sengupta, Sujan
Seshadri, Sridhar
Shahul Hameed, Mohin
Shastri, Prajval
Shevgaonkar, R.
Shukla, K.
Shukre, C.

Bhardwaj, Anil
Bhat, Narayana
Bhatia, Prem
Bhatia, Vishnu
Bhatnagar, K.
Bhatt, H.
Bhattacharjee, Pijush
Bhattacharya, Dipankar
Bhattacharyya, J.
Bhattacharyya, Sudip
Biswas, Sukumar
Boddapati, Anandarao
Bondal, Krishna
Buti, Bimla
Chakrabarti, Sandip
Chakraborty, Deo
Chandra, Suresh
Chandrasekhar, Thyagarajan
Chaoudhuri, Arnab
Chaubey, Uma
Chengalur, Jayaram
Chitre, Shashikumar
Chokshi, Arati
Choudhury, Tirthankar
Dadhich, Naresh
Das, Mrinal
Das, P.
Degaonkar, S.
Desai, Jyotindra
Deshpande, M.
Deshpande, Avinash
Dhurandhar, Sanjeev
Duari, Debiprosad
Duorah, Hira
Dwarakanath, K.
Dwivedi, Bhola
Gaur, V.
Ghosh, Swarna
Ghosh, P.
Giridhar, Sunetra
Gokhale, Moreshwar
Gopala, Rao
Goswami, J.
Goyal, A.
Goyal, Ashok
Gupta, Ranjan
Mangalam, Arun
Manoharan, P.
Marar, T.
Mitra, Abhas
Mohan, Anita
Mohan, Chander
Mohan, Vijay
Murthy, Jayant
Nagendra, K.
Nair, Sunita
Namboodiri, P.
Narain, Udit
Naranan, S.
Narasimha, Delampady
Narayana, J.
Narlikar, Jayant
Nath, Biman
Nath, Mishra
Nayar, S.R.Prabhakaran
Nityananda, Rajaram
Ojha, Devendra
Padalia, T.
Padmanabhan, Janardhan
Padmanabhan, Thanu
Pande, Girish
Pandey, A.
Pandey, S.
Pandey, Uma
Parthasarathy, Mudumba
Pati, Ashok
Paul, Biswajit
Peraiah, Annamaneni
Prabhu, Tushar
Prasanna, A.
Radhakrishnan, V.
Rajamohan, R.
Raju, P.
Raju, Vasundhara
Ramadurai, Souriraja
Ramamurthy, Swaminathan
Ramana, Murthy
Rangarajan, K.
Rao, Arikkala
Rao, Ramachandra
Rao, A.
Rao, N.
Singal, Ashok
Singh, Mahendra
Singh, Jagdev
Singh, Harinder
Singh, Kulinder
Singh, Markandey
Sinha, Krishnanand
Sivaram, C.
Somasundaram, Seetha
Sonti, Sreedhar
Souradeep, Tarun
Sreekantan, B.
Sreekumar, Parameswaran
Srianand, Roghunathan
Srinivasan, Ganesan
Srivastava, J.
Srivastava, Ram
Srivastava, Dhruwa
Stephens, S.
Subrahmanya, C.
Subrahmanyan, Ravi
Subramaniam, Annapurni
Subramanian, Kandaswamy
Subramanian, K.
Subramanian, Prasad
Swarup, Govind
Talwar, Satya
Tandon, S.
Tonwar, Suresh
Tripathi, B.
Udaya, Shankar
Uddin, Wahab
Vahia, Mayank
Vaidya, P.
Vardya, M.
Varma, Ram
Vasu-Mallik, Sushma
Vats, Hari
Venkatakrishnan, P.
Verma, R.
Verma, V.
Vinod, S.
Vishveshwara, C.
Vivekanand, M.
Vivekananda, Rao
Yadav, Jagdish

Indonesia

Ardi, Eliani
Dawanas, Djoni
Djamaluddin, Thomas
Herdiwijaya, Dhani
Hidayat, Taufiq
Hidayat, Bambang
Ibrahim, Jorga
Kunjaya, Chatief
Malasan, Hakim
Premadi, Premana
Radiman, Iratio
Raharto, Moedji
Ratag, Mezak
Siregar, Suryadi
Sukartadiredja, Darsa
Wiramihardja, Suhardja

Iran

Adjabshirizadeh, Ali
Ardebili, M.
Asareh, Habibolah
Bordbar, Gholam
Dehghani, Mohammad
Edalati Sharbaf, Mohammad
Ghanbari, Jamshid
Hamedivafa, Hashem
Jalali, Mir Abbas
Jassur, Davoud
Kalafi, Manoucher
Khalesseh, Bahram
Khesali, Ali
Kiasatpour, Ahmad
Malakpur, Iradj
Mansouri, Reza
Nasiri, Sadollah
Nasr-Esfahani, Bahram
Nozari, Kourosh
Rahvar, Sohrab
Riazi, Nematollah
Samimi, Jalal
Sobouti, Yousef

Ireland

Breslin, Ann
Butler, Raymond
Callanan, Paul
Carroll, P.
Cawley, Michael
Coffey, Deirdre
Cunniffe, John
Devaney, Martin
Downes, Turlough
Drury, Luke
Elliott, Ian
Espey, Brian
Fegan, David
Florides, Petros
Gabuzda, Denise
Gillanders, Gerard
Golden, Aaron
Hanlon, Lorraine
Haywood, J.
Hoey, Michael
Kennedy, Eugene
Lang, Mark
McAteer, R.T. James
McBreen, Brian
McKeith, Niall
McKenna, Lawlor
Meurs, Evert
Murphy, John
Norci, Laura
O'Connor, Seamus
O'Mongain, Eon
O'Sullivan, Denis
O'Sullivan, Creidhe
Podio, Linda
Quinn, John
Ray, Thomas
Redfern, Michael
Redman, Matthew
Shadmehri, Mohsen
Shearer, Andrew
Smith, Niall
Trappe, Neil
Whelan, Emma
Wrixon, Gerard

Israel

Alexander, Tal
Almoznino, Elhanan
Bar-Nun, Akiva
Barkana, Rennan
Barkat, Zalman
Bekenstein, Jacob
Braun, Arie
Brosch, Noah
Cuperman, Sami
Dekel, Avishai
Eichler, David
Eppelbaum, Lev
Ershkovich, Alexander
Eviatar, Aharon
Finzi, Arrigo
Formiggini, Lilliana
Gedalin, Michael
Glasner, Shimon
Goldman, Itzhak
Goldsmith, S.
Griv, Evgeny
Harpaz, Amos
Heller, Ana
Horesh, Assaf
Horwitz, Gerald
Ibbetson, Peter
Israelevich, Peter
Jacob, Uri
Joséph, Joachim
Kaspi, Shai
Katz, Joséph
Kovetz, Attay
Kozlovsky, Ben
Laor, Ari
Leibowitz, Elia
Levinson, Amir
Lyubarsky, Yury
Maoz, Dan
Marco, Shmulik
Mazeh, Tsevi
Meidav, Meir
Mekler, Yuri
Merman, Hirsh G.
Naoz, Smadar
Netzer, Nathan
Netzer, Hagai
Nusser, Adi
Ofir, Aviv
Ohring, George
Piran, Tsvi
Podolak, Morris
Prialnik-Kovetz, Dina
Price, Colin
Rakavy, Gideon
Regev, Oded
Rephaeli, Yoel
Ribak, Erez
Sack, Noam
Segaluvitz, Alexander
Seidov, Zakir
Shaviv, Giora
Shiryaev, Alexander
Soker, Noam
Steinitz, Raphael
Trakhtenbrot, Benny
Tuchman, Ytzhak
Usov, Vladimir
Vager, Zeev
Vazan-Shukrun, Allona
Vidal, Nissim
Waxman, Eli
Woo, Joanna
Yeivin, Y.
Zitrin, Adi
Zucker, Shay

Italy

Abbas, Ummi
Aiello, Santi
Albanese, Lara
Alcal, Juan Manuel
Altamore, Aldo
Amati, Lorenzo
Ambrosini, Roberto
Amendola, Luca
Andreani, Paola
Andreon, Stefano
Andretta, Vincenzo
Andreuzzi, Gloria
Angeletti, Lucio
Anile, Angelo
Antonelli, Lucio Angelo
Antonello, Elio
Antonucci, Ester
Antonuccio-Delogu, Vincenzo
Arcidiacono, Carmelo
Auriemma, Giulio
Badolati, Ennio
Baffa, Carlo
Balbi, Amedeo
Baldinelli, Luigi
Bandiera, Rino
Banni, Aldo
Baratta, Giuseppe
Baratta, Giovanni
Barbaro, Guido
Barberis, Bruno
Barbieri, Cesare
Barbon, Roberto
Bardelli, Sandro
Barletti, Raffaele
Barone, Fabrizio
Bartolini, Corrado
Battinelli, Paolo
Battistini, Pierluigi
Bazzano, Angela
Becciani, Ugo
Bedogni, Roberto
Belinski, Vladimir
Bellazzini, Michele
Belloni, Tomaso
Belvedere, Gaetano
Bemporad, Alessandro
Benacchio, Leopoldo
Benetti, Stefano
Bernacca, Pierluigi
Bernardi, Fabrizio
Berrilli, Francesco
Berta, Stefano
Bertelli, Gianpaolo
Bertin, Giuseppe
Bertola, Francesco
Bettoni, Daniela
de Zotti, Gianfranco
Del Santo, Melania
Del Zanna, Luca
Delbo, Marco
Dell' Oro, Aldo
Della Ceca, Roberto
Della Valle, Massimo
Di Cocco, Guido
Di Fazio, Alberto
Di Martino, Mario
Di Mauro, Maria Pia
di Serego Alighieri, Sperello
Diaferio, Antonaldo
Dotto, Elisabetta
Drimmel, Ronald
Einaudi, Giorgio
Elia, Davide
Emanuele, Alessandro
Ermolli, Ilaria
Ettori, Stefano
Facondi, Silvia
Falciani, Roberto
Falomo, Renato
Fanti, Carla
Fanti, Roberto
Faraggiana, Rosanna
Fasano, Giovanni
Fedeli, Cosimo
Federici, Luciana
Felli, Marcello
Feretti, Luigina
Ferluga, Steno
Ferrari, Attilio
Ferreri, Walter
Ferrini, Federico
Filacchione, Gianrico
Focardi, Paola
Fofi, Massimo
Foschini, Luigi
Franceschini, Alberto
Franchini, Mariagrazia
Frasca, Antonio
Fulle, Marco
Fusco-Femiano, Roberto
Fusi-Pecci, Flavio
Gai, Mario
Galeotti, Piero
Galletta, Giuseppe
Galletto, Dionigi
Galli, Daniele
Gallino, Roberto
Garilli, Bianca
Gaudenzi, Silvia
Gavazzi, Giuseppe
Gemmo, Alessandra
Gervasi, Massimo
Nipoti, Carlo
Nobili, L.
Nobili, Anna
Nocera, Luigi
Noci, Giancarlo
Nucita, Achille
Oliva, Ernesto
Olmi, Luca
Omizzolo, Alessandro
Origlia, Livia
Orio, Marina
Orlandini, Mauro
Orlando, Salvatore
Ortolani, Sergio
Ostorero, Luisa
Pacini, Franco
Pagano, Isabella
Palla, Francesco
Palumbo, Maria Elisabetta
Palumbo, Giorgio
Panessa, Francesca
Pannunzio, Renato
Paolicchi, Paolo
Paresce, Francesco
Parma, Paola
Pasian, Fabio
Pastori, Livio
Patern, Lucio
Patriarchi, Patrizio
Pellegrini, Silvia
Peres, Giovanni
Perola, Giuseppe
Persi, Paolo
Pettini, Marco
Pian, Elena
Picca, Domenico
Piccioni, Adalberto
Pigatto, Luisa
Piotto, Giampaolo
Pipino, Antonio
Piro, Luigi
Pirronello, Valerio
Pizzella, Alessandro
Pizzella, Guido
Pizzichini, Graziella
Plainaki, Christina
Poggianti, Bianca
Polcaro, V.
Poletto, Giannina
Politi, Romolo
Polletta, Maria del Carmen
Poma, Angelo
Porceddu, Ignazio
Poretti, Ennio
Prandoni, Isabella
Preite Martinez, Andrea

Bianchi, Simone
Bianchini, Antonio
Biazzo, Katia
Bignami, Giovanni
Biviano, Andrea
Blanco, Carlo
Blanco, Armando
Boccaletti, Dino
Bocchino, Fabrizio
Bodo, Gianluigi
Bolzonella, Micol
Bombaci, Ignazio
Bonanno, Alfio
Bonanno, Giovanni
Bondi, Marco
Bono, Giuseppe
Bnoli, Fabrizio
Bonometto, Silvio
Borgani, Stefano
Braccesi, Alessandro
Bragaglia, Angela
Branchesi, Marca
Brand, Jan
Brescia, Massimo
Bressan, Alessandro
Brocato, Enzo
Broglia, Pietro
Brunetti, Gianfranco
Bruno, Roberto
Bucciarelli, Beatrice
Buonanno, Roberto
Burderi, Luciano
Busà, Innocenza
Busarello, Giovanni
Buson, Lucio
Busso, Maurizio
Buzzoni, Alberto
Caccianiga, Alessandro
Cacciari, Carla
Calamai, Giovanni
Caloi, Vittoria
Calura, Francesco
Calvani, Massimo
Cantiello, Michele
Capaccioli, Massimo
Capaccioni, Fabrizio
Cappellaro, Enrico
Cappi, Alberto
Cappi, Massimo
Capria, Maria
Caputo, Filippina
Capuzzo Dolcetta, Roberto
Caraveo, Patrizia
Cardini, Daniela
Carollo, Daniela
Carpino, Mario
Carraro, Giovanni
Carretta, Eugenio
Carretti, Ettore
Ghia, Piera Luisa
Ghirlanda, Giancarlo
Ghisellini, Gabriele
Giallongo, Emanuele
Giani, Elisabetta
Giannone, Pietro
Giannuzzi, Maria
Gioia, Isabella
Giordano, Silvio
Giovanardi, Carlo
Giovannelli, Franco
Giovannini, Gabriele
Girardi, Marisa
Girardi, Leo
Giroletti, Marcello
Gitti, Myriam
Godoli, Giovanni
Gomez, Maria
Granato, Gian Luigi
Gratton, Raffaele
Greggio, Laura
Gregorini, Loretta
Gregorio, Anna
Gronchi, Giovanni
Grueff, Gavril
Guarnieri, Adriano
Guerrero, Gianantonio
Guzzo, Luigi
Hack, Margherita
Held, Enrico
Hunt, Leslie
Iijima, Takashi
Iovino, Angela
Janssen, Katja
Keheyan, Yeghis
La Barbera, Francesco
La Franca, Fabio
La Padula, Cesare
La Spina, Alessandra
Lai, Sebastiana
Lanciano, Nicoletta
Landi, Degli Innocenti
Landi, Simone
Landini, Massimo
Landolfi, Marco
Lanza, Antonino
Lanzafame, Alessandro
Lattanzi, Mario
Lazzarin, Monica
Leone, Franco
Leschiutta, S.
Leto, Giuseppe
Ligori, Sebastiano
Limongi, Marco
Lisi, Franco
Londrillo, Pasquale
Longo, Giuseppe
Longo, Giuseppe
Lucatello, Sara
Proverbio, Edoardo
Pucillo, Mauro
Pugliano, Antonio
Pulone, Luigi
Rafanelli, Piero
Raimondo, Gabriella
Raiteri, Claudia
Ramella, Massimo
Rampazzo, Roberto
Ranalli, Piero
Randich, Sofia
Ranieri, Marcello
Reale, Fabio
Reardon, Kevin
Righini, Alberto
Robba, Natale
Romano, Giuliano
Romano, Patrizia
Romoli, Marco
Rossi, Lucio
Rossi, Corinne
Rossi, Alessandro
Ruffini, Remo
Russo, Guido
Sabbadin, Franco
Saggion, Antonio
Salinari, Piero
Salvati, Marco
Sancisi, Renzo
Sandrelli, Stefano
Santin, Paolo
Saracco, Paolo
Sarasso, Maria
Sasso, Clementina
Sbarufatti, Boris
Scappini, Flavio
Scaramella, Roberto
Scardia, Marco
Schneider, Raffaella
Sciortino, Salvatore
Scodeggio, Marco
Scuderi, Salvo
Secco, Luigi
Sedmak, Giorgio
Selvelli, Pierluigi
Setti, Giancarlo
Severgnini, Paola
Severino, Giuseppe
Shore, Steven
Silva, Laura
Silvestro, Giovanni
Silvotti, Roberto
Sironi, Giorgio
Smaldone, Luigi
Smart, Richard
Smriglio, Filippo
Sozzetti, Alessandro
Spadaro, Daniele
Spagna, Alessandro

Carusi, Andrea
Casasola, Viviana
Caselli, Paola
Cassatella, Angelo
Castelli, Fiorella
Catalano, Franco
Catanzaro, Giovanni
Cauzzi, Gianna
Cavaliere, Alfonso
Cavallini, Fabio
Cazzola, Paolo
Cecchi-Pellini, Cesare
Celletti, Alessandra
Cellino, Alberto
Celotti, Anna Lisa
Centurin Martin, Miriam
Ceppatelli, Guido
Cerroni, Priscilla
Cerruti Sola, Monica
Cester, Bruno
Cevolani, Giordano
Chiappetti, Lucio
Chiappini Moraes Leite, Cristina
Chincarini, Guido
Chinnici, Ileana
Chiosi, Cesare
Chiuderi, Claudio
Chiuderi-Drago, Franca
Chlistovsky, Franca
Ciliegi, Paolo
Cinzano, Pierantonio
Ciotti, Luca
Ciroi, Stefano
Citterio, Oberto
Claudi, Riccardo
Clementini, Gisella
Codella, Claudio
Colafrancesco, Sergio
Colangeli, Luigi
Comastri, Andrea
Comoretto, Giovanni
Conconi, Paolo
Coradini, Angioletta
Corbelli, Edvige
Corsini, Enrico
Cosentino, Rosario
Cosmovici, Cristiano
Costa, Enrico
Cottini, Valeria
Covino, Elvira
Covone, Giovanni
Cremonese, Gabriele
Cristiani, Stefano
Cugusi, Leonino
Curir, Anna
Cusumano, Giancarlo
Cutispoto, Giuseppe
D' Odorico, Valentina
D'Amico, Nicolo'
Lucchesi, David
Maccacaro, Tommaso
Maccagni, Dario
Maccarone, Maria Concetta
Maceroni, Carla
Mack, Karl-Heinz
Magazz, Antonio
Maggio, Antonio
Magni, Gianfranco
Maiolino, Roberto
Malagnini, Maria
Malara, Francesco
Manara, Alessandro
Mancuso, Santi
Mandolesi, Nazzareno
Mangano, Vanessa
Mannucci, Filippo
Mantegazza, Luciano
Mantovani, Franco
Marano, Bruno
Maraschi, Laura
Marchi, Simone
Marconi, Alessandro
Marconi, Marcella
Mardirossian, Fabio
Marilli, Ettore
Maris, Michele
Marmolino, Ciro
Marulli, Federico
Marzari, Francesco
Marziani, Paola
Masani, A.
Massaglia, Silvano
Massardi, Marcella
Matt, Giorgio
Matteucci, Francesca
Mazzali, Paolo
Mazzei, Paola
Mazzitelli, Italo
Mazzoni, Massimo
Mazzotta Epifani, Elena
Mazzucconi, Fabrizio
Meneghetti, Massimo
Mennella, Vito
Mercurio, Amata
Mereghetti, Sandro
Merighi, Roberto
Merluzzi, Paola
Messerotti, Mauro
Messina, Antonio
Messina, Sergio
Mezzetti, Marino
Micela, Giuseppina
Milani, Andrea
Milano, Leopoldo
Mineo, Teresa
Missana, Marco
Molaro, Paolo
Molinari, Emilio
Spinoglio, Luigi
Stalio, Roberto
Stanga, Ruggero
Stanghellini, Carlo
Stirpe, Giovanna
Strafella, Francesco
Straus, Thomas
Strazzulla, Giovanni
Tagliaferri, Gianpiero
Tantalo, Rosaria
Tanzella-Nitti, Giuseppe
Tavecchio, Fabrizio
Tempesti, Piero
Ternullo, Maurizio
Terranegra, Luciano
Testi, Leonardo
Tofani, Gianni
Tomasella, Lina
Tommei, Giacomo
Torelli, M.
Tormen, Giuseppe
Tornambe, Amedeo
Torricelli, Guidetta
Tosi, Monica
Tosi, Federico
Tozzi, Gian
Tozzi, Paolo
Tozzi, Andrea
Trevese, Dario
Trigilio, Corrado
Trinchieri, Ginevra
Trussoni, Edoardo
Tuccari, Gino
Turatto, Massimo
Turolla, Roberto
Ubertini, Pietro
Umana, Grazia
Uras, Silvano
Uslenghi, Michela
Vagnetti, Fausto
Vallenari, Antonella
Valsecchi, Giovanni
Velli, Marco
Ventura, Rita
Ventura, Paolo
Venturi, Tiziana
Vercellone, Stefano
Vettolani, Giampaolo
Viel, Matteo
Vietri, Mario
Vignali, Cristian
Villata, Massimo
Viotti, Roberto
Virgopia, Nicola
Vittone, Alberto
Vittorio, Nicola
Vladilo, Giovanni
Walmsley, C.
Wolter, Anna

D'Antona, Francesca
D'Onofrio, Mauro
Dall'Ora, Massimo
Dall-Oglio, Giorgio
Dalla Bontà, Elena
Dallacasa, Daniele
Danese, Luigi
Danziger, I.
De Bernardis, Paolo
de Biase, Giuseppe
de Felice, Fernando
de Martino, Domitilla
de Petris, Marco
de Ruiter, Hans
de Sanctis, Giovanni
de Sanctis, M.
Monaco, Pierluigi
Morbidelli, Lorenzo
Morbidelli, Roberto
Morossi, Carlo
Moscadelli, Luca
Moscardini, Lauro
Motta, Santo
Mucciarelli, Paola
Mulas, Giacomo
Munari, Ulisse
Mureddu, Leonardo
Napolitano, Nicola
Natta, Antonella
Nesci, Roberto
Nicastro, Luciano
Zacchei, Andrea
Zaggia, Simone
Zamorani, Giovanni
Zampieri, Luca
Zanichelli, Alessandra
Zaninetti, Lorenzo
Zanini, Valeria
Zannoni, Mario
Zappal, Rosario
Zappal, Vincenzo
Zavatti, Franco
Zitelli, Valentina
Zlobec, Paolo
Zucca, Elena
Zuccarello, Francesca

Japan

Agata, Hidehiko
Aikawa, Toshiki
Aikawa, Yuri
Akabane, Kenji
Akio, Inoue
Akita, Kyo
Akiyama, Masayuki
Ando, Hiroyasu
Aoki, Shinko
Aoki, Wako
Arafune, Jiro
Arai, Kenzo
Arakida, Hideyoshi
Arimoto, Nobuo
Asai, Ayumi
Awaki, Hisamitsu
Aya, Bamba
Aya, Yamauchi
Ayani, Kazuya
Azuma, Takahiro
Baba, Hajime
Baba, Naoshi
Chiba, Masashi
Chiba, Takeshi
Chikada, Yoshihiro
Chikawa, Michiyuki
Daisaku, Nogami
Daishido, Tsuneaki
Daisuke, Iono
Deguchi, Shuji
Dobashi, Kazuhito
Doi, Mamoru
Dotani, Tadayasu
Ebisuzaki, Toshikazu
Ebizuka, Noboru
Enoki, Motohiro
Enome, Shinzo
Eriguchi, Yoshiharu
Ezawa, Hajime
Koda, Jin
Kodaira, Keiichi
Kodama, Hideo
Kodama, Tadayuki
Kogure, Tomokazu
Kohno, Kotaro
Koide, Shinji
Koike, Chiyoe
Koji, Mori
Kojima, Masayoshi
Kojima, Yasufumi
Kokubo, Eiichiro
Komiyama, Yutaka
Kondo, Masaaki
Kondo, Masayuki
Kosai, Hiroki
Koshiba, Masatoshi
Koshiishi, Hideki
Koyama, Katsuji
Kozai, Yoshihide
Kozasa, Takashi
Kubota, Jun
Kudoh, Takahiro
Kumagai, Shiomi
Kumai, Yasuki
Kunieda, Hideyo
Kuno, Nario
Kurokawa, Hiroki
Kúsano, Kanya
Kusunose, Masaaki
Machida, Mami
Maeda, Keiichi
Maeda, Kei-ichi
Maeda, Koitiro
Maehara, Hideo
Maihara, Toshinori
Makino, Fumiyoshi
Makino, Junichiro
Makishima, Kazuo
Sato, Koichi
Sato, Shinji
Sato, Shuji
Satoshi, Honda
Satoshi, Nozawa
Sawa, Takeyasu
Sawada-Satoh, Satoko
Seki, Munezo
Sekido, Mamoru
Sekiguchi, Kazuhiro
Sekiguchi, Maki
Sekiguchi, Naosuke
Sekiguchi, Tomohiko
Sekimoto, Yutaro
Shibahashi, Hiromoto
Shibai, Hiroshi
Shibasaki, Kiyoto
Shibata, Katsunori
Shibata, Kazunari
Shibata, Masaru
Shibata, Shinpei
Shibata, Yukio
Shibazaki, Noriaki
Shigeyama, Toshikazu
Shimasaku, Kazuhiro
Shimojo, Masumi
Shimura, Toshiya
Shin, Watanabe
Shin'Ichi, Nagata
Shioya, Yasuhiro
Sho, Sasaki
Simoda, Mahiro
Sofue, Yoshiaki
Sma, Mitsuru
Sorai, Kazuo
Suda, Kazuo
Suematsu, Yoshinori
Sugai, Hajime
Suganuma, Masahiro

Fharo, Toshihiro
Fujimoto, Masa-Katsu
Fujimoto, Masayuki
Fujimoto, Shin-ichiro
Fujishita, Mitsumi
Fujita, Mitsutaka
Fujita, Yoshio
Fujita, Yutaka
Fujiwara, Akira
Fujiwara, Takao
Fujiwara, Tomoko
Fukuda, Ichiro
Fukuda, Naoya
Fukue, Jun
Fukugita, Masataka
Fukui, Takao
Fukui, Yasuo
Fukushige, Toshiyuki
Fukushima, Toshio
Funato, Yoko
Furusho, Reiko
George, Kosugi
Goto, Tomotsugu
Gouda, Naoteru
Gunji, Shuichi
Habe, Asao
Hachisu, Izumi
Hamabe, Masaru
Hamajima, Kiyotoshi
Hanami, Hitoshi
Hanaoka, Yoichiro
Hanawa, Tomoyuki
Handa, Toshihiro
Hara, Hirohisa
Hara, Tetsuya
Hasegawa, Ichiro
Hasegawa, Tetsuo
Hashimoto, Masa-aki
Hashimoto, Osamu
Hatsukade, Isamu
Hattori, Makoto
Hayasaki, Kimitake
Hayashi, Chushiro
Hideki, Asada
Hideko, Nomura
Hidenori, Kokubun
Hiei, Eijiro
Hirabayashi, Hisashi
Hirai, Masanori
Hirao, Takanori
Hirashita, Hiroyuki
Hirata, Ryuko
Hirayama, Tadashi
Hiroki, Chihara
Hiroko, Nagahara
Hiromoto, Norihisa
Hiroshi, Akitaya
Hiroshi, Yoshida
Hirota, Tomoya
Makita, Mitsugu
Makiuti, Sin'itirou
Manabe, Seiji
Masahiro, Tsujimoto
Masai, Kuniaki
Masaki, Yoshimitsu
Masuda, Satoshi
Masumichi, Seta
Matsuda, Takuya
Matsuhara, Hideo
Matsui, Takafumi
Matsumoto, Hironori
Matsumoto, Katsura
Matsumoto, Ryoji
Matsumoto, Tomoaki
Matsumoto, Toshio
Matsumura, Masafumi
Matsuo, Hiroshi
Matsuoka, Masaru
Matsuura, Shuji
Mikami, Takao
Mineshige, Shin
Misato, Fukagawa
Mitsuda, Kazuhisa
Mitsunobu, Kawada
Miyaji, Shigeki
Miyama, Syoken
Miyamoto, Masanori
Miyata, Emi
Miyawaki, Ryosuke
Miyazaki, Atsushi
Miyoshi, Makoto
Miyoshi, Shigeru
Mizumoto, Yoshihiko
Mizuno, Akira
Mizuno, Norikazu
Mizuno, Shun
Mizuno, Takao
Mizutani, Kohei
Momose, Muntake
Mori, Masaki
Mori, Masao
Morimoto, Masaki
Morita, Kazuhiko
Morita, Koh-ichiro
Morita, Satoshi
Moriyama, Fumio
Motizuki, Yuko
Motohara, Kentaro
Mukai, Tadashi
Munetoshi, Tokumaru
Murakami, Hiroshi
Murakami, Hiroshi
Murakami, Izumi
Murakami, Toshio
Murayama, Takashi
Nagase, Fumiaki
Nagata, Tetsuya
Nagataki, Shigehiro
Sugawa, Chikara
Sugimoto, Daiichiro
Suginohara, Tatsushi
Sugitani, Koji
Sugiyama, Naoshi
Sunada, Kazuyoshi
Suto, Yasushi
Suzuki, Hideyuki
Suzuki, Takeru
Suzuki, Tomoharu
Tachihara, Kengo
Tagoshi, Hideyuki
Takaba, Hiroshi
Takada-Hidai, Masahide
Takahara, Fumio
Takahara, Mariko
Takahashi, Junko
Takahashi, Koji
Takahashi, Masaaki
Takahashi, Tadayuki
Takakubo, Keiya
Takami, Hideki
Takano, Shuro
Takano, Toshiaki
Takase, Bunshiro
Takashi, Hasegawa
Takata, Masao
Takato, Naruhisa
Takayanagi, Kazuo
Takayoshi, Kohmura
Takeda, Hidenori
Takeda, Yoichi
Takei, Yoh
Takenouchi, Tadao
Takeuchi, Tsutomu
Takeuti, Mine
Takizawa, Motokazu
Tamenaga, Tatsuo
Tamura, Motohide
Tamura, Shin'ichi
Tanabe, Hiroyoshi
Tanabe, Kenji
Tanabe, Toshihiko
Tanaka, Masuo
Tanaka, Riichiro
Tanaka, Wataru
Tanaka, Yasuo
Tanaka, Yutaka
Taniguchi, Yoshiaki
Tanikawa, Kiyotaka
Taruya, Atsushi
Tashiro, Makoto
Tatehiro, Mihara
Tatekawa, Takayuki
Tatematsu, Ken'ichi
Tawara, Yuzuru
Terashima, Yuichi
Terashita, Yoichi
Tohru, Nagao

Honma, Mareki
Horaguchi, Toshihiro
Hori, Genichiro
Horiuchi, Ritoku
Hosokawa, Mizuhiko
Hozumi, Shunsuke
Hurukawa, Kiitiro
Ichikawa, Shin-ichi
Ichikawa, Takashi
Ichimaru, Setsuo
Iguchi, Osamu
Iguchi, Satoru
Iijima, Shigetaka
Ikeuchi, Satoru
Imai, Hiroshi
Imanishi, Masatoshi
Inada, Naohisa
Inagaki, Shogo
Inatani, Junji
Inoue, Hajime
Inoue, Makoto
Inutsuka, Shu-ichiro
Ioka, Kunihito
Iriyama, Jun
Ishida, Keiichi
Ishida, Manabu
Ishida, Toshihito
Ishiguro, Masato
Ishihara, Hideki
Ishii, Takako
Ishimaru, Yuhri
Ishizaka, Chiharu
Ishizawa, Toshiaki
Ishizuka, Toshihisa
Ita, Yoshifusa
Ito, Kensai
Ito, Takashi
Ito, Yutaka
Itoh, Hiroshi
Itoh, Masayuki
Itoh, Naoki
Iwamuro, Fumihide
Iwasaki, Kyosuke
Iwata, Ikuru
Iwata, Takahiro
Iye, Masanori
Izumiura, Hideyuki
Jeong, Woong-Seob
Jugaku, Jun
Kaburaki, Osamu
Kaifu, Norio
Kajino, Toshitaka
Kakinuma, Takakiyo
Kakuta, Chuichi
Kamaya, Hideyuki
Kambe, Eiji
Kameya, Osamu
Kamijo, Fumio
Kanamitsu, Osamu

Nakada, Yoshikazu
Nakagawa, Akiharu
Nakagawa, Naoya
Nakagawa, Takao
Nakagawa, Yoshitsugu
Nakai, Naomasa
Nakajima, Hiroshi
Nakajima, Junichi
Nakajima, Koichi
Nakajima, Tadashi
Nakamichi, Akika
Nakamoto, Taishi
Nakamura, Akiko
Nakamura, Fumitaka
Nakamura, Takashi
Nakamura, Takuji
Nakamura, Tsuko
Nakamura, Yasuhisa
Nakanishi, Hiroyuki
Nakanishi, Kouichiro
Nakano, Makoto
Nakano, Syuichi
Nakano, Takenori
Nakao, Yasushi
Nakasato, Naohito
Nakayama, Kunji
Nakazawa, Kiyoshi
Nambu, Yasusada
Nariai, Kyoji
Narita, Shinji
Nishi, Keizo
Nishi, Ryoichi
Nishida, Minoru
Nishida, Mitsugu
Nishikawa, Jun
Nishimura, Jun
Nishimura, Masayoshi
Nishimura, Osamu
Nishimura, Shiro
Nishio, Masanori
Nitta, Shin-ya
Nobuyuki, Iwamoto
Noguchi, Kunio
Noguchi, Masafumi
Nomoto, Ken'ichi
Obi, Shinya
Ogawa, Hideo
Ogawara, Yoshiaki
Ogura, Katsuo
Ohashi, Takaya
hashi, Yukio
Ohishi, Masatoshi
Ohki, Kenichiro
Ohnishi, Kouji
Ohta, Kouji
Ohtani, Hiroshi
Ohtsubo, Junji
Ohtsuki, Keiji
Ohyama, Noboru

Tomimatsu, Akira
Tomisaka, Kohji
Tomita, Akihiko
Tomita, Kenji
Tosa, Makoto
Tosaki, Tomoka
Toshifumi, Shimizu
Toshihiro, Kasuga
Toshinobu, Takagi
Totani, Tomonori
Totsuka, Yoji
Toyama, Kiyotaka
Tsubaki, Tokio
Tsuboi, Masato
Tsuchiya, Toshio
Tsuguya, Naito
Tsuji, Takashi
Tsujimoto, Takuji
Tsunemi, Hiroshi
Tsuneta, Saku
Tsuru, Takeshi
Tsutsumi, Takahiro
Uchida, Juichi
Ueda, Yoshihiro
Uemura, Makoto
Ueno, Munetaka
Ueno, Sueo
Uesugi, Akira
Ukita, Nobuharu
Umeda, Hideyuki
Umemoto, Tomofumi
Umemura, Masayuki
Unno, Wasaburo
Utsumi, Kazuhiko
Wada, Keiichi
Wada, Takehiko
Wakamatsu, Ken-Ichi
Wako, Kojiro
Wanajo, Shinya
Washimi, Haruichi
Watanabe, Jun-ichi
Watanabe, Noriaki
Watanabe, Takashi
Watanabe, Tetsuya
Watarai, Hidneori
Watarai, Kenya
Watari, Shinichi
Yabushita, Shin
Yagi, Masafumi
Yahagi, Hideki
Yamada, Masako
Yamada, Toru
Yamada, Yoshiyuki
Yamagata, Tomohiko
Yamamoto, Masayuki
Yamamoto, Satoshi
Yamamoto, Tetsuo
Yamamoto, Yoshiaki
Yamamura, Issei

Kaneda, Hidehiro
Kaneko, Noboru
Karoji, Hiroshi
Kashikawa, Nobunari
Kasuga, Takashi
Kato, Ken-ichi
Kato, Mariko
Kato, Shoji
Kato, Taichi
Kato, Takako
Kato, Tsunehiko
Kato, Yoshiaki
Kawabata, Kinaki
Kawabata, Kiyoshi
Kawabata, Koji
Kawabata, Shusaku
Kawabe, Ryohei
Kawaguchi, Ichiro
Kawaguchi, Kentarou
Kawai, Nobuyuki
Kawakatu, Nozomu
Kawakita, Hideyo
Kawamura, Akiko
Kawara, Kimiaki
Kawasaki, Masahiro
Kawata, Yoshiyuki
Kenji, Nakamura
Kentaro, Matsuda
Kiguchi, Masayoshi
Kimura, Toshiya
Kinoshita, Hiroshi
Kinugasa, Kenzo
Kitai, Reizaburo
Kitamoto, Shunji
Kitamura, Masatoshi
Kiyoaki, Wajima
Kiyoshi, Hayashida
Kiyotomo, Ichiki
Kobayashi, Eisuke
Kobayashi, Hideyuki
Kobayashi, Naoto
Kobayashi, Yukiyasu
Okamoto, Isao
Okamura, Sadanori
Okazaki, Akira
Okazaki, Atsuo
Okuda, Haruyuki
Okuda, Toru
Okumura, Sachiko
Okumura, Shin-ichiro
Omukai, Kazuyuki
Onaka, Takashi
Onishi, Toshikazu
Oohara, Ken-ichi
Oowaki, Naoaki
Osaki, Yoji
Oyabu, Shinki
Ozaki, Masanobu
Ozeki, Hiroyuki
Sabano, Yutaka
Sadakane, Kozo
Saigo, Kazuya
Saijo, Keiichi
Saio, Hideyuki
Saito, Kuniji
Saito, Mamoru
Saito, Sumisaburo
Saito, Takao
Saitoh, Takayuki
Sakai, Junichi
Sakamoto, Seiichi
Sakao, Taro
Sakashita, Shiro
Sakurai, Kunitomo
Sakurai, Takashi
Sasaki, Minoru
Sasaki, Misao
Sasaki, Shin
Sasao, Tetsuo
Sato, Fumio
Sato, Humitaka
Sato, Isao
Sato, Jun'ichi
Sato, Katsuhiko
Yamaoka, Hitoshi
Yamasaki, Atsuma
Yamasaki, Noriko
Yamashita, Kojun
Yamashita, Takuya
Yamashita, Yasumasa
Yamauchi, Makoto
Yamauchi, Shigeo
Yanagisawa, Masahisa
Yano, Hajime
Yano, Taihei
Yasuda, Haruo
Yasuda, Naoki
Yasuhiro, Koyama
Yasuhiro, Murata
Yoichiro, Suzuki
Yokosawa, Masayoshi
Yokoyama, Jun-ichi
Yokoyama, Koichi
Yonehara, Atsunori
Yonekura, Yoshinori
Yoneyama, Tadaoki
Yoshida, Atsumasa
Yoshida, Fumi
Yoshida, Haruo
Yoshida, Junzo
Yoshida, Michitoshi
Yoshida, Shigeomi
Yoshida, Shin'ichirou
Yoshida, Takashi
Yoshii, Yuzuru
Yoshikawa, Kohji
Yoshikawa, Makoto
Yoshimura, Hirokazu
Yoshioka, Kazuo
Yoshioka, Satoshi
Yoshizawa, Masanori
Yuasa, Manabu
Yui, Yukari
Yuji, Ikeda
Yumiko, Oasa
Yutaka, Shiratori

Republic of Korea

Ahn, Youngsook
Ann, Hong
Baek, Chang Hyun
Byun, Yong-Ik
Cha, Seung-Hoon
Chae, Jongchul
Chae, Kyu-Hyun
Chang, Heon-Young
Chang, Kyongae
Cho, Kyung Suk
Cho, Se-Hyung
Choe, Gwangson
Choe, Seung
Choi, Chul-Sung
Choi, Kyu-Hong
Kim, Hyun-Goo
Kim, Jongsoo
Kim, Kap-sung
Kim, Kwang-tae
Kim, Sang Joon
Kim, Seung-Lee
Kim, Sug-Whan
Kim, Sungeun
Kim, Sungsoo
Kim, Tu-Whan
Kim, Yong-Cheol
Kim, Yonggi
Kim, Yongha
Kim, Yoo Jea
Kim, Young-Soo
Mkrtichian, David
Moon, Shin
Moon, Yong-Jae
Nha, Il-Seong
Noh, Hyerim
Noh, Hyerim
Oh, Kap-Soo
Oh, Kyu
Pak, Soojong
Park, Byeong-Gon
Park, Changbom
Park, Hong
Park, Il-Hung
Park, Jang-Hyun
Park, Myeong-gu

Choi, Minho
Chou, Kyong
Chun, Mun-suk
Chung, Hyun
Han, Cheongho
Han, Wonyong
Hong, Seung
Hwang, Jai-chan
Hyung, Siek
Im, Myungshin
Ishiguro, Masateru
Jang, Minwhan
Jeon, Young-Beom
Jeong, Jang
Jung, Jae
Kang, Hyesung
Kang, Yong
Kang, Young-Woon
Kim, Bong
Kim, Chulhee
Kim, Chun
Koo, Bon-Chul
Lee, Chang-Won
Lee, Dae-Hee
Lee, Dong-Hun
Lee, Dong-Wook
Lee, Eun-Hee
Lee, Hee-Won
Lee, Hyung
Lee, Jae-Woo
Lee, Kang Hwan
Lee, Myung Gyoon
Lee, Sang
Lee, Seong-Jae
Lee, Sungho
Lee, Woo-baik
Lee, Yong Bok
Lee, Yong-Sam
Lee, Young
Lee, Younguung
Minh, Young
Minn, Young
Park, Pil-Ho
Park, Seok
Park, Yong-Sun
Park, Young-Deuk
Rey, Soo-Chang
Rhee, Myung-Hyun
Ryu, Dongsu
Seon, Kwang-IL
Sohn, Young-Jong
Song, Doo
Song, In-Ok
Suh, Kyung-Won
Sung, Hwankyung
Yang, Hong-Jin
Yang, Jongmann
Yi, Sukyoung
Yi, Yu
Yim, Hong-Suh
Yoon, Suk-Jin
Yoon, Tae
Yun, Hong-Sik

Latvia

Abele, Maris
Alksnis, Andrejs
Balklavs-Grinhofs, Arturs
Daube-Kurzemniece, Ilga
Docenko, Dmitrijs
Dzervitis, Uldis
Eglitis, Ilgmars
Freimanis, Juris
Gills, Martins
Grasberg, Ernest
Lapushka, Kazimirs
Ryabov, Boris
Salitis, Antonijs
Shmeld, Ivar
Vilks, Ilgonis
Začs, Laimons
Zhagars, Youris

Lebanon

Bittar, Jamal
Touma, Jihad
El Eid, Mounib
Sabra, Bassem

Lithuania

Bartašiute, Stanislava
Bartkevicius, Antanas
Bogdanovicius, Pavelas
Cernis, Kazimieras
Kazlauskas, Algirdas
Kupliauskiene, Alicija
Lazauskaite, Romualda
Meištas, Edmundas
Rudzikas, Zenonas
Sperauskas, Julius
Straizys, Vytautas
Jokubas Sudžius, Jokubas
Tautvaišiene, Gražina
Vansevicius, Vladas
Zdanavicius, Kazimeras

Malaysia

Abdul Aziz, Ahmad
Abu Kassim, Hasan
Ibrahim, Zainol Abidin
Ilyas, Mohammad
Mahat, Rosli
Majid, Abdul
Mohd, Zambri
Othman, Mazlan
Zainal Abidin, Zamri

Mexico

Aguilar, Luis A.
Allen, Christine
Andernach, Heinz
Arellano Ferro, Armando
Aretxaga, Itziar
Echevarria, Juan
Escalante, Vladimir
Fierro, Julieta
Franco, José
Galindo-Trejo, Jesús
Obregón Daz, Octavio
Page, Dany
Peimbert, Manuel
Pena, Miriam
Peña Saint-Martin, José

Arthur, Jane
Avila Foucault, Remy
Avila-Reese, Vladimir
Ballesteros-Paredes, Javier
Bentez, Erika
Bertone, Emanuele
Binette, Luc
Bravo-Alfaro, Hector
Canto, Jorge
Cardona, Octavio
Caretta, Cesar
Carigi, Leticia
Carramiñana, Alberto
Carrasco, Bertha
Carrasco, Luis
Carrillo, René
Chatterjee, Tapan
Chavarria-K, Carlos
Chavez-Dagostino, Miguel
Chavushyan, Vahram
Coln, Pedro
Contreras, Maria
Cornejo, Alejandro
Costero, Rafael
Coziol, Roger
Cruz-González, Irene
Cuevas, Salvador
Curiel, Salvador
Daltabuit, Enrique
de Diego Onsurbe, José
de La Herran, José
Dultzin-Hacyan, Deborah
Garcia-Barreto, José
Gomez, Yolanda
González, J.
González, Alejando
Guichard, José
Henney, William
Hernández, Xavier
Hiriart, David
Hughes, David
Jeyakumar, Solai
Kemp, Simon
Klapp, Jaime
Koenigsberger, Gloria
Kurtz, Stanley
Lee, William
Lekht, Evgeni
Lizano, Susana
Loinard, Laurent
López Cruz, Omar
López Garcia, José
Malacara, Daniel
Martinez, Mario
Martnez Bravo, Oscar
Martos, Marco
Mayya, Divakara
Mendez, Emmanuel
Mendoza, V.
Mendoza-Torress, José-Eduardo
Miyaji, Takamitsu
Moreno, Edmundo
Moreno-Corral, Marco
Mújica, Raul
Perez de Tejada, Hector
Perez-Peraza, Jorge
Phillips, John
Porras, Bertha
Poveda, Arcadio
Ivnio, Ivanio
Recillas-Cruz, Elsa
Rodriguez, Luis
Rodrguez, Monica
Rohrmann, Rene
Rosa González, Daniel
Rosado, Margarita
Ruelas-Mayorga, R.
Santillan, Alfredo
Sarmiento-Galan, Antonio
Schuster, William
Serrano, Alfonso
Tapia, Mauricio
Tenorio-Tagle, Guillermo
Torres-Peimbert, Silvia
Tovmassian, Gaghik
Tovmassian, Hrant
Trinidad, Miguel
Valdes Parra, José
Valdés-Sada, Pedro
Vázquez, Roberto
Vázquez-Semadeni, Enrique
Velázquez, Pablo
Wall, William
Warman, Josef
Wilkin, Francis
Zharikov, Sergey

Morocco

Benkhaldoun, Zouhair
El, Bakkali
Najid, Nour-Eddine
Chamcham, Khalil
Kadiri, Samir
Siher, El Arbi
Darhmaoui, Hassane
Lazrek, Mohamed
Touma, Hamid

Netherlands

Achterberg, Abraham
Baan, Willem
Barthel, Peter
Baryshev, Andrey
Baud, Boudewijn
Begeman, Kor
Beintema, Douwe
Bennett, Kevin
Bijleveld, Willem
Blaauw, Adriaan
Bleeker, Johan
Bloemen, Hans
Böker, Torsten
Boland, Wilfried
Bontekoe, Romke
Boonstra, Albert
Borgman, Jan
Hoekstra, Hendrik
Hörandel, Jörg
Houdebine, Eric
Hovenier, J.
Hoyng, Peter
Hubenet, Henri
Hulsbosch, A.
Icke, Vincent
in't Zand, Johannes
Israel, Frank
Ives, John
Jaffe, Walter
Jakobsen, Peter
Jeffers, Sandra
Jonker, Peter
Jourdain de Muizon, Marie
Kaastra, Jelle
Savonije, Gerrit
Schaye, Joop
Scheepmaker, Anton
Schrijver, Johannes
Schulz, Rita
Schutte, Willem
Schwarz, Ulrich
Schwehm, Gerhard
Shipman, Russell
Smit, J.
Snellen, Ignas
Spaans, Marco
Spoelstra, T.
Stam, Daphne
Stark, Ronald
Strom, Richard
Stuik, Remko

Bos, Albert
Bosma, Pieter
Brandl, Bernhard
Bregman, Jacob
Brentjens, Michiel
Brinchmann, Jarle
Brinkman, Bert
Brouw, Willem
Brown, Anthony
Campbell, Robert
Clavel, Jean
Constantini, Elisa
De Bruijne, Jos
de Bruyn, A.
de Geus, Eugene
de Jager, Cornelis
de Jong, Teije
de Korte, Pieter
de Koter, Alex
de Lange, Gert
De Marchi, Guido
de Vries, Cornelis
Dekker, E.
den Herder, Jan-Willem
Deul, Erik
Dewi, Jasinta
Dieleman, Pieter
Dominik, Carsten
Douglas, Nigel
Ehrenfreund, Pascale
Foing, Bernard
Foley, Anthony
Franx, Marijn
Fridlund, Malcolm
Fritzova-Svestka, L.
Garrett, Michael
Gathier, Roel
Goedbloed, Johan
Gondoin, Philippe
Groot, Paul
Gurvits, Leonid
Habing, Harm
Hammerschlag, Robert
Hammerschlag-Hensberge, Godelieve
Haverkorn, Marijke
Heald, George
Heise, John
Helmi, Amina
Helmich, Frank
Henrichs, Hubertus
Heras, Ana
Hermsen, Willem
Heske, Astrid
Hessels, Hessels
Kahlmann, Hans
Kamp, Inga
Kaper, Lex
Katgert, Peter
Katgert-Merkelijn, J.
Keller, Christoph
Koopmans, Leon
Koornneef, Jan
Kuijken, Koen
Kuijpers, H.
Kuiper, Lucien
Kuperus, Max
Kwee, K.
Lamers, Henny
Langer, Norbert
Larsen, Soeren
Laureijs, Rene
Le Poole, Rudolf
Levin, Yuri
Linnartz, Harold
Lub, Jan
Markoff, Sera
Martens, Petrus
Méndez, Mariano
Miley, George
Morales Rueda, Luisa
Morganti, Raffaella
Namba, Osamu
Nelemans, Gijs
Nieuwenhuijzen, Hans
Ödman, Carolina
Ollongren, A.
Olnon, Friso
Olthof, Henk
Oosterloo, Thomas
Paragi, Zsolt
Parmar, Arvind
Peacock, Anthony
Pel, Jan
Peletier, Reynier
Perryman, Michael
Pogrebenko, Sergei
Polatidis, Antonios
Pols, Onno
Portegies Zwart, Simon
Pottasch, Stuart
Prusti, Timo
Raassen, Ion
Raimond, Ernst
Roelfsema, Peter
Roettgering, Huub
Roos, Nicolaas
Rutten, Robert
Sanders, Robert
Svestka, Zdenek
Szomoru, Arpad
Tauber, Jan
te Lintel Hekkert, Peter
Thé, Pik-Sin
Tinbergen, Jaap
Tjin-a-Djie, Herman
Tolstoy, Eline
Trager, Scott
Valentijn, Edwin
van Albada, Tjeerd
van Bueren, Hendrik
van de Stadt, Herman
van de Weygaert, Rien
van den Heuvel, Edward
van den Oord, Bert
van der Hucht, Karel A.
van der Hulst, Jan
van der Klis, Michiel
van der Kruit, Pieter
van der Laan, Harry
van der Tak, Floris
van der Werf, Paul
van Dishoeck, Ewine
van Duinen, R.
van Genderen, Arnoud
van Gent, Robert
van Haarlem, Michiel
van Houten-Groeneveld, Ingrid
van Langevelde, Huib
van Leeuwen, Joeri
van Woerden, Hugo
Verbunt, Franciscus
Verdoes Kleijn, Gijsbert
Verheijen, Marc
Vermeulen, Rene
Vink, Jacco
Waters, Laurens
Watts, Anna
Wesselius, Paul
Wijers, Ralph
Wijnands, Rudy
Wijnbergen, Jan
Wild, Wolfgang
Winkler, Christoph
Wise, Michael
Zaroubi, Saleem
Zayer, Igor

New Zealand

Adams, Jenni
Albrow, Michael
Allen, William
Andrews, Frank
Arnold, Richard
Baggaley, William
Blow, Graham
Bond, Ian
Budding, Edwin
Carter, Brian
Christie, Grant
Cottrell, Peter
Craig, Ian
Dodd, Richard
Gilmore, Alan
Gulyaev, Sergei
Hearnshaw, John
Hill, Graham
Jones, Albert
Kerr, Roy
Kilmartin, Pamela
Kitaeff, Vyacheslav
Pollard, Karen
Sullivan, Denis
Sweatman, Winston
Taylor, Andrew
Walker, William
Yock, Philip

Nigeria

Akujor, Chidi
Okeke, Pius
Okpala, Kingsley
Urama, Johnson
Chukwude, Augustine
Okeke, Francisca
Rabiu, Akeem
Eze, Romanus
Okoye, Samuel
Schmitter, Edward

Norway

Aksnes, Kaare
Andersen, Bo Nyborg
Brekke, Pål
Brynildsen, Nils
Carlsson, Mats
Dahle, Haakon
Elgaroy, Oystein
Engvold, Oddbjørn
Eriksen, Hans Kristian
Esser, Ruth
Gudiksen, Boris
Hansen, Frode
Hansteen, Viggo
Haugan, Stein Vidar
Havnes, Ove
Jaunsen, Andreas
Kjeldseth-Moe, Olav
Leer, Egil
Lie-Svendsen, Oystein
Lilje, Per
Lin, Yong
Mota, David
Mysen, Eirik
Pettersen, Bjørn
Puetzfeld, Dirk
Rouppe van der Voort, Luc
Skjæraasen, Olaf
Solheim, Jan
Stabell, Rolf
Trulsen, Jan
Wiik Toutain, Jun
Wikstol, Oivind
Wold, Margrethe

Peru

Aguilar, Maria
Ishitsuka, Mutsumi
Reyes, Rafael

Philippines

Celebre, Cynthia
Vinluan, Renato
Soriano, Bernardo
Torres, Jesus Rodrigo

Poland

Bajtlik, Stanislaw
Baran, Andrzej
Bartkiewicz, Anna
Bem, Jerzy
Berlicki, Arkadiusz
Borkowski, Kazimierz
Breiter, Slawomir
Brzezinski, Aleksander
Bulik, Tomasz
Bzowski, Maciej
Chodorowski, Michal
Choloniewsski, Jacek
Kijak, Jaroslaw
Kiraga, Marcin
Kluzniak, Wlodzimiere
Kolaczek, Barbara
Kołomański, Sylwester
Konacki, Maciej
Kopacki, Grzegorz
Kosek, Wieslaw
Kozlowski, Maciej
Krasinski, Andrzej
Kreiner, Jerzy
Krelowski, Jacek
Pokrzywka, Bartlomiej
Pres, Pawek
Rompolt, Bogdan
Rozyczka, Michal
Rudak, Bronislaw
Rudawy, Pawel
Rudnicki, Konrad
Rys, Stanislaw
Sarna, Marek
Schillak, Stanislaw
Schreiber, Roman
Schwarzenberg-Czerny, Alex

Chyzy, Krzysztof Tadeusz
Ciurla, Tadeusz
Cugier, Henryk
Czern, Bozena
Daszyńska-Daszkiewicz, Jadwiga
Demianski, Marek
Drozyner, Andrzej
Dybczynski, Piotr
Dziembowski, Wojciech
Essai, Essai
Falewicz, Robert
Flin, Piotr
Gaska, Stanislaw
Gesicki, Krzysztof
Giersz, Miroslav
Gil, Janusz
Godlowski, Wlodzimierz
Gorgolewski, Stanislaw
Goździewski, Krzysztof
Grabowski, Boleslaw
Gronkowski, Piotr
Grzedzielski, Stanislaw
Haensel, Pawel
Hanasz, Jan
Hanasz, Michal
Heller, Michael
Hurnik, Hieronim
Iwaniszewska, Cecylia
Jahn, Krzysztof
Jakimiec, Jerzy
Janiuk, Agnieszka
Jaranowski, Piotr
Jaroszynski, Michal
Jerzykiewicz, Mikolaj
Jopek, Tadeusz
Juszkiewicz, Roman
Kaluzny, Janusz
Katarzyński, Krzysztof
Krempec-Krygier, Janina
Krolikowska-Soltan, Malgorzata
Kruszewski, Andrzej
Krygier, Bernard
Kryszczyn'ska, Agnieszka
Krzesinski, Jerzy
Kubiak, Marcin
Kunert-Bajraszewska, Magdalena
Kurpinska-Winiarska, M.
Kurzynska, Krystyna
Kus, Andrzej
Kwiatkowski, Tomasz
Lehmann, Marek
Machalski, Jerzy
Maciejewski, Andrzej
Madej, Jerzy
Marecki, Andrzej
Maslowski, Jozef
Michalec, Adam
Michalek, Grzegorz
Michalowski, Tadeusz
Mietelski, Jan
Mikołajewska, Joanna
Mikołajewski, Maciej
Moderski, Rafal
Molenda-Zakowicz, Joanna
Moskalik, Pawel
Nastula, Jolanta
Niedzielski, Andrzej
Niemczura, Ewa
Nikołajuk, Marek
Ogłoza, Waldemar
Opolski, Antoni
Ostrowski, Michal
Otmianowska-Mazur, Katarzyna
Pietrzynski, Grzegorz
Pigulski, Andrzej
Pojmanski, Grzegorz
Semeniuk, Irena
Sienkiewicz, Ryszard
Sikora, Marek
Sikorski, Jerzy
Sitarski, Grzegorz
Smak, Jozef
Sokolowski, Lech
Soltan, Andrzej
Soszyński, Igor
Stawikowski, Antoni
Stepień, Kazimierz
Strobel, Andrzej
Sylwester, Janusz
Sylwester, Barbara
Szczerba, Ryszard
Szutowicz, Slawomira
Szydlowski, Marek
Szymanski, Michal
Szymczak, Marian
Tomczak, Michal
Tomov, Toma
Turlo, Zygmunt
Tylenda, Romuald
Udalski, Andrzej
Urbanik, Marek
Usowics, Jerzy
Winiarski, Maciej
Wnuk, Edwin
Woszczyk, Andrzej
Woszczyna, Andrzej
Wytrzyszczak, Iwona
Zdunik, Julian
Zdziarski, Andrzej
Zieba, Stanislaw
Ziolkowski, Janusz
Ziolkowski, Krzysztof
Zola, Stanislaw

Portugal

Anton, Sonia
Augusto, Pedro
Avelino, Pedro
Bazot, Michael
Braga da Costa Campos, L.
Cardoso Santos, Nuno
Costa, Vitor
Cunha, Margarida
da Costa, Antonio
De Avillez, Miguel
Doran, Rosa
Fernandes, Amadeu
Fernandes, João
Ferreira, João
Gameiro, Jorge
Garcia, Paulo
Gizani, Nectaria
Lago, Maria
Lima, João
Lobo, Catarina
Luz, David
Machado Folha, Daniel
Magalhães, Antonio
Marques, Manuel
Martins, Carlos
Melendez, Jorge
Mendes, Virgilio
Moitinho, André
Monteiro, Mario João
Moreira Morais, Maria
Mourão, Ana Maria
Osório, Isabel
Osório, José
Pascoal, Antonio
Pedrosa, António
Peixinho, Nuno
Roos-Serote, Maarten
Santos, Filipe
Santos Agostinho, Rui
Serote Roos, Margarida
Viana, Pedro
Vicente, Raimundo
Yun, João

Romania

Badescu, Octavian
Barbosu, Mihail
Blaga, Christina
Botez, Elvira
Breahna, Iulian
Burs, Lucian
Carsmaru, Maria
Dinulescu, Simona
Dumitrache, Cristiana
Dumitrescu, Alexandru
Imbroane, Alexandru
Lungu, Nicolaie
Marcu, Alexandru
Marilena, Mierla
Maris, Georgeta
Mihaila, Ieronim
Mioc, Vasile
Oprescu, Gabriela
Oproiu, Tiberiu
Parv, Bazil
Pop, Alexandru
Pop, Vasile
Popescu, Nedelia
Popescu, Petre
Pricopi, Dumitru
Roman, Rodica
Rusu, I.
Stanila, George
Stavinschi, Magda
Suran, Marian
Szenkovits, Ferenc
Todoran, Ioan
Toro, Tibor
Ureche, Vasile
Vass, Gheorghe

Russian Federation

Abalakin, Viktor
Afanas'ev, Viktor
Akim, Efraim
Altyntsev, Alexandre
Antipin, Sergei
Antipova, Lyudmila
Antokhin, Igor
Antonov, Vadim
Arefiev, Vadim
Arkharov, Arkadij
Arkhipova, Vera
Artamonov, Boris
Artyukh, Vadim
Babadzhahianc, Mikhail
Bagrov, Alexander
Bajkova, Anisa
Bakhtigaraev, Nail
Balega, Yurij
Barabanov, Sergey
Baranov, Alexander
Barkin, Yuri
Baryshev, Yurij
Batrakov, Yurij
Baturin, Vladimir
Belkovich, Oleg
Berdnikov, Leonid
Beskin, Gregory
Beskin, Vasily
Bikmaev, Ilfan
Bisikalo, Dmitrij
Bisnovatyi-Kogan, Gennadij
Blinnikov, Sergey
Bobylev, Vadim
Bochkarev, Nikolai
Bogod, Vladimir
Bondarenko, Lyudmila
Borisov, Nikolay
Borovik, Valery
Boyarchuk, Alexander
Burba, George
Burdyuzha, Vladimir
Karachentsev, Igor
Kardashev, Nicolay
Karitskaya, Evgeniya
Karpov, Sergey
Kascheev, Rafael
Kastel', Galina
Katsova, Maria
Kazarovets, Elena
Khaikin, Vladimir
Khaliullin, Khabibrachman
Kholshevnikov, Konstantin
KholtYgin, Alexander
Khovritchev, Maxim
Kilpio, Elena
Kim, Iraida
Kiselev, Aleksej
Kislyakov, Albert
Kisseleva, Tamara
Kitchatinov, Leonid
Klochkova, Valentina
Kocharov, Grant
Kocharovsky, Vitalij
Kolesov, Aleksandr
Komberg, Boris
Kompaneets, Dmitrij
Kondratiev, Vladislav
Kononovich, Edvard
Kopylov, Aleksandr
Kopylova, Yulia
Kornilov, Viktor
Korzhavin, Anatolij
Koshelyaevsky, Nikolay
Kostyakova, Elena
Kovalev, Yuri
Kovaleva, Dana
Krasinsky, George
Ksanfomality, Leonid
Kudryavtsev, Dmitry
Kuimov, Konstantin
Kulikova, Nelli
Kumkova, Irina
Rizvanov, Naufal
Rodin, Alexander
Rodionova, Zhanna
Romanov, Andrey
Romanyuk, Iosif
Rudenko, Valentin
Rudnitskij, Georgij
Ruskol, Evgeniya
Ryabchikova, Tatiana
Ryabov, Yurij
Ryabova, Galina
Rykhlova, Lidiya
Rzhiga, Oleg
Sachkov, Mikhail
Sagdeev, Roald
Sakhibullin, Nail
Samodurov, Vladimir
Samus', Nikolaj
Sazhin, Mikhail
Sazonov, Sergey
Serber, Alexander
Shakht, Natalia
Shakura, Nikolaj
Shaposhnikov, Vladimir
Shapovalova, Alla
Sharina, Margarita
Shatsky, Nicolai
Shchekinov, Yuri
Sheffer, Evgenij
Shefov, Nikolaj
Shematovich, Valerij
Shenavrin, Victor
Shevchenko, Vladislav
Shevchenko, Ivan
Shibanov, Yuri
Shingareva, Kira
Shishov, Vladimir
Sholukhova, Olga
Shor, Viktor
Shustov, Boris
Shuygina, Nadia

Burenin, Rodion
Bychkov, Konstantin
Bykov, Andrei
Chashei, Igor
Chechetkin, Valerij
Chentsov, Eugene
Cherepashchuk, Anatolij
Chernetenko, Yulia
Chernin, Artur
Chernov, Gennadij
Chertok, Ilya
Chertoprud, Vadim
Chugai, Nikolaj
Churazov, Evgenij
Dagkesamansky, Rustam
Dambis, Andrei
Danilov, Vladimir
Denisenkov, Pavel
Devyatkin, Aleksandr
Dluzhnevskaya, Olga
Dodonov, Sergej
Dokuchaev, Vyacheslav
Dokuchaeva, Olga
Doubinskij, Boris
Dudorov, Alexander
Efremov, Yurij
Emelianov, Nikolaj
Emel'yanenko, Vacheslav
Eroshkin, Georgij
Esipov, Valentin
Fabrika, Sergei
Fadeyev, Yurij
Fatkhullin, Timur
Fedotov, Leonid
Finkelstein, Andrej
Fokin, Andrei
Fomichev, Valerij
Fominov, Aleksandr
Fridman, Aleksej
Galeev, Albert
Gayazov, Iskander
Gelfreikh, Georgij
Gilfanov, Marat
Glagolevskij, Yurij
Glebova, Nina
Glushkova, Elena
Glushneva, Irina
Gnedin, Yurij
Gontcharov, George
Gorschkov, Aleksandr
Gorshanov, Denis
Gosachinskij, Igor
Grachev, Stanislav
Grebenev, Sergej
Grebenikov, Evgenij
Grechnev, Victor
Grib, Sergey
Grigorjev, Viktor
Kurilchik, Vladimir
Kurt, Vladimir
Kutuzov, Sergej
Kuznetsov, Eduard
Kuznetsov, Vladimir
Lamzin, Sergei
Larionov, Mikhail
Larionov, Valeri
Lavrov, Mikhail
Leushin, Valerij
Lipunov, Vladimir
Livshits, Mikhail
Loktin, Alexhander
Loskutov, Viktor
Lotova, Natalja
Loukitcheva, Maria
Lozinskaya, Tatjana
Lukash, Vladimir
Lukashova, Marina
Lutovinov, Alexander
Makalkin, Andrei
Makarov, Dmitry
Makarova, Lidia
Malkin, Zinovy
Malkov, Oleg
Malofeev, Valery
Mardyshkin, Vyacheslav
Marov, Mikhail
Mashonkina, Lyudmila
Maslennikov, Kirill
Matveenko, Leonid
Medvedev, Yurij
Mikhailov, Andrey
Miletsky, Eugeny
Mingaliev, Marat
Mironov, Aleksey
Mishurov, Yuri
Mogilevskij, Emmanuil
Moiseenko, Sergey
Moiseev, Alexei
Nadyozhin, Dmitrij
Nagirner, Dmitrij
Nagnibeda, Valerij
Nagovitsyn, Yurij
Nefed'ev, Yury
Nefedeva, Antonina
Nikiforov, Igor
Obridko, Vladimir
Orlov, Victor
Pakhomov, Yury
Pamyatnykh, Aleksej
Panchuk, Vladimir
Papushev, Pavel
Parfinenko, Leonid
Parijskij, Yurij
Pashchenko, Mikhail
Pavlinsky, Mikhail
Petrov, Gennadij
Sidorenkov, Nikolaj
Sil'chenko, Olga
Silantev, Nikolaj
Skripnichenko, Vladimir
Skulachov, Dmitrij
Smirnov, Grigorij
Smolentsev, Sergej
Smolkov, Gennadij
Snegirev, Sergey
Snezhko, Leonid
Sobolev, Andrej
Soboleva, Natalja
Sokolov, Leonid
Sokolov, Viktor
Sokolov, Vladimir
Solovaya, Nina
Soloviev, Alexandr
Somov, Boris
Sorochenko, Roman
Sorokin, Nikolaj
Sotnikova, Natalia
Stankevich, Kazimir
Stepanov, Aleksander
Strukov, Igor
Struminsky, Alexei
Suleimanov, Valery
Sunyaev, Rashid
Surdin, Vladimir
Svechnikov, Marij
Sveshnikov, Mikhail
Taranova, Olga
Tatevyan, Suriya
Tchouikova, Nadezhda
Teplitskaya, Raisa
Terekhov, Oleg
Terentjeva, Aleksandra
Tikhonov, Nikolai
Tlatov, Andrej
Tokarev, Yurij
Trushkin, Sergej
Tsvetkov, Dmitry
Tsygan, Anatolij
Tutukov, Aleksandr
Tyul'bashev, Sergei
Udaltsov, Vyacheslav
Ugolnikov, Oleg
Utrobin, Victor
Vainshtein, Leonid
Valtts, Irina
Valeev, Sultan
Valyaev, Valerij
Varshalovich, Dmitrij
Vashkov'yak, Michael
Vashkovyak, Sofja
Vassiliev, Nikolaj
Vereshchagin, Sergej
Verkhodanov, Oleg
Vikhlinin, Alexey
Vityazev, Veniamin

Grishchuk, Leonid
Gubanov, Vadim
Gubchenko, Vladimir
Gulyaev, Rudolf
Gusev, Alexander
Guseva, Irina
Hagen-Thorn, Vladimir
Idlis, Grigorij
Ikhsanov, Robert
Ikhsanov, Nazar
Il'in, Vladimir
Illarionov, Andrei
Ilyasov, Yuri
Imshennik, Vladimir
Ipatov, Aleksandr
Ivanov, Evgenij
Ivanov, Vsevolod
Ivanov-Kholodny, Gor
Kaidanovski, Mikhail
Kaigorodov, Pavel
Kalenskii, Sergei
Kaltman, Tatyana
Kanayev, Ivan
Petrov, Sergei
Petrovskaya, Margarita
Pipin, Valery
Piskunov, Anatolij
Pitjev, Nikolaj
Pitjeva, Elena
Pogodin, Mikhail
Poliakov, Evgenij
Polyachenko, Evgeny
Polyakhova, Elena
Popov, Mikhail
Popov, Sergey
Porfir'ev, Vladimir
Postnov, Konstantin
Pozanenko, Alexei
Prokhorov, Mikhail
Pushkin, Sergej
Pustilnik, Simon
Rahimov, Ismail
Rastorguev, Aleksej
Razin, Vladimir
Reshetnikov, Vladimir
Revnivtsev, Mikhail
Vityazev, Andrej
Vlasyuk, Valerij
Volkov, Evgeni
Voloshina, Irina
Voshchinnikov, Nikolai
Wiebe, Dmitri
Yakovlev, Dmitrij
Yakovleva, Valerija
Yudin, Ruslan
Yungelson, Lev
Yushkin, Maxim
Zabolotny, Vladimir
Zagretdinov, Renat
Zaitsev, Valerij
Zakharova, Polina
Zasov, Anatolij
Zharkov, Vladimir
Zharov, Vladimir
Zheleznyakov, Vladimir
Zhitnik, Igor
Zhugzhda, Yuzef
Zinchenko, Igor
Zlotnik, Elena

Saudi Arabia

Al-Malki, Mohammed
Basurah, Hassan
Elmhamdi, Abouazza
Malawi, Abdulrahman
Al-Mostafa, Zaki
Bukhari, Fadel
Goharji, Adan
Niazy, Adnan
Almleaky, Yasseen
Eker, Zeki
Kordi, Ayman
Saleh, Magdy

Serbia

Angelov, Trajko
Arsenijević, Jelisaveta
Atanackovic-Vukmanović, Olga
Čadež, Vladimir
Ćirković Milan
Cvetkovi'c, Zorica
Dačić, Miodrag
Damljanović, Goran
Dimitrijević, Milan
Djurasević, Gojko
Djurović, Dragutin
Ignjatović, Ljubinko
Jankov, Slobodan
Jovanović, Predrag
Knežević, Zoran
Kovačević, Andjelka
Kubičela, Aleksandar
Kuzmanoski, Mike
Lazović, Jovan
Lukacević, Ilija
Mihajlov, Anatolij
Milogradov-Turin, Jelena
Nikolić, Silvana
Ninković, Slobodan
Pakvor, Ivan
Pejović, Nadezda
Popović, Georgije
Popović, Luka
Sadzakov, Sofija
Samurović, Srdjan
Segan, Stevo
Urošević, Dejan
Vince, Ištvan
Vukićević-Karabin, Mirjana

Slovakia

Budaj, Jan
Chochol, Drahomir
Dorotovič, Ivan
Dzifcáková, Elena
Gajdoš, Štefan
Galád, Adrian
Ga'lis, Rudolf
Hajdukova, Maria
Hajduková, Jr., Maria
Hefty, Jan
Hric, Ladislav
Jakubík, Marián
Kapisinsky, Igor
Klačka, Jozef
Klocok, Lubomir
Kocifaj, Miroslav
Komžík, Richard
Koza, Julius
Kučera, Aleš
Minarovjech, Milan
Neslušan, Luboš
Palůš, Pavel
Parimucha, Stefan
Pittich, Eduard
Porubčan, Vladimír
Pribulla, Theodor
Rušin, Vojtech
Rybák, Jan
Rybanský, Milan
Saniga, Metod
Skopal, Augustin
Svoreň, Ján
Sykora, Julius
Tóth, Juraj
Tremko, Jozef
Vaňko, Martin
Žižňovský, Jozef
Zverko, Juraj

South Africa

Balona, Luis
Barway, Sudhashu
Bassett, Bruce
Beesham, Aroonkumar
Block, David
Booth, Roy
Bouchard, Antoine
Buckley, David
Charles, Philip
Cooper, Timothy
Crawford, Steven
Cress, Catherine
Cunow, Barbara
de Jager, Gerhard
de Jager, Ocker
Dunsby, Peter
Ellis, George
Fanaroff, Bernard
Feast, Michael
Flanagan, Claire
Fraser, Brian
Gaylard, Michael
Glass, Ian
Goedhart, Sharmila
Govender, Kevindran
Govinder, Keshlan
Gulbis, Amanda
Hashimoto, Yasuhiro
Hellaby, Charles
Hers, Jan
Hughes, Arthur
Jonas, Justin
Kilkenny, David
Kniazev, Alexei
Koen, Marthinus
Kraan-Korteweg, Renée
Laney, Clifton
Leeuw, Lerothodi
Loaring, Nicola
Maharaj, Sunil
Martinez, Peter
Medupe, Rodney
Meintjes, Petrus
Menzies, John
Mokhele, Khotso
Nicolson, George
O'Donoghue, Darragh
Olivier, Enrico
Poole, Graham
Potter, Stephen
Raubenheimer, Barend
Rijsdijk, Case
Roche, David
Romero-Colmenero, Encarnacion
Schröder, Anja
Sefako, Ramotholo
Smits, Derck
Soltynski, Maciej
Still, Martin
Stoker, Pieter
Tiplady, Adrian
Väisänen, Petri
van der Heyden, Kurt
van der Walt, Diederick
Viollier, Raoul
Wainwright, John
Warner, Brian
Whitelock, Patricia
Williams, William
Winkler, Hartmut
Woudt, Patrick

Spain

Abad Medina, Alberto
Abia, Carlos
Acosta Pulido, José
Alberdi, Antonio
Alcobé, Santiago
Alcolea, Javier
Aldaya, Victor
Alfaro, Emilio
Alvarez, Pedro
Alves, Joao
Amado González, Pedro
Andrade, Manuel
Anglada, Guillem
Aparicio, Antonio
Argüeso, Francisco
Arribas Mocoroa, Santiago
Atrio Barandela, Fernando
Bachiller, Rafael
Balcells, Marc
Ballester, José
Ballesteros, Ezequiel
Barcia, Alberto
Barcons, Xavier
Barrado Navascués, David
Battaner, Eduardo
Beckman, John
Belizon, Fernando
Bellot Rubio, Luis
Belmonte Avilés, Juan Antonio
Gallego, Jesús
García, López
Garcia, Domingo
Garca de la Rosa, Ignacio
Garca-Berro, Enrique
Garcia-Burillo, Santiago
Garcia-Lario, Pedro
Garcia-Lorenzo, Maria
Garcia-Pelayo, Jose
Garrido, Rafael
Garzón, Francisco
Getino Fernandez, Juan
Gimenez, Alvaro
Gomez, González
Gomez, Monique
Gómez de Castro, Ana
Gómez Fernndez, Jose
Gómez Rivero, José
Gonzáles-Alfonso, Eduardo
González, Serrano
González Camacho, Antonio
González de Buitrago, Jesús
González Delgado, Rosa
González Martinez Pais, Ignacio
González-Riestra, R.
Gorgas, Garcia
Grieger, Bjoern
Gutiérrez, Carlos
Hernández-Pajares, Manuel
Najarro de la Parra, Francisco
Negueruela, Ignacio
Núñez, Jorge
Oliver, Ramn
Ortiz, Jose
Oscoz, Alejandro
Pallé, Pere
Pallé Bagó, Enric
Paredes Poy, Josep
Perea-Duarte, Jaime
Perez, Enrique
Perez, Fournon
Perez Hernandez, Fernando
Perez-Torres, Miguel
Planesas, Pere
Prieto, Mercedes
Prieto, Cristina
Prieto, Almudena
Quintana, José
Rebolo, Rafael
Reglero Velasco, Victor
Rego, Fernandez
Régulo, Clara
Ribas, Ignasi
Ribó, Marc
Rioja, Maria
José, Jose
Roca Cortés, Teodoro
Rodrigo, Rafael

Benn, Chris
Bernabeu, Guillermo
Betancor Rijo, Juan
Boloix, Rafael
Bonet, Jose
Boschin, Walter
Bravo, Eduardo
Bujarrabal, Valentin
Calvo, Manuel
Camarena Badia, Vicente
Canal, Ramon
Caon, Nicola
Carbonell, Marc
Casares, Jorge
Castañeda, Héctor
Castro-Tirado, Alberto
Cepa, Jordi
Cernicharo, Jose
Clement, Rosa Maria
Codina, Vidal
Colina, Luis
Collados, Manuel
Colomer, Francisco
Coma, Juan
Cornelisse, Remon
Cornide, Manuel
Corradi, Romano
Costa, Victor
Cubarsi, Rafael
Cuesta Crespo, Luis
de Castro, Elisa
de Vicente, Pablo
Deeg, Hans
Del Olmo, Orozco
Del Rio, Gerardo
Del Toro Iniesta, Jose
Delgado, Antonio
Diaz, Angeles
Diaz Trigo, Maria
Djupvik, Anlaug
Docobo, Jose A.
Domingo, Vicente
Dominguez, Inma
Duffard, Rene
Ehle, Matthias
Eiroa, Carlos
Elipe, Sánchez
Elizalde, Emilio
Escapa, Alberto
Estalella, Robert
Esteban, César
Fabregat, Juan
Fabricius, Claus
Falcón Barroso, Jesus
Fernández, David
Fernandez-Figueroa, M.
Ferrandiz, Jose
Ferrer, Martinez
Ferriz Mas, Antonio
Hernanz, Margarita
Herrero, Davó
Hidalgo, Miguel
Isern, Jorge
Israelian, Garik
Jimenez, Mancebo
Jiménez-Vicente, Jorge
Jordi, Carme
José, Jordi
Kessler, Martin
Kidger, Mark
Knapen, Johan
Kretschmar, Peter
Küppers, Michael
Labay, Javier
Lahulla, J.
Lara, Luisa
Lara, Martin
Lázaro Hernando, Carlos
Licandro, Javier
Ling, Josefina
Lodieu, Nicolas
Loiseau, Nora
López de Coca, M. D. P.
López Aguerri, Jose Alfonso
López Arroyo, M.
López Corredoira, Martin
López González, Maria
López Hermoso, Maria
López Moratalla, Teodoro
López Moreno, Jose
López Puertas, Manuel
López Valverde, M.
Luri, Xavier
Mahoney, Terence
Mampaso, Antonio
Manchado, Arturo
Manrique, Alberto
Manteiga Outeiro, Minia
Marcaide, Juan-Maria
Marco, Amparo
Maria de Garcia, J.M.
Marquez, Isabel
Marti, Josep
Martín, Eduardo
Martin Diaz, Carlos
Martin-Pintado, Jesus
Martínez-Delgado, David
Martinez Fiorenzano, Aldo
Martinez Garcia, Vicent
Martínez Pillet, Valentin
Martinez Roger, Carlos
Martinez-Frias, Jesus
Martinez-González, Enrique
Masegosa, Gallego
Massaguer, Josep
Mediavilla, Evencio
Medina, José
Mern Martn, Bruno
Rodríguez, Eloy
Rodríguez Espinosa, Jose
Rodríguez Hidalgo, Inés
Rodríguez-Eillamil, R.
Rodríguez-Franco, Arturo
Rolland, Angel
Romero Pérez, Maria
Ros, Rosa
Rossello, Gaspar
Rozas, Maite
Rubino-Martin, Jose Alberto
Ruiz Cobo, Basilio
Ruiz-Lapuente, Mara
Rutten, René
Sabau-Graziati, Lola
Sala, Ferran
Salazar, Antonio
Salvador-Sole, Eduardo
Sanahuja Parera, Blai
Sánchez, Francisco
Sánchez, Almeida
Sánchez, Filomeno
Sánchez Béjar, Victor
Sánchez Doreste, Néstor
Sánchez-Lavega, Agustin
Sánchez-Saavedra, M.
Sanroma, Manuel
Sansaturio, Maria
Santos-Lleó, Maria
Sanz, Jose
Sanz, Jaume
Schartel, Norbert
Schoedel, Rainer
Sein-Echaluce, M.
Sempere, Maria
Sequeiros, Juan
Serra Ricart, Miquel
Sevilla, Miguel
Shahbaz, Tariq
Simó, Carles
Skillen, Ian
Socas-Navarro, Hector
Solanes, Josep
Sulentic, Jack
Tafalla, Mario
Talavera, Antonio
Tamazian, Vakhtang
Toffolatti, Luigi
Torra, Jordi
José Miguel Torrejón, Jose Miguel
Torrelles, Jose
Torres, Diego
Trigo-Rodrguez, Josep
Trujillo Bueno, Javier
Trujillo Cabrera, Ignacio
Trullols, I.
Unger, Stephen
Vallejo, Miguel
Valtchanov, Ivan

Figueras, Francesca
Floria Peralta, Luis
Florido, Estrella
Folgueira, Marta
Fonseca González, Maria
Fors, Octavi
Fuensalida, Jiménez
Fuente, Asuncion
Gabriel, Carlos
Galadi Enriquez, David
Galan, Maximino
Gallart, Carme
Gallego, Juan Daniel
Metcalfe, Leo
Miralda-Escudé, Jordi
Moles Villamate, Mariano
Molina, Antonio
Mollá, Mercedes
Montes, David
Montesinos Comino, Benjamn
Morales-Duran, Carmen
Moreno, Fernando
Moreno Insertis, Fernando
Moreno Lupiañez, Manuel
Muinos Haro, José
Muñoz Tuñón, Casiana
Varela Perez, Antonia
Vazdekis, Alexandre
Vázquez, Manuel
Verdes-Montenegro, Lourdes
Verdugo, Eva
Vila, Samuel
Vilchez, Jose
Villaver, Eva
Viñuales Gavín, Ederlinda
Vives, Teodoro
Watson, Robert
Zamorano, Jaime
Zapatero-Osorio, Maria Rosa

Sweden

Aalto, Suzanne
Abramowicz, Marek
Andersen, Torben
Ardeberg, Arne
Artymowicz, Pawel
Bååth, Lars
Barklem, Paul
Bergman, Per
Bergström, Lars
Bergvall, Nils
Björnsson, Claes-Ingvar
Black, John
Brandenburg, Axel
Carlqvist, Per
Carlson, Per
Cato, B.
Conway, John
Davidsson, Björn
Dravins, Dainis
Edsjö, Joakim
Edvardsson, Bengt
Ellder, Joel
Elvius, Aina
Eriksson, Kjell
Faelthammar, Carl
Fathi, Kambiz
Feltzing, Sofia
Fransson, Claes
Fredga, Kerstin
Freytag, Bernd
Frisk, Urban
Gahm, Gösta
Gleisner, Hans
Goobar, Ariel
Gustafsson, Bengt
Hartman, Henrik
Heikkilä, Arto
Heiter, Ulrike
Hjalmarson, Ake
Hobbs, David
Höfner, Susanne
Hoeglund, Bertil
Högbom, Jan
Holm, Nils
Holmberg, Gustav
Horellou, Cathy
Hulth, Per
Justtanont-Liseau, Kay
Kiselman, Dan
Kochukhov, Oleg
Kollberg, Erik
Korn, Andreas
Kozma, Cecilia
Kristiansson, Krister
Lagerkvist, Claes-Ingvar
Larsson, Stefan
Larsson-Leander, Gunnar
Lauberts, Andris
Lehnert, Bo
Lindblad, Bertil
Lindblad, Per
Linde, Peter
Lindegren, Lennart
Lindqvist, Michael
Liseau, René
Lodén, Kerstin
Lofdahl, Mats
Lundqvist, Peter
Lundstedt, Henrik
Lundstrom, Ingemar
Magnusson, Per
Mellema, Garrelt
Muller, Sebastien
Nilsson, Hampus
Nordh, Lennart
Nummelin, Albert
Oja, Tarmo
Olberg, Michael
Olofsson, Goeran
Olofsson, Hans
Olofsson, Kjell
Östlin, Göran
Owner-Petersen, Mette
Pearce, Mark
Pellinen-Wannberg, Asta
Persson, Carina
Piskunov, Nikolai
Raadu, Michael
Rickman, Hans
Roennaeng, Bernt
Romeo, Alessandro
Roslund, Curt
Rosquist, Kjell
Rydbeck, Gustaf
Ryde, Felix
Ryde, Nils
Sandqvist, Aage
Scharmer, Goeran
Schöier, Fredrik
Sinnerstad, Ulf
Söderhjelm, Staffan
Sollerman, Jesper
Stenholm, Björn
Sundin, Maria
Sundman, Anita
Thomasson, Magnus
Torkelsson, Ulf
van Groningen, Ernst
Wallinder, Fredrik
Winnberg, Anders
Wramdemark, Stig

Switzerland

Altwegg, Kathrin
Audard, Marc
Behrend, Raoul
Benz, Arnold
Grenon, Michel
Güdel, Manuel
Hauck, Bernard
Huber, Martin
Ramelli, Renzo
Rauscher, Thomas
Revaz, Yves
Rothacher, Markus

Benz, Willy
Beutler, Gerhard
Bianda, Michele
Binggeli, Bruno
Bochsler, Peter
Burki, Gilbert
Buser, Roland
Carollo, Marcella
Charbonnel, Corinne
Cherchneff, Isabelle
Courbin, Frédéric
Courvoisier, Thierry
Csillaghy, Andre
Dessauges-Zavadsky, Miroslava
Dubath, Pierre
Eggenberger, Patrick
Finsterle, Wolfgang
Fluri, Dominique
Froehlich, Claus
Gautschy, Alfred
Geiss, Johannes
Golay, Marcel
Jablonka, Pascale
Jetzer, Philippe
Liebendörfer, Matthias
Lilly, Simon
Locher, Kurt
Maeder, Andre
Martinet, Louis
Mayor, Michel
Megevand, Denis
Meylan, Georges
Meynet, Georges
Moore, Ben
Mowlavi, Nami
Nicolet, Bernard
North, Pierre
Nussbaumer, Harry
Paltani, Stéphane
Pepe, Francesco
Pfenniger, Daniel
Produit, Nicolas
Queloz, Didier
Schaerer, Daniel
Schildknecht, Thomas
Schmid, Hans
Schmutz, Werner
Ségransan, Damien
Simoniello, Rosaria
Steinlin, Uli
Stenflo, Jan
Straumann, Norbert
Tammann, Gustav
Thielemann, Friedrich-Karl
Thomas, Nicolas
Tuerler, Marc
Udry, Stéphane
Verdun, Andreas
von Steiger, Rudolf
Walder, Rolf
Walter, Roland
Woltjer, Lodewijk
Wurz, Peter
Zelenka, Antoine

Tajikistan

Babadjanov, Pulat
Ibadinov, Khursand
Kokhirova, Gulchehra
Minnikulov, Nasriddin
Sahibov, Firuz
Subhon, Ibadov

Thailand

Aungwerojwit, Amornrat
Channok, Chanruangrit
Eungwanichayapant, Anant
Gasiprong, Nipon
Gumjudpai, Burin
Khumlemlert, Thiranee
Komonjinda, Siramas
Kramer, Busaba
Muanwong, Orrarujee
Pacharin-Tanakun, P.
Ruffolo, David
Sáiz, Alejandro
Songsathaporn, Ruangsak
Soonthornthum, Boonrucksar

Turkey

Akan, Mustafa
Alpar, Mehmet
Aslan, Zeki
Atac, Tamer
Bolcal, Cetin
Demircan, Osman
Derman, I.
Enginol, Turan
Ercan, Enise
Esenoğlu, Hasan
Eskioglu, A.
Evren, Serdar
Gulmen, Omur
Gulsecen, Hulusi
Gün, Gulnur
Ibanoglu, Cafer
Kalemci, Emrah
Kalomeni, Belinda
Karaali, Salih
Keskin, Varol
Kirbiyik, Halil
Kiziloglu, Nilgun
Kiziloglu, Umit
Kocer, Dursun
Mentese, Huseyin
Özel, Mehmet
Özgü, Atila
Ozkan, Mustafa
Pekünlü, E.
Saygac, Ahmet
Selam, Selim
Sezer, Cengiz
Taş, Günay
Tektunali, H.
Topaktas, Latif
Tunca, Zeynel
Yakut, Kadri
Yeşilyurt, Ibrahim
Yilmaz, Fatma
Yüce, Kutluay

Ukraine

Akimov, Leonid
Alexandrov, Alexander
Alexandrov, Yuri
Khmil, Sergiy
Khoda, Oleg
Kiselev, Nikolai
Pronik, Iraida
Pronik, Vladimir
Protsyuk, Yuri

Andrienko, Dmitry
Andrievsky, Sergei
Andronov, Ivan
Antonyuk, Kirill
Babin, Arthur
Baranovsky, Edward
Baransky, Olexander
Belskaya, Irina
Berczik, Peter
Bolotin, Sergei
Bolotina, Olga
Borysenko, Sergiy
Bruns, Andrey
Butkovskaya, Varvara
Chubko, Larysa
Churyumov, Klim
Danylevsky, Vassyl
Denishchik, Yurii
Dlugach, Zhanna
Dorokhova, Tetyana
Doroshenko, Valentina
Dragunova, Alina
Dudinov, Volodymyr
Duma, Dmitrij
Efimenko, Volodymyr
Efimov, Yuri
Elyiv, Andrii
Epishev, Vitali
Fedorov, Petro
Fedorova, Elena
Fomin, Piotr
Fomin, Valery
Gershberg, R.
Glazunova, Ljudmila
Golovatyj, Volodymyr
Gopasyuk, Olga
Gopka, Vera
Gorbanev, Jury
Gozhy, Adam
Grinin, Vladimir
Guseva, Natalia
Guziy, Sergiy
Halevin, Alexander
Hnatyk, Bohdan
Hudkova, Ludmila
Ivanchuk, Victor
Ivanova, Aleksandra
Ivantsov, Anatoliy
Ivchenko, Vasily
Izotov, Yuri
Izotova, Iryna
Kablak, Nataliya
Kalenichenko, Valentin
Karachentseva, Valentina
Karetnikov, Valentin
Karpov, Nikolai
Kazantsev, Anatolii
Kazantseva, Liliya
Kharchenko, Nina
Kharin, Arkadiy
Kislyuk, Vitalij
Klymyshyn, I.
Kolesnikov, Sergey
Kolomiyets, Svitlana
Kondrashova, Nina
Konovalenko, Alexander
Kontorovich, Victor
Korotin, Sergey
Korsun, Alla
Korsun, Pavlo
Koshkin, Nikolay
Kostik, Roman
Kotov, Valery
Koval, I.
Kovalchuk, George
Kovtyukh, Valery
Kozak, Lyudmyla
Kozak, Pavlo
Kravchuk, Sergei
Kruchinenko, Vitaliy
Krugly, Yurij
Kryshtal, Alexander
Kryvdyk, Volodymyr
Kryvodubskyj, Valery
Kudashkina, Larisa
Kudrya, Yury
Kurochka, Evgenia
Kuzkov, Vladimir
Lazorenko, Peter
Leiko, Uliana
Litvinenko, Leonid
Lozitskij, Vsevolod
Lukyanyk, Igor
Lupishko, Dmitrij
Lyubchik, Yuri
Lyubimkov, Leonid
Maigurova, Nadiia
Makarenko, Ekaterina
Marsakova, Vladislava
Mel'nik, Valentin
Melnyk, Olga
Men, Anatolij
Minakov, Anatoliy
Miroshnichenko, Alla
Mishenina, Tamara
Morozhenko, A.
Nazarenko, Victor
Novosyadlyj, Bohdan
Parnovsky, Sergei
Pavlenko, Elena
Pavlenko, Yakov
Petrov, G.
Petrov, Peter
Petrova, Svetlana
Pilyugin, Leonid
Pinigin, Gennadiy
Plachinda, Sergei
Polosukhina-Chuvaeva, Nina
Pozhalova, Zhanna
Prokof'eva, Valentina
Psaryov, Volodymyr
Pugach, Alexander
Rachkovsky, D.
Rashkovskij, Sergey
Reshetnyk, Volodymyr
Romanchuk, Pavel
Romanov, Yuri
Romanyuk, Yaroslav
Rosenbush, Vera
Rosenbush, Alexander
Rostopchina, Alla
Ryabov, Michael
Sergeev, Aleksandr
Sergeev, Sergey
Shakhov, Boris
Shakhovskaya, Nadejda
Shakhovskoj, Nikolay
Shchukina, Nataliya
Sheminova, Valentina
Shevchenko, Vasilij
Shkuratov, Yurii
Shulga, Oleksandr
Shulga, Valery
Silich, Sergey
Sizonenko, Yuri
Skulskyj, Mychajlo
Sobolev, Yakov
Sodin, Leonid
Sokolov, Konstantin
Stepanian, Natali
Steshenko, N.
Stodilka, Myroslav
Tarady, Vladimir
Tarashchuk, Vera
Tarasov, Anatolii
Tarasova, Taya
Terebizh, Valery
Tsap, Teodor
Tsap, Yuri
Tugay, Anatoliy
Udovichenko, Sergei
Ulyanov, Oleg
Usenko, Igor
Vavilova, Iryna
Vidmachenko, Anatoliy
Voitsekhovska, Anna
Voloschuk, Yuri
Volvach, Alexander
Volyanskaya, Margarita
Voroshilov, Volodymyr
Yanovitskij, Edgard
Yatsenko, Anatolij
Yatskiv, Yaroslav
Yukhimuk, Adam
Yushchenko, Alexander
Zakhozhaj, Volodimir
Zhdanov, Valery
Zheleznyak, Alexander
Zhilyaev, Boris

United Kingdom

Aarseth, Sverre
Adamson, Andrew
Ade, Peter
Afram, Nadine
Aggarwal, Kanti
Aitken, David
Albinson, James
Alexander, Paul
Allan, Peter
Allen, Anthony
Allington-Smith, Jeremy
Almaini, Omar
Alsabti, Abdul Athem
Andrews, David
Aragón-Salamanca, Alfonso
Archontis, Vasilis
Ardavan, Houshang
Argyle, Robert
Asher, David
Baddiley, Christopher
Bagnulo, Stefano
Bailey, Mark
Baldry, Ivan
Baldwin, John
Balikhin, Michael
Bamford, Steven
Barber, Robert
Barclay, Charles
Barlow, Michael
Barocas, Vinicio
Barrow, John
Barstow, Martin
Baruch, John
Baskill, Darren
Bastin, John
Bates, Brian
Bayet, Estelle
Beale, John
Bell, Steven
Bell Burnell, Jocelyn
Bennett, Jim
Bentley, Robert
Berger, Mitchell
Bersier, David
Beurle, Kevin
Bewsher, Danielle
Bingham, Robert
Binney, James
Birkinshaw, Mark
Blackman, Clinton
Blackwell, Donald
Blundell, Katherine
Bode, Michael
Boksenberg, Alec
Boles, Thomas
Bonnor, W.
Botha, Gert
Grainge, Keith
Gray, Norman
Gray, Meghan
Green, Robin
Green, Simon
Green, David
Green, Anne
Griffin, Roger
Griffiths, William
Gull, Stephen
Gunn, Alastair
Guthrie, Bruce
Haehnelt, Martin
Haniff, Christopher
Hardcastle, Martin
Harper, David
Harra, Louise
Harris-Law, Stella
Harrison, Richard
Hartquist, Thomas
Hassall, Barbara
Haswell, Carole
Hatchell, Jennifer
Hawarden, Timothy
Hayward, John
Heavens, Alan
Heggie, Douglas
Hellier, Coel
Hendry, Martin
Hewett, Paul
Hewish, Antony
Hide, Raymond
Hilditch, Ronald
Hills, Richard
Hingley, Peter
Hirschi, Raphael
Hoare, Melvin
Hohenkerk, Catherine
Holloway, Nigel
Hood, Alan
Horne, Keith
Horner, Jonathan
Houdek, Gunter
Hough, James
Howarth, Ian
Hughes, David
Hughes, David
Hunt, Garry
Hütsi, Gert
Hysom, Edmund
Irwin, Michael
Irwin, Patrick
Ivison, Robert
Jackson, John
Jackson, Neal
James, John
James, Richard
Osborne, John
Osborne, Julian
Oudmaijer, Rene
Padman, Rachael
Page, Clive
Page, Mathew
Palmer, Philip
Papaloizou, John
Parker, Neil
Parkinson, John
Parnell, Clare
Peach, Gillian
Peach, John
Peacock, John
Pearce, Frazer
Pearce,Gillian
Pedlar, Alan
Penny, Alan
Penston, Margaret
Petford, A.
Petkaki, Panagiota
Pettini, Max
Phillipps, Steven
Phillips, Kenneth
Pijpers, Frank
Pilkington, John
Pillinger, Colin
Pittard, Julian
Podsiadlowski, Philipp
Pollacco, Don
Ponman, Trevor
Ponsonby, John
Pont, Frdric
Pooley, Guy
Pounds, Kenneth
Priest, Eric
Pringle, James
Prinja, Raman
Proctor, Michael
Pye, John
Quenby, John
Raine, Derek
Rawlings, Jonathan
Raychaudhury, Somak
Rees, Martin
Regnier, Stéphane
Richer, John
Rijnbeek, Richard
Riley, Julia
Roberts, Timothy
Robertson, John
Robertson, Norna
Robson, Ian
Rowan-Robinson, Michael
Rowson, Barrie
Roxburgh, Ian
Roy, Archie

Boyd, David
Boyle, Stephen
Brand, Peter
Branduardi-Raymont, Graziella
Bridle, Sarah
Brinks, Elias
Bromage, Barbara
Bromage, Gordon
Brooke, John
Brookes, Clive
Brown, John
Browne, Ian
Browning, Philippa
Bunker, Andrew
Bureau, Martin
Burgess, David
Burton, William
Buscher, David
Butchins, Sydney
Butler, Christopher
Cameron, Andrew
Cappellari, Michele
Cargill, Peter
Carr, Bernard
Carson, T.
Carswell, Robert
Carter, David
Catchpole, Robin
Chakrabarty, Dalia
Chapman, Sandra
Christou, Apostolos
Cioni, Maria-Rosa
Clark, David
Clarke, David
Clegg, Peter
Clegg, Robin
Clowes, Roger
Clube, S.
Coe, Malcolm
Cole, Shaun
Coles, Peter
Conway, Andrew
Conway, Robin
Cooper, Nicholas
Couper, Heather
Crawford, Carolin
Crawford, Ian
Cropper, Mark
Crowther, Paul
Cruise, Adrian
Culhane, John
Dalla, Silvia
Damian, Audley
Davidson, William
Davies, John
Davies, Melvyn
Davies, Rodney
Davies, Roger
Davis, A. E. L.
Jameson, Richard
Jardine, Moira
Jeffery, Christopher
Jenkins, Charles
Johnstone, Roderick
Jones, Barrie
Jones, Derek
Jones, Janet
Jordan, Carole
Jorden, Paul
Kaiser, Christian
Kennicutt, Robert
Khan, J
Khosroshahi, Habib
King, Andrew
Kingston, Arthur
Knapp, Johannes
Kobayashi, Shiho
Kolb, Ulrich
Kollerstrom, Nicholas
Kontar, Eduard
Krajnovic, Davor
Kramer, Michael
Kroto, Harold
Kuin, Nicolaas
Kunz, Martin
Kurtz, Donald
Labrosse, Nicolas
Lacey, Cedric
Lahav, Ofer
Lapington, Jonathan
Lasenby, Anthony
Leahy, J.
Liddle, Andrew
Lloyd, Christopher
Lloyd, Myfanwy
Lloyd Evans, Thomas
Longair, Malcolm
Love, Gordon
Loveday, Jon
Lovell, Bernard
Lucey, John
Lucy, Leon
Lynden-Bell, Donald
Lyne, Andrew
Lyon, Ian
Maartens, Roy
MacCallum, Malcolm
Maccarone, Thomas
MacDonald, Geoffrey
Maciejewski, Witold
Mackay, Duncan
MacKinnon, Alexander
Maddison, Ronald
Maddox, Stephen
Magorrian, Stephen
Major, John
Mann, Robert
Mao, Shude
Ruffert, Maximilian
Ruggles, Clive
Russel, Sara
Ryan, Sean
Sahln, Martin
Sakano, Masaaki
Sansom, Anne
Sarre, Peter
Satterthwaite, Gilbert
Schilizzi, Richard
Schroeder, Klaus
Schwartz, Steven
Scott, Paul
Serjeant, Stephen
Seymour, P.
Shanklin, Jonathan
Shanks, Thomas
Sharples, Ray
Shone, David
Shukurov, Anvar
Silk, Joseph
Simnett, George
Simpson, Allen
Sims, Mark
Skilling, John
Sleath, John
Smail, Ian
Smalley, Barry
Smith, Michael
Smith, Robert
Smith, Rodney
Smyth, Michael
Somerville, William
Sorensen, Soren-Aksel
Spencer, Ralph
Stallard, Thomas
Steel, Duncan
Stephenson, F.
Stevens, Ian
Steves, Bonita
Stewart, John
Stewart, Paul
Stickland, David
Stolyarov, Vladislav
Suda, Takuma
Sylvester, Roger
Tavakol, Reza
Taylor, Donald
Taylor, Fredric
Taylor, John
Tedds, Jonathan
Teixeira, Teresa
Tennyson, Jonathan
Terlevich, Elena
Terlevich, Roberto
Thomas, David
Thomas, Peter
Thomasson, Peter
Thomson, Robert

de Grijs, Richard
Dent, William
Dewhirst, David
Dhillon, Vikram
Diamond, Philip
Diego, Francisco
Dipper, Nigel
Disney, Michael
Diver, Declan
Donnison, John
Dormand, John
Doyle, John
Drew, Janet
Driver, Simon
Duffett-Smith, Peter
Dufton, Philip
Dunlop, Storm
Dworetsky, Michael
Eales, Stephen
Edge, Alastair
Edmunds, Michael
Efstathiou, George
Elliott, Kenneth
Elsworth, Yvonne
Emerson, James
Erdelyi, Robert
Evans, Aneurin
Evans, Christopher
Evans, Dafydd
Evans, Michael
Evans, Wyn
Eyres, Stewart
Fabian, Andrew
Falle, Samuel
Ferguson, Annette
Ferreras, Ignacio
Field, J.
Fitzsimmons, Alan
Fletcher, Lyndsay
Flower, David
Fludra, Andrzej
Fong, Richard
Fraser, Helen
Frenk, Carlos
Fuller, Gary
Furniss, Ian
Fyfe, Duncan
Gnsicke, Boris
Gibson, Brad
Gill, Peter
Gilmore, Gerard
Gledhill, Timothy
Glencross, William
Godwin, Jon
González-Solares, Eduardo
Goodwin, Simon
Gough, Douglas
Grady, Monica
Graffagnino, Vito
Marek, John
Marsh, Thomas
Marshall, Kevin
Masheder, Michael
Mason, Helen
Mason, Keith
Mason, John
Massey, Robert
Mathioudakis, Mihalis
Matsuura, Mikako
Mattila, Seppo
Mccombie, June
McDonnell, J.
McHardy, Ian
McMahon, Richard
McNally, Derek
Meadows, A.
Meikle, William
Meiksin, Avery
Merrifield, Michael
Mestel, Leon
Miles, Howard
Millar, Thomas
Miller, Steven
Miller, John
Mitton, Simon
Mitton, Jacqueline
Moffatt, Henry
Moore, Patrick
Moore, Daniel
Morison, Ian
Morris, David
Morris, Rhys
Morris, Simon
Morrison, Leslie
Moss, Christopher
Moss, David
Murdin, Paul
Murray, John
Murray, Carl
Murray, Andrew
Murtagh, Fionn
Muxlow, Thomas
Nakariakov, Valery
Napier, William
Napiwotzki, Ralf
Naylor, Tim
Nelson, Alistair
Nelson, Richard
Neukirch, Thomas
Newton, Gavin
Nicolson, Iain
Norton, Andrew
O'Brien, Paul
O'Brien, Tim
Oliveira, Joana
Oliver, Sebastian
Onuora, Lesley
Orford, Keith
Tobias, Steven
Tout, Christopher
Tritton, Keith
Tritton, Susan
Trotta, Roberto
Tsiklauri, David
Tworkowski, Andrzej
van Leeuwen, Floor
van Loon, Jacco
Vaughan, Simon
Vekstein, Gregory
Verwichte, Erwin
Vink, Jorick
Viti, Serena
Waddington, Ian
Walker, David
Walker, Edward
Walker, Helen
Wall, Jasper
Wallace, Patrick
Wallis, Max
Walton, Nicholas
Ward, Martin
Ward-Thompson, Derek
Warner, Peter
Warwick, Robert
Watson, Michael
Watt, Graeme
Webster, Adrian
Weiss, Nigel
Wheatley, Peter
White, Glenn
Whiting, Alan
Whitworth, Anthony
Wickramasinghe, N.
Wilkins, George
Wilkinson, Althea
Wilkinson, Peter
Williams, David
Williams, Iwan
Williams, Peredur
Williams, Robin
Willis, Allan
Willmore, A.
Willstrop, Roderick
Wilson, Lionel
Woan, Graham
Wolfendale, Arnold
Wolstencroft, Ramon
Wood, Roger
Woolfson, Michael
Worrall, Diana
Worswick, Susan
Wright, Ian
Yallop, Bernard
Zane, Silvia
Zarnecki, John
Zharkova, Valentina
Zijlstra, Albert

United States

Aannestad, Per
Abbett, William
Abbott, David
Ables, Harold
Abramenko, Valentina
Abt, Helmut
Accomazzi, Alberto
Acton, Loren
Adams Jr., James
Adams, Fred
Adams, Mark
Adams, Thomas
Adelman, Saul
Adler, David
A'Hearn, Michael
Ahluwalia, Harjit
Ahmad, Imad
Aizenman, Morris
Ake III, Thomas
Akeson, Rachel
Albers, Henry
Alexander, David
Alexander, Joseph
Allan, David
Allen Jr., John
Allen, Lori
Allen, Ronald
Allende Prieto, Carlos
Aller, Hugh
Aller, Margo
Alley, Carrol
Allison, Michael
Aloisi, Alessandra
Altrock, Richard
Altschuler, Daniel
Altschuler, Martin
Altunin, Valery
Alvarez del Castillo, Elizabeth
Alvarez, Manuel
Ambruster, Carol
Anand, S.
Andersen, Morten
Anderson, Christopher
Anderson, Kinsey
Anderson, Kurt
Andersson, Bengt
Angel, J.
Angione, Ronald
Anosova, Joanna
Anthony-Twarog, Barbara
Antiochos, Spiro
Aoki, Kentaro
Apai, Daniel
Appleby, John
Appleton, Philip
Archinal, Brent
Ardila, David
Arion, Douglas
Haghighipour, Nader
Hagyard, Mona
Haisch Jr., Karl
Haisch, Bernard
Hajian, Arsen
Hakkila, Jon
Hale, Alan
Hall, Donald
Hall, Douglas
Hallam, Kenneth
Hamilton, Andrew
Hamilton, Douglas
Hammel, Heidi
Hammond, Gordon
Hanisch, Robert
Hankins, Timothy
Hanner, Martha
Hansen, Carl
Hansen, Richard
Hanson, Margaret
Hanson, Robert
Hapke, Bruce
Hardebeck, Ellen
Hardee, Philip
Harmer, Charles
Harmer, Dianne
Harms, Richard
Harnden Jr., Frank
Harnett, Julienne
Harper, Graham
Harrington, J.
Harris, Alan
Harris, Daniel
Harris, Hugh
Hart, Michael
Harten, Ronald
Hartigan, Patrick
Hartkopf, William
Hartmann, Dieter
Hartmann, Lee
Hartmann, William
Hartoog, Mark
Harvel, Christopher
Harvey, Gale
Harvey, John
Harvey, Paul
Harwit, Martin
Hasan, Hashima
Haschick, Aubrey
Hathaway, David
Hatzes, Artie
Hauser, Michael
Havlen, Robert
Hawley, John
Hawley, Suzanne
Hayashi, Masahiko
Hayashi, Saeko
Hayes, Donald
Patten, Brian
Pauls, Thomas
Pavlidou, Vasiliki
Pavlov, George
Payne, David
Peale, Stanton
Pearson, Kevin
Pearson, Timothy
Peebles, P.
Peery, Benjamin
Pellerin Jr., Charles
Pence, William
Pendleton, Yvonne
Penzias, Arno
Perez, Mario
Perkins, Francis
Perley, Richard
Perrin, Marshall
Perry, Peter
Pesch, Peter
Peters, Geraldine
Peters, William
Peterson, Bradley
Peterson, Charles
Peterson, Deane
Peterson, Laurence
Peterson, Ruth
Petre, Robert
Petrie, Gordon
Petro, Larry
Petrosian, Vahe
Pettengill, Gordon
Petuchowski, Samuel
Pevtsov, Alexei
Pfeiffer, Raymond
Phelps, Randy
Philip, A.G.
Phillips, Thomas
Pickles, Andrew
Pier, Jeffrey
Pierce, David
Pihlstrm, Ylva
Pihlstrm, Ylva
Pilachowski, Catherine
Pilcher, Carl
Pines, David
Pinsonneault, Marc
Pinto, Philip
Pipher, Judith
Pirzkal, Norbert
Pisano, Daniel
Pittichova, Jana
Platais, Imants
Plavchan, Jr., Peter
Plavec, Zdenka
Pneuman, Gerald
Podesta, John
Pogge, Richard

Armandroff, Taft
Armstrong, John
Arnett, W.
Arnold, James
Arny, Thomas
Arons, Jonathan
Arthur, David
Aschwanden, Markus
Ashby, Neil
Aspin, Colin
Assousa, George
Athay, R.
Atkinson, David
Atreya, Sushil
Auer, Lawrence
Augason, Gordon
Avrett, Eugene
Axon, David
Ayres, Thomas
Backer, Donald
Bagnuolo Jr., William
Bagri, Durgadas
Bahcall, Neta
Bailey Mathae, Katherine
Bailyn, Charles
Baird, Scott
Baker, Andrew
Baker, Joanne
Bakker, Eric

Balachandran, Suchitra
Balasubramaniam, K.
Balbus, Steven
Baldwin, Jack
Baldwin, Ralph
Balick, Bruce
Baliunas, Sallie
Ball, John
Bally, John
Balonek, Thomas
Balser, Dana
Bandermann, L.
Bangert, John
Bania, Thomas
Bardeen, James
Barden, Samuel
Barger, Amy
Baring, Matthew
Barker, Edwin
Barker, Timothy
Barlow, Nadine
Barnbaum, Cecilia
Barnes III, Thomas
Barnes, Aaron
Barnes, Graham
Barnes, Sydney
Baron, Edward
Barrett, Paul
Barsony, Mary
Barth, Aaron

Haymes, Robert
Haynes, Martha
Heacox, William
Heap, Sara
Heasley, James
Hecht, James
Heckathorn, Harry
Heckman, Timothy
Heeschen, David
Heger, Alexander
Hegyi, Dennis
Heiles, Carl
Hein, Righini
Heiser, Arnold
Helfand, David
Helfer, H.
Heller, Clayton
Helmken, Henry
Helou, George
Hemenway, Mary
Hemenway, Paul
Henden, Arne
Hendrix, Amanda
Henning, Patricia
Henriksen, Mark
Henry, Richard B. C.
Henry, Richard Conn
Herbig, George
Herbst, Eric
Herbst, William
Herczeg, Tibor
Hernquist, Lars
Hershey, John
Hertz, Paul
Hewitt, Adelaide
Hewitt, Anthony
Heyer, Mark
Hibbard, John
Hibbs, Albert
Hicks, Amalia
Hildebrand, Roger
Hildner, Ernest
Hill, Frank
Hill, Grant
Hill, Henry
Hillenbrand, Lynne
Hilliard, Ron
Hillier, Desmond
Hills, Jack
Hillwig, Todd
Hilton, James
Hindsley, Robert
Hinkle, Kenneth
Hinners, Noel
Hintz, Eric
Hintzen, Paul
Ho, Luis
Ho, Paul
Hoard, Donald
Hobbs, Lewis

Poland, Arthur
Polidan, Ronald
Pompea, Stephen
Pontoppidan, Klaus
Pott, Jorg-Uwe
Potter, Andrew
Pottschmidt, Katja
Pound, Marc
Pouquet, Annick
Pradhan, Anil
Prasad, Sheo
Pravdo, Steven
Press, William
Preston, George
Preston, Robert
Price, Michael
Price, R.
Price, Stephan
Pritzl, Barton
Probstein, R.
Prochaska, Jason
Proffitt, Charles
Protheroe, William
Pryor, Carlton
Puche, Daniel
Puetter, Richard
Puschell, Jeffery
Pyper Smith, Diane
Quirk, William
Rabin, Douglas
Radford, Simon
Radick, Richard
Rafert, James
Rafferty, Theodore
Rajagopal, Jayadev
Ralchenko, Yuri
Ramsey, Lawrence
Rand, Richard
Rank, David
Rankin, Joanna
Ransom, Scott
Rasio, Frederic
Rasmussen, Jesper
Ratcliff, Stephen
Ray, James
Ray, Paul
Raymond, John
Reach, William
Readhead, Anthony
Reames, Donald
Reasenberg, Robert
Rebull, Luisa
Rector, Travis
Reed, Bruce
Reichert, Gail
Reid, Iain
Reid, Mark
Reinard, Alysha
Reipurth, Bo
Reitsema, Harold

Barton, Elizabeth
Barvainis, Richard
Basart, John
Bash, Frank
Basri, Gibor
Bastian, Timothy
Basu, Sarbani
Batchelor, David
Batson, Raymond
Bauer, Carl
Bauer, James
Bauer, Wendy
Baum, Stefi
Baum, William
Baustian, W.
Bautz, Laura
Baym, Gordon
Beasley, Anthony
Beavers, Willet
Bechtold, Jill
Becker, Peter
Becker, Robert A.
Becker, Robert H.
Becker, Stephen
Beckers, Jacques
Becklin, Eric
Beckwith, Steven
Beebe, Herbert
Beebe, Reta
Beer, Reinhard
Beers, Timothy
Begelman, Mitchell
Beiersdorfer, Peter
Bell III, James
Bell, Barbara
Bell, Jeffrey
Bell, Roger
Belserene, Emilia
Belton, Michael
Bender, Peter
Benedict, George
Benevolenskaya, Elena
Benford, Gregory
Bennett, Charles
Bennett, David
Benson, James
Bentz, Misty
Berendzen, Richard
Berg, Richard
Berge, Glenn
Bergin, Edwin
Bergstrahl, Jay
Berman, Robert
Berman, Vladimir
Bernat, Andrew
Bertschinger, Edmund
Bettis, Dale
Bhavsar, Suketu
Bianchi, Luciana
Bidelman, William

Hobbs, Robert
Hockey, Thomas
Hodapp, Klaus-Werner
Hodge, Paul
Hoeflich, Peter
Hoeksema, Jon
Hoessel, John
Hoff, Darrel
Hoffman, Jeffrey
Hofner, Peter
Hogan, Craig
Hogg, David
Holberg, Jay
Hollenbach, David
Hollis, Jan
Hollowell, David
Hollweg, Joseph
Holman, Gordon
Holt, Stephen
Holz, Daniel
Holzer, Thomas
Honeycutt, R.
Hora, Joseph
Horch, Elliott
Horner, Scott
Hornschemeier, Ann
Horowitz, Paul
Houck, James
Houdashelt, Mark
Houk, Nancy
House, Lewis
Howard, Andrew
Howard, Robert
Howard, Sethanne
Howard, W.
Howard, William
Howell, Ellen
Howell, Steve
Hrivnak, Bruce
Hu, Esther
Huang, Jiasheng
Hubbard, William
Hubeny, Ivan
Huchra, John
Hudson, Hugh
Hudson, Reggie
Huebner, Walter
Huenemoerder, David
Huggins, Patrick
Hughes, John
Hughes, Philip
Huguenin, G.
Humphreys, Elizabeth
Humphreys, Roberta
Hunten, Donald
Hunter, Christopher
Hunter, Deidre
Hunter, James
Hunter, Todd
Huntress, Wesley

Rense, William
Rettig, Terrence
Revelle, Douglas
Reyes, Francisco
Reynolds, Ronald
Reynolds, Stephen
Rhoads, James
Rhodes Jr., Edward
Rich, Robert
Richards, Mercedes
Richardson, R.
Richstone, Douglas
Rickard, James
Rickard, Lee
Ricker, George
Ricotti, Massimo
Riddle, Anthony
Ridgway, Stephen
Ridgway, Susan
Riegel, Kurt
Riegler, Guenter
Riley, Pete
Rindler, Wolfgang
Ringwald, Frederick
Rivolo, Arthur
Roark, Terry
Roberge, Wayne
Roberts Jr., Lewis
Roberts Jr., William
Roberts, David
Roberts, Douglas
Roberts, Mallory
Roberts, Morton
Robertson, Douglas
Robinson Jr., Richard
Robinson, Andrew
Robinson, Edward
Robinson, I.
Robinson, Leif
Robinson, Lloyd
Roddier, Claude
Roddier, Francois
Rodman, Richard
Roeder, Robert
Roellig, Thomas
Roemer, Elizabeth
Rogers, Alan
Rogers, Forrest
Rogerson, John
Rogstad, David
Roman, Nancy
Romani, Roger
Romanishin, William
Romano-Diaz, Emilio
Romer, Anita
Roming, Peter
Romney, Jonathan
Rood, Herbert
Rood, Robert
Roosen, Robert

Bieging, John
Bignell, R.
Billingham, John
Binzel, Richard
Biretta, John
Bjorkman, Jon
Bjorkman, Karen
Black, Adam
Blades, John
Blair, Guy
Blair, William
Blakeslee, John
Blanco, Victor
Blandford, Roger
Blasius, Karl
Bless, Robert
Blitz, Leo
Bloemhof, Eric
Blondin, John
Blum, Robert
Blumenthal, George
Boboltz, David
Bobrowsky, Matthew
Bock, Douglas
Boden, Andrew
Bodenheimer, Peter
Boesgaard, Ann
Boeshaar, Gregory
Bogdan, Thomas
Boggess, Albert
Boggess, Nancy
Bohannan, Bruce
Bohlin, J.
Bohlin, Ralph
Bohm, Karl-Heinz
Bohm-Vitense, Erika
Boice, Daniel
Bolatto, Alberto
Boley, Forrest
Bond, Howard
Bonsack, Walter
Book, David
Bookbinder, Jay
Bookmyer, Beverly
Booth, Andrew
Bopp, Bernard
Bord, Donald
Borderies, Nicole
Borkowski, Kazimierz
Borne, Kirk
Bornmann, Patricia
Boroson, Bram
Borozdin, Konstantin
Boss, Alan
Bourke, Tyler
Bowell, Edward
Bowen, David
Bowen, George
Bower, Gary
Bower, Geoffrey

Hurford, Gordon
Hurley, Kevin
Hurwitz, Mark
Hut, Piet
Hutter, Donald
Huynh, Minh
Hyun, Jong-June
Ianna, Philip
Iben Jr., Icko
Ibrahim, Alaa
Ignace, Richard
Illing, Rainer
Illingworth, Garth
Ilya, Mandel
Imamura, James
Imhoff, Catherine
Impey, Christopher
Ingerson, Thomas
Inglis, Michael
Ipatov, Sergei
Ipser, James
Iro, Nicolas
Irvine, William
Ivezic, Zeljko
Jackson, Bernard
Jackson, James
Jackson, Peter
Jackson, William
Jacobs, Kenneth
Jacobson, Robert
Jacoby, George
Jaffe, Daniel
Janes, Kenneth
Janiczek, Paul
Jannuzi, Buell
Janssen, Michael
Jarrett, Thomas
Jastrow, Robert
Jedicke, Robert
Jefferies, Stuart
Jefferys, William
Jenkins, Edward
Jenkner, Helmut
Jenner, David
Jenniskens, Petrus
Jevremovic, Darko
Jewell, Philip
Johnson, Donald
Johnson, Fred
Johnson, Hollis
Johnson, Hugh
Johnson, Jennifer
Johnson, Thomas
Johnson, Torrence
Johnston, Kenneth
Jokipii, J.
Joner, Michael
Jones, Barbara
Jones, Burton
Jones, Christine

Rose, James
Rose, William
Rosen, Edward
Rosendhal, Jeffrey
Rosner, Robert
Ross, Dennis
Rothberg, Barry
Rots, Arnold
Rountree, Janet
Rouse, Carl
Routly, Paul
Roy, Jean-René
Rubenstein, Eric
Rubin, Robert
Rubin, Vera
Rudnick, Lawrence
Rugge, Hugo
Rule, Bruce
Russell, Christopher
Russell, Jane
Rust, David
Ruszkowski, Mateusz
Rybicki, George
Sackmann, Ingrid
Sadun, Alberto
Safko, John
Sage, Leslie
Saha, Abhijit
Sahai, Raghvendra
Sahu, Kailash
Saito, Yoshihiko
Sakai, Shoko
Salama, Farid
Salas, Luis
Salisbury, J.
Salstein, David
Salter, Christopher
Salzer, John
Samarasinha, Nalin
Samec, Ronald
Sampson, Russell
Sandage, Allan
Sandell, Goran
Sanders, David
Sanders, Walt
Sanders, Wilton
Sandford, Maxwell
Sandford, Scott
Sandmann, William
Sanyal, Ashit
Sarajedini, Ata
Sarazin, Craig
Sargent, Annelia
Sargent, Wallace
Sarma, Anuj
Sartori, Leo
Sasaki, Toshiyuki
Saslaw, William
Sasselov, Dimitar
Savage, Blair

Bowers, Phillip
Bowyer, C.
Boyce, Peter
Boyle, Richard
Boynton, Paul
Bradley, Arthur
Bradley, Paul
Bradstreet, David
Branch, David
Brandt, John
Brandt, William
Branscomb, L.
Braun, Douglas
Breakiron, Lee
Brecher, Aviva
Brecher, Kenneth
Breckinridge, James
Bregman, Joel
Brickhouse, Nancy
Bridle, Alan
Britt, Daniel
Broadfoot, A.
Broderick, John
Brodie, Jean
Brosius, Jeffrey
Broucke, Roger
Brown, Alexander
Brown, Douglas
Brown, Robert
Brown, Robert
Brown, Thomas
Brown, Warren
Brownlee, Donald
Brownlee, Robert
Brucato, Robert
Bruenn, Stephen
Brugel, Edward
Bruhweiler, Frederick
Bruner, Marilyn
Bruning, David
Brunk, William
Buchler, J.
Buff, James
Buhl, David
Buie, Marc
Bunner, Alan
Buote, David
Buratti, Bonnie
Burbidge, Eleanor
Burbidge, Geoffrey
Burgasser, Adam
Burke, Bernard
Burkhead, Martin
Burlaga, Leonard
Burns Jr., Jack
Burns, Joseph
Burrows, Adam
Burrows, David
Burstein, David
Burton, W.

Jones, Dayton
Jones, Eric
Jones, Frank
Jones, Harrison
Jones, Thomas
Jordan, Stuart
Jorgensen, Anders
Jorgensen, Inger
Jørgensen, Jes
Joselyn, Jo
Joseph, Charles
Joseph, Robert
Joss, Paul
Joy, Marshall
Judge, Philip
Junkkarinen, Vesa
Junor, William
Jura, Michael
Jurgens, Raymond
Kafatos, Menas
Kaftan, May
Kahler, Stephen
Kaiser, Mary
Kaitchuck, Ronald
Kaler, James
Kalirai, Jason
Kalkofen, Wolfgang
Kammeyer, Peter
Kamp, Lucas
Kane, Sharad
Kane, Stephen
Kanekar, Nissim
Kaplan, George
Kaplan, Lewis
Karovska, Margarita
Karp, Alan
Karpen, Judith
Kassim, Namir
Katz, Jonathan
Kaufman, Michele
Kawaler, Steven
Kay, Laura
Kaye, Anthony
Keel, William
Keene, Jocelyn
Keil, Klaus
Keil, Stephen
Keller, Charles
Keller, Geoffrey
Kellermann, Kenneth
Kellogg, Edwin
Kelly, Brandon
Kemball, Athol
Kent, Stephen
Kenyon, Scott
Khare, Bishun
Kielkopf, John
Killen, Rosemary
Kilston, Steven

Savedoff, Malcolm
Savin, Daniel
Sawyer, Constance
Scalo, John
Scargle, Jeffrey
Schaefer, Bradley
Schaefer, Martha
Schaller, Emily
Schatten, Kenneth
Schechter, Paul
Scheeres, Daniel
Scherb, Frank
Scherrer, Philip
Schild, Rudolph
Schiller, Stephen
Schlegel, Eric
Schleicher, David
Schlesinger, Barry
Schloerb, F.
Schmahl, Edward
Schmalberger, Donald
Schmelz, Joan
Schmidt, Edward
Schmidt, Maarten
Schmidtke, Paul
Schmitt, Henrique
Schmitz, Marion
Schneider, Donald
Schneider, Glenn
Schneider, Nicholas
Schneps, Matthew
Schnopper, Herbert
Schoolman, Stephen
Schou, Jesper
Schreier, Ethan
Schrijver, Karel
Schroeder, Daniel
Schucking, Engelbert
Schuler, Simon
Schulte, D.
Schulte-Ladbeck, Regina
Schultz, Alfred
Schultz, David
Schulz, Norbert
Schutz, Bob
Schwartz, Daniel
Schwartz, Philip
Schwartz, Richard
Schweizer, Franois
Sconzo, Pasquale
Scott, Eugene
Scott, John
Scoville, Nicholas
Seaman, Rob
Searle, Leonard
Seeds, Michael
Seeger, Philip
Seidelmann, P.
Seielstad, George
Seigar, Marcus

Busko, Ivo
Buta, Ronald
Butler, Bryan
Butler, Dennis
Butler, Paul
Butterworth, Paul
Buzasi, Derek
Byard, Paul
Byrd, Gene
Cahn, Julius
Cai, Kai
Caillault, Jean
Caldwell, John
Calvin, William
Calzetti, Daniela
Cameron, Winifred
Campbell, Belva
Campbell, Donald
Campbell, Murray
Campins, Humberto
Canfield, Richard
Canizares, Claude
Cannon, John
Canzian, Blaise
Capen, Charles
Capriotti, Eugene
Carbon, Duane
Carilli, Christopher
Carleton, Nathaniel
Carlson, John
Carney, Bruce
Caroff, Lawrence
Carpenter, Kenneth
Carpenter, Lloyd
Carr, John
Carr, Thomas
Carruthers, George
Carter, William
Casertano, Stefano
Cash Jr., Webster
Cassinelli, Joseph
Castelaz, Micheal
Castelli, John
Castor, John
Caton, Daniel
Catura, Richard
Cefola, Paul
Centrella, Joan
Cersosimo, Juan
Chaboyer, Brian
Chaffee, Frederic
Chaisson, Eric
Chamberlain, Joseph
Chambers, John
Chambliss, Carlson
Chaname, Julio
Chance, Kelly
Chandler, Claire
Chandra, Subhash
Chapman, Clark
Kim, Dong-Woo
Kimble, Randy
Kinemuchi, Karen
King, David
King, Ivan
Kinman, Thomas
Kinney, Anne
Kiplinger, Alan
Kirby, Kate
Kirkpatrick, Joseph
Kirkpatrick, Ronald
Kirshner, Robert
Kisseleva-Eggleton, Ludmila
Kissell, Kenneth
Kitiashvili, Irina
Klein, Richard
Kleinmann, Douglas
Klemola, Arnold
Klemperer, Wilfred
Klepczynski, William
Klimchuk, James
Klinglesmith III, Daniel
Klinglesmith, Daniel
Kliore, Arvydas
Klock, Benny
Knacke, Roger
Knapp, Gillian
Knezek, Patricia
Kniffen, Donald
Knoelker, Michael
Knowles, Stephen
Ko, Hsien
Koch, David
Koch, Robert
Koekemoer, Anton
Kofman, Lev
Kohl, John
Kolb, Edward
Kolokolova, Ludmilla
Kondo, Yoji
Kong, Albert
Konigl, Arieh
Koo, David
Kopeikin, Sergei
Kopp, Greg
Kopp, Roger
Koratkar, Anuradha
Korchagin, Vladimir
Kormendy, John
Kosovichev, Alexander
Koupelis, Theodoros
Kouveliotou, Chryssa
Kovalev, Yuri
Kovar, N.
Kovar, Robert
Kowal, Charles
Kraft, Robert
Kramer, Kh
Kramida, Alexander
Kraushaar, William
Seitzer, Patrick
Sekanina, Zdenek
Sellgren, Kristen
Sellwood, Jerry
Sembach, Kenneth
Serabyn, Eugene
Severson, Scott
Seward, Frederick
Shaffer, David
Shafter, Allen
Shaham, Jacob
Shandarin, Sergei
Shane, William
Shank, Michael
Shankland, Paul
Shao, Cheng-yuan
Shapero, Donald
Shapiro, Irwin
Shapiro, Maurice
Shapiro, Stuart
Sharma, A.
Sharp, Christopher
Sharp, Nigel
Sharpless, Stewart
Shaw, James
Shaw, John
Shaw, Richard
Shawl, Stephen
Shaya, Edward
Shea, Margaret
Sheeley, Neil
Sheikh, Suneel
Sheinis, Andrew
Shelus, Peter
Shen, Benjamin
Shepherd, Debra
Shetrone, Matthew
Shields, Gregory
Shields, Joseph
Shine, Richard
Shinnaga, Hiroko
Shipman, Harry
Shivanandan, Kandiah
Shlosman, Isaac
Shore, Bruce
Shostak, G.
Shull, John
Shull, Peter
Sigurdsson, Steinn
Silva, David
Silverberg, Eric
Simkin, Susan
Simon, George
Simon, Michal
Simon, Norman
Simon, Theodore
Simon-Miller, Amy
Simonson, S.
Sinha, Rameshwar
Sion, Edward

Chapman, Gary
Chapman, Robert
Chartas, George
Chen, Hsiao-Wen
Cheng, Kwang
Chesley, Steven
Cheung, Cynthia
Chevalier, Roger
Chitre, Dattakumar
Chiu, Hong
Chiu, Liang-Tai
Chodas, Paul
Choudhary, Debi Prasad
Christian, Carol
Christiansen, Wayne
Christodoulou, Dimitris
Christy, James
Christy, Robert
Chu, You-Hua
Chubb, Talbot
Chupp, Edward
Churchwell, Edward
Ciardi, David
Ciardullo, Robin
Clark Jr., Alfred
Clark, Barry
Clark, Frank
Clark, George
Clarke, John
Clarke, Tracy
Claussen, Mark
Clayton, Donald
Clayton, Geoffrey
Clegg, Andrew
Clem, James
Clemens, Dan
Clifton, Kenneth
Cline, Thomas
Cliver, Edward
Clowe, Douglas
Cochran, Anita
Cochran, William
Cocke, William
Coffeen, David
Coffey, Helen
Cohen, Jeffrey
Cohen, Judith
Cohen, Leon
Cohen, Marshall
Cohen, Martin
Cohen, Richard
Cohen, Ross
Cohn, Haldan
Colbert, Edward
Colburn, David
Cole, David
Coleman, Paul
Coletti, Donna
Colgate, Stirling
Collins, George

Kreidl, Tobias
Krieger, Allen
Krisciunas, Kevin
Kriss, Gerard
Krogdahl, W.
Krolik, Julian
Kron, Richard
Krucker, Sam
Krumholz, Mark
Krupp, Edwin
Kudritzki, Rolf-Peter
Kuhi, Leonard
Kuhn, Jeffery
Kuiper, Thomas
Kulkarni, Shrinivas
Kulsrud, Russell
Kumar, C.
Kumar, Shiv
Kundu, Arunav
Kundu, Mukul
Kurfess, James
Kurtz, Michael
Kurucz, Robert
Kutner, Marc
Kutter, G.
Kuzio de Naray, Rachel
Kwitter, Karen
Lacy, Claud
Lacy, John
Lada, Charles
Laird, John
Lamb Jr., Donald
Lamb, Frederick
Lamb, Richard
Lamb, Susan
Lambert, David
Lampton, Michael
Lande, Kenneth
Landecker, Peter
Landman, Donald
Landolt, Arlo
Lane, Adair
Lane, Arthur
Lang, Kenneth
Langer, William
Langhoff, Stephanie
Langston, Glen
Lanz, Thierry
Lario, David
LaRosa, Theodore
Larsen, Jeffrey
Larson, Harold
Larson, Richard
Larson, Stephen
Lasala Jr., Gerald
Lasher, Gordon
Latham, David
Latter, William
Lattimer, James
Lauroesch, James

Sitko, Michael
Sivaraman, Koduvayur
Sjogren, William
Skalafuris, Angelo
Skillman, Evan
Skinner, Gerald
Skinner, Stephen
Skumanich, Andrew
Slade, Martin
Slane, Patrick
Slater, Timothy
Sloan, Gregory
Slovak, Mark
Smale, Alan
Smecker-Hane, Tammy
Smith, Barham
Smith, Bradford
Smith, Bruce
Smith, Dean
Smith, Eric
Smith, Graeme
Smith, Haywood
Smith, Horace
Smith, Howard
Smith, J.
Smith, Linda
Smith, Myron
Smith, Peter
Smith, Randall
Smith, Tracy
Smith, Verne
Smith, William
Smoot III, George
Sneden, Chris
Snell, Ronald
Snow, Theodore
Snyder, Lewis
Soberman, Robert
Sobieski, Stanley
Soderblom, David
Soderblom, Larry
Sofia, Sabatino
Sofia, Ulysses
Soifer, Baruch
Sonett, Charles
Song, Inseok
Sonneborn, George
Soon, Willie
Sovers, Ojars
Sowell, James
Spahr, Timothy
Sparke, Linda
Sparks, Warren
Sparks, William
Spencer, John
Spencer, John
Spergel, David
Spicer, Daniel
Spiegel, E.
Spinrad, Hyron

Combi, Michael
Comins, Neil
Condon, James
Conklin, Edward
Connolly, Leo
Conrad, Albert
Conselice, Christopher
Conti, Peter
Cook, John
Cook, Kem
Cooray, Asantha
Corbally, Christopher
Corbet, Robin
Corbin, Brenda
Corbin, Michael
Corbin, Thomas
Corcoran, Michael
Cordes, James
Cordova, France
Corliss, C.
Corwin Jr., Harold
Cotton Jr., William
Coulson, Iain
Counselman, Charles
Couvidat, Sebastien
Couvidat, Sebastien
Cowan, John
Cowie, Lennox
Cowley, Anne
Cowley, Charles
Cowsik, Ramanath
Cox, Arthur
Cox, Donald
Craine, Eric
Crane, Patrick
Crane, Philippe
Crannell, Carol
Crawford, David
Crawford, Fronefield
Crenshaw, Daniel
Crocker, Deborah
Crotts, Arlin
Cruikshank, Dale
Crutcher, Richard
Cudworth, Kyle
Cuillandre, Jean-Charles
Culver, Roger
Cuntz, Manfred
Czyzak, Stanley
Dahn, Conard
Dalgarno, Alexander
Daly, Ruth
Danby, J.
Danchi, William
Danford, Stephen
Danks, Anthony
Danly, Laura
Danner, Rolf
Dappen, Werner
Dasyra, Kalliopi
Lautman, D.
Lawlor, Timothy
Lawrence, Charles
Lawrence, G.
Lawrence, John
Lawrie, David
Lawson, Peter
Layden, Andrew
Layzer, David
Lazarian, Alexandre
Lazio, Joseph
Lea, Susan
Leacock, Robert
Lebofsky, Larry
Lebohec, Stephan
Lebovitz, Norman
Lebron, Mayra
Lecar, Myron
Leckrone, David
Lee, Paul
Leggett, Sandy
Leibacher, John
Leighly, Karen
Leisawitz, David
Leitherer, Claus
Leka, Kimberly
Lemmon, Mark
Leonard, Peter
Lepine, Sebastien
Lepp, Stephen
Lester, Daniel
Leung, Chun
Leung, Kam
Levine, Randolph
Levine, Stephen
Levison, Harold
Levreault, Russell
Levy, Eugene
Lewin, Walter
Lewis, Brian
Lewis, J.
Li, Hong-Wei
Liang, Edison
Libbrecht, Kenneth
Liddell, U.
Liebert, James
Lieske, Jay
Likkel, Lauren
Lilley, Edward
Lillie, Charles
Lin, Chia
Lin, Douglas
Lindsey, Charles
Lingenfelter, Richard
Linke, Richard
Linnell, Albert
Linsky, Jeffrey
Lippincott Zimmerman, Sarah
Lipschutz, Michael
Lis, Dariusz
Sprague, Ann
Squires, Gordon
Sramek, Richard
Sridharan, Tirupati
Stacey, Gordon
Stachnik, Robert
Stahler, Steven
Stahr-Carpenter, M.
Stancil, Philip
Standish, E.
Stanford, Spencer
Stanghellini, Letizia
Stanimirovic, Snezana
Stanley, G.
Stapelfeldt, Karl
Stark, Antony
Stark, Glen
Starrfield, Sumner
Statler, Thomas
Stauffer, John
Stebbins, Robin
Stecher, Theodore
Stecker, Floyd
Steele, John
Stefanik, Robert
Steiger, W.
Steigman, Gary
Steiman-Cameron, Thomas
Stein, John
Stein, Robert
Stein, Wayne
Steinolfson, Richard
Stellingwerf, Robert
Stencel, Robert
Stepinski, Tomasz
Stern, Robert
Stern, S.
Stiavelli, Massimo
Stier, Mark
Stinebring, Daniel
Stocke, John
Stockman Jr., Hervey
Stockton, Alan
Stokes, Grant
Stone, Edward
Stone, James
Stone, R.
Stone, Remington
Storrie-Lombardi, Lisa
Storrs, Alexander
Strachan Jr., Leonard
Strauss, Michael
Strelnitski, Vladimir
Stringfellow, Guy
Strittmatter, Peter
Strobel, Darrell
Strohmayer, Tod
Strom, Karen
Strom, Robert
Strom, Stephen

Datlowe, Dayton
David, Laurence
Davidson, Kris
Davies, Ashley
Davies, Paul
Davila, Joseph
Davis Jr., Cecil
Davis, Christopher
Davis, Donald
Davis, Gary
Davis, Marc
Davis, Michael
Davis, Morris
Davis, Robert
de Frees, Douglas
de Jong, Roelof
de Jonge, J.
de Marco, Orsola
de Mello, Duilia
de Pater, Imke
de Toma, Giuliana
de Vincenzi, Donald
de Young, David
Dearborn, David
Deeming, Terence
DeGioia-Eastwood, Kathleen
Deliyannis, Constantine
Dell'Antonio, Ian
Delsemme, Armand
Demarque, Pierre
Deming, Leo
Dempsey, Robert
Dennis, Brian
Dennison, Edwin
Dent, William
Dere, Kenneth
Dermer, Charles
Dermott, Stanley
Despain, Keith
Deustua, Susana
Deutschman, William
Devinney, Edward
DeVorkin, David
Devost, Daniel
Dewey, Rachel
Dewitt-Morette, Cecile
Dhawan, Vivek
Dick, Steven
Dickel, Helene
Dickel, John
Dickey, Jean
Dickinson, Dale
Dickman, Robert
Dickman, Steven
Didkovsky, Leonid
Dieter Conklin, Nannielou
Dietrich, Joerg
Dietrich, Matthias
Digel, Seth
Dikova, Smilyana

Lissauer, Jack
Lisse, Carey
Lister, Matthew
Liszt, Harvey
Little-Marenin, Irene
Littleton, John
Litvak, Marvin
Liu, Michael
Liu, Sou-Yang
Liu, Yang
Livingston, William
Livio, Mario
Lo, Kwok-Yung
Lochner, James
Lockman, Felix
Lockwood, G.
Lodders, Katharina
Long, Knox
Longmore, Andrew
Lonsdale, Carol
Lopes-Gautier, Rosaly
Lord, Steven
Loren, Robert
Lovas, Francis
Lovelace, Richard
Low, Boon
Lu, Limin
Lu, Phillip
Lubin, Lori
Lubowich, Donald
Luck, R.
Lucke, Peter
Lugger, Phyllis
Lundquist, Charles
Luttermoser, Donald
Lutz, Barry
Lutz, Julie
Luu, Jane
Luzum, Brian
Lynch, David
Lynds, Beverly
Lynds, Roger
Ma, Chopo
Mac Clain, Edward
Mac Low, Mordecai-Mark
Macalpine, Gordon
Macchetto, Ferdinando
MacConnell, Darrell
MacDonald, James
Mack, Peter
Macquart, Jean-Pierre
MacQueen, Robert
Macri, Lucas
Macy, William
Madau, Piero
Madore, Barry
Magee-Sauer, Karen
Magnani, Loris
Majid, Walid
Makarov, Valeri

Strong, Ian
Strong, Keith
Struble, Mitchell
Struck-Marcell, Curtis
Stryker, Linda
Sturrock, Peter
Suess, Steven
Sukumar, Sundarajan
Sullivan, III, Woodruff
Sutton, Edmund
Svalgaard, Leif
Swade, Daryl
Swank, Jean
Sweigart, Allen
Sweitzer, James
Swenson Jr., George
Swerdlow, Noel
Sykes, Mark
Synnott, Stephen
Szalay, Alex
Szkody, Paula
Taam, Ronald
Tademaru, Eugene
Takata, Tadafumi
Talbot Jr., Raymond
Tandberg-Hanssen, Einar
Tapley, Byron
Tarnstrom, Guy
Tarter, C.
Tarter, Jill
Tassis, Konstantinos
Tayal, Swaraj
Taylor, Donald
Taylor, Gregory
Taylor, Joseph
Taylor, Keith
Teays, Terry
Tedesco, Edward
Telesco, Charles
ten Brummelaar, Theo
Terrell, Dirk
Terrell, James
Terrile, Richard
Terzian, Yervant
Teske, Richard
Teuben, Peter
Thaddeus, Patrick
Tholen, David
Thomas, John
Thomas, Roger
Thompson, A.
Thompson, Ian
Thompson, Rodger
Thonnard, Norbert
Thorne, Kip
Thornley, Michele
Thorsett, Stephen
Thorstensen, John
Thronson Jr., Harley
Thuan, Trinh

Dinerstein, Harriet
Dinescu-Casetti, Dana
Dixon, Robert
Djorgovski, Stanislav
Doherty, Lowell
Dolan, Joseph
Donahue, Megan
Donahue, Robert
Donn, Bertram
Doppmann, Gregory
Doschek, George
Downes, Ronald
Downs, George
Doyle, Laurance
Draine, Bruce
Drake, Frank
Drake, Jeremy
Drake, Stephen
Dreher, John
Dressel, Linda
Dressler, Alan
Drever, Ronald
Drilling, John
Dryer, Murray
Dubin, Maurice
Dufour, Reginald
Dukes Jr., Robert
Duncan, Douglas
Duncombe, Raynor
Dunham, David
Dupree, Andrea
Dupuy, David
Durisen, Richard
Durney, Bernard
Durrance, Samuel
Durrell, Patrick
Duthie, Joseph
Duvall Jr., Thomas
Dwarkadas, Vikram
Dwek, Eli
Dyck, Melvin
Dyson, Freeman
Eaton, Joel
Eddy, John
Edelson, Rick
Edwards, Alan
Edwards, Suzan
Efroimsky, Michael
Egan, Michael
Eggleton, Peter
Eicher, David
Eilek, Jean
Eisner, Josh
El Baz, Farouk
Elitzur, Moshe
Elliot, James
Ellis, Richard
Elmegreen, Bruce
Elmegreen, Debra
Elste, Gunther
Malhotra, Renu
Malhotra, Sageeta
Malina, Roger
Malitson, Harriet
Malkamaeki, Lauri
Malkan, Matthew
Malville, J.
Mamajek, Eric
Mangum, Jeffrey
Manset, Nadine
Maran, Stephen
Marcialis, Robert
Marcy, Geoffrey
Marengo, Massimo
Margon, Bruce
Margot, Jean-Luc
Margrave Jr., Thomas
Mariska, John
Markworth, Norman
Marley, Mark
Marochnik, L.
Marr, Jonathon
Marschall, Laurence
Marscher, Alan
Marsden, Brian
Marshall, Herman
Marston, Anthony
Martin, Christopher
Martin, Crystal
Martin, Donn
Martin, Rene
Martin, Robert
Martin, William
Martini, Paul
Martins, Donald
Marvel, Kevin
Marvin, Ursula
Marzo, Giuseppe
Mason, Brian
Mason, Glenn
Mason, Paul
Massa, Derck
Massey, Philip
Masson, Colin
Matese, John
Mather, John
Mathews, William
Mathieu, Robert
Mathis, John
Matsakis, Demetrios
Matson, Dennis
Matthews, Clifford
Matthews, Henry
Matthews, Lynn
Matthews, Thomas
Mattox, John
Matz, Steven
Matzner, Richard
Mauche, Christopher
Max, Claire
Tifft, William
Tilanus, Remo
Timothy, J.
Tipler, Frank
Tiscareno, Matthew
Title, Alan
Tody, Douglas
Tohline, Joel
Tokunaga, Alan
Tolbert, Charles
Toller, Gary
Tomasko, Martin
Tonry, John
Toomre, Alar
Toomre, Juri
Torres Dodgen, Ana
Torres, Guillermo
Tothill, Nicholas
Townes, Charles
Townsend, Richard
Trafton, Laurence
Trammell, Susan
Traub, Wesley
Treffers, Richard
Tremaine, Scott
Treu, Tommaso
Trimble, Virginia
Tripathy, Sushanta
Tripicco, Michael
Troland, Thomas
Trujillo, Chadwick
Truran Jr., James
Tsuruta, Sachiko
Tsvetanov, Zlatan
Tucker, Wallace
Tull, Robert
Tully, Richard
Turner, Edwin
Turner, Jean
Turner, Kenneth
Turner, Michael
Turner, Nils
Turnshek, David
Twarog, Bruce
Tycner, Christopher
Tyler Jr., G.
Tylka, Allan
Tyson, John
Tytler, David
Ulich, Bobby
Ulmer, Melville
Ulrich, Roger
Ulvestad, James
Underwood, James
Unwin, Stephen
Uomoto, Alan
Upgren, Arthur
Upson, Walter
Upton, E.
Urban, Sean

Elston, Wolfgang
Elvis, Martin
Emerson, Darrel
Emslie, A.
Endal, Andrew
Epps, Harland
Epstein, Eugene
Epstein, Gabriel
Epstein, Richard
Eshleman, Von
Eskridge, Paul
Esposito, F.
Esposito, Larry
Etzel, Paul
Eubanks, Thomas
Evans, Daniel
Evans, Ian
Evans, J.
Evans, Nancy
Evans, Neal
Evans, W.
Ewen, Harold
Ewing, Martin
Eyer, Laurent
Fabbiano, Giuseppina
Faber, Sandra
Fabricant, Daniel
Fahey, Richard
Falco, Emilio
Falk Jr., Sydney
Fall, S.
Faller, James
Fallon, Frederick
Fan, Yuhong
Fang, Li-zhi
Fanselow, John
Farnham, Tony
Fassnacht, Christopher
Faulkner, John
Fay, Theodore
Fazio, Giovanni
Federman, Steven
Feigelson, Eric
Fekel, Francis
Feldman, Uri
Felten, James
Ferland, Gary
Fesen, Robert
Fey, Alan
Feynman, Joan
Fiala, Alan
Fichtel, Carl
Fiedler, Ralph
Field, George
Fienberg, Richard
Filippenko, Alexei
Fink, Uwe
Finn, Lee
Fischel, David
Fischer, Jacqueline
Maxwell, Alan
Mayfield, Earle
Mazurek, Thaddeus
Mazzarella, Joseph
McAlister, Harold
McCabe, Marie
McCammon, Dan
McCarthy, Dennis
McClintock, Jeffrey
McCluskey Jr., George
McCollum, Bruce
McCord, Thomas
McCray, Richard
McCrosky, Richard
McCullough, Peter
McDavid, David
McDonald, Frank
McDonough, Thomas
McElroy, M.
McFadden, Lucy
McGaugh, Stacy
McGimsey Jr., Ben
McGrath, Melissa
McGraw, John
McIntosh, Patrick
McKee, Christopher
McKinnon, William
McLaren, Robert
McLean, Brian
McLean, Ian
McMahan, Robert
McMillan, Robert
McMullin, Joseph
McNamara, Delbert
McNeil, Stephen
McSwain, Mary
Mead, Jaylee
Meech, Karen
Meeks, M.
Megeath, S.
Meier, David
Meier, Robert
Meisel, David
Meixner, Margaret
Melbourne, William
Melia, Fulvio
Melnick, Gary
Melott, Adrian
Mendillo, Michael
Menndez-Delmestre, Karin
Merline, William
Mertz, Lawrence
Messmer, Peter
Meszaros, Peter
Metcalfe, Travis
Metevier, Anne
Meyer, Angela
Meyer, David
Meyers, Karie
Michel, F.
Urry, Claudia
Usher, Peter
Uson, Juan
Vaccaro, Todd
Valenti, Jeff
van Altena, William
van Breugel, Wil
van Citters, Gordon
van der Marel, Roeland
van der Veen, Wilhelmus
van Dorn, Bradt
van Dyk, Schuyler
van Gorkom, Jacqueline
van Hamme, Walter
van Horn, Hugh
van Hoven, Gerard
van Moorsel, Gustaaf
van Putten, Maurice
van Riper, Kenneth
Van Zee, Liese
VandenBout, Paul
Vandervoort, Peter
Vaughan, Arthur
Veeder, Glenn
Veillet, Christian
Veilleux, Sylvain
Velusamy, T.
Venugopal, V.
Verner, Ekaterina
Verschuur, Gerrit
Verter, Frances
Vesecky, J.
Vesperini, Enrico
Vestergaard, Marianne
Veverka, Joseph
Vidali, Gianfranco
Vijh, Uma
Vilas, Faith
Vogel, Stuart
Vogt, Steven
Volk, Kevin
von Braun, Kaspar
von Hippel, Theodore
Vorpahl, Joan
Vrba, Frederick
Vrtilek, Jan
Vrtilek, Saeqa
Wachter, Stefanie
Waddington, C.
Wade, Richard
Wagner, Raymond
Wagner, Robert
Wagner, William
Wagoner, Robert
Wahlgren, Glenn
Wainscoat, Richard
Wakker, Bastiaan
Walborn, Nolan
Walker, Alta
Walker, Constance

Fisher, George
Fisher, J.
Fisher, Philip
Fisher, Richard
Fishman, Gerald
Fitch, Walter
Fitzpatrick, Edward
Fix, John
Flannery, Brian
Fleck, Bernhard
Fleck, Robert
Fleischer, Robert
Fleming, Thomas
Fliegel, Henry
Florkowski, David
Fogarty, William
Folkner, William
Foltz, Craig
Fomalont, Edward
Fontenla, Juan
Forbes, Terry
Ford Jr., W.
Ford, Holland
Forman, William
Forrest, William
Forster, James
Foster, Roger
Foukal, Peter
Fox, Kenneth
Frail, Dale
Frank, Juhan
Franklin, Fred
Franz, Otto
Frazier, Edward
Fredrick, Laurence
Freedman, Wendy
French, Richard
Friberg, Per
Friedlander, Michael
Friedman, Scott
Friel, Eileen
Frisch, Priscilla
Frogel, Jay
Frost, Kenneth
Fruchter, Andrew
Fruscione, Antonella
Frye, Glenn
Ftaclas, Christ
Fuhr, Jeffrey
Fulbright, Jon
Fullerton, Alexander
Funes, Jos
Furlanetto, Steven
Furuya, Ray
Fuse, Tetsuharu
Gaetz, Terrance
Gaisser, Thomas
Gallagher III, John
Gallet, Roger
Gallimore, Jack
Michel, Raul
Mickelson, Michael
Migenes, Victor
Mighell, Kenneth
Mihalas, Barbara
Mihalas, Dimitri
Mikesell, Alfred
Milkey, Robert
Millan Gabet, Rafael
Miller, Guy
Miller, Hugh
Miller, Joseph
Miller, Michael
Miller, Neal
Miller, Richard
Milligan, J.
Millikan, Allan
Millis, Robert
Minchin, Robert
Mink, Douglas
Minter, Anthony
Mintz Blanco, Betty
Miralles, Mari
Misawa, Toru
Misconi, Nebil
Misner, Charles
Mitchell, Kenneth
Miyazaki, Satoshi
Mo, Houjun
Mo, Jinger
Modali, Sarma
Modisette, Jerry
Moellenbrock III, George
Moffett, David
Moffett, Thomas
Mohr, Joseph
Molnar, Michael
Momjian, Emmanuel
Monet, Alice
Monet, David
Monnier, John
Moody, Joseph
Mook, Delo
Moore, Elliott
Moore, Marla
Moore, Ronald
Moos, Henry
Morabito, David
Moran, James
Morgan, John
Morgan, Thomas
Morino, Jun-Ichi
Morris, Charles
Morris, Mark
Morris, Steven
Morrison, David
Morrison, Nancy
Morton, G.
Motz, Lloyd
Mouschovias, Telemachos
Walker, Merle
Walker, Robert C.
Walker, Robert M. A.
Wallace, James
Wallace, Lloyd
Wallace, Richard
Waller, William
Wallerstein, George
Wallin, John
Walter, Fabian
Walter, Frederick
Walterbos, Rene
Wampler, E.
Wang, Haimin
Wang, Q. Daniel
Wang, Yi-ming
Wang, Zhong
Wannier, Peter
Ward, Richard
Ward, William
Wardle, John
Warner, John
Warren Jr., Wayne
Warwick, James
Wasserman, Lawrence
Wasson, John
Watanabe, Ken
Watson, William
Wdowiak, Thomas
Weaver, Harold
Weaver, Kimberly
Weaver, Thomas
Weaver, William
Webb, David
Webber, John
Webbink, Ronald
Weber, Stephen
Webster, Zodiac
Weedman, Daniel
Weekes, Trevor
Wegner, Gary
Wehinger, Peter
Wehrle, Ann
Wei, Mingzhi
Weidenschilling, S.
Weiler, Edward
Weiler, Kurt
Weill, Gilbert
Weinberg, Jerry
Weinberg, Steven
Weis, Edward
Weisberg, Joel
Weisheit, Jon
Weisskopf, Martin
Weissman, Paul
Weistrop, Donna
Welch, William
Weller, Charles
Wells, Donald
Wells, Eddie

Galloway, Duncan
Galvin, Antoinette
Ganguly, Rajib
Gaposchkin, Edward
Garcia, Michael
Gardner, Jonathan
Garlick, George
Garmany, Katy
Garmire, Gordon
Garnett, Donald
Gary, Dale
Gary, Gilmer
Gaskell, C.
Gatewood, George
Gatley, Ian
Gaume, Ralph
Gauss, Stephen
Geballe, Thomas
Gebbie, Katharine
Gehrels, Neil
Gehrels, Tom
Gehrz, Robert
Gelderman, Richard
Geldzahler, Bernard
Geller, Margaret
Genet, Russel
Gerakines, Perry
Gergely, Tomas
Germain, Marvin
Gerola, Humberto
Gezari, Daniel
Ghez, Andrea
Ghigo, Francis
Ghosh, Kajal
Ghosh, Tapashi
Giampapa, Mark
Gibson, David
Gibson, James
Gibson, Sarah
Gibson, Steven
Gierasch, Peter
Gies, Douglas
Gigas, Detlef
Gilliland, Ronald
Gillingham, Peter
Gilman, Peter
Gilra, Daya
Gingerich, Owen
Giorgini, Jon
Giovane, Frank
Giovanelli, Riccardo
Gizis, John
Glaser, Harold
Glaspey, John
Glass, Billy
Glassgold, Alfred
Glatzmaier, Gary
Glinski, Robert
Goddi, Ciriaco

Moustakas, Leonidas
Mozurkewich, David
Mueller, Beatrice
Mueller, Ivan
Muench, Guido
Mufson, Stuart
Mukai, Koji
Mukherjee, Krishna
Mulchaey, John
Mullan, Dermott
Muller, Richard
Mumford, George
Mumma, Michael
Mundy, Lee
Munro, Richard
Murdock, Thomas
Murphy, Brian
Murphy, Robert
Murray, Stephen D.
Murray, Stephen S.
Musen, Peter
Mushotzky, Richard
Musielak, Zdzislaw
Musman, Steven
Mutel, Robert
Mutschlecner, J.
Myers, Philip
Nacozy, Paul
Nahar, Sultana
Narayan, Ramesh
Nather, R.
Nave, Gillian
Neff, James
Neff, Susan
Neidig, Donald
Nelson, Burt
Nelson, George
Nelson, Jerry
Nelson, Robert
Nelson, Robert
Nemiroff, Robert
Ness, Norman
Neugebauer, Gerry
Neupert, Werner
Newberg, Heidi
Newburn Jr., Ray
Newhall, X.
Newman, Michael
Newsom, Gerald
Nice, David
Nichols, Joy
Nicolas, Kenneth
Nicoll, Jeffrey
Nicollier, Claude
Niedner, Malcolm
Niell, Arthur
Nielsen, Krister
Nilsson, Carl
Ninkov, Zoran
Nishikawa, Ken-Ichi

Welsh, William
Wentzel, Donat
Westerhout, Gart
Weymann, Ray
Wheeler, J.
Whipple, Arthur
Whitaker, Ewen
Whitcomb, Stanley
White, James
White, Nathaniel
White, Nicholas
White, Oran
White, R.
White, Raymond
White, Richard E.
White, Richard L.
White, Stephen
Whitmore, Bradley
Whitney, Charles
Whittet, Douglas
Whittle, D.
Wickramasinghe, Thulsi
Widing, Kenneth
Wiese, Wolfgang
Wiita, Paul
Wiklind, Tommy
Wilcots, Eric
Wilkening, Laurel
Wilkes, Belinda
Will, Clifford
Williamon, Richard
Williams, Barbara
Williams, Carol
Williams, Gareth
Williams, Glen
Williams, James
Williams, John
Williams, Jonathan
Williams, Robert
Williams, Theodore
Williams, Thomas
Willner, Steven
Wills, Beverley
Wills, Derek
Willson, Lee Anne
Willson, Robert
Wilner, David
Wilson, Albert
Wilson, Curtis
Wilson, Richard E.
Wilson, Robert W.
Wilson, Robert
Windhorst, Rogier
Winebarger, Amy
Wing, Robert
Winget, Donald
Winkler, Gernot
Winkler, Karl-Heinz
Winkler, Paul
Winn, Joshua

Goebel, John
Golap, Kumar
Goldman, Martin
Goldreich, Peter
Goldsmith, Donald
Goldsmith, Paul
Goldstein, Richard
Goldwire Jr., Henry
Golub, Leon
González, Guillermo
Goode, Philip
Goodman, Alyssa
Goodrich, Robert
Goody, Richard
Gopalswamy, Natchimuthuk
Gordon, Courtney
Gordon, Kurtiss
Gordon, Mark
Gorenstein, Marc
Gorenstein, Paul
Gorkavyi, Nikolai
Gosling, John
Goss, W. Miller
Gott, J.
Gottesman, Stephen
Gotthelf, Eric
Gottlieb, Carl
Gould, Robert
Graboske Jr., Harold
Gradie, Jonathan
Grady, Carol
Graham, Eric
Graham, John
Grandi, Steven
Grasdalen, Gary
Grauer, Albert
Grav, Tommy
Gray, Peter
Gray, Richard
Grayzeck, Edwin
Green, Daniel
Green, Elizabeth
Green, Jack
Green, Richard
Greenberg, Richard
Greenhill, Lincoln
Greenhouse, Matthew
Greenstein, George
Gregg, Michael
Gregory, Stephen
Greisen, Eric
Greve, Thomas
Greyber, Howard
Griest, Kim
Griffin, Ian
Griffiths, Richard
Grindlay, Jonathan
Grinspoon, David
Gronwall, Caryl
Gross, Peter

Nishimura, Tetsuo
Noerdlinger, Peter
Nolan, Michael
Noll, Keith
Noonan, Thomas
Noriega-Crespo, Alberto
Norman, Colin
Norman, Dara
Noumaru, Junichi
Noyes, Robert
Nulsen, Paul
Nuth, Joseph
O'Brian, Thomas
O'Connell, Robert
O'Connell, Robert
O'Dea, Christopher
Odell, Andrew
O'Dell, Charles
O'Dell, Stephen
Odenwald, Sten
Oegerle, William
Oemler Jr., Augustus
Oey, Sally
Ofman, Leon
Ogelman, Hakki
O'Handley, Douglas
Ohashi, Nagayoshi
Ojha, Roopesh
Oka, Takeshi
Oliver, John
Olivier, Scot
Olling, Robert
Olowin, Ronald
Olsen, Kenneth
Olsen, Knut
Olson, Edward
O'Neal, Douglas
O'Neil, Karen
Onello, Joseph
Opendak, Michael
Ord, Stephen
Orlin, Hyman
Ormes, Jonathan
Orton, Glenn
Osborn, Wayne
Osmer, Patrick
Osten, Rachel
Ostriker, Eve
Ostriker, Jeremiah
Oswalt, Terry
Ott, Juergen
Ouchi, Masami
Owen Jr., William
Owen, Frazer
Owen, Tobias
Owocki, Stanley
Ozsvath, I.
Pacholczyk, Andrzej
Paciesas, William
Paerels, Frederik

Withbroe, George
Witt, Adolf
Witten, Louis
Wofford, Aida
Woillez, Julien
Wolfe, Arthur
Wolff, Michael
Wolff, Sidney
Wolfire, Mark
Wolfson, C.
Wolfson, Richard
Wolszczan, Alexander
Wood, Douglas
Wood, H.
Wood, Matthew
Wooden, Diane
Wooden, William
Woodward, Paul
Woolf, Neville
Woosley, Stanford
Wootten, Henry
Worden, Simon
Wouterloot, Jan
Wray, James
Wright, Edward
Wright, James
Wright, Melvyn
Wright, Nicholas
Wrobel, Joan
Wu, Chi
Wu, Nailong
Wu, Shi
Wu, Yanling
Wlser, Jean-Pierre
Wyckoff, Susan
Wyse, Rosemary
Xanthopoulos, Emily
Yahil, Amos
Yanamandra-Fisher, Padma
Yaplee, B.
Yau, Kevin
Yeh, Tyan
Yeomans, Donald
Yin, Qi-Feng
Yoder, Charles
York, Donald
Yorke, Harold
Yoshino, Kouichi
Yoss, Kenneth
Young, Andrew
Young, Arthur
Young, Judith
Young, Lisa
Young, Louise
Yu, Yan
Yusef-Zadeh, Farhad
Zacharias, Norbert
Zare, Khalil
Zarro, Dominic
Zavala, Jr., Robert

Gross, Richard
Grossman, Allen
Grossman, Lawrence
Grundy, William
Grupe, Dirk
Gudehus, Donald
Guetter, Harry
Guhathakurta, Madhulika
Guidice, Donald
Guinan, Edward
Gulkis, Samuel
Gull, Theodore
Gunn, James
Gurman, Joseph
Gurshtein, Alexander
Gustafson, Bo
Guzik, Joyce
Gwinn, Carl
Habbal, Shadia
Haberreiter, Margit
Hackman, Christine
Hackwell, John
Palmer, Patrick
Pan, Xiao-Pei
Panagia, Nino
Panek, Robert
Pang, Kevin
Pankonin, Vernon
Pannuti, Thomas
Pantoja, Carmen
Pap, Judit
Papaliolios, Costas
Parhi, Shyamsundar
Parise, Ronald
Parker, Eugene
Parkinson, Truman
Parkinson, William
Parrish, Allan
Parsons, Sidney
Partridge, Robert
Pasachoff, Jay
Pascu, Dan
Pascucci, Ilaria
Patel, Nimesh
Zeilik, Michael
Zellner, Benjamin
Zepf, Stephen
Zezas, Andreas
Zhang, Cheng-Yue
Zhang, Er-Ho
Zhang, Qizhou
Zhang, Sheng-Pan
Zhang, Shouzhong
Zhang, William
Zhang, Xiaolei
Zhao, Jun
Zhao, Junwei
Zheng, Wei
Zinn, Robert
Zirin, Harold
Zirker, Jack
Ziurys, Lucy
Zombeck, Martin
Zuckerman, Ben

Uruguay

Fernández, Julio
Gallardo Castro, Carlos
Tancredi, Gonzalo

Vatican City State

Consolmagno, Guy
Coyne, George

Venezuela

Abad Hiraldo, Carlos
Briceño, Cesar
Falcon Veloz, Nelson
Hernández, Jesús
Mendoza, Claudio
Rengel, Miriam
Vivas, Anna
Bautista, Manuel
Bruzual, Gustavo
Ferrin, Ignacio
Ibanez, S.
Parravano, Antonio
Rosenzweig-Levy, Patrica
Bongiovanni, Angel
Calvet, Nuria
Fuenmayor, Francisco
Magris, Gladis
Ramrez, José
Sigalotti, Leonardo

Viet Nam

Darriulat, Pierre
Nguyen, Mau
Dinh, Van
Huan, Nguyen

Individual Membership by Non-Adhering Country

Armenia

Hafizi, Mimoza

Algeria

Abdelatif, Toufik	Irbah, Abdanour	Makhlouf, Amar

Andorra

Miralles, Joan-Marc

Angola

Vilinga, Jaime

Azerbaijan

Aslanov, I.	Asvarov, Abdul	Babayev, Elchin
Eminzade, T.	Guseinov, O.	Gusejnov, Ragim
Ismailov, Nariman	Kasumov, Fikret	Sultanov, G.

Colombia

Brieva, Eduardo	de Greiff, J.

Ecuador

Lopez, Ericson

Ethiopia

Kebede, Legesse

Georgia

Bartaya, R.	Kalandadze, N.	Kurtanidze, Omar
Borchkhadze, Tengiz	Khatisashvili, Alfez	Lominadze, Jumber
Chkhikvadze, Iakob	Khetsuriani, Tsiala	Salukvadze, G.
Dolidze, Madona	Kiladze, Rolan	Shvelidze, Teimuraz
Dzhapiashvili, Victor	Kogoshvili, Natela	Toroshlidze, Teimuraz
Dzigvashvili, R.	Kumsiashvily, Mzia	

Iraq

Abdulla, Shaker
Jabbar, Sabeh
Jabir, Niama
Kadouri, Talib
Mohammed, Ali
Sadik, Aziz
Younis, Saad

Kazakhstan

Denisyuk, Edvard
Genkin, Igor
Karygina, Zoya
Kharitonov, Andrej
Omarov, Tuken
Rozhkovskij, Dimitrij
Tejfel, Viktor
Vilkoviskij, Emmanuil

Korea, DPR

Baek, Chang-Ryong
Cha, Gi-Ung
Choe, Won-Chol
Kang, Jin-Sok
Kang, Gon-Ik
Kim, Jik
Ri, Son-Jae

Macedonia

Apostolovska, Gordana

Malta

Mallia, Edward

Mauritius

Goelbasi, Orhan
Heeralall-Issur, Nalini
Somanah, Radhakhrishna

Pakistan

Quamar, Jawaid

Singapore

Snowden, Michael
Walsh, Wilfred

Slovenia

Čadež, Andrej
Dintinjana, Bojan
Dominko, Fran
Gomboc, Andreja
Kilar, Bogdan
Slosar, Anze
Zwitter, Tomaž

Sri Lanka

de Silva, Lindamulage
Maheswaran, Murugesapillai
Ratnatunga, Kavan

Trinidad & Tobago

Haque Copilah, Shirin

United Arab Emirates

Al-Naimiy, Hamid
Guessoum, Nidhal
Mohamed Ali, Alaa Eldin

Uzbekistan

Ehgamberdiev, Shuhrat
Hojaev, Alisher
Ilyasov, Sabit
Kalmykov, A.
Latypov, A.
Mamadazimov, Mamadmuso
Muminov, Muydinjon
Nuritdinov, Salakhutdin
Sattarov, Isroil
Yuldashbaev, Taimas
Zakirov, Mamnum

Transactions IAU, Volume XXVIIB
Proc. XXVII IAU General Assembly, August 2009
Ian F. Corbett, ed.

doi:10.1017/S1743921310005430

CHAPTER XII

INDIVIDUAL MEMBERSHIP BY COMMISSION

Division I Commission 4 Ephemerides

President Kaplan, George
Vice-President Hohenkerk, Catherine

Organizing Committee

Arlot, Jean-Eudes
Bangert, John
Bell, Steven
Folkner, William
Lara, Martin
Pitjeva, Elena
Urban, Sean
Vondrák, Jan

Members

Abalakin, Viktor
Ahn, Youngsook
Aoki, Shinko
Arakida, Hideyoshi
Bandyopadhyay, A.
Capitaine, Nicole
Chapront, Jean
Chapront-Touze, Michelle
Coma, Juan
Cooper, Nicholas
de Greiff, J.
Di, Xiaohua
Dickey, Jean
Duncombe, Raynor
Dunham, David
Eroshkin, Georgij
Fiala, Alan
Fienga, Agnès
Fominov, Aleksandr
Fu, Yanning
Giorgini, Jon
Glebova, Nina
Harper, David
He, Miao-fu
Hilton, James
Howard, Sethanne
Ilyas, Mohammad
Ivantsov, Anatoliy
Janiczek, Paul
Johnston, Kenneth
Kinoshita, Hiroshi
Klepczynski, William
Kolaczek, Barbara
Krasinsky, George
Laskar, Jacques
Lehmann, Marek
Lieske, Jay
Lopez Moratalla, Teodoro
Lukashova, Marina
Madsen, Claus
Majid, Abdul
Mallamaci, Claudio
Emelyanenko, Yoshimitsu
Morrison, Leslie
Mueller, Ivan
Newhall, X.
Nguyen, Mau
O'Handley, Douglas
Olivier, Enrico
Reasenberg, Robert
Rodin, Alexander
Rossello, Gaspar
Salazar, Antonio
Schwan, Heiner
Seidelmann, P.
Shapiro, Irwin
Shiryaev, Alexander
Shuygina, Nadia
Simon, Jean-Louis
Sôma, Mitsuru
Standish, E.
Thuillot, William
Ting, Yeou-Tswen
Vilinga, Jaime
Wang, Xiao-bin
Wielen, Roland
Wilkins, George
Williams, Carol
Williams, James
Winkler, Gernot
Wytrzyszczak, Iwona
Yallop, Bernard

Division XII Commission 5 Documentation & Astronomical Data

President Ohishi, Masatoshi
Vice-President Hanisch, Robert

Organizing Committee

Andernach, Heinz
Bishop, Marsha
Griffin, R.
Kembhavi, Ajit
Murphy, Tara
Pasian, Fabio

Members

A'Hearn, Michael
Abalakin, Viktor
Abt, Helmut
Accomazzi, Alberto
Aizenman, Morris
Alvarez, Pedro
Antonelli, Lucio Angelo
Banhatti, Dilip
Beckmann, Volker
Benacchio, Leopoldo
Benn, Chris
Berthier, Jerôme
Bessell, Michael
Bond, Ian
Borde, Suzanne
Bouška, Jiří
Boyce, Peter
Brinchmann, Jarle
Brouw, Willem
Calabretta, Mark
Caretta, Cesar
Chang, Hsiang-Kuang
Chapman, Jacqueline
Cheung, Cynthia
Chiappetti, Lucio
Christlieb , Norbert
Chu, Yaoquan
Coletti, Donna
Corbin, Brenda
Crézé, Michel
Cui, Chenzhou
Cunniffe, John
Dalla, Silvia
Davis, Morris
Davis, Robert
de Boer, Klaas
Dewhirst, David
Dickel, Helene
Dixon, Robert
Dluzhnevskaya, Olga
Dobrzycki, Adam
Dubois, Pascal
Ducati, Jorge
Duncombe, Raynor
Durand, Daniel
Griffin, Roger
Grosbol, Preben
Guibert, Jean
Guo, Hongfang
Harvel, Christopher
Hauck, Bernard
Hefele, Herbert
Helou, George
Hopkins, Andrew
Hudkova, Ludmila
Jenkner, Helmut
Kalberla, Peter
Kaplan, George
Kelly, Brandon
Kharin, Arkadiy
Kleczek, Josip
Kong, Xu
Kovaleva, Dana
Kuin, Nicolaas
Kunz, Martin
Lequeux, James
Lesteven, Soizick
Linde, Peter
Lonsdale, Carol
Lu, Fangjun
Madore, Barry
Mann, Robert
Mason, Brian
Matz, Steven
McLean, Brian
McNally, Derek
McNamara, Delbert
Mead, Jaylee
Meadows, A.
Mein, Pierre
Michel, Laurent
Mink, Douglas
Mitton, Simon
Morris, Rhys
Murtagh, Fionn
Nakajima, Koichi
Nefed'ev, Yury
Nishimura, Shiro
Schlüter, Franois
Pakhomov, Yury
Quintana, Hernan
Raimond, Ernst
Ratnatunga, Kavan
Reardon, Kevin
Renson, P.
Riegler, Guenter
Roman, Nancy
Rots, Arnold
Russo, Guido
Samodurov, Vladimir
Sarasso, Maria
Schade, David
Schilbach, Elena
Schlesinger, Barry
Schlueter, A.
Schmadel, Lutz
Schneider, Jean
Schröder, Anja
Seaman, Rob
Serrano, Alfonso
Sharp, Nigel
Shaw, Richard
Song, Liming
Spite, Franois
Su, Cheng-yue
Tedds, Jonathan
Terashita, Yoichi
Teuben, Peter
Tody, Douglas
Tritton, Susan
Tsvetkov, Milcho
Turner, Kenneth
Uesugi, Akira
Valeev, Sultan
Wallace, Patrick
Wang, Jian-Min
Warren Jr., Wayne
Weidemann, Volker
Weilbacher, Peter
Wells, Donald
Wenger, Marc
Westerhout, Gart
Whiting, Matthew
Wielen, Roland
Wilkins, George

Egret, Daniel	Pamyatnykh, Aleksej	Wise, Michael
Elia, Davide	Paturel, Georges	Wright, Alan
Elyiv, Andrii	Pecker, Jean-Claude	Yang, Hong-Jin
Fyfe, Duncan	Philip, A.G.	Yang, Xiaohu
Gabriel, Carlos	Piskunov, Anatolij	Zacchei, Andrea
Gomez, Monique	Pizzichini, Graziella	Zhang, Yanxia
Green, David	Polechova, Pavla	Zhao, Yongheng
Greisen, Eric	Pucillo, Mauro	Zhao, Jun Liang
		Zhou, Jianfeng

Composition of Division XII Commission 6 Astronomical Telegrams

President Samus, Nikolaj
Vice-President Yamaoka, Hitoshi

Organizing Committee

Aksnes, Kaare	Nakano, Syuichi	Williams, Gareth
Green, Daniel	Spahr, Timothy	
Marsden, Brian	Tichá, Jana	

Members

Apostolovska, Gordana	Gilmore, Alan	Phillips, Mark
Baransky, Olexander	Grindlay, Jonathan	Roemer, Elizabeth
Bazzano, Angela	Hers, Jan	Seaman, Rob
Bouchard, Antoine	Kastel', Galina	Sôma, Mitsuru
Coletti, Donna	Kouveliotou, Chryssa	Tholen, David
Corbin, Brenda	Mattila, Seppo	Tsvetkov, Milcho
Esenoğlu, Hasan	Nakamura, Tsuko	West, Richard
Filippenko, Alexei	Paragi, Zsolt	Williams, Gareth

Division I Commission 7 Celestial Mechanics & Dynamical Astronomy

President Knežević, Zoran
Vice-President Morbidelli, Alessandro

Organizing Committee

Athanassoula, Evangelia
Laskar, Jacques
Malhotra, Renu
Mikkola, Seppo
Peale, Stanton
Roig, Fernando

Members

Abad Medina, Alberto
Abalakin, Viktor
Ahmed, Mostafa
Akim, Efraim
Aksnes, Kaare
Andrade, Manuel
Anosova, Joanna
Antonacopoulos, Gregory
Aoki, Shinko
Arakida, Hideyoshi
Archinal, Brent
Balmino, Georges
Barabanov, Sergey
Barberis, Bruno
Barbosu, Mihail
Barkin, Yuri
Batrakov, Yurij
Benest, Daniel
Bettis, Dale
Beutler, Gerhard
Bhatnagar, K.
Boccaletti, Dino
Bois, Eric
Borderies, Nicole
Boss, Alan
Bouchard, Antoine
Bozis, George
Branham, Richard
Breiter, Slawomir
Brieva, Eduardo
Brookes, Clive
Broucke, Roger
Brož, Miroslav
Brunini, Adrian
Cai, Michael
Calame, Odile
Caranicolas, Nicholas
Carpino, Mario
Carruba, Valerio
Cefola, Paul
Celletti, Alessandra
Chakrabarty, Dalia
Chambers, John
Chapanov, Yavor
Chapront, Jean
Chapront-Touze, Michelle
Haghighipour, Nader
Hallan, Prem
Hamid, S.
Hamilton, Douglas
Hanslmeier, Arnold
Hau, George
He, Miao-fu
Heggie, Douglas
Helali, Yhya
Henon, Michel
Hori, Genichiro
Horner, Jonathan
Hu, Xiaogong
Huang, Tianyi
Huang, Cheng
Hurley, Jarrod
Hut, Piet
Ipatov, Sergei
Ismail, Mohamed
Ito, Takashi
Ivanova, Violeta
Ivantsov, Anatoliy
Jakubík, Marian
Janiczek, Paul
Jefferys, William
Ji, Jianghui
Jiang, Ing-Guey
Kalvouridis, Tilemachos
Kammeyer, Peter
Kholshevnikov, Konstantin
Kim, Sungsoo
Kim, Yoo Jea
King, Sun-Kun
Kinoshita, Hiroshi
Kitiashvili, Irina
Klioner, Sergei
Klocok, Lubomir
Klokocnik, Jaroslav
Kokubo, Eiichiro
Korchagin, Vladimir
Koshkin , Nikolay
Kovavčević, Andjelka
Kovalevsky, Jean
Kozai, Yoshihide
Krasinsky, George
Krivov, Alexander
Polyakhova, Elena
Puetzfeld, Dirk
Rodin, Alexander
Rodriguez-Eillamil, R.
Roman, Rodica
Rossi, Alessandro
Roy, Archie
Ryabov, Yurij
Saad, Abdel-naby
Sansaturio, Maria
Scheeres, Daniel
Scholl, Hans
Schubart, Joachim
Sconzo, Pasquale
Segan, Stevo
Sehnal, Ladislav
Seidelmann, P.
Sein-Echaluce, M.
Shankland, Paul
Shapiro, Irwin
Sheng, Wan
Shevchenko, Ivan
Sima, Zdislav
Simó, Carles
Simon, Jean-Louis
Skripnichenko, Vladimir
Soffel, Michael
Sokolov, Leonid
Sokolov, Viktor
Sorokin, Nikolaj
Souchay, Jean
Standish, E.
Stellmacher, Irène
Steves, Bonita
Sultanov, G.
Sun, Yisui
Sun, Fuping
Sweatman, Winston
Szenkovits, Ferenc
Tao, Jin-he
Tatevyan, Suriya
Tawadros, Maher
Taylor, Donald
Thiry, Yves
Thuillot, William
Tiscareno, Matthew

Choi, Kyu-Hong
Christou, Apostolos
Cionco, Rodolfo
Colin, Jacques
Conrad, Albert
Contopoulos, George
Cooper, Nicholas
Counselman, Charles
Danby, J.
Davis, Morris
Deleflie, Florent
Descamps, Pascal
Di Sisto, Romina
Dikova, Smilyana
Dong, Xiaojun
Dormand, John
Dourneau, Gerard
Drozyner, Andrzej
Du, Lan
Duncombe, Raynor
Duriez, Luc
Dvorak, Rudolf
Efroimsky, Michael
El, Bakkali
Elipe, Sánchez
Emelianov, Nikolaj
Emelyanenko, Vacheslav
Fernández, Silvia
Ferraz Mello, Sylvio
Ferrer, Martinez
Fiala, Alan
Floria Peralta, Luis
Fong, Chugang
Froeschlé, Claude
Fukushima, Toshio
Galletto, Dionigi
Gaposchkin, Edward
Gaska, Stanislaw
Geng, Lihong
Giordano, Claudia
Giuliatti Winter, Silvia
Goldreich, Peter
Gomes, Rodney
Goodwin, Simon
Goudas, Constantine
Goz'dziewski, Krzysztof
Grebenikov, Evgenij
Greenberg, Richard
Gronchi, Giovanni
Gusev, Alexander
Hadjidemetriou, John
Kuznetsov, Eduard
La Spina, Alessandra
Lala, Petr
Lazovic, Jovan
Lecavelier des Etangs, Alain
Lega, Elena
Levine, Stephen
Liao, Xinhao
Lieske, Jay
Lin, Douglas
Lissauer, Jack
Lu, BenKui
Lucchesi, David
Lundquist, Charles
Ma, Jingyuan
Ma, Lihua
Makhlouf, Amar
Marchal, Christian
Marsden, Brian
Martinet, Louis
Masaki, Yoshimitsu
Matas, Vladimir
Mavraganis, Anastasios
Melbourne, William
Merman, Hirsh G.
Metris, Gilles
Michel, Patrick
Mignard, Franois
Mioc, Vasile
Moreira Morais, Maria
Musen, Peter
Muzzio, Juan
Mysen, Eirik
Nacozy, Paul
Namouni, Fathi
Navone, Hugo
Nefed'ev, Yury
Nobili, Anna
O'Handley, Douglas
Ollongren, A.
Omarov, Tuken
Orellana, Rosa
Orlov, Victor
Oscare Giacaglia, Giorgio
Osório, Isabel
Osório, José
Parv, Bazil
Pauwels, Thierry
Petit, Jean-Marc
Petrovskaya, Margarita
Pilat-Lohinger, Elke
Tommei, Giacomo
Tremaine, Scott
Tsuchida, Masayoshi
Valsecchi, Giovanni
Valtonen, Mauri
Varvoglis, Harry
Vashkovyak, Sofja
Vassiliev, Nikolaj
Veillet, Christian
Vieira, Martins
Vieira Neto, Ernesto
Vienne, Alain
Vilhena, Rodolpho
Vinet, Jean-Yves
Virtanen, Jenni
Vondrák, Jan
Walch, Jean-Jacques
Watanabe, Noriaki
Whipple, Arthur
Wiegert, Paul
Williams, Carol
Winter, Othon
Wnuk, Edwin
Wu, Lianda
Wytrzyszczak, Iwona
Xiong, Jianning
Yang, Xuhai
Yano, Taihei
Yi, Zhaohua
Yokoyama, Tadashi
Yong, Zhou
Yoshida, Junzo
Yoshida, Haruo
Yseboodt, Marie
Yuasa, Manabu
Zafiropoulos, Basil
Zare, Khalil
Zhang, Sheng-Pan
Zhang, Wei
Zhang, Xiaoxiang
Zhang, Wei
Zhao, Changyin
Zhao, You
Zhao, Haibin
Zhdanov, Valery
Zheng, Xuetang
Zheng, Jia-Qing
Zheng, Yong
Zhou, Hongnan
Zhu, Wenyao

Division I Commission 8 Astrometry

President Evans, Dafydd
Vice-President Zacharias, Norbert

Organizing Committee

Andrei, Alexandre
Brown, Anthony
Gouda, Naoteru
Kumkova, Irina
Popescu, Petre
Souchay, Jean
Unwin, Stephen
Zhu, Zi

Members

Abad Hiraldo, Carlos
Abbas, Ummi
Ahmed, Abdel-aziz
Arenou, Frédéric
Argyle, Robert
Arias, Elisa
Arlot, Jean-Eudes
Assafin, Marcelo
Babusiaux, Carine
Bacham, Eswar
Backer, Donald
Badescu, Octavian
Bakhtigaraev, Nail
Ballabh, Goswami
Bangert, John
Barkin, Yuri
Bastian, Ulrich
Belizon, Fernando
Benedict, George
Benevides Soares, Paulo
Bernstein, Hans-Heinrich
Bien, Reinhold
Boboltz, David
Bouchard, Antoine
Bougeard, Mireille
Bradley, Arthur
Branham, Richard
Brosche, Peter
Brouw, Willem
Bucciarelli, Beatrice
Capitaine, Nicole
Carrasco, Guillermo
Chakrabarty, Dalia
Chapanov, Yavor
Chen, Li
Chen, Alfred
Cioni, Maria-Rosa
Cooper, Nicholas
Corbin, Thomas
Costa, Edgardo
Crézé, Michel
Crifo, Françoise
Cudworth, Kyle
Dahn, Conard
Danylevsky, Vassyl
De Bruijne, Jos
Dejaiffe, René
Helmer, Leif
Hemenway, Paul
Hering, Roland
Heudier, Jean-Louis
Hill, Graham
Hobbs, David
Høg, Erik
Hong, Zhang
Ianna, Philip
Irwin, Michael
Ivantsov, Anatoliy
Jackson, Paul
Jahreiss, Hartmut
Jefferys, William
Jin, WenJing
Johnston, Kenneth
Jones, Burton
Jones, Derek
Jordi, Carme
Kalomeni, Belinda
Kanayev, Ivan
Kaplan, George
Kazantseva, Liliya
Kharchenko, Nina
Kharin, Arkadiy
Kislyuk, Vitalij
Klemola, Arnold
Klioner, Sergei
Klock, Benny
Kovalevsky, Jean
Kuimov, Konstantin
Kurzynska, Krystyna
Lattanzi, Mario
Latypov, A.
Lazorenko, Peter
Le Poole, Rudolf
Lenhardt, Helmut
Lepine, Sebastien
Li, Zhigang
Li, Qi
Lindegren, Lennart
Lopez, José
Lopez, Carlos
Lu, Phillip
Lu, Chunlin
Ma, Wenzhang
MacConnell, Darrell
Platais, Imants
Poma, Angelo
Poppe, Paulo
Pourbaix, Dimitri
Protsyuk, Yuri
Proverbio, Edoardo
Prusti, Timo
Pugliano, Antonio
Rafferty, Theodore
Raimond, Ernst
Reffert, Sabine
Reynolds, John
Rizvanov, Naufal
Rodin, Alexander
Roemer, Elizabeth
Roeser, Siegfried
Russell, Jane
Sadzakov, Sofija
Sanders, Walt
Sarasso, Maria
Sato, Koichi
Schilbach, Elena
Schildknecht, Thomas
Scholz, Ralf-Dieter
Schwan, Heiner
Schwekendiek, Peter
Ségransan, Damien
Seidelmann, P.
Shelus, Peter
Shen, Kaixian
Shen, Zhiqiang
Shulga, Oleksandr
Smart, Richard
Söderhjelm, Staffan
Solarić, Nikola
Sôma, Mitsuru
Sovers, Ojars
Sozzetti, Alessandro
Spoljaric, Drago
Standish, E.
Stein, John
Steinert, Klaus
Steinmetz, Matthias
Suganuma, Masahiro
Sun, Fuping
Tang, Zheng
Tedds, Jonathan

Del Santo, Melania
Delmas, Christian
Devyatkin, Aleksandr
Dick, Wolfgang
Dick, Steven
Dinescu-Casetti, Dana
Dommanget, Jean
Du, Lan
Ducourant, Christine
Duma, Dmitrij
Duncombe, Raynor
Emilio, Marcelo
Fabricius, Claus
Fan, Yu
Fey, Alan
Firneis, Maria
Fomin, Valery
Fors, Octavi
Franz, Otto
Fredrick, Laurence
Fresneau, Alain
Froeschlé, Michel
Fujishita, Mitsumi
Fukushima, Toshio
Gatewood, George
Gaume, Ralph
Gauss, Stephen
Geffert, Michael
Germain, Marvin
Goddi, Ciriaco
Gontcharov, George
Goyal, A.
Guibert, Jean
Guseva, Irina
Hajian, Arsen
Hanson, Robert
Hartkopf, William
Maigurova, Nadiia
Makarov, Valeri
Mallamaci, Claudio
Marschall, Laurence
Martin, Vera
Mason, Brian
Mavridis, Lyssimachos
McAlister, Harold
McLean, Brian
Mignard, François
Mink, Douglas
Mioc, Vasile
Miyamoto, Masanori
Monet, David
Morbidelli, Roberto
Muinos Haro, José
Murray, Andrew
Nakagawa, Akiharu
Nakajima, Koichi
Nefed'ev, Yury
Nefedeva, Antonina
Nikoloff, Ivan
Noël, Fernando
Nunez, Jorge
Ohnishi, Kouji
Oja, Tarmo
Olsen, Hans
Osborn, Wayne
Osório, José
Pakvor, Ivan
Pannunzio, Renato
Pascu, Dan
Pauwels, Thierry
Penna, Jucira
Perryman, Michael
Ping, Jinsong
Pinigin, Gennadiy
Teixeira, Ramachrisna
Teixeira, Paula
ten Brummelaar, Theo
Thuillot, William
Tsujimoto, Takuji
Turon, Catherine
Upgren, Arthur
Urban, Sean
Vallejo, Miguel
van Altena, William
van Leeuwen, Floor
Vass, Gheorghe
Vilkki, Erkki
Volyanskaya, Margarita
Wallace, Patrick
Walter, Hans
Wang, Jiaji
Wang, Zhengming
Wasserman, Lawrence
Westerhout, Gart
White, Graeme
Wielen, Roland
Xia, Yifei
Xu, Jiayan
Yamada, Yoshiyuki
Yang, Tinggao
Yano, Taihei
Yasuda, Haruo
Yatsenko, Anatolij
Yatskiv, Yaroslav
Ye, Shuhua
Yoshizawa, Masanori
Zacchei, Andrea
Zhang, Wei
Zheng, Yong

Division II Commission 10 Solar Activity

President van Driel-Gesztelyi, Lidia
Vice-President Schrijver, Karel

Organizing Committee

Charbonneau, Paul
Fletcher, Lyndsay
Hasan, S. Sirajul
Hudson, Hugh
Kusano, Kanya
Mandrini, Cristina
Peter, Hardi
Vršnak, Bojan
Yan, Yihua

Members

Abbett, William
Abdelatif, Toufik
Aboudarham, Jean
Abraham, Peter
Abramenko, Valentina
Afram, Nadine
Ahluwalia, Harjit
Ai, Guoxiang
Akita, Kyo
Alissandrakis, Costas
Almleaky, Yasseen
Altrock, Richard
Altschuler, Martin
Altyntsev, Alexandre
Aly, Jean-Jacques
Ambastha, Ashok
Ambroz, Pavel
Anastasiadis, Anastasios
Andersen, Bo Nyborg
Anderson, Kinsey
Andretta, Vincenzo
Andries, Jesse
Antiochos, Spiro
Antonucci, Ester
Anzer, Ulrich
Aschwanden, Markus
Atac, Tamer
Athay, R.
Aurass, Henry
Babin, Arthur
Bagala, Liria
Bagare, S.
Balasubramaniam, K.
Balikhin, Michael
Ballester, José
Bao, Shudong
Baranovsky, Edward
Barrow, Colin
Bárta, Miroslav
Basu, Sarbani
Batchelor, David
Beckers, Jacques
Bedding, Timothy
Beebe, Herbert
Bell, Barbara
Bellot Rubio, Luis
Belvedere, Gaetano
Hagyard, Mona
Hammer, Reiner
Hanaoka, Yoichiro
Hanasz, Jan
Hansen, Richard
Hanslmeier, Arnold
Hara, Hirohisa
Harra, Louise
Harvey, John
Hasan, S. Sirajul
Hathaway, David
Haugan, Stein Vidar
Hayward, John
Heinzel, Petr
Henoux, Jean-Claude
Herdiwijaya, Dhani
Hiei, Eijiro
Hildebrandt, Joachim
Hildner, Ernest
Hochedez, Jean-François
Hoeksema, Jon
Hohenkerk, Catherine
Hollweg, Joseph
Holman, Gordon
Holzer, Thomas
Hood, Alan
Houdebine, Eric
Howard, Robert
Hoyng, Peter
Hudson, Hugh
Hughes, David
Hurford, Gordon
Ishii, Takako
Ishitsuka, Mutsumi
Isliker, Heinz
Ivanchuk, Victor
Ivanov, Evgenij
Ivchenko, Vasily
Jackson, Bernard
Jain, Rajmal
Jakimiec, Jerzy
Janssen, Katja
Jardine, Moira
Ji, Haisheng
Jiang, Yun
Jiang, Aimin
Jimenez, Mancebo
Pevtsov, Alexei
Pflug, Klaus
Phillips, Kenneth
Pick, Monique
Pipin, Valery
Plainaki, Christina
Pneuman, Gerald
Podesta, John
Poedts, Stefaan
Pohjolainen, Silja
Poland, Arthur
Poquerusse, Michel
Preka-Papadema, Panagiota
Pres, Pawek
Priest, Eric
Proctor, Michael
Prokakis, Theodore
Raadu, Michael
Ramelli, Renzo
Rao, A.
Raoult, Antoinette
Raulin, Jean-Pierre
Rayrole, Jean
Reale, Fabio
Rees, David
Reeves, Hubert
Regnier, Stéphane
Régulo, Clara
Reinard, Alysha
Rendtel, Juergen
Rengel, Miriam
Reshetnyk, Volodymyr
Ri, Son-Jae
Riehokainen, Aleksandr
Rijnbeek, Richard
Riley, Pete
Robinson Jr., Richard
Roca Cortés, Teodoro
Roemer, Max
Romanchuk, Pavel
Romoli, Marco
Rompolt, Bogdan
Roša, Dragan
Roudier, Thierry
Rovira, Marta
Roxburgh, Ian
Rozelot, Jean-Pierre

Bemporad, Alessandro
Benevolenskaya, Elena
Benz, Arnold
Berger, Mitchell
Bergeron, Jacqueline
Berghmans, David
Berrilli, Francesco
Bewsher, Danielle
Bianda, Michèle
Bingham, Robert
Bobylev, Vadim
Bocchia, Romeo
Bogdan, Thomas
Bommier, Veronique
Bondal, Krishna
Bornmann, Patricia
Botha, Gert
Bothmer, Volker
Bouchard, Antoine
Bougeret, Jean-Louis
Boyer, René
Brajša, Roman
Brandenburg, Axel
Brandt, Peter
Braun, Douglas
Bray, Robert
Brekke, Pål
Bromage, Barbara
Brooke, John
Brosius, Jeffrey
Brown, John
Browning, Philippa
Brun, Allan
Bruner, Marilyn
Bruno, Roberto
Bruns, Andrey
Brynildsen, Nils
Buccino, Andrea
Buchlin, Eric
Buecher, Alain
Büchner, Jörg
Bumba, Václav
Busà, Innocenza
Čadež, Vladimir
Cally, Paul
Cane, Hilary
Carbonell, Marc
Cargill, Peter
Cauzzi, Gianna
Chae, Jongchul
Chambe, Gilbert
Chandra, Suresh
Chang, Heon-Young
Channok, Chanruangrit
Chaoudhuri, Arnab
Chapman, Gary
Charbonneau, Paul
Chen, Peng Fei
Jing, Hairong
Jockers, Klaus
Jones, Harrison
Jordan, Stuart
Joselyn, Jo
Jurčák, Jan
Kaburaki, Osamu
Kahler, Stephen
Kallenbach, Reinald
Kalman, Béla
Kaltman, Tatyana
Kane, Sharad
Kang, Jin-Sok
Käpylä, Petri
Karlický, Marian
Karpen, Judith
Kašparová, Jana
Katsova, Maria
Kaufmann, Pierre
Keppens, Rony
Khan, J
Khumlemlert, Thiranee
Kim, Iraida
Kim, Kap-sung
Kiplinger, Alan
Kitai, Reizaburo
Kitchatinov, Leonid
Kitiashvili, Irina
Kjeldseth-Moe, Olav
Kleczek, Josip
Klein, Karl
Kliem, Bernhard
Klvana, Miroslav
Kołomański, Sylwester
Kondrashova, Nina
Kontar, Eduard
Kopylova, Yulia
Kostik, Roman
Kotrč, Pavel
Koutchmy, Serge
Kovacs, Agnes
Koza, Julius
Kozlovsky, Ben
Krimigis, Stamatios
Krucker, Sam
Kryshtal, Alexander
Kryvodubskyj, Valery
Kubota, Jun
Kučera, Aleš
Kulhánek, Petr
Kundu, Mukul
Kuperus, Max
Kurochka, Evgenia
Kurokawa, Hiroki
Kusano, Kanya
Kuznetsov, Vladimir
Labrosse, Nicolas
Landi, Simone
Rudawy, Pawel
Rúdiger, Gúnther
Ruffolo, David
Ruiz Cobo, Basilio
Rušin, Vojtech
Rust, David
Ruždjak, Vladimir
Ruzickova-Topolova, B.
Rybák, Jan
Rybanský, Milan
Sahal-Brechot, Sylvie
Saito, Kuniji
Sáiz, Alejandro
Sakao, Taro
Sakurai, Kunitomo
Sánchez, Almeida
Saniga, Metod
Sasso, Clementina
Satoshi, Nozawa
Sattarov, Isroil
Sawyer, Constance
Schindler, Karl
Schlichenmaier, Roef
Schmahl, Edward
Schmelz, Joan
Schmidt, H.
Schmieder, Brigitte
Schober, Hans
Schússler, Manfred
Schwartz, Pavol
Schwenn, Rainer
Semel, Meir
Shea, Margaret
Sheeley, Neil
Shibasaki, Kiyoto
Shimojo, Masumi
Shin'Ichi, Nagata
Shine, Richard
Sigalotti, Leonardo
Silva, Adriana
Simnett, George
Simon, Guy
Sinha, Krishnanand
Smaldone, Luigi
Smith, Dean
Smolkov, Gennadij
Snegirev, Sergey
Sobotka, Michal
Socas-Navarro, Hector
Solanki, Sami
Soliman, Mohamed
Soloviev, Alexandr
Somov, Boris
Spadaro, Daniele
Spicer, Daniel
Spruit, Henk
Steiner, Oskar
Stellmacher, Götz

Chen, Zhiyuan
Chernov, Gennadij
Chertok, Ilya
Chertoprud, Vadim
Chiuderi-Drago, Franca
Chiueh, Tzihong
Cho, Kyung Suk
Choe, Gwangson
Choudhary, Debi Prasad
Chupp, Edward
Cliver, Edward
Coffey, Helen
Collados, Manuel
Conway, Andrew
Cook, John
Correia, Emilia
Costa, Joaquim
Craig, Ian
Cramer, Neil
Crannell, Carol
Culhane, John
Curdt, Werner
Dalla, Silvia
Dasso, Sergio
Datlowe, Dayton
Davila, Joseph
de Groof, Anik
de Jager, Cornelis
de Toma, Giuliana
Dechev, Momchil
Del Toro Iniesta, José
Démoulin, Pascal
Deng, YuanYong
Dennis, Brian
Dere, Kenneth
Deubner, Franz-Ludwig
Dialetis, Dimitris
Ding, Mingde
Dinulescu, Simona
Dobler, Wolfgang
Dobrzycka, Danuta
Dollfus, Audouin
Dorch, Søren Bertil
Dorotovic, Ivan
Dryer, Murray
Dubau, Jacques
Dubois, Marc
Duchlev, Peter
Duldig, Marc
Dumitrache, Cristiana
Dwivedi, Bhola
Eddy, John
Efimenko, Volodymyr
Elste, Gunther
Emslie, A.
Engvold, Oddbjørn
Enome, Shinzo
Erdelyi, Robert
Landman, Donald
Lang, Kenneth
Lawrence, John
Lazrek, Mohamed
Leibacher, John
Leiko, Uliana
Leka, Kimberly
Leroy, Jean-Louis
Leroy, Bernard
Li, Wei
Li, Hui
Li, Kejun
Li, Zhi
Lie-Svendsen, Oystein
Lima, João
Lin, Yong
Liritzis, Ioannis
Liu, Xinping
Liu, Yang
Liu, Yu
Livshits, Mikhail
Lopez Fuentes, Marcelo
Loukitcheva, Maria
Low, Boon
Lozitskij, Vsevolod
Lundstedt, Henrik
Luo, Xianhan
Machado, Marcos
Mackay, Duncan
MacKinnon, Alexander
MacQueen, Robert
Makita, Mitsugu
Malara, Francesco
Malherbe, Jean-Marie
Malitson, Harriet
Malville, J.
Manabe, Seiji
Mann, Gottfried
Marcu, Alexandru
Maričić, Darije
Marilena, Mierla
Maris, Georgeta
Mariska, John
Marková, Eva
Martens, Petrus
Mason, Glenn
Masuda, Satoshi
Matsuura, Oscar
Mattig, W.
Maxwell, Alan
McAteer, R.T. James
McCabe, Marie
McIntosh, Patrick
McKenna, Lawlor
McLean, Donald
Mein, Pierre
Melnik, Valentin
Mendes, Da
Stenflo, Jan
Stepanian, Natali
Stepanov, Aleksander
Steshenko, N.
Stewart, Ronald
Stix, Michael
Stoker, Pieter
Strong, Keith
Struminsky, Alexei
Sturrock, Peter
Subramanian, K.
Subramanian, Prasad
Sukartadiredja, Darsa
Suzuki, Takeru
Švestka, Zdeněk
Sykora, Julius
Sylwester, Janusz
Sylwester, Barbara
Szalay, Alex
Takano, Toshiaki
Talon, Raoul
Tamenaga, Tatsuo
Tandberg-Hanssen, Einar
Tang, Yuhua
Tapping, Kenneth
Tarashchuk, Vera
Ternullo, Maurizio
Teske, Richard
Thomas, John
Thomas, Roger
Tikhomolov, Evgeniy
Tlamicha, Antonin
Tlatov, Andrej
Tobias, Steven
Tomczak, Michal
Toshifumi, Shimizu
Treumann, Rudolf
Tripathy, Sushanta
Tritakis, Basil
Trottet, Gerard
Tsap, Yuri
Tsinganos, Kanaris
Tuominen, Ilkka
Uddin, Wahab
Underwood, James
Usoskin, Ilya
van den Oord, Bert
van der Heyden, Kurt
van der Linden, Ronald
van Hoven, Gerard
van't Veer, Frans
Vaughan, Arthur
Vekstein, Gregory
Velli, Marco
Venkatakrishnan, P.
Ventura, Rita
Vergez, Madeleine
Verheest, Frank

Ermolli, Ilaria
Esenoğlu, Hasan
Falciani, Roberto
Falewicz, Robert
Fan, Yuhong
Fang, Cheng
Fárník, František
Fernandes, Francisco
Ferreira, João
Ferriz Mas, Antonio
Fisher, George
Fluri, Dominique
Foing, Bernard
Fontenla, Juan
Forbes, Terry
Forgács-Dajka, Emese
Fossat, Eric
Fu, Hsieh-Hai
Gabriel, Alan
Gaizauskas, Victor
Galal, A.
Galloway, David
Galsgaard, Klaus
Gan, Weiqun
García de la Rosa, Ignacio
Gary, Gilmer
Gelfreikh, Georgij
Gergely, Tomas
Ghizaru, Mihai
Gibson, David
Gill, Peter
Gilliland, Ronald
Gilman, Peter
Giménez de Castro, Carlos
Glatzmaier, Gary
Gleisner, Hans
Godoli, Giovanni
Goedbloed, Johan
Gokhale, Moreshwar
Gómez, Daniel
Gontikakis, Constantin
Goossens, Marcel
Gopasyuk, Olga
Graffagnino, Vito
Grandpierre, Attila
Gray, Norman
Grechnev, Victor
Gregorio, Anna
Grib, Sergey
Gudiksen, Boris
Guhathakurta, Madhulika
Gupta, Surendra
Gurman, Joseph
Györi, Lajos
Messerotti, Mauro
Messmer, Peter
Meszárosová, Hana
Michalek, Gregory
Miletsky, Eugeny
Miralles, Mari
Mogilevskij, Emmanuil
Mohan, Anita
Moreno Insertis, Fernando
Morita, Satoshi
Moriyama, Fumio
Motta, Santo
Muller, Richard
Musielak, Zdzislaw
Nakajima, Hiroshi
Nakariakov, Valery
Namba, Osamu
Narain, Udit
Neidig, Donald
Neukirch, Thomas
Neupert, Werner
Nickeler, Dieter
Ning, Zongjun
Nishi, Keizo
Nocera, Luigi
Noens, Jacques-Clair
Noyes, Robert
Nussbaumer, Harry
Obridko, Vladimir
Ofman, Leon
Ohki, Kenichiro
Oliver, Ramón
Orlando, Salvatore
özgüç, Atila
Padmanabhan, Janardhan
Paletou, Frederic
Pallé, Pere
Pallé Bagó, Enric
Palus, Pavel
Pan, Liande
Pap, Judit
Parfinenko, Leonid
Pariat, Etienne
Park, Young-Deuk
Parkinson, John
Parkinson, William
Parnell, Clare
Pasachoff, Jay
释Paternò, Lucio
Peres, Giovanni
Petrie, Gordon
Petrosian, Vahe
Petrov, Nikola
Petrovay, Kristof
Verma, V.
Verwichte, Erwin
Vial, Jean-Claude
Vieytes, Mariela
Vilinga, Jaime
Vilmer, Nicole
Vinod, S.
Voitenko, Yuriy
Wang, Huaning
Wang, Yi-ming
Wang, Haimin
Wang, Min
Wang, Dongguang
Wang, Shujuan
Wang, Jingyu
Webb, David
Wentzel, Donat
White, Stephen
Wiehr, Eberhard
Wikstol, Oivind
Wilson, Peter
Winebarger, Amy
Wittmann, Axel
Wöhl, Hubertus
Wolfson, Richard
Woltjer, Lodewijk
Wu, De Jin
Wu, Shi
Xie, Xianchun
Xu, Aoao
Xu, Jun
Xu, Zhi
Yan, Yihua
Yang, Zhiliang
Yang, Hong-Jin
Yang, Lei
Yeh, Tyan
Yeşilyurt, Ibrahim
Yi, Yu
Yoshimura, Hirokazu
Yu, Dai
Yun, Hong-Sik
Zachariadis, Theodosios
Zappalà, Rosario
Zelenka, Antoine
Zhang, Mei
Zhang, Jun
Zharkova, Valentina
Zhitnik, Igor
Zhou, Daoqi
Zhou, Guiping
Zhugzhda, Yuzef
Zirin, Harold
Zlobec, Paolo

Division II Commission 12 Solar Radiation & Structure

President Kosovichev, Alexander
Vice-President Cauzzi, Gianna

Organizing Committee

Asplund, Martin
Bogdan, Thomas
Brandenburg, Axel
Cauzzi, Gianna
Christensen-Dalsgaard, Jørgen
Cram, Lawrence
Gan, Weiqun
Gizon, Laurent
Heinzel, Petr
Kuznetsov, Vladimir
Rovira, Marta
Shchukina, Nataliya
Venkatakrishnan, P.
Warren Jr., Wayne

Members

Abbett, William
Aboudarham, Jean
Acton, Loren
Ai, Guoxiang
Aime, Claude
Alissandrakis, Costas
Altrock, Richard
Altschuler, Martin
Andersen, Bo Nyborg
Ando, Hiroyasu
Andretta, Vincenzo
Ansari, S.M.
Antia, H.
Arnaud, Jean-Paul
Artzner, Guy
Asai, Ayumi
Athay, R.
Ayres, Thomas
Babayev, Elchin
Baliunas, Sallie
Balthasar, Horst
Bárta, Miroslav
Basu, Sarbani
Baturin, Vladimir
Beckers, Jacques
Beckman, John
Beebe, Herbert
Beiersdorfer, Peter
Bemporad, Alessandro
Benford, Gregory
Bhardwaj, Anil
Bhattacharyya, J.
Bi, Shao
Bianda, Michèle
Bingham, Robert
Blackwell, Donald
Blamont, Jacques-Emile
Bocchia, Romeo
Bommier, Veronique
Bonnet, Roger
Book, David
Bornmann, Patricia
Borovik, Valery
Bougeret, Jean-Louis
Glatzmaier, Gary
Godoli, Giovanni
Goldman, Martin
Gomez, Maria
Gopalswamy, Natchimuthuk
Grevesse, Nicolas
Guhathakurta, Madhulika
Hagyard, Mona
Hamedivafa, Hashem
Hammer, Reiner
Harvey, John
Hein, Righini
Hejna, Ladislav
Hiei, Eijiro
Hildner, Ernest
Hill, Frank
Hoang, Binh
House, Lewis
Howard, Robert
Hoyng, Peter
Huang, Guangli
Illing, Rainer
Ivanov, Evgenij
Jabbar, Sabeh
Jackson, Bernard
Janssen, Katja
Jefferies, Stuart
Jones, Harrison
Jordan, Carole
Jordan, Stuart
Jurčák, Jan
Kalkofen, Wolfgang
Kalman, Béla
Kaltman, Tatyana
Käpylä, Petri
Karlický, Marian
Karpen, Judith
Kaufmann, Pierre
Keil, Stephen
Khan, J
Khetsuriani, Tsiala
Khumlemlert, Thiranee
Kim, Yong-Cheol
Kim, Iraida
Petrovay, Kristof
Pflug, Klaus
Phillips, Kenneth
Picazzio, Enos
Poquerusse, Michel
Priest, Eric
Prokakis, Theodore
Qu, Zhong Quan
Radick, Richard
Ramelli, Renzo
Raoult, Antoinette
Reardon, Kevin
Rees, David
Régulo, Clara
Rengel, Miriam
Riehokainen, Aleksandr
Roca Cortés, Teodoro
Roddier, François
Rodríguez Hidalgo, Inés
Roudier, Thierry
Rouppe van der Voort, Luc
Rušin, Vojtech
Rutten, Robert
Ryabov, Boris
Rybanský, Milan
Sakai, Junichi
Samain, Denys
Sánchez, Almeida
Saniga, Metod
Sasso, Clementina
Satoshi, Nozawa
Sauval, A.
Schleicher, Helmold
Schmahl, Edward
Schmidt, Wolfgang
Schmieder, Brigitte
Schmitt, Dieter
Schober, Hans
Schou, Jesper
Schüssler, Manfred
Schwartz, Steven
Schwartz, Pavol
Schwenn, Rainer
Semel, Meir

Brandt, Peter
Bray, Robert
Breckinridge, James
Brosius, Jeffrey
Bruls, Jo
Brun, Allan
Bruner, Marilyn
Bruning, David
Buchlin, Eric
Bumba, Václav
Čadež, Vladimir
Cavallini, Fabio
Ceppatelli, Guido
Chambe, Gilbert
Chan, Kwing
Chapman, Gary
Chertok, Ilya
Choe, Gwangson
Clark, Thomas
Clette, Frederic
Collados, Manuel
Cook, John
Couvidat, Sebastien

Cox, Arthur
Craig, Ian
Cramer, Neil
Dara, Helen
Dasso, Sergio
de Jager, Cornelis
de Toma, Giuliana
Dechev, Momchil
Degenhardt, Detlev
Del Toro Iniesta, José
Deliyannis, Jean
Demarque, Pierre
Deming, Leo
Deubner, Franz-Ludwig
Di Mauro, Maria Pia
Ding, Mingde
Diver, Declan
Donea, Alina
Dravins, Dainis
Dumont, Simone
Duvall Jr., Thomas
Ehgamberdiev, Shuhrat
Einaudi, Giorgio
Elliott, Ian
Elste, Gunther
Epstein, Gabriel
Ermolli, Ilaria
Esser, Ruth
Evans, J.
Falciani, Roberto
Falewicz, Robert
Fang, Cheng
Feldman, Uri
Fernandes, Francisco
Fiala, Alan

Kitiashvili, Irina
Klein, Karl
Kneer, Franz
Knoelker, Michael
Kołomański, Sylwester
Kononovich, Edvard
Kopylova, Yulia
Kostik, Roman
Kotov, Valery
Kotrč, Pavel
Koutchmy, Serge
Koza, Julius
Krivova, Natalie
Kryvodubskyj, Valery
Kubičela, Aleksandar
Kučcera, Aleš
Kundu, Mukul
Kuperus, Max
Labrosse, Nicolas
Landi, Degli Innocenti
Landman, Donald
Landolfi, Marco
Lanzafame, Alessandro
Leibacher, John
Leroy, Jean-Louis
Linsky, Jeffrey
Livingston, William
Locke, Jack
Lopez Arroyo, M.
Lopez Fuentes, Marcelo
Loukitcheva, Maria
Lüst, Reimar
Makita, Mitsugu
Mandrini, Cristina
Maričić, Darije
Marilli, Ettore
Marmolino, Ciro
Mattig, W.
McAteer, R.T. James
McKenna, Lawlor
Mein, Pierre
Melrose, Donald
Meszárosová, Hana
Meyer, Friedrich
Michard, Raymond
Mihalas, Dimitri
Milkey, Robert
Monteiro, Mario João
Moore, Ronald
Morabito, David
Moreno Insertis, Fernando
Morita, Satoshi
Moriyama, Fumio
Mouradian, Zadig
Muller, Richard
Munro, Richard
Namba, Osamu
Neckel, Heinz
Nesis, Anastasios

Severino, Giuseppe
Shchukina, Nataliya
Sheeley, Neil
Sheminova, Valentina
Shine, Richard
Sigalotti, Leonardo
Sigwarth, Michael
Simon, George
Simon, Guy
Singh, Jagdev
Sinha, Krishnanand
Sivaraman, Koduvayur
Skumanich, Andrew
Smith, Peter
Solanki, Sami
Soloviev, Alexandr
Spicer, Daniel
Stathopoulou, Maria
Staude, Juergen
Stebbins, Robin
Steffen, Matthias
Steiner, Oskar
Stenflo, Jan
Stix, Michael
Stodilka, Myroslav
Straus, Thomas
Suematsu, Yoshinori
Švestka, Zdeněk
Tandberg-Hanssen, Einar
Tarashchuk, Vera
Teplitskaya, Raisa
Thomas, John
Torelli, M.
Trujillo Bueno, Javier
Tsap, Teodor
Tsiklauri, David
Tsiropoula, Georgia
Usoskin, Ilya
van Hoven, Gerard
Vaughan, Arthur
Ventura, Paolo
Vial, Jean-Claude
Vilmer, Nicole
Voitsekhovska, Anna
Volonte, Sergio
von der Lühe, Oskar
Vukićević-Karabin, Mirjana
Wang, Jingxiu
Wang, Shujuan
Warwick, James
Weiss, Nigel
Wentzel, Donat
Wiik Toutain, Jun
Wilson, Peter
Wittmann, Axel
Wöhl, Hubertus
Worden, Simon
Wu, Hsin-Heng
Xie, Xianchun

Fisher, George
Fleck, Bernhard
Fluri, Dominique
Fofi, Massimo
Fomichev, Valerij
Fontenla, Juan
Forgács-Dajka, Emese
Fossat, Eric
Foukal, Peter
Frazier, Edward
Frölich, Claus
Gabriel, Alan
Gaizauskas, Victor
Garcia, Rafael
García-Berro, Enrique
Nicolas, Kenneth
Nishi, Keizo
Nordlund, Aake
Noyes, Robert
Ossendrijver, Mathieu
Owocki, Stanley
Padmanabhan, Janardhan
Pallé, Pere
Palus, Pavel
Papathanasoglou, Dimitrios
Parkinson, William
Pasachoff, Jay
Pecker, Jean-Claude
Petrie, Gordon
Petrov, Nikola
Xu, Zhi
Yoichiro , Suzuki
Yoshimura, Hirokazu
Youssef, Nahed
Yun, Hong-Sik
Zampieri, Luca
Zarro, Dominic
Zelenka, Antoine
Zhang, Jun
Zhao, Junwei
Zharkova, Valentina
Zhou, Daoqi
Zhugzhda, Yuzef
Zirin, Harold
Zirker, Jack

Division XII Commission 14 Atomic & Molecular Data

President Wahlgren, Glenn
Vice-President van Dishoeck, Ewine

Organizing Committee

Beiersdorfer, Peter
Dimitrijevic, Milan
Jorissen, Alain
Mashonkina, Lyudmila
Nilsson, Hampus
Salama, Farid
Tennyson, Jonathan

Members

Adelman, Saul
Afram, Nadine
Aggarwal, Kanti
Allard, Nicole
Allard, France
Allen Jr., John
Allende Prieto, Carlos
Arduini-Malinovsky, Monique
Arion, Douglas
Artru, Marie-Christine
Balanca, Christian
Barber, Robert
Barklem, Paul
Barnbaum, Cecilia
Bartaya, R.
Bautista, Manuel
Bayet, Estelle
Bely-Dubau, Françoise
Bhardwaj, Anil
Biémont, Emile
Black, John
Boechat-Roberty, Heloisa
Bommier, Veronique
Borysow, Aleksandra
Branscomb, L.
Bromage, Gordon
Buchlin, Eric
Carbon, Duane
Carroll, P.
Casasola, Viviana
Chance, Kelly
Chen, Huei-Ru
Chen, Guoming
Christlieb , Norbert
Cichowolski, Silvina
Corliss, C.
Cornille, Marguerite
Czyzak, Stanley
d'Hendecourt, Louis
Dalgarno, Alexander
de Frees, Douglas
de Kertanguy, Amaury
Delsemme, Armand
Désesquelles, Jean
Diercksen, Geerd
Dubau, Jacques
Dufay, Maurice
Hoang, Binh
Homeier, Derek
Horáček, JiříJiri
Hörandel, Jörg
House, Lewis
Huber, Martin
Huebner, Walter
Ignjatovic, Ljubinko
Iliev, Ilian
Irwin, Alan
Irwin, Patrick
Jamar, Claude
Johnson, Donald
Johnson, Fred
Joly, François
Jordan, Carole
Jørgensen, Uffe
Jørgensen, Henning
Kamp, Inga
Kanekar, Nissim
Kato, Takako
Kennedy, Eugene
Kerber, Florian
Kielkopf, John
Kim, Sang Joon
Kingston, Arthur
Kipper, Tonu
Kirby, Kate
Kohl, John
Kramida, Alexander
Kroto, Harold
Kuan, Yi-Jehng
Kupka, Friedrich
Kurucz, Robert
Lambert, David
Landman, Donald
Langhoff, Stephanie
Launay, Jean-Michel
Launay, Françoise
Lawrence, G.
Layzer, David
Le Bourlot, Jacques
Le Floch, André
Leach, Sydney
Léger, Alain
Lemaire, Jean-louis
Linnartz, Harold
Orton, Glenn
Ozeki, Hiroyuki
Palmeri, Patrick
Park, Yong-Sun
Parkinson, William
Paron, Sergio
Peach, Gillian
Pei, Chunchuan
Petrini, Daniel
Pettini, Marco
Piacentini, Ruben
Pradhan, Anil
Querci, François
Quinet, Pascal
Ralchenko, Yuri
Ramírez, José
Rastogi, Shantanu
Redman, Matthew
Rogers, Forrest
Rostas, François
Roueff, Evelyne
Ruder, Hanns
Rudzikas, Zenonas
Ryabchikova, Tatiana
Sahal-Brechot, Sylvie
Sarre, Peter
Savin, Daniel
Schrijver, Johannes
Schultz, David
Sharp, Christopher
Shi, Jianrong
Shore, Bruce
Sinha, Krishnanand
Smith, William
Smith, Peter
Somerville, William
Song, In-Ok
Spielfiedel, Annie
Stancil, Philip
Stark, Glen
Stehlé, Chantal
Strachan, Leonard Jr
Strelnitski, Vladimir
Sutherland, Ralph
Swings, Jean-Pierre
Takayanagi, Kazuo
Tatum, Jeremy

Dulieu, François
Eidelsberg, Michèle
Epstein, Gabriel
Feautrier, Nicole
Federici, Luciana
Filacchione, Gianrico
Fillion, Jean-Hugues
Fink, Uwe
Flower, David
Fluri, Dominique
Fraser, Helen
Fuhr, Jeffrey
Gabriel, Alan
Gallagher III, John
Gargaud, Muriel
Glagolevskij, Yurij
Glinski, Robert
Glover , Simon
Goddi, Ciriaco
Goldbach, Claudine
Grevesse, Nicolas
Hartman, Henrik
Hesser, James
Hillier, Desmond
Liu, Sheng-Yuan
Loulergue, Michelle
Lovas, Francis
Lutz, Barry
Maillard, Jean-Pierre
Martin, William
Mason, Helen
Mickelson, Michael
Mihajlov, Anatolij
Morton, Donald
Mumma, Michael
Nahar, Sultana
Nave, Gillian
Newsom, Gerald
Nicholls, Ralph
Nielsen, Krister
Nollez, Gerard
Nussbaumer, Harry
O'Brian, Thomas
Obi, Shinya
Oetken, L.
Oka, Takeshi
Omont, Alain
Tayal, Swaraj
Tchang-Brillet, Lydia
Tozzi, Gian
TranMinh, Nguyet
Trefftz, Eleonore
Ulyanov, Oleg
van Rensbergen, Walter
Varshalovich, Dmitrij
Völk, Heinrich
Volonte, Sergio
Vujnovic, Vladis
Wang, Junxian
Wang, Feilu
Wiese, Wolfgang
Winnewisser, Gisbert
Wunner, Guenter
Yang, Changgen
Yoshino, Kouichi
Young, Louise
Yu, Yan
Zeippen, Claude
Zeng, Qin
Zhao, Gang
Zirin, Harold

Division III Commission 15 Physical Studies of Comets & Minor Planets

President Cellino, Alberto
Vice-President Bockelee-Morvan, Dominique

Organizing Committee

Bockelee-Morvan, Dominique
Davidsson, Björn
Dotto, Elisabetta
Fitzsimmons, Alan
Jenniskens, Petrus
Lupishko, Dmitrij
Mothé-Diniz, Thais
Tancredi, Gonzalo
Wooden, Diane

Members

Agata, Hidehiko
A'Hearn, Michael
Allegre, Claude
Altwegg, Kathrin
Andrienko, Dmitry
Angeli, Claudia
Archinal, Brent
Arnold, James
Arpigny, Claude
Axford, W.
Babadjanov, Pulat
Bailey, Mark
Barabanov, Sergey
Baransky, Olexander
Barber, Robert
Barker, Edwin
Bar-Nun, Akiva
Barriot, Jean-Pierre
Barucci, Maria
Bell, Jeffrey
Belskaya, Irina
Belton, Michael
Bemporad, Alessandro
Bendjoya, Philippe
Bhardwaj, Anil
Bingham, Robert
Binzel, Richard
Birch, Peter
Birlan, Mirel
Biver, Nicolas
Blamont, Jacques-Emile
Blanco, Armando
Boehnhardt, Hermann
Bonev, Tanyu
Borysenko, Sergiy
Bouška, Jiří
Bowell, Edward
Brandt, John
Brecher, Aviva
Britt, Daniel
Brown, Robert
Brownlee, Donald
Brunk, William
Buie, Marc
Buratti, Bonnie
Burlaga, Leonard
Burns, Joseph
Gil-Hutton, Ricardo
Giovane, Frank
Gounelle, Matthieu
Gradie, Jonathan
Grady, Monica
Green, Daniel
Green, Simon
Greenberg, Richard
Gronkowski, Piotr
Grossman, Lawrence
Grün, Eberhard
Grundy, William
Gulbis, Amanda
Gustafson, Bo
Hadamcik, Edith
Halliday, Ian
Hanner, Martha
Hapke, Bruce
Harris, Alan
Hartmann, William
Harwit, Martin
Haupt, Hermann
Hestroffer, Daniel
Howell, Ellen
Hughes, David
Huntress, Wesley
Ibadinov, Khursand
Ip, Wing-Huen
Irvine, William
Irwin, Patrick
Israelevich, Peter
Ivanova, Aleksandra
Ivanova, Violeta
Ivezic, Zeljko
Jackson, William
Jakubík, Marian
Jedicke, Robert
Jockers, Klaus
Johnson, Torrence
Jorda, Laurent
Kaasalainen, Mikko
Karatekin, Özgür
Käufl, Hans Ulrich
Kavelaars, JJ.
Kawakita, Hideyo
Keay, Colin
Keil, Klaus
Nakamura, Akiko
Nakamura, Tsuko
Napier, William
Neukum, Gerhard
Newburn Jr., Ray
Niedner, Malcolm
Ninkov, Zoran
Nolan, Michael
Noll, Keith
O'Dell, Charles
Ortiz, José
Paolicchi, Paolo
Parisot, Jean-Paul
Peixinho, Nuno
Pellas, Paul
Pendleton, Yvonne
Perez de Tejada, Hector
Picazzio, Enos
Piironen, Jukka
Pilcher, Carl
Pillinger, Colin
Pittich, Eduard
Pittichová, Jana
Prialnik-Kovetz, Dina
Proisy, Paul
Revelle, Douglas
Rickman, Hans
Roemer, Elizabeth
Roig, Fernando
Rosenbush , Vera
Rossi, Alessandro
Rousselot, Philippe
Russel, Sara
Russell, Kenneth
Sagdeev, Roald
Saito, Takao
Salitis, Antonijs
Samarasinha, Nalin
Schaller, Emily
Schleicher, David
Schloerb, F.
Schmidt, H.
Schmidt, Maarten
Schober, Hans
Scholl, Hans
Sekanina, Zdeněk
Sekiguchi, Tomohiko

Butler, Bryan
Campins, Humberto
Capaccioni, Fabrizio
Capria, Maria
Carruba, Valerio
Carruthers, George
Carsenty, Uri
Carusi, Andrea
Carvano, Jorge
Ceplecha, Zdeněk
Cerroni, Priscilla
Chandrasekhar, Thyagarajan
Chapman, Clark
Chapman, Robert
Chen, Daohan
Chubko, Larysa
Clairemidi, Jacques
Clayton, Donald
Clayton, Geoffrey
Clube, S.
Cochran, Anita
Cochran, William
Colom, Pierre
Combi, Michael
Connors, Martin
Conrad, Albert
Consolmagno, Guy
Cosmovici, Cristiano
Cremonese, Gabriele
Crovisier, Jacques
Cruikshank, Dale
Cuypers, Jan
da Silveira, Enio
Danks, Anthony
Davidsson, Björn
Davies, John
de Almeida, Amaury
de Pater, Imke
de Sanctis, Giovanni
de Sanctis, M.
Delbo, Marco
Dell' Oro, Aldo
Delsanti, Audrey
Delsemme, Armand
Dermott, Stanley
Deutschman, William
Di Martino, Mario
Donn, Bertram
Doressoundiram, Alain
Dotto, Elisabetta
Dryer, Murray
Duffard, René
Duncan, Martin
Durech, Josef
Dzhapiashvili, Victor
Encrenaz, Therésè
Erard, Stéphane
Ershkovich, Alexander
Eviatar, Aharon
Keller, Horst
Kidger, Mark
Kim, Bong
King, Sun-Kun
Kiselev, Nikolai
Klačka, Jozef
Kliem, Bernhard
Knacke, Roger
Knežević, Zoran
Koeberl, Christian
Kohoutek, Lubos
Korsun, Pavlo
Koshkin , Nikolay
Kowal, Charles
Kozasa, Takashi
Krimigis, Stamatios
Krishna, Swamy
Kristensen, Leif
Krugly, Yurij
Kryszczynska, Agnieszka
Kuan, Yi-Jehng
Küppers, Michael
La Spina, Alessandra
Lagerkvist, Claes-Ingvar
Lamy, Philippe
Lane, Arthur
Lara, Luisa
Larson, Harold
Larson, Stephen
Lazzarin, Monica
Lebofsky, Larry
Lee, Thyphoon
Levasseur-Regourd, Anny-Chantal
Licandro, Javier
Liller, William
Lillie, Charles
Lindsey, Charles
Lipschutz, Michael
Lissauer, Jack
Lisse, Carey
Lo Curto, Gaspare
Lodders, Katharina
Lopes-Gautier, Rosaly
Lukyanyk, Igor
Lumme, Kari
Lutz, Barry
Luu, Jane
Lyon, Ian
Magee-Sauer, Karen
Magnusson, Per
Makalkin, Andrei
Maran, Stephen
Marchi, Simone
Marcialis, Robert
Maris, Michèle
Marsden, Brian
Marzari, Francesco
Matson, Dennis
Matsuura, Oscar
Sen, Asoke
Serra Ricart, Miquel
Shanklin, Jonathan
Sharma, A.
Sharp, Christopher
Shevchenko, Vasilij
Shinsuke, Abe
Shkodrov, Vladimir
Sho, Sasaki
Shor, Viktor
Sims, Mark
Sivaraman, Koduvayur
Sizonenko, Yuri
Smith, Bradford
Snyder, Lewis
Šolc, Martin
Spinrad, Hyron
Steel, Duncan
Stern, S.
Subhon, Ibadov
Surdej, Jean
Svoreň, Ján
Swade, Daryl
Sykes, Mark
Szego, Karoly
Szutowicz, Slawomira
Tacconi-Garman, Lowell
Takeda, Hidenori
Tanabe, Hiroyoshi
Tanga, Paolo
Tao, Jun
Tarashchuk, Vera
Tatum, Jeremy
Tedesco, Edward
Terentjeva, Aleksandra
Tholen, David
Thomas, Nicolas
Tiscareno, Matthew
Tosi, Federico
Toth, Imre
Tozzi, Gian
Trujillo, Chadwick
Ugolnikov, Oleg
Valdés-Sada, Pedro
Vázquez, Roberto
Veeder, Glenn
Veverka, Joseph
Vilas, Faith
Voelzke, Marcos
Wallis, Max
Wang, Xiao-bin
Wasson, John
Watanabe, Jun-ichi
Wdowiak, Thomas
Weaver, Harold
Wehinger, Peter
Weidenschilling, S.
Weissman, Paul
Wells, Eddie

Farnham, Tony
Feldman, Paul
Fernández, Julio
Ferrin, Ignacio
Filacchione, Gianrico
Fitzsimmons, Alan
Fornasier, Sonia
Foryta, Dietmar
Fraser, Helen
Froeschlé, Christiane
Fujiwara, Akira
Fulchignoni, Marcello
Furusho, Reiko
Gajdoš, Štefan
Galád, Adrian
Gammelgaard, Peter
Gehrels, Tom
Geiss, Johannes
Gerakines, Perry
Gerard, Eric
Gibson, James
Mazzotta Epifani, Elena
McCord, Thomas
McCrosky, Richard
McDonnell, J.
McFadden, Lucy
McKenna, Lawlor
Meech, Karen
Meisel, David
Mendillo, Michael
Merline, William
Michalowski, Tadeusz
Milani, Andrea
Millis, Robert
Moehlmann, Diedrich
Moore, Elliott
Morrison, David
Mothé-Diniz, Thais
Mueller, Thomas
Muinonen, Karri
Mukai, Tadashi
Mumma, Michael
West, Richard
Wilkening, Laurel
Williams, Iwan
Wooden, Diane
Woolfson, Michael
Woszczyk, Andrzej
Wyckoff, Susan
Yabushita, Shin
Yanagisawa, Masahisa
Yang, Jongmann
Yang, Xiaohu
Yeomans, Donald
Yi, Yu
Yoshida, Fumi
Zappalà, Vincenzo
Zarnecki, John
Zellner, Benjamin
Zhang, Jun
Zhang, Xiaoxiang
Zhao, Haibin
Zhu, Jin

Division III Commission 16 Physical Study of Planets & Satellites

President McGrath, Melissa
Vice-President Lemmon, Mark

Organizing Committee

Kim, Sang Joon
Ksanfomality, Leonid
Lara, Luisa
Morrison, David
Tejfel, Viktor
Yanamandra-Fisher, Padma

Members

Akimov, Leonid
Alexandrov, Yuri
Appleby, John
Archinal, Brent
Arthur, David
Atkinson, David
Atreya, Sushil
Balikhin, Michael
Barkin, Yuri
Barlow, Nadine
Barrow, Colin
Batson, Raymond
Battaner, Eduardo
Baum, William
Beebe, Reta
Beer, Reinhard
Bell III, James
Belton, Michael
Bender, Peter
Ben-Jaffel, Lofti
Berge, Glenn
Bergstralh, Jay
Bertaux, Jean-Loup
Beurle, Kevin
Bezard, Bruno
Bhardwaj, Anil
Billebaud, Françoise
Binzel, Richard
Blamont, Jacques-Emile
Blanco, Armando
Bondarenko, Lyudmila
Bosma, Pieter
Boss, Alan
Boyce, Peter
Brahic, André
Brecher, Aviva
Broadfoot, A.
Brown, Robert
Brunk, William
Buie, Marc
Buratti, Bonnie
Burba, George
Burns, Joseph
Calame, Odile
Caldwell, John
Cameron, Winifred
Campbell, Donald
Goody, Richard
Gorenstein, Paul
Gor'kavyi, Nikolai
Goudas, Constantine
Gounelle, Matthieu
Grav, Tommy
Green, Jack
Grieger, Bjoern
Grossman, Lawrence
Gulkis, Samuel
Gurshtein, Alexander
Halliday, Ian
Hammel, Heidi
Hänninen, Jyrki
Harris, Alan
Harris, Alan
Hasegawa, Ichiro
Hide, Raymond
Holberg, Jay
Horedt, Georg
Hovenier, J.
Hubbard, William
Hunt, Garry
Hunten, Donald
Irvine, William
Irwin, Patrick
Iwasaki, Kyosuke
Johnson, Torrence
Jordán, Andrés
Jurgens, Raymond
Kascheev, Rafael
Käufl, Hans Ulrich
Kiladze, Rolan
Killen, Rosemary
Kim, Sang Joon
Kim, Yongha
Kim, Yoo Jea
Kislyuk, Vitalij
Kley, Wilhelm
Kowal, Charles
Kozak, Lyudmyla
Krimigis, Stamatios
Kumar, Shiv
Kurt, Vladimir
Lane, Arthur
Larson, Harold
Ness, Norman
Neukum, Gerhard
Noll, Keith
Ohtsuki, Keiji
Ortiz, José
Owen, Tobias
Pang, Kevin
Paolicchi, Paolo
Park, Yong-Sun
Peixinho, Nuno
Pérez, Mario
Petit, Jean-Marc
Pettengill, Gordon
Pillinger, Colin
Ping, Jinsong
Politi, Romolo
Potter, Andrew
Psaryov, Volodymyr
Rao, M.
Rodionova, Zhanna
Rodrigo, Rafael
Roos-Serote, Maarten
Roques, Françoise
Rosenbush , Vera
Rossi, Alessandro
Ruskol, Evgeniya
Russo, Pedro
Saissac, Joseph
Sampson, Russell
Sánchez-Lavega, Agustín
Schaller, Emily
Schleicher, David
Schloerb, F.
Schneider, Nicholas
Sergis, Nick
Shapiro, Irwin
Shevchenko, Vladislav
Shi, Huli
Shkuratov, Yurii
Sho, Sasaki
Sicardy, Bruno
Sims, Mark
Sjogren, William
Smith, Bradford
Snellen, Ignas
Soderblom, Larry
Sonett, Charles

Capria, Maria
Carsmaru, Maria
Cecconi, Baptiste
Chapman, Clark
Chen, Daohan
Chevrel, Serge
Clairemidi, Jacques
Cochran, Anita
Combi, Michael
Connes, Janine
Coradini, Angioletta
Cottini, Valeria
Counselman, Charles
Courtin, Régis
Coustenis, Athena
Cruikshank, Dale
Davies, Ashley
Davis, Gary
de Bergh, Catherine
de Pater, Imke
Dermott, Stanley
Dickel, John
Dickey, Jean
Dlugach, Zhanna
Dollfus, Audouin
Doressoundiram, Alain
Drake, Frank
Drossart, Pierre
Duffard, René
Durrance, Samuel
Dzhapiashvili, Victor
El Baz, Farouk
Elliot, James
Elston, Wolfgang
Encrenaz, Therésè
Epishev, Vitali
Eshleman, Von
Esposito, Larry
Evans, Michael
Ferrari, Cecile
Filacchione, Gianrico
Fink, Uwe
Fox, Kenneth
Fujiwara, Akira
Gautier, Daniel
Gehrels, Tom
Geiss, Johannes
Gérard, Jean-Claude
Gierasch, Peter
Goldreich, Peter
Goldstein, Richard
Larson, Stephen
Lellouch, Emmanuel
Lemmon, Mark
Lewis, J.
Licandro, Javier
Lichtenegger, Herbert
Lineweaver, Charles
Lissauer, Jack
Lo Curto, Gaspare
Lockwood, G.
Lodders, Katharina
Lopes-Gautier, Rosaly
Lopez Moreno, José
Lopez Puertas, Manuel
Lopez Valverde, M.
Lumme, Kari
Lutz, Barry
Luz, David
Lyon, Ian
Makalkin, Andrei
Marchi, Simone
Marcialis, Robert
Margot, Jean-Luc
Marov, Mikhail
Martinez-Frias, Jesùs
Marzo, Giuseppe
Matson, Dennis
Matsui, Takafumi
Mazzotta Epifani, Elena
McCord, Thomas
McCullough, Peter
McElroy, M.
McKinnon, William
Meadows, A.
Mendillo, Michael
Mickelson, Michael
Mikhail, Joseph
Millis, Robert
Mills, Franklin
Möhlmann, Diedrich
Molina, Antonio
Moncuquet, Michel
Montmessin, Franck
Moore, Patrick
Moreno, Fernando
Morozhenko, A.
Mosser, Benot
Mumma, Michael
Murphy, Robert
Nakagawa, Yoshitsugu
Nelson, Richard
Sprague, Ann
Stallard, Thomas
Stam, Daphne
Stern, S.
Sterzik, Michael
Stoev, Alexey
Stone, Edward
Strobel, Darrell
Strom, Robert
Synnott, Stephen
Tanga, Paolo
Taylor, Fredric
Tchouikova, Nadezhda
Tedds, Jonathan
Terrile, Richard
Tholen, David
Thomas, Nicolas
Tiscareno, Matthew
Tosi, Federico
Trafton, Laurence
Tran-Minh, Françoise
Trujillo, Chadwick
Tyler Jr., G.
Veiga, Carlos
Veverka, Joseph
Vidmachenko, Anatoliy
Walker, Alta
Walker, Robert
Wallace, Lloyd
Wasserman, Lawrence
Wasson, John
Weidenschilling, S.
Wells, Eddie
Whitaker, Ewen
Williams, Iwan
Williams, James
Woolfson, Michael
Woszczyk, Andrzej
Wu, Yanqin
Wurz, Peter
Yanamandra-Fisher, Padma
Yang, Xiaohu
Yi, Yu
Yoder, Charles
Young, Andrew
Young, Louise
Zarka, Philippe
Zhang, Jun
Zharkov, Vladimir

Division I Commission 19 Rotation of the Earth

President Schuh, Harald
Vice-President Huang, Cheng-Li
Secretary Seitz, Florian

Organizing Committee

Bizouard, Christian
Chao, Benjamin
Gross, Richard
Kosek, Wieslaw
Malkin, Zinovy
Richter, Bernd
Salstein, David
Titov, Oleg

Members

Arabelos, Dimitrios
Archinal, Brent
Arias, Elisa
Banni, Aldo
Barkin, Yuri
Barlier, François
Beutler, Gerhard
Bolotin, Sergei
Bolotina, Olga
Boucher, Claude
Bougeard, Mireille
Boytel, Jorge
Brentjens, Michiel
Brosche, Peter
Capitaine, Nicole
Cazenave, Anny
Chao, Benjamin
Chapanov, Yavor
Damljanovic, Goran
De Biasi, Maria
de Viron, Olivier
Debarbat, Suzanne
Dejaiffe, René
Deleflie, Florent
Dick, Wolfgang
Dickman, Steven
Djurovic, Dragutin
Du, Lan
El Shahawy, Mohamad
Eppelbaum, Lev
Escapa, Alberto
Fernández, Laura
Ferrándiz, José
Fliegel, Henry
Folgueira, Marta
Fong, Chugang
Fujishita, Mitsumi
Fukushima, Toshio
Gambis, Daniel
Gao, Buxi
Gaposchkin, Edward
Gayazov, Iskander
Getino Fernandez, Juan
Gontier, Anne-Marie
Gozhy, Adam
Hugentobler, Urs
Iijima, Shigetaka
Jin, WenJing
Johnson, Thomas
Kakuta, Chuichi
Kameya, Osamu
Khoda, Oleg
Klepczynski, William
Knowles, Stephen
Kolaczek, Barbara
Korsun, Alla
Kosteleck, Jan
Kouba, Jan
Lehmann, Marek
Li, Jinling
Li, Yong
Liao, Dechun
Lieske, Jay
Liu, Ciyuan
Luzum, Brian
Ma, Lihua
Malkin, Zinovy
Manabe, Seiji
Masaki, Yoshimitsu
McCarthy, Dennis
Meinig, Manfred
Melbourne, William
Merriam, James
Monet, Alice
Morgan, Peter
Morrison, Leslie
Mueller, Ivan
Mysen, Eirik
Nastula, Jolanta
Newhall, X.
Niemi, Aimo
Nothnagel, Axel
Paquet, Paul
Park, Pil-Ho
Pejovic, Nadezda
Pešek, Ivan
Petit, Gérard
Petrov, Sergei
Picca, Domenico
Pilkington, John
Ron, Cyril
Roosbeek, Fabian
Rothacher, Markus
Ruder, Hanns
Rusu, M V
Rykhlova, Lidiya
Sadzakov, Sofija
Sasao, Tetsuo
Sato, Koichi
Schillak, Stanislaw
Schutz, Bob
Seitz, Florian
Sekiguchi, Naosuke
Shapiro, Irwin
Shelus, Peter
Shi, Huli
Shuygina, Nadia
Sidorenkov, Nikolaj
Soffel, Michael
Stanila, George
Stephenson, F.
Sugawa, Chikara
Sun, Fuping
Tapley, Byron
Tarady, Vladimir
Titov, Oleg
Tsao, Mo
Veillet, Christian
Vicente, Raimundo
Wallace, Patrick
Wang, Zhengming
Wang, Kemin
Weber, Robert
Williams, James
Wilson, P.
Wooden, William
Wu, Bin
Wu, Shouxian
Xiao, Naiyuan
Xu, Jiayan
Yang, Fumin
Yatskiv, Yaroslav
Ye, Shuhua
Yokoyama, Koichi
Yu, Nanhua

Groten, Erwin	Poma, Angelo	Zhang, Zhongping
Guinot, Bernard	Proverbio, Edoardo	Zharov, Vladimir
Han, Tianqi	Pugliano, Antonio	Zheng, Yong
Han, Yanben	Ray, James	Zhong, Min
Hefty, Jan	Richter, Bernd	Zhou, Yonghong
Huang, Cheng	Robertson, Douglas	Zhu, Yaozhong

Division III Commission 20 Positions & Motions of Minor Planets, Comets & Satellites

President **Yoshikawa, Makoto**

Organizing Committee t.b.d.

Members

A'Hearn, Michael
Abalakin, Viktor
Aikman, G.
Aksnes, Kaare
Arlot, Jean-Eudes
Babadjanov, Pulat
Baggaley, William
Bailey, Mark
Baransky, Alexander
Batrakov, Yurij
Behrend, Raoul
Benest, Daniel
Bernardi, Fabrizio
Berthier, Jerôme
Bien, Reinhold
Blanco, Carlo
Blow, Graham
Boerngen, Freimut
Bowell, Edward
Branham, Richard
Burns, Joseph
Calame, Odile
Carpino, Mario
Carusi, Andrea
Chapront-Touze, Michelle
Chodas, Paul
Cooper, Nicholas
Darhmaoui, Hassane
de Sanctis, Giovanni
Delbo, Marco
Delsemme, Armand
Di Sisto, Romina
Dollfus, Audouin
Donnison, John
Dourneau, Gerard
Doval, Jorge M.
Dunham, David
Dvorak, Rudolf
Dybczynski, Piotr
Elliot, James
Elst, Eric
Emelianov, Nikolaj
Emelyanenko, Vacheslav
Epishev, Vitali
Evans, Michael
Fernández, Julio
Ferraz Mello, Sylvio
Ferreri, Walter
Hudkova, Ludmila
Hurnik, Hieronim
Hurukawa, Kiitiro
Ianna, Philip
Ivanova, Violeta
Ivantsov, Anatoliy
Jacobson, Robert
Jakubík, Marián
Kablak, Nataliya
Kazantsev, Anatolii
Khatisashvili, Alfez
Kilmartin, Pamela
Kim, Sang Joon
Kinoshita, Hiroshi
Kisseleva, Tamara
Klemola, Arnold
Knežević, Zoran
Kohoutek, Lubos
Kosai, Hiroki
Koshkin, Nikolay
Kowal, Charles
Kozai, Yoshihide
Krasinsky, George
Kristensen, Leif
Krolikowska-Soltan, Malgorzata
Krugly, Yurij
Kulikova, Nelli
Lagerkvist, Claes-Ingvar
Larsen, Jeffrey
Laurin, Denis
Lematre, Anne
Li, Guangyu
Li, Yong
Lieske, Jay
Lomb, Nicholas
Lovas, Miklos
Manara, Alessandro
Marsden, Brian
Matese, John
Maury, Alain
McCrosky, Richard
McMillan, Robert
McNaught, Robert
Medvedev, Yurij
Melita, Mario
Milani, Andrea
Millis, Robert
Mintz Blanco, Betty
Rapaport, Michel
Reitsema, Harold
Rickman, Hans
Roemer, Elizabeth
Roeser, Siegfried
Rossi, Alessandro
Rui, Qi
Russell, Kenneth
Sato, Isao
Schmadel, Lutz
Schober, Hans
Scholl, Hans
Schubart, Joachim
Schuster, William
Seidelmann, P.
Sekanina, Zdeněk
Shanklin, Jonathan
Shelus, Peter
Shen, Kaixian
Shi, Huli
Shkodrov, Vladimir
Shor, Viktor
Sitarski, Grzegorz
Solovaya, Nina
Sôma, Mitsuru
Standish, E.
Steel, Duncan
Stellmacher, Irène
Stokes, Grant
Sultanov, G.
Svoreň, Ján
Synnott, Stephen
Szutowicz, Slawomira
Tancredi, Gonzalo
Tatum, Jeremy
Taylor, Donald
Thuillot, William
Tiscareno, Matthew
Torres, Carlos
Trujillo, Chadwick
Tsuchida, Masayoshi
Tuccari, Gino
van Houten-Groeneveld, Ingrid
Veillet, Christian
Vieira, Martins
Vienne, Alain
Virtanen, Jenni
Voelzke, Marcos

Fors, Octavi
Franklin, Fred
Fraser, Brian
Freitas Mouro, Ronaldo
Froeschlé, Claude
Fuse, Tetsuharu
Gehrels, Tom
Gibson, James
Giorgini, Jon
Gorshanov, Denis
Green, Daniel
Greenberg, Richard
Hahn, Gerhard
Harper, David
Harris, Alan
Hasegawa, Ichiro
Haupt, Hermann
He, Miao-fu
Hemenway, Paul
Hers, Jan
Heudier, Jean-Louis
Hol, Pedro
Monet, Alice
Moravec, Zdeněk
Morris, Charles
Murray, Carl
Nacozy, Paul
Nakamura, Tsuko
Nakano, Syuichi
Neslušan, Luboš
Nobili, Anna
Owen Jr., William
Pandey, A.
Pascu, Dan
Pauwels, Thierry
Pierce, David
Ping, Jinsong
Pittich, Eduard
Polyakhova, Elena
Porubčan, Vladimír
Pozhalova, Zhanna
Qiao, Rongchuan
Rajamohan, R.
Raju, Vasundhara
Wasserman, Lawrence
Weissman, Paul
West, Richard
Whipple, Arthur
Williams, Gareth
Williams, Iwan
Williams, James
Xiong, Jianning
Yabushita, Shin
Yeomans, Donald
Yim, Hong-Suh
Yuasa, Manabu
Zagretdinov, Renat
Zappalà, Vincenzo
Zhang, Qiang
Zhang, Jiaxiang
Zhang, Xiaoxiang
Zhang, Wei
Zhao, Haibin
Ziolkowski, Krzysztof

Division IX Commission 21 Galactic and Extragalactic Background Radiation

President Witt, Adolf
Vice-President Murthy, Jayant

Organizing Committee

Baggaley, William
Dwek, Eli
Gustafson, Bo
Levasseur-Regourd, Anny-Chantal
Mann, Ingrid
Mattila, Kalevi
Watanabe, Jun-ichi

Members

Angione, Ronald
Belkovich, Oleg
Blamont, Jacques-Emile
Bowyer, C.
Broadfoot, A.
Clairemidi, Jacques
d'Hendecourt, Louis
Dermott, Stanley
Dodonov, Sergej
Dubin, Maurice
Dufay, Maurice
Dumont, René
Feldman, Paul
Fujiwara, Akira
Giovane, Frank
Gounelle, Matthieu
Grün, Eberhard
Hanner, Martha
Harwit, Martin
Hauser, Michael
Hecht, James
Henry, Richard
Hofmann, Wilfried
Hong, Seung
Hurwitz, Mark
Ivanov-Kholodny, Gor
Jackson, Bernard
James, John
Joubert, Martine
Karygina, Zoya
Kopylov, Aleksandr
Koutchmy, Serge
Kozak, Lyudmyla
Kramer, Busaba
Kulkarni, Prabhakar
Lamy, Philippe
Léger, Alain
Leinert, Christoph
Lemke, Dietrich
Lillie, Charles
Lopez Gonzalez, Maria
Lopez Moreno, José
Lopez Puertas, Manuel
Lumme, Kari
Maihara, Toshinori
Martin, Donn
Mather, John
Matsumoto, Toshio
Maucherat, Jean
McDonnell, J.
Mikhail, Joseph
Misconi, Nebil
Muinonen, Karri
Mukai, Tadashi
Nakamura, Akiko
Nawar, Samir
Nishimura, Tetsuo
Paresce, Francesco
Perrin, Jean-Marie
Pfleiderer, Jorg
Reach, William
Renard, Jean-Baptiste
Robley, Robert
Rodrigo, Rafael
Rozhkovskij, Dimitrij
Sánchez, Francisco
Sánchez-Saavedra, M.
Saxena, P.
Schlosser, Wolfhard
Schuh, Harald
Schwehm, Gerhard
Shefov, Nikolaj
Smith, Robert
Soberman, Robert
Sparrow, James
Staude, Hans
Sykes, Mark
Tanabe, Hiroyoshi
Toller, Gary
Toroshlidze, Teimuraz
Tyson, John
Ueno, Munetaka
Ugolnikov, Oleg
Vrtilek, Jan
Wallis, Max
Weinberg, Jerry
Wesson, Paul
Wheatley, Peter
Wilson, P.
Wolstencroft, Ramon
Woolfson, Michael
Yamamoto, Tetsuo
Yamashita, Kojun
Yang, Xiaohu

Division III Commission 22 Meteors, Meteorites & Interplanetary Dust

President Watanabe, Jun-ichi
Vice-President Jenniskens, Petrus

Organizing Committee

Borovicka, Jirí
Campbell-Brown, Margaret
Consolmagno, Guy
Jenniskens, Petrus
Jopek, Tadeusz
Vaubaillon, Jérémie
Williams, Iwan

Members

Alexandrov, Alexander
Apai, Daniel
Asher, David
Babadjanov, Pulat
Belkovich, Oleg
Bhandari, N.
Brownlee, Donald
Campbell-Brown, Margaret
Carusi, Andrea
Ceplecha, Zdeněk
Cevolani, Giordano
Clifton, Kenneth
Clube, S.
Cooper, Timothy
Dieleman, Pieter
Djorgovski, Stanislav
Dubin, Maurice
Duffard, René
Elford, William
Gajdoš, Štefan
Glass, Billy
Gorbanev, Jury
Goswami, J.
Gounelle, Matthieu
Grady, Monica
Grün, Eberhard
Gustafson, Bo
Hajduková, Mária
Hajduková, Jr., Mária
Halliday, Ian
Hanner, Martha
Harvey, Gale
Hasegawa, Ichiro
Hawkes, Robert
Hiroko, Nagahara
Hodge, Paul
Hong, Seung
Hughes, David
Jakubík, Marián
Jones, James
Jopek, Tadeusz
Kalenichenko, Valentin
Kapisinsky, Igor
Karakas, Amanda
Keay, Colin
Khovritchev, Maxim
Koeberl, Christian
Kokhirova, Gulchehra
Kolomiyets, Svitlana
Koten, Pavel
Kozak, Pavlo
Kramer, Kh
Kruchinenko, Vitaliy
Lamy, Philippe
Lemaire, Joseph
Levasseur-Regourd, Anny-Chantal
Lodders, Katharina
Lovell, Sir Bernard
Lugaro, Maria
Lyon, Ian
Makalkin, Andrei
Mann, Ingrid
Maris, Michèle
Martinez-Frias, Jesùs
Marvin, Ursula
Mason, John
McCrosky, Richard
McDonnell, J.
McIntosh, Bruce
Meisel, David
Miles, Howard
Misconi, Nebil
Murray, Carl
Murray, Andrew
Nakamura, Takuji
Nakazawa, Kiyoshi
Napier, William
Newburn Jr., Ray
Nuth, Joseph
Pecina, Petr
Pillinger, Colin
Plavec, Zdenka
Politi, Romolo
Poole, Graham
Rendtel, Juergen
Revelle, Douglas
Rickman, Hans
Ripken, Hartmut
Ryabova, Galina
Sekanina, Zdeněk
Shao, Cheng-yuan
Shinsuke, Abe
Sho, Sasaki
Šimek, Miloš
Soberman, Robert
Steel, Duncan
Švestka, Jiří
Svoreň, Ján
Tatum, Jeremy
Taylor, Andrew
Tedesco, Edward
Terentjeva, Aleksandra
Toshihiro, Kasuga
Tosi, Federico
Tóth, Juraj
Trigo-Rodríguez, Josep
Valsecchi, Giovanni
Vaubaillon, Jérémie
Voloschuk, Yuri
Webster, Alan
Weinberg, Jerry
Woolfson, Michael
Yamamoto, Masayuki
Yeomans, Donald
Zhang, Xiaoxiang
Zhao, Haibin
Zhu, Jin

Division IX Commission 25 Stellar Photometry & Polarimetry

President Milone, Eugene
Vice-President Walker, Alistair

Organizing Committee

Anthony-Twarog, Barbara
Bastien, Pierre
Knude, Jens
Kurtz, Donald
Menzies, John
Mironov, Aleksey
Qian, Shengbang

Members

Ables, Harold
Adelman, Saul
Ahumada, Javier
Albrecht, Rudolf
Alecian, Evelyne
Anandaram, Mandayam
Andreuzzi, Gloria
Angel, J.
Angione, Ronald
Arnaud, Jean-Paul
Arsenijevic, Jelisaveta
Ashok, N.
Aspin, Colin
Aungwerojwit, Amornrat
Axon, David
Baldinelli, Luigi
Baliyan, Kiran
Balona, Luis
Baran, Andrzej
Barnes III, Thomas
Barrett, Paul
Barrientos, Luis
Baume, Gustavo
Bellazzini, Michele
Berdyugin, Andrei
Bessell, Michael
Bjorkman, Jon
Blanco, Victor
Bookmyer, Beverly
Borgman, Jan
Borra, Ermanno
Breger, Michel
Brown, Douglas
Brown, Thomas
Buser, Roland
Cantiello, Michele
Carney, Bruce
Carter, Brian
Castelaz, Micheal
Celis, Leopoldo
Chadid, Merieme
Chen, Wen Ping
Chen, An-Le
Cioni, Maria-Rosa
Grenon, Michel
Grewing, Michael
Grundahl, Frank
Guetter, Harry
Gulbis, Amanda
Gutierrez-Moreno, A.
Hall, Douglas
Hauck, Bernard
Hayes, Donald
Heck, Andre
Hensberge, Herman
Hilditch, Ronald
Hiroshi, Akitaya
Hubrig, Swetlana
Huovelin, Juhani
Hyland, Harry
Irwin, Alan
Ivezic, Zeljko
Jeffers, Sandra
Jerzykiewicz, Mikolaj
Joshi, Umesh
Kawara, Kimiaki
Kazlauskas, Algirdas
Kebede, Legesse
Keller, Stefan
Kentaro, Matsuda
Kepler, S.
Kilkenny, David
Kim, Seung-Lee
King, Ivan
Koch, Robert
Kornilov, Viktor
Kulkarni, Prabhakar
Kunkel, William
Landstreet, John
Laskarides, Paul
Lazauskaite, Romualda
Lemke, Michael
Lenzen, Rainer
Leroy, Jean-Louis
Li, Li
Linde, Peter
Lockwood, G.
Lub, Jan
Parimucha, Štefan
Pavani, Daniela
Pedreros, Mario
Pel, Jan
Penny, Alan
Perrin, Marshall
Petit, Pascal
Pfeiffer, Raymond
Philip, A.G.
Piirola, Vilppu
Platais, Imants
Pokrzywka, Bartlomiej
Pulone, Luigi
Rao, Pasagada
Raveendran, A.
Rawlings, Mark
Reglero Velasco, Victor
Reshetnyk, Volodymyr
Robb, Russell
Robinson, Edward
Rodrigues, Claudia
Romanyuk, Yaroslav
Roslund, Curt
Rostopchina, Alla
Santos Agostinho, Rui
Schuster, William
Sekiguchi, Kazuhiro
Sen, Asoke
Shakhovskoj, Nikolay
Shankland, Paul
Shawl, Stephen
Smith, J.
Smyth, Michael
Snowden, Michael
Steinlin, Uli
Stetson, Peter
Stockman Jr., Hervey
Stone, Remington
Straizys, Vytautas
Stritzinger, Maximilian
Subramaniam, Annapurni
Sudžius, Jokubas
Sullivan, Denis
Szkody, Paula

Clem, James
Connolly, Leo
Coyne, George
Crawford, David
Cuypers, Jan
Dachs, Joachim
Dahn, Conard
Dall,, Nogami
Danford, Stephen
Deshpande, M.
Dolan, Joseph
Dubout, Renée
Ducati, Jorge
Ducourant, Christine
Dzervitis, Uldis
Edwards, Paul
Efimov, Yuri
Elmhamdi, Abouazza
Fabregat, Juan
Fabrika, Sergei
Feinstein, Alejandro
Fernández Lajús, Eduardo
Fernie, J.
Fluri, Dominique
Forte, Juan
Galadi Enriquez, David
Gehrz, Robert
Genet, Russel
Gerbaldi, Michèle
Ghosh, Swarna
Gilliland, Ronald
Giorgi, Edgard
Glass, Ian
Golay, Marcel
Graham, John
Grauer, Albert
Luna, Homero
Maitzen, Hans
Manfroid, Jean
Manset, Nadine
Markkanen, Tapio
Marraco, Hugo
Marsden, Stephen
Martinez Roger, Carlos
Masani, A.
Maslennikov, Kirill
Mason, Paul
Mathys, Gautier
Mayer, Pavel
McDavid, David
McLean, Ian
Mendoza, V.
Metcalfe, Travis
Mianes, Pierre
Miller, Joseph
Mintz Blanco, Betty
Moffett, Thomas
Moitinho, André
Mourard, Denis
Mumford, George
Munari, Ulisse
Naylor, Tim
Neiner, Coralie
Nicolet, Bernard
Noguchi, Kunio
Notni, Peter
Oblak, Edouard
Östreicher, Roland
Orsatti, Ana
Page, Arthur
Pak, Soojong
Szymanski, Michal
Tandon, S.
Taranova, Olga
Taş, Günay
Tedds, Jonathan
Tinbergen, Jaap
Todoran, Ioan
Tokunaga, Alan
Tolbert, Charles
Townsend, Richard
Umeda, Hideyuki
Ureche, Vasile
Uslenghi, Michela
Vaughan, Arthur
Verma, R.
Voloshina, Irina
Vrba, Frederick
Walker, William
Warren Jr., Wayne
Weiss, Werner
Weistrop, Donna
Wesselius, Paul
Wheatley, Peter
White, Nathaniel
Wielebinski, Richard
Willstrop, Roderick
Winiarski, Maciej
Wramdemark, Stig
Yamashita, Yasumasa
Yao, Yongqiang
Young, Andrew
Yudin, Ruslan
Yuji, Ikeda
Zhu, Liying
Žižňovský, Jozef

Division IV Commission 26 Double & Multiple Stars

President Docobo, José A.
Vice-President Mason, Brian

Organizing Committee

Arenou, Frédéric
Balega, Yurij
Oswalt, Terry
Pourbaix, Dimitri
Scardia, Marco
Scarfe, Colin
Tamazian, Vakhtang

Members

Abt, Helmut
Ahumada, Javier
Andrade, Manuel
Anosova, Joanna
Arenou, Frédéric
Argyle, Robert
Armstrong, John
Aungwerojwit, Amornrat
Bagnuolo Jr., William
Bailyn, Charles

Batten, Alan
Beavers, Willet
Bernacca, Pierluigi
Boden, Andrew
Bonneau, Daniel
Brosche, Peter
Budaj, Ján
Cester, Bruno
Chen, Wen Ping
Couteau, Paul
Culver, Roger
Cvetković, Zorica
Daisaku, Nogami
Davis, John
De Cat, Peter
Dominis Prester, Dijana
Dommanget, Jean
Dukes Jr., Robert
Dunham, David
Falceta-Gonçalves, Diego
Fekel, Francis
Fernandes, João
Ferrer, Osvaldo
Fletcher, J.
Fors, Octavi
Franz, Otto
Fredrick, Laurence
Freitas Mouro, Ronaldo
Gatewood, George
Gaudenzi, Silvia
Geyer, Edward
Ghez, Andrea
Gonçalves, Denise
Goodwin, Simon
Gün, Gulnur
Hakkila, Jon
Halbwachs, Jean-Louis
Hartigan, Patrick

Horch, Elliott
Hummel, Christian
Hummel, Wolfgang
Ianna, Philip
Ireland, Michael
Jahreiss, Hartmut
Jassur, Davoud
Jeon, Young-Beom
Johnston, Helen
Jurdana-Šepić, Rajka

Kazantseva, Liliya
Kilpio, Elena
Kiselev, Aleksej
Kisseleva-Eggleton, Ludmila
Kitsionas, Spyridon
Kley, Wilhelm
Köhler, Rainer
Kroupa, Pavel
Kubát, Jiří
Lampens, Patricia
Latham, David
Lattanzi, Mario
Lee, William
Leinert, Christoph
Lepine, Sebastien
Lim, Jeremy
Ling, Josefina
Lippincott Zimmerman, Sarah
Liu, Michael
Loden, Kerstin
Lyubchik, Yuri
Maddison, Sarah
Malkov, Oleg
Marsakova, Vladislava
Martín, Eduardo
Mason, Brian
Mathieu, Robert
McAlister, Harold
McDavid, David
Mennickent, Ronald
Mikkola, Seppo
Mikotajewski, Maciej
Mohan, Chander
Morbey, Christopher
Morbidelli, Roberto
Morel, Pierre-Jacques
Negueruela, Ignacio
Neuhäuser, Ralph

Petr-Gotzens, Monika
Pollacco, Don
Popović Georgije
Poveda, Arcadio
Prieto, Cristina
Prieur, Jean-Louis
Rakos, Karl
Reipurth, Bo
Roberts Jr., Lewis
Rodrigues de Oliveira Filho, Irapuan
Russell, Jane
Sagar, Ram
Salukvadze, G.
Scardia, Marco
Schmidtke, Paul
Schöller, Markus
Shakht, Natalia
Shatsky, Nicolai
Simon, Michaël
Sinachopoulos, Dimitris
Skokos, Charalambos
Smak, Józef
Smith, J.
Söderhjelm, Staffan
Sowell, James
Stein, John
Sterzik, Michael
Szabados, Laśzló
Tamazian, Vakhtang
Tango, William
Tarasov, Anatolii
Teixeira, Paula
ten Brummelaar, Theo
Terquem, Caroline
Tokovinin, Andrei
Tomasella, Lina
Torres, Guillermo
Trimble, Virginia
Tsay, Wean-Shun
Turner, Nils
Upgren, Arthur
Valtonen, Mauri
van Altena, William
van der Hucht, Karel
van Dessel, Edwin
Vaňko, Martin
Vaz, Luiz Paulo
Vennes, Stéphane

He, Jinhua
Heacox, William
Hershey, John
Hidayat, Bambang
Hill, Graham
Hillwig, Todd
Hindsley, Robert
Nürnberger, Dieter
Orlov, Victor
Pannunzio, Renato
Parimucha, Štefan
Pauls, Thomas
Pereira, Claudio
Peterson, Deane
Wang, Jiaji
Weis, Edward
Wen, Linqing
Zavala, Jr., Robert
Zheleznyak, Alexander
Zhu, Liying
Zinnecker, Hans

Division V Commission 27 Variable Stars

President Handler, Gerald
Vice-President Pollard, Karen

Organizing Committee

Bedding, Timothy
Catelán, Márcio
Cunha, Margarida
Eyer, Laurent
Jeffery, Christopher
Kepler, S.
Kolenberg, Katrien
Mkrtichian, David
Olah, Katalin

Members

Aerts, Conny
Aizenman, Morris
Albinson, James
Albrow, Michael
Alencar, Silvia
Alfaro, Emilio
Allan, David
Alpar, Mehmet
Amado Gonzalez, Pedro
Ando, Hiroyasu
Andrievsky, Sergei
Antipin, Sergei
Antipova, Lyudmila
Antonello, Elio
Antonyuk, Kirill
Antov, Alexandar
Arellano Ferro, Armando
Arentoft, Torben
Arias, Maria
Arkhipova, Vera
Arsenijevic, Jelisaveta
Asteriadis, Georgios
Aungwerojwit, Amornrat
Avgoloupis, Stavros
Baade, Dietrich
Baglin, Annie
Balona, Luis
Baran, Andrzej
Baransky, Alexander
Barban, Caroline
Barnes III, Thomas
Bartolini, Corrado
Barway, Sudhashu
Barwig, Heinz
Baskill, Darren
Bastien, Pierre
Bauer, Wendy
Bazot, Michael
Beaulieu, Jean-Philippe
Bedogni, Roberto
Belmonte Avilés, Juan Antonio
Belserene, Emilia
Belvedere, Gaetano
Benkoe, Jozsef
Berdnikov, Leonid
Bersier, David
Berthomieu, Gabrielle
Bessell, Michael
Günthardt, Guillermo
Guo, Jianheng
Guzik, Joyce
Hackwell, John
Haefner, Reinhold
Haisch, Bernard
Halbwachs, Jean-Louis
Hall, Douglas
Hamdy, M.
Hansen, Carl
Hao, Jinxin
Harmanec, Petr
Hawley, Suzanne
Heiser, Arnold
Hempelmann, Alexander
Henden, Arne
Herbig, George
Hers, Jan
Hesser, James
Hill, Henry
Hintz, Eric
Hojaev, Alisher
Horner, Scott
Houdek, Gunter
Houk, Nancy
Howell, Steve
Huenemoerder, David
Humphreys, Elizabeth
Hutchings, John
Iben Jr., Icko
Iijima, Takashi
Ireland, Michael
Ishida, Toshihito
Ismailov, Nariman
Ita, Yoshifusa
Ivezic, Zeljko
Jablonski, Francisco
Jankov, Slobodan
Jeffers, Sandra
Jeon, Young-Beom
Jerzykiewicz, Mikolaj
Jetsu, Lauri
Jewell, Philip
Jiang, Biwei
Jin, Zhenyu
Joner, Michael
Jones, Albert
Jurcsik, Johanna
Olivier, Enrico
Opolski, Antoni
stensen, Roy
Oswalt, Terry
Panei, Jorge
Papaloizou, John
Paparo, Margit
Parimucha, Štefan
Park, Byeong-Gon
Parsamyan, Elma
Parthasarathy, Mudumba
Patat, Ferdinando
Paternò, Lucio
Pavlovski, Kresimir
Pearson, Kevin
Percy, John
Perez Hernandez, Fernando
Petersen, J.
Petit, Pascal
Petrov, Peter
Pettersen, Bjørn
Piirola, Vilppu
Pijpers, Frank
Plachinda, Sergei
Plavchan, Jr., Peter
Pollacco, Don
Pont, Frédéric
Pop, Alexandru
Pop, Vasile
Pricopi, Dumitru
Pringle, James
Pritzl, Barton
Provost, Janine
Pugach, Alexander
Pustõnski, Vladislav-Veniamin
Rakos, Karl
Ransom, Scott
Rao, N.
Ratcliff, Stephen
Reale, Fabio
Reiners, Ansgar
Reinsch, Klaus
Renson, P.
Rey, Soo-Chang
Rivinius, Thomas
Robinson, Edward
Rodriguez, Eloy
Romano, Giuliano

Bianchini, Antonio
Bjorkman, Karen
Bochonko, D.
Bolton, Charles
Bond, Howard
Bopp, Bernard
Bowen, George
Boyd, David
Bradley, Paul
Breger, Michel
Briquet, Maryline
Brown, Douglas
Bruch, Albert
Bruntt, Hans
Buccino, Andrea
Buchler, J.
Burki, Gilbert
Burwitz, Vadim
Busà, Innocenza
Busko, Ivo
Butkovskaya, Varvara
Butler, Christopher
Butler, Dennis
Buzasi, Derek
Cacciari, Carla
Caldwell, John
Cameron, Andrew
Cao, Huilai
Carrier, Fabien
Casares, Jorge
Catchpole, Robin
Catelán, Márcio
Chadid, Merieme
Chen, An-Le
Chen, Alfred
Cherchneff, Isabelle
Cherepashchuk, Anatolij
Chou, Yi
Christensen-Dalsgaard, Jørgen
Christie, Grant
Christy, Robert
Cioni, Maria-Rosa
Clementini, Gisella
Cohen, Martin
Connolly, Leo
Contadakis, Michael
Cook, Kem
Córsico, Alejandro
Costa, Vitor
Cottrell, Peter
Coulson, Iain
Coutts-Clement, Christine
Cox, Arthur
Cutispoto, Giuseppe
Cuypers, Jan
D'Amico, Nicolò
Daisaku, Nogami
Dall'Ora, Massimo
Kadouri, Talib
Käufl, Hans Ulrich
Kalomeni, Belinda
Kambe, Eiji
Kanamitsu, Osamu
Kanyo, Sandor
Karitskaya, Evgeniya
Karovska, Margarita
Karp, Alan
Katsova, Maria
Kaufer, Andreas
Kawaler, Steven
Kaye, Anthony
Kazarovets, Elena
Keller, Stefan
Khaliullin, Khabibrachman
Kilkenny, David
Kilpio, Elena
Kim, Tu-Whan
Kim, Chulhee
Kim, Seung-Lee
Kim, Young-Soo
Kiplinger, Alan
Kippenhahn, Rudolf
Kiss, Laśzló
Kjeldsen, Hans
Kjurkchieva, Diana
Kochukhov, Oleg
Koen, Marthinus
Koevari, Zsolt
Kollath, Zoltan
Komžík, Richard
Konstantinova-Antova, Renada
Kopacki, Grzegorz
Korhonen, Heidi
Kraft, Robert
Krautter, Joachim
Kreiner, Jerzy
Krisciunas, Kevin
Krzeminski, Wojciech
Krzesinski, Jerzy
Kubiak, Marcin
Kuhi, Leonard
Kunjaya, Chatief
Kunkel, William
Kurtz, Donald
Lago, Maria
Lampens, Patricia
Landolt, Arlo
Laney, Clifton
Lanza, Antonino
Larionov, Valeri
Laskarides, Paul
Lawlor, Timothy
Lawson, Warrick
Lázaro Hernando, Carlos
Le Bertre, Thibaut
Lebzelter, Thomas
Romanov, Yuri
Rosenbush, Alexander
Rountree, Janet
Russev, Ruscho
Sachkov, Mikhail
Sadik, Aziz
Saha, Abhijit
Samus, Nikolaj
Sandmann, William
Sanyal, Ashit
Sareyan, Jean-Pierre
Sasselov, Dimitar
Schaefer, Bradley
Schlegel, Eric
Schmidt, Edward
Schmidtobreick, Linda
Schuh, Sonja
Schwartz, Philip
Schwarzenberg-Czerny, Alex
Schwope, Axel
Scuflaire, Richard
Seeds, Michael
Selam, Selim
Shahbaz, Tariq
Shahul Hameed, Mohin
Shakhovskaya, Nadezhda
Sharma, Dharma
Shenavrin, Victor
Sherwood, William
Shobbrook, Robert
Silvotti, Roberto
Skinner, Stephen
Smak, Józef
Smeyers, Paul
Smith, Myron
Smith, Horace
Soliman, Mohamed
Somasundaram, Seetha
Soszyński, Igor
Srivastava, Ram
Starrfield, Sumner
Stellingwerf, Robert
Stello, Dennis
Stępień, Kazimierz
Sterken, Christiaan
Strassmeier, Klaus
Stringfellow, Guy
Strom, Stephen
Strom, Karen
Szabados, Laśzló
Szabó, Robert
Szatmary, Karoly
Szécsényi-Nagy, Gábor
Szeidl, Béla
Szkody, Paula
Takata, Masao
Takeuti, Mine
Tammann, Gustav

Danford, Stephen
Daszyńska-Daszkiewicz, Jadwiga
De Cat, Peter
de Ridder, Joris
Delgado, Antonio
Demers, Serge
Deng, LiCai
Deupree, Robert
Di Mauro, Maria Pia
Donahue, Robert
Dorokhova, Tetyana
Downes, Ronald
Dukes Jr., Robert
Dunlop, Storm
Dupuy, David
Dziembowski, Wojciech
Edwards, Paul
Edwards, Suzan
Efremov, Yurij
Eggenberger, Patrick
El Basuny, Ahmed
Esenoğlu, Hasan
Eskioğlu, A. Nihat
Evans, Dafydd
Evans, Nancy
Evans, Aneurin
Evren, Serdar
Fadeyev, Yurij
Feast, Michael
Ferland, Gary
Fernández Lajús, Eduardo
Fernie, J.
Fitch, Walter
Fokin, Andrei
Formiggini, Lilliana
Friedjung, Michael
Fu, Jianning
Fu, Hsieh-Hai
Fujiwara, Tomoko
Gahm, Goesta
Galis, Rudolf
Gameiro, Jorge
Gamen, Roberto
Garrido, Rafael
Gascoigne, S.
Genet, Russel
Gershberg, R.
Geyer, Edward
Gieren, Wolfgang
Gies, Douglas
Gillet, Denis
Glagolevskij, Yurij
Godoli, Giovanni
Gondoin, Philippe
Gosset, Eric
Gough, Douglas
Goupil, Marie-Jo
Graham, John
Grasberg, Ernest
Lee, Myung Gyoon
Lee, Jae-Woo
Leung, Kam
Li, Yan
Li, Zhiping
Little-Marenin, Irene
Lloyd, Christopher
Lockwood, G.
Longmore, Andrew
López de Coca Castañer, Pilar
Lorenz-Martins, Silvia
Lub, Jan
Machado Folha, Daniel
Macri, Lucas
Madore, Barry
Maeder, André
Mahmoud, Farouk
Makarenko, Ekaterina
Mantegazza, Luciano
Marchev, Dragomir
Marconi, Marcella
Margrave Jr., Thomas
Markoff, Sera
Marsakova, Vladislava
Martic, Milena
Martinez, Peter
Masani, A.
Mason, Paul
Mathias, Philippe
Matsumoto, Katsura
Matthews, Jaymie
Mauche, Christopher
Mavridis, Lyssimachos
McGraw, John
McNamara, Delbert
McSaveney, Jennifer
Melikian, Norair
Mennickent, Ronald
Messina, Sergio
Michel, Eric
Mikotajewski, Maciej
Milone, Eugene
Milone, Luis
Minnikulov, Nasriddin
Moffett, Thomas
Mohan, Chander
Monteiro, Mário João
Morales Rueda, Luisa
Morrison, Nancy
Moskalik, Pawel
Mukai, Koji
Mumford, George
Murdin, Paul
Nather, R.
Naylor, Tim
Neiner, Coralie
Ngeow, Chow Choong
Niarchos, Panayiotis
Niemczura, Ewa
Tamura, Shin'ichi
Tarasova, Taya
Taş, Günay
Taylor, John
Teixeira, Teresa
Teixeira, Paula
Tempesti, Piero
Terzan, Agop
Tjin-a-Djie, Herman
Tomov, Toma
Townsend, Richard
Traulsen, Iris
Tremko, Jozef
Tsvetkov, Milcho
Tsvetkova, Katja
Turner, David
Tutukov, Aleksandr
Tylenda, Romuald
Udovichenko, Sergei
Uemura, Makoto
Usher, Peter
Uslenghi, Michela
Utrobin, Victor
Vaccaro, Todd
Valtier, Jean-Claude
van Genderen, Arnoud
van Hoolst, Tim
Ventura, Rita
Viotti, Roberto
Vivas, Anna
Vogt, Nikolaus
Voloshina, Irina
von Braun, Kaspar
Votruba, Viktor
Waelkens, Christoffel
Walker, Merle
Walker, William
Walker, Edward
Wallerstein, George
Wang, Xunhao
Warner, Brian
Watson, Robert
Webbink, Ronald
Wehlau, Amelia
Weis, Kerstin
Weiss, Werner
Welch, Douglas
Wheatley, Peter
Whitelock, Patricia
Williamon, Richard
Willson, Lee Anne
Wilson, Lionel
Wing, Robert
Wittkowski, Markus
Wood, Peter
Xiong, Da Run
Yakut, Kadri
Yüce, Kutluay
Yuji, Ikeda

Grasdalen, Gary
Green, Daniel
Grinin, Vladimir
Groenewegen, Martin
Grygar, Jiří
Guerrero, Gianantonio
Guinan, Edward
Gün, Gulnur
Nikolov, Andrej
Nikolov, Nikola
Nugis, Tiit
O'Donoghue, Darragh
O'Toole, Simon
Ogłoza, Waldemar
Oliveira, Alexandre
Zamanov, Radoslav
Zhang, Xiaobin
Zhu, Liying
Zijlstra, Albert
Zola, Stanislaw
Zsoldos, Endre
Zuckerman, Ben

Division VIII Commission 28 Galaxies

President Davies, Roger
Vice-President Gallagher III, John

Organizing Committee

Courteau, Stéphane	Jog, Chanda	Tacconi, Linda
Dekel, Avishai	Nakai, Naomasa	Terlevich, Elena
Franx, Marijn	Rubio, Monica	

Members

Aalto, Suzanne
Ables, Harold
Abrahamian, Hamlet
Adler, David
Afanassiev, Viktor
Aguero, Estela
Aguilar, Luis A.
Ahmad, Farooq
Akio, Inoue
Akiyama, Masayuki
Alcaino, Gonzalo
Aldaya, Victor
Alexander, Tal
Alladin, Saleh
Allen, Ronald
Allington-Smith, Jeremy
Alloin, Danielle
Almaini, Omar
Aloisi, Alessandra
Alonso, Maria
Alonso, Maria
Amram, Philippe
Andernach, Heinz
Andrillat, Yvette
Ann, Hong
Anosova, Joanna
Anton, Sonia
Antonelli, Lucio Angelo
Aoki, Kentaro
Aparicio, Antonio
Aragón-Salamanca, Alfonso
Ardeberg, Arne
Aretxaga, Itziar
Argo, Megan
Arkhipova, Vera
Arnaboldi, Magda
Artamonov, Boris
Athanassoula, Evangelia
Aussel, Herve
Avila-Reese, Vladimir
Aya, Yamauchi
Ayani, Kazuya
Azzopardi, Marc
Bacham, Eswar
Bachev, Rumen
Baddiley, Christopher
Graham, John
Graham, Alister
Granato, Gian Luigi
Gray, Meghan
Grebel, Eva
Gregg, Michael
Greve, Thomas
Griffiths, Richard
Griv, Evgeny
Gronwall, Caryl
Grove, Lisbeth
Grupe, Dirk
Gu, Qiusheng
Gunn, James
Günthardt, Guillermo
Guseva, Natalia
Gutiérrez, Carlos
Gyulbudaghian, Armen
Hagen-Thorn, Vladimir
Hamabe, Masaru
Hambaryan, Valeri
Hammer, François
Han, Cheongho
Hanami, Hitoshi
Hara, Tetsuya
Hardy, Eduardo
Harms, Richard
Harnett, Julienne
Hasan, Hashima
Hashimoto, Yasuhiro
Hattori, Makoto
Hau, George
Haugbølle, Troels
He, XiangTao
Heald, George
Heckman, Timothy
Heidt, Jochen
Held, Enrico
Helou, George
Henning, Patricia
Henry, Richard
Hensler, Gerhard
Héraudeau, Philippe
Hernández, Xavier
Hewitt, Anthony
Hewitt, Adelaide
Perez, Fournon
Perez, Maria
Perez-Torres, Miguel
Péroux, Céline
Peters, William
Peterson, Charles
Petit, Jean-Marc
Petrosian, Artaches
Petrov, Georgi
Petuchowski, Samuel
Pfenniger, Daniel
Philipp, Sabine
Phillipps, Steven
Phillips, Mark
Pihlström, Ylva
Pihlström, Ylva
Pikichian, Hovhannes
Pipino, Antonio
Pirzkal, Norbert
Pisano, Daniel
Pizzella, Alessandro
Plana, Henri
Pogge, Richard
Poggianti, Bianca
Polletta, Maria del Carmen
Polyachenko, Evgeny
Pompei, Emanuela
Popescu, Cristina
Popović Luka
Portinari, Laura
Poveda, Arcadio
Prabhu, Tushar
Pracy, Michael
Prandoni, Isabella
Press, William
Prevot-Burnichon, Marie-Louise
Prieto, Almudena
Prires Martins, Lucimara
Pritchet, Christopher
Proctor, Robert
Pronik, Iraida
Pronik, Vladimir
Proust, Dominique
Prugniel, Philippe
Ivnio, Ivanio
Pustil'nik, Simon

Baes, Maarten
Bailey, Mark
Bajaja, Esteban
Baker, Andrew
Baldwin, Jack
Balkowski-Mauger, Chantal
Ballabh, Goswami
Balogh, Michael
Bamford, Steven
Banhatti, Dilip
Barbon, Roberto
Barcons, Xavier
Barnes, David
Barrientos, Luis
Barth, Aaron
Barthel, Peter
Barton, Elizabeth
Barway, Sudhashu
Bassino, Lilia
Basu, Baidyanath
Battaner, Eduardo
Battinelli, Paolo
Baum, Stefi
Baum, William
Bautista, Manuel
Bayet, Estelle
Beaulieu, Sylvie
Beck, Rainer
Beckmann, Volker
Begeman, Kor
Bender, Ralf
Benedict, George
Benetti, Stefano
Benítez, Erika
Bensby, Thomas
Bentz, Misty
Berczik, Peter
Bergeron, Jacqueline
Berkhuijsen, Elly
Berman, Vladimir
Berta, Stefano
Bertola, Francesco
Bettoni, Daniela
Bian, Yulin
Bianchi, Simone
Biermann, Peter
Bignall, Hayley
Bijaoui, Albert
Binette, Luc
Binggeli, Bruno
Binney, James
Biretta, John
Birkinshaw, Mark
Björnsson, Claes-Ingvar
Blakeslee, John
Bland-Hawthorn, Jonathan
Blitz, Leo
Block, David
Blumenthal, George
Hicks, Amalia
Hickson, Paul
Hintzen, Paul
Hirashita, Hiroyuki
Hjalmarson, Ake
Hjorth, Jens
Ho, Luis
Hodge, Paul
Hoekstra, Hendrik
Holz, Daniel
Hong, Wu
Hopkins, Andrew
Hopp, Ulrich
Horellou, Cathy
Hornschemeier, Ann
Hornstrup, Allan
Hou, Jinliang
Houdashelt, Mark
Hough, James
Hu, Fuxing
Hua, Chon Trung
Huang, Keliang
Huang, Jiasheng
Huchra, John
Huchtmeier, Walter
Hüttemeister, Susanne
Hughes, David
Humphreys, Roberta
Humphreys, Elizabeth
Hunstead, Richard
Hunt, Leslie
Hunter, Christopher
Hunter, James
Huynh, Minh
Hwang, Chorng-Yuan
Ichikawa, Takashi
Ichikawa, Shin-ichi
Idiart, Thais
Illingworth, Garth
Im, Myungshin
Imanishi, Masatoshi
Impey, Christopher
Infante, Leopoldo
Irwin, Judith
Ishimaru, Yuhri
Israel, Frank
Issa, Issa
Ivezic, Zeljko
Ivison, Robert
Iwamuro, Fumihide
Iwata, Ikuru
Iye, Masanori
Izotov, Yuri
Izotova, Iryna
Jablonka, Pascale
Jaffe, Walter
Jahnke, Knud
Jang, Minwhan
Jarrett, Thomas
Qin, Yi-Ping
Quinn, Peter
Quintana, Hernan
Rafanelli, Piero
Raiteri, Claudia
Rampazzo, Roberto
Ranalli, Piero
Rand, Richard
Rasmussen, Jesper
Ravindranath, Swara
Raychaudhury, Somak
Reboul, Henri
Recchi, Simone
Rector, Travis
Reichert, Gail
Rejkuba, Marina
Rephaeli, Yoel
Reshetnikov, Vladimir
Reunanen, Juha
Revaz, Yves
Revnivtsev, Mikhail
Rey, Soo-Chang
Reynolds, Cormac
Ribeiro, Andre Luis
Richer, Harvey
Richstone, Douglas
Richter, Gotthard
Richter, Philipp
Ridgway, Susan
Rix, Hans-Walter
Robert, Carmelle
Roberts, Morton
Roberts, Timothy
Roberts Jr., William
Rodrigues de Oliveira Filho, Irapuan
Rödiger, Elke
Roeser, Hermann-Josef
Romano, Patrizia
Romeo, Alessandro
Romero-Colmenero, Encarnacion
Roos, Nicolaas
Rosa, Michael
Rosa González, Daniel
Rosado, Margarita
Rose, James
Rothberg, Barry
Rots, Arnold
Rozas, Maite
Rubin, Vera
Rudnicki, Konrad
Ružička, Adam
Ryder, Stuart
Sackett, Penny
Sadat, Rachida
Sadun, Alberto
Sahibov, Firuz
Saitoh, Takayuki
Sáiz, Alejandro
Sakai, Shoko

Böker, Torsten
Boissier, Samuel
Boisson, Catherine
Boksenberg, Alec
Boles, Thomas
Bolzonella, Micol
Bomans, Dominik
Bongiovanni, Angel
Borchkhadze, Tengiz
Borne, Kirk
Bosma, Albert
Bottinelli, Lucette
Bowen, David
Bower, Gary
Braccesi, Alessandro
Braine, Jonathan
Braun, Robert
Bravo-Alfaro, Hector
Brecher, Kenneth
Bressan, Alessandro
Bridges, Terry
Briggs, Franklin
Brinchmann, Jarle
Brinkmann, Wolfgang
Brinks, Elias
Brodie, Jean
Brosch, Noah
Brouillet, Nathalie
Brown, Thomas
Brown, Michael
Bruzual, Gustavo
Bryant, Julia
Buat, Véronique
Buote, David
Burbidge, Eleanor
Burbidge, Geoffrey
Bureau, Martin
Burgarella, Denis
Burkert, Andreas
Burns Jr., Jack
Burstein, David
Busarello, Giovanni
Buta, Ronald
Butcher, Harvey
Byrd, Gene
Byun, Yong-Ik
Cai, Michael
Calderón, Jess
Calura, Francesco
Calzetti, Daniela
Campusano, Luis
Cannon, Russell
Cannon, John
Cantiello, Michele
Canzian, Blaise
Cao, Xinwu
Cao, Li
Caon, Nicola
Capaccioli, Massimo
Jerjen, Helmut
Jiang, Ing-Guey
Jiménez-Vicente, Jorge
Johansson, Peter
Johnston, Helen
Joly, Monique
Jones, Paul
Jones, Thomas
Jones, Christine
Jordán, Andrés
Jørgensen, Inger
Joshi, Umesh
Jovanovic, Predrag
Joy, Marshall
Jugaku, Jun
Jungwiert, Bruno
Junkes, Norbert
Junkkarinen, Vesa
Junor, William
Kalloglian, Arsen
Kandalyan, Rafik
Kanekar, Nissim
Kaneko, Noboru
Karachentsev, Igor
Karoji, Hiroshi
Kashikawa, Nobunari
Kaspi, Shai
Kassim, Namir
Katgert, Peter
Katsiyannis, Athanassios
Kauffmann, Guinevere
Kaufman, Michèle
Kawakatu, Nozomu
Keel, William
Kellermann, Kenneth
Kelly, Brandon
Kemp, Simon
Kennicutt, Robert
Khachikian, Edward
Khanna, Ramon
Khare, Pushpa
Khosroshahi, Habib
Kilborn, Virginia
Kim, Dong-Woo
Kim, Sungsoo
King, Ivan
Kinman, Thomas
Kirshner, Robert
Kissler-Patig, Markus
Klein, Ulrich
Knapen, Johan
Knezek, Patricia
Kniazev, Alexei
Knudsen, Kirsten
Ko, Chung-Ming
Kobayashi, Chiaki
Kochhar, Rajesh
Kodaira, Keiichi
Kodama, Tadayuki
Sala, Ferran
Salvador-Sole, Eduardo
Samurović, Srdjan
Sanahuja Parera, Blai
Sancisi, Renzo
Sanders, Robert
Sanders, David
Sanroma, Manuel
Sansom, Anne
Santiago, Basilio
Santos-Lleó, Maria
Sapre, Ashok
Saracco, Paolo
Sarazin, Craig
Sargent, Wallace
Sasaki, Toshiyuki
Sasaki, Minoru
Saslaw, William
Sastry, Shankara
Savage, Ann
Saviane, Ivo
Sawa, Takeyasu
Scaramella, Roberto
Schaerer, Daniel
Schaye, Joop
Schechter, Paul
Schmidt, Maarten
Schmitt, Henrique
Schmitz, Marion
Schröder, Anja
Schucking, Engelbert
Schwarz, Ulrich
Schweizer, François
Scodeggio, Marco
Scorza de Appl, Cecilia
Scoville, Nicholas
Searle, Leonard
Seigar, Marcus
Sellwood, Jerry
Semelin, Benoit
Sempere, Maria
Seon, Kwang-Il
Sergeev, Sergey
Serjeant, Stephen
Serote Roos, Margarida
Seshadri, Sridhar
Setti, Giancarlo
Severgnini, Paola
Shan, Hongguang
Shapovalova, Alla
Sharp, Nigel
Sharples, Ray
Shaver, Peter
Shaya, Edward
Shen, Zhiqiang
Sherwood, William
Shields, Gregory
Shields, Joseph
Shimasaku, Kazuhiro

Cappellari, Michele
Caproni, Anderson
Caretta, Cesar
Carigi, Leticia
Carollo, Marcella
Carrillo, René
Carswell, Robert
Carter, David
Casasola, Viviana
Casoli, Fabienne
Cattaneo, Andrea
Cayatte, Veronique
Cellone, Sergio
Cepa, Jordi
Cha, Seung-Hoon
Chae, Kyu-Hyun
Chakrabarti, Sandip
Chakrabarty, Dalia
Chamaraux, Pierre
Chang, Ruixiag
Charmandaris, Vassilis
Chatterjee, Tapan
Chatzichristou, Eleni
Chavushyan, Vahram
Chen, Jiansheng
Chen, Yang
Chen, Lin-wen
Chiappini Moraes Leite, Cristina
Chiba, Masashi
Chincarini, Guido
Chou, Chih-Kang
Choudhury, Tirthankar
Chu, Yaoquan
Chugai, Nikolaj
Chun, Sun
Cid Fernandes Jr., Roberto
Cinzano, Pierantonio
Cioni, Maria-Rosa
Ciotti, Luca
Ciroi, Stefano
Clavel, Jean
Clementini, Gisella
Cohen, Ross
Colbert, Edward
Colina, Luis
Comte, Georges
Conselice, Christopher
Contopoulos, George
Cook, Kem
Corbin, Michael
Corsini, Enrico
Corwin Jr., Harold
Côté, Stéphanie
Côte, Patrick
Couch, Warrick
Courbin, Frederic
Courtes, Georges
Courvoisier, Thierry

Kogoshvili, Natela
Kollatschny, Wolfram
Komiyama, Yutaka
Kong, Xu
Kontizas, Evangelos
Kontorovich, Victor
Koo, David
Koopmans, Leon
Koratkar, Anuradha
Koribalski, Bärbel
Kormendy, John
Kotilainen, Jari
Krajnovic, Davor
Krause, Marita
Krishna, Gopal
Kron, Richard
Krumholz, Mark
Kumai, Yasuki
Kunchev, Peter
Kunert-Bajraszewska, Magdalena
Kuno, Nario
Kunth, Daniel
Kuntschner, Harald
Kunz, Martin
Kuzio de Naray, Rachel
La Barbera, Francesco
La Franca, Fabio
Koji,, Guilaine
Lal, Dharam
Lançon, Ariane
Lanfranchi, Gustavo
Larsen, Søren
Larson, Richard
Laurikainen, Eija
Layzer, David
Leacock, Robert
Leão, João Rodrigo
Lebron, Mayra
Lee, Myung Gyoon
Leeuw, Lerothodi
Lefevre, Olivier
Lehnert, Matthew
Lehto, Harry
Lequeux, James
Levine, Robyn
Li, Qibin
Li, Li-Xin
Li, Ji
Liang, Yanchun
Lilly, Simon
Lim, Jeremy
Lima Neto, Gastao
Lin, Chia
Lin, Weipeng
Lindblad, Per
Linden-Vørnle, Michael
Lo, Kwok-Yung
Lobo, Catarina

Shostak, G.
Shukurov, Anvar
Siebenmorgen, Ralf
Sigurdsson, Steinn
Sil'chenko, Olga
Sillanpaa, Aimo
Silva, David
Silva, Laura
Simkin, Susan
Singh, Kulinder
Siopis, Christos
Skillman, Evan
Slezak, Eric
Smail, Ian
Smecker-Hane, Tammy
Smith, Malcolm
Smith, Haywood
Smith, Eric
Soares, Domingos
Sobouti, Yousef
Sohn, Young-Jong
Soltan, Andrzej
Song, Liming
Sorai, Kazuo
Sparks, William
Spinoglio, Luigi
Spinrad, Hyron
Srinivasan, Ganesan
Statler, Thomas
Staveley-Smith, Lister
Stavrev, Konstantin
Steiman-Cameron, Thomas
Steinbring, Eric
Stiavelli, Massimo
Stirpe, Giovanna
Stoehr, Felix
Stone, Remington
Storchi-Bergmann, Thaisa
Strauss, Michael
Strom, Richard
Strom, Robert
Stuik, Remko
Su, Cheng-yue
Subramaniam, Annapurni
Sugai, Hajime
Sulentic, Jack
Sullivan, III, Woodruff
Sundin, Maria
Sutherland, Ralph
Tacconi-Garman, Lowell
Tagger, Michel
Takashi, Hasegawa
Takata, Tadafumi
Takato, Naruhisa
Takeuchi, Tsutomu
Takizawa, Motokazu
Tamm, Antti
Tammann, Gustav

Couto da Silva, Telma
Cowsik, Ramanath
Coziol, Roger
Crane, Philippe
Crawford, Carolin
Cress, Catherine
Croton, Darren
Cui, Wen-Yuan
Cunniffe, John
Cunow, Barbara
Cypriano, Eduardo
d'Odorico, Sandro
D'Onofrio, Mauro
da Costa, Luiz
Da Rocha, Cristiano
Daisuke, Iono
Dalla Bontaà, Elena
Dallacasa, Daniele
Danks, Anthony
Dantas, Christine
Dasyra, Kalliopi
Davidge, Timothy
Davies, Rodney
Davis, Marc
De Blok, Erwin
de Boer, Klaas
de Bruyn, A.
de Carvalho, Reinaldo
de Diego Onsurbe, José
de Grijs, Richard
de Jong, Roelof
de Mello, Duilia
de Propris, Roberto
de Rijcke, Sven
de Zeeuw, Pieter
Dejonghe, Herwig
Demers, Serge
Deng, Zugan
Dennefeld, Michel
Dessauges-Zavadsky, Miroslava
Dettmar, Ralf-Jürgen
Devost, Daniel
Diaferio, Antonaldo
Diaz, Angeles
Diaz, Ruben
Dickey, John
Dietrich, Matthias
Doi, Mamoru
Dokuchaev, Vyacheslav
Domínguez, Mariano
Dominis Prester, Dijana
Donas, José
Donea, Alina
Dong, Xiao-Bo
Donner, Karl
Donzelli, Carlos
Dopita, Michael
Doroshenko, Valentina
Dottori, Horacio
Londrillo, Pasquale
Longo, Giuseppe
Lopez, Ericson
Lopez Aguerri, José Alfonso
López Cruz, Omar
Lopez Hermoso, Maria
López-Sánchez, Angel
Lord, Steven
Loup, Cecile
Lu, Limin
Lugger, Phyllis
Luminet, Jean-Pierre
Luo, Ali
Lutz, Dieter
Lynden-Bell, Donald
Lynds, Beverly
Lynds, Roger
Ma, Jun
Macalpine, Gordon
Maccagni, Dario
Maccarone, Thomas
Macchetto, Ferdinando
Maciejewski, Witold
Mackie, Glen
Macquart, Jean-Pierre
Madden, Suzanne
Madore, Barry
Magorrian, Stephen
Magris, Gladis
Mahtessian, Abraham
Mainieri, Vincenzo
Maiolino, Roberto
Makarov, Dmitry
Makarova, Lidia
Malagnini, Maria
Malhotra, Sageeta
Mann, Robert
Mannucci, Filippo
Marcelin, Michel
Marco, Olivier
Marconi, Alessandro
Markoff, Sera
Marquez, Isabel
Marr, Jonathon
Marston, Anthony
Martin, René
Martin, Maria
Martin, Crystal
Martinet, Louis
Martinez Garcia, Vicent
Martini, Paul
Marziani, Paola
Masegosa, Gallego
Massardi, Marcella
Mathewson, Donald
Matthews, Lynn
Mattila, Seppo
Mauersberger, Rainer
Maurice, Eric
Tanaka, Yutaka
Taniguchi, Yoshiaki
Tantalo, Rosaria
Taylor, James
Telles, Eduardo
Temporin, Sonia
Tenjes, Peeter
Terlevich, Roberto
Terzian, Yervant
Theis, Christian
Thomasson, Magnus
Thonnard, Norbert
Thornley, Michèle
Thuan, Trinh
Tifft, William
Tikhonov, Nikolai
Tilanus, Remo
Tissera, Patricia
Tisserand,Patrick
Toft, Sune
Tohru , Nagao
Tolstoy, Eline
Tomita, Akihiko
Toomre, Alar
Toshinobu , Takagi
Tovmassian, Hrant
Toyama, Kiyotaka
Traat, Peeter
Trager, Scott
Tremaine, Scott
Treu, Tommaso
Trimble, Virginia
Trinchieri, Ginevra
Trujillo Cabrera, Ignacio
Tsuchiya, Toshio
Tsvetkov, Dmitry
Tuffs, Richard
Tugay, Anatoliy
Tully, Richard
Turner, Edwin
Tyson, John
Tyulbashev, Sergei
Ulrich, Marie-Helene
Urbanik, Marek
Uslenghi, Michela
Utrobin, Victor
Valcheva, Antoniya
Valdés Parra, José
Valentijn, Edwin
Vallenari, Antonella
Valotto, Carlos
Valtchanov, Ivan
Valtonen, Mauri
van Albada, Tjeerd
van den Bergh, Sidney
van der Hulst, Jan
van der Kruit, Pieter
van der Laan, Harry
van der Marel, Roeland

Dovciak, Michal
Doyon, René
Dressel, Linda
Dressler, Alan
Drinkwater, Michael
Driver, Simon
Duc, Pierre-Alain
Dufour, Reginald
Dultzin-Hacyan, Deborah
Dumont, Anne-Marie
Durret, Florence
Duval, Marie-France
Eales, Stephen
Edelson, Rick
Edmunds, Michael
Efstathiou, George
Ehle, Matthias
Einasto, Jaan
Ekers, Ronald
Ellis, Simon
Elmegreen, Debra
Elvis, Martin
Elvius, Aina
Elyiv, Andrii
Emsellem, Eric
English, Jayanne
Enoki, Motohiro
Espey, Brian
Evans, Robert
Fabbiano, Giuseppina
Faber, Sandra
Fabricant, Daniel
Falceta-Gonçalves, Diego
Falco, Emilio
Falcón Barroso, Jesùs
Fall, S.
Famaey, Benoit
Fan, Junhui
Fasano, Giovanni
Fathi, Kambiz
Feain, Ilana
Feast, Michael
Feinstein, Carlos
Feitzinger, Johannes
Ferguson, Annette
Ferland, Gary
Ferrarese, Laura
Ferreras, Ignacio
Ferrini, Federico
Fharo, Toshihiro
Field, George
Filippenko, Alexei
Flin, Piotr
Florido, Estrella
Florsch, Alphonse
Foltz, Craig
Forbes, Duncan
Ford, Holland
Ford Jr., W.
Mayya, Divakara
Mazzarella, Joseph
McBreen, Brian
McGaugh, Stacy
McNeil, Stephen
Mediavilla, Evencio
Mehlert, Dörte
Meier, David
Meikle, William
Meisenheimer, Klaus
Melnyk, Olga
Mendes de Oliveira, Claudia
Menon, T.
Mercurio, Amata
Merluzzi, Paola
Merrifield, Michael
Metevier, Anne
Meusinger, Helmut
Meyer, Martin
Meyer, Angela
Meza, Andres
Mihov, Boyko
Miley, George
Miller, Joseph
Miller, Hugh
Miller, Richard
Miller, Neal
Milvang-Jensen, Bo
Ming, Bai
Mirabel, Igor
Miroshnichenko, Alla
Misawa, Toru
Mitsunobu, Kawada
Mizuno, Takao
Moiseev, Alexei
Moles Villamate, Mariano
Molinari, Emilio
Mollá, Mercedes
Monaco, Pierluigi
Moody, Joseph
Mori, Masao
Moss, Christopher
Motohara, Kentaro
Mould, Jeremy
Mourão, Ana Maria
Müller, Volker
Dorotovič,, Andreas
Mjica, Raul
Mulchaey, John
Muller, Erik
Muller, Sebastien
Muñoz Tuñón, Casiana
Muratorio, Gerard
Murayama, Takashi
Murray, Stephen
Mushotzky, Richard
Muzzio, Juan
Nair, Sunita
Nakanishi, Kouichiro
van Driel, Wim
van Gorkom, Jacqueline
van Moorsel, Gustaaf
van Woerden, Hugo
Van Zee, Liese
Vansevicius, Vladas
Varma, Ram
Vaughan, Simon
Vauglin, Isabelle
Vavilova, Iryna
Vazdekis, Alexandre
Veilleux, Sylvain
Vercellone, Stefano
Verdes-Montenegro, Lourdes
Verdoes Kleijn, Gijsbert
Vermeulen, René
Véron, Philippe
Véron, Marie-Paule
Viel, Matteo
Vigroux, Laurent
Villata, Massimo
Vivas, Anna
Vlasyuk, Valerij
Vollmer, Bernd
Vrtilek, Jan
Wada, Keiichi
Wagner, Stefan
Wagner, Alexander
Wakamatsu, Ken-Ichi
Walker, Mark
Walter, Fabian
Walterbos, René
Wanajo, Shinya
Wang, Yiping
Wang, Tinggui
Wang, Huiyuan
Wang, Hong-Guang
Ward, Martin
Weedman, Daniel
Wei, Jianyan
Weilbacher, Peter
Weiler, Kurt
Welch, Gary
Westmeier, Tobias
White, Simon
Whiting, Matthew
Whitmore, Bradley
Wielebinski, Richard
Wielen, Roland
Wiita, Paul
Wilcots, Eric
Wild, Wolfgang
Williams, Robert
Williams, Theodore
Williams, Barbara
Wills, Beverley
Wills, Derek
Windhorst, Rogier
Winkler, Hartmut

Foschini, Luigi
Fouqué, Pascal
Fraix-Burnet, Didier
Francis, Paul
Freedman, Wendy
Freeman, Kenneth
Fricke, Klaus
Fried, Josef
Fritze, Klaus
Frogel, Jay
Fuchs, Burkhard
Fujita, Yutaka
Fukugita, Masataka
Funato, Yoko
Funes, José
Furlanetto, Steven
Gaensler, Bryan
Gallagher, Sarah
Gallart, Carme
Gallego, Jess
Galletta, Giuseppe
Gallimore, Jack
Gamaleldin, Abdulla
Ganguly, Rajib
Gao, Yu
Garcia-Lorenzo, Maria
Gardner, Jonathan
Garilli, Bianca
Gascoigne, S.
Gavignaud, Isabelle
Gelderman, Richard
Geller, Margaret
Georgiev, Tsvetan
Gerhard, Ortwin
Ghigo, Francis
Ghosh, P.
Giacani, Elsa
Giani, Elisabetta
Gibson, Brad
Gigoyan, Kamo
Giovanardi, Carlo
Giovanelli, Riccardo
Giroletti, Marcello
Gitti, Myriam
Glass, Ian
Glazebrook, Karl
Godlowski, Wlodzimierz
Gonzalez, Serrano
Gonzalez Delgado, Rosa
Gonzalez-Solares, Eduardo
Goodrich, Robert
Gorgas, Garcia
Goss, W. Miller
Goto, Tomotsugu
Gottesman, Stephen
Gouguenheim, Lucienne
Nakanishi, Hiroyuki
Nakos, Theodoros
Namboodiri, P.
Napolitano, Nicola
Navarro, Julio
Nedialkov, Petko
Nikołajuk, Marek
Ninkovic, Slobodan
Nipoti, Carlo
Nishikawa, Ken-Ichi
Nityananda, Rajaram
Noguchi, Masafumi
Noonan, Thomas
Norman, Colin
Nucita, Achille
Nulsen, Paul
O'Connell, Robert
O'Dea, Christopher
Ocvirk, Pierre
Oemler Jr., Augustus
Oey, Sally
Ohta, Kouji
Okamura, Sadanori
Olling, Robert
Olofsson, Kjell
Omizzolo, Alessandro
Oosterloo, Thomas
Origlia, Livia
Osman, Anas
östlin, Göran
Ostorero, Luisa
Ostriker, Eve
Ott, Juergen
Ouchi, Masami
Ovcharov, Evgeni
Oyabu, Shinki
Pacholczyk, Andrzej
Page, Mathew
Pak, Soojong
Palmer, Philip
Palumbo, Giorgio
Panessa, Francesca
Pannuti, Thomas
Papayannopoulos, Theodoros
Park, Jang-Hyun
Parker, Quentin
Pastoriza, Miriani
Paturel, Georges
Pearce, Frazer
Pedrosa, Susana
Peimbert, Manuel
Peletier, Reynier
Pellegrini, Silvia
Pello, Roser
Peng, Qingyu
Perea-Duarte, Jaime
Wise, Michael
Wise, Michael
Wisotzki, Lutz
Wlérick, Gérard
Wofford, Aida
Wold, Margrethe
Woosley, Stanford
Worrall, Diana
Woudt, Patrick
Wozniak, Hervé
Wrobel, Joan
Wu, Xue-bing
Wu, Wentao
Wu, Jianghua
Wu, Yanling
Wünsch, Richard
Wynn-Williams, C.
Xanthopoulos, Emily
Xia, Xiao-Yang
Xilouris, Emmanouel
Xu, Dawei
Xue, Suijian
Yagi, Masafumi
Yahagi, Hideki
Yakovleva, Valerija
Yamada, Yoshiyuki
Yamada, Toru
Yamagata, Tomohiko
Yi, Sukyoung
Yonehara, Atsunori
Yoshida, Michitoshi
Yoshikawa, Kohji
Young, Judith
Zaggia, Simone
Zamorano, Jaime
Zaroubi, Saleem
Zasov, Anatolij
Zavatti, Franco
Zeilinger, Werner
Zepf, Stephen
Zezas, Andreas
Zhang, Xiaolei
Zhang, Yang
Zhang, JiangShui
Zhang, Jingyi
Zhang, Bo
Zhang, Fenghui
Zhou, Youyuan
Zhou, Xu
Zhou, Hongyan
Zhou, Jianjun
Ziegler, Bodo
Ziegler, Harald
Zinn, Robert
Zou, Zhenlong
Zwaan, Martin

Division IV Commission 29 Stellar Spectra

President Piskunov, Nikolai
Vice-President Cunha, Katia

Organizing Committee

Aoki, Wako
Asplund, Martin
Bohlender, David
Carpenter, Kenneth
Melendez, Jorge
Rossi, Silvia
Smith, Verne
Soderblom, David
Wahlgren, Glenn

Members

Abia, Carlos
Abt, Helmut
Adelman, Saul
Afram, Nadine
Aikman, G.
Ake III, Thomas
Alcalá, Juan Manuel
Alecian, Georges
Alencar, Silvia
Allende Prieto, Carlos
Ambruster, Carol
Andretta, Vincenzo
Andreuzzi, Gloria
Andrillat, Yvette
Annuk, Kalju
Aoki, Wako
Appenzeller, Immo
Ardila, David
Arias, Maria
Arkharov, Arkadij
Artru, Marie-Christine
Asplund, Martin
Atac, Tamer
Audard, Marc
Baade, Dietrich
Bacham, Eswar
Bagnulo, Stefano
Bakker, Eric

Baliunas, Sallie
Ballereau, Dominique
Banerjee, Dipankar
Baratta, Giovanni
Barber, Robert
Barbuy, Beatriz
Baron, Edward
Basri, Gibor
Batalha, Celso
Bauer, Wendy
Beckman, John
Beers, Timothy
Beiersdorfer, Peter
Bellas-Velidis, Ioannis
Bensby, Thomas
Bertone, Emanuele
Bessell, Michael
Biazzo, Katia
Bikmaev, Ilfan
Gomboc, Andreja
Gonzalez, Guillermo
Gopka, Vera
Grady, Carol
Gratton, Raffaele
Gray, David
Griffin, R.
Griffin, Roger
Gu, Sheng-hong
Gustafsson, Bengt
Guthrie, Bruce
Hack, Margherita
Hanson, Margaret
Hanuschik, Reinhard
Harmer, Charles
Harmer, Dianne
Hartman, Henrik
Hartmann, Lee
Hashimoto, Osamu
Hearnshaw, John
Heber, Ulrich
Heiter, Ulrike
Henrichs, Hubertus
Herbig, George
Heske, Astrid
Hessman, Frederic
Hill, Grant
Hill, Vanessa
Hillier, Desmond
Hinkle, Kenneth
Hirai, Masanori
Hirata, Ryuko
Hoeflich, Peter
Horaguchi, Toshihiro
Houk, Nancy
Houziaux, Leo
Hron, Josef
Hu, Zhong wen
Hubert-Delplace, Anne-Marie
Hubrig, Swetlana
Huenemoerder, David
Hyland, Harry
Israelian, Garik
Ivans, Inese
Izumiura, Hideyuki
Jankov, Slobodan
Jehin, Emmanuel
Johnson, Hollis
Owocki, Stanley
Pakhomov, Yury
Parsons, Sidney
Pavani, Daniela
Pavlenko, Yakov
Pedoussaut, André
Peery, Benjamin
Peters, Geraldine
Peterson, Ruth
Petit, Pascal
Pilachowski, Catherine
Pintado, Olga
Plez, Bertrand
Polcaro, V.
Polidan, Ronald
Polosukhina-Chuvaeva, Nina
Pompeia, Luciana
Porto de Mello, Gustavo
Primas, Francesca
Prinja, Raman
Prires Martins, Lucimara
Querci, François
Querci, Monique
Raassen, Ion
Rao, N.
Rashkovskij, Sergey
Rastogi, Shantanu
Rautela, B.
Rauw, Gregor
Rawlings, Mark
Rebolo, Rafael
Rego, Fernandez
Reimers, Dieter
Reiners, Ansgar
Rettig, Terrence
Ringuelet, Adela
Rivinius, Thomas
Romanyuk, Iosif
Rose, James
Rossi, Lucio
Rossi, Corinne
Rutten, Robert
Ryan, Sean
Ryde, Nils
Sachkov, Mikhail
Sadakane, Kozo
Saffe, Carlos
Sahade, Jorge

Boehm, Torsten
Boesgaard, Ann
Boggess, Albert
Bohlender, David
Bond, Howard
Bonifacio, Piercarlo
Bonsack, Walter
Bopp, Bernard
Bouvier, Jerôme
Bragaglia, Angela
Brandi, Elisande
Breysacher, Jacques
Brickhouse, Nancy
Briot, Danielle
Brown, Douglas
Bruhweiler, Frederick
Bruning, David
Bruntt, Hans
Bues, Irmela
Burkhart, Claude
Busà, Innocenza
Butkovskaya, Varvara
Butler, Keith
Carney, Bruce
Carretta, Eugenio
Carter, Bradley
Catala, Claude
Catanzaro, Giovanni
Catchpole, Robin
Cayrel, Roger
Cayrel de Strobel, Giusa
Chadid, Merieme
Chavez-Dagostino, Miguel
Chen, Alfred
Chen, Yuqin
Cidale, Lydia
Claudi, Riccardo
Climenhaga, John
Conti, Peter
Corbally, Christopher
Cornide, Manuel
Cottrell, Peter
Cowley, Anne
Cowley, Charles
Crowther, Paul
Cui, Wen-Yuan
Curé, Michel
da Silva, Licio
Dačic, Miodrag
Daflon, Simone
Daisaku, Nogami
Dall, Thomas
Damineli Neto, Augusto
Dawanas, Djoni
de Araujo, Francisco
de Castro, Elisa
de Laverny, Patrick
del Peloso, Eduardo
Divan, Lucienne
Johnson, Jennifer
Jordan, Carole
Josselin, Eric
Jugaku, Jun
Käufl, Hans Ulrich
Kawka, Adela
Kipper, Tonu
Klochkova, Valentina
Kochukhov, Oleg
Kodaira, Keiichi
Kogure, Tomokazu
Kolka, Indrek
Kordi, Ayman
Korn, Andreas
Korotin, Sergey
Kotnik-Karuza, Dubravka
Koubsky, Pavel
Kovachev, Bogomil
Kovtyukh, Valery
Kraft, Robert
Krempec-Krygier, Janina
Kwok, Sun
Lago, Maria
Lagrange, Anne-Marie
Laird, John
Lambert, David
Lamers, Henny
Lamontagne, Robert
Landstreet, John
Lanz, Thierry
Le Contel, Jean-Michel
Leão, João Rodrigo
Lebre, Agnes
Leckrone, David
Lee, Jae-Woo
Leedjärv, Laurits
Lester, John
Leushin, Valerij
Levato, Orlando
Li, Ji
Liang, Yanchun
Liebert, James
Little-Marenin, Irene
Liu, Michael
Lodders, Katharina
Lopes, Dalton
Lubowich, Donald
Lucatello, Sara
Luck, R.
Lugaro, Maria
Lundstrom, Ingemar
Luo, Ali
Lyubimkov, Leonid
Magain, Pierre
Magazz, Antonio
Maillard, Jean-Pierre
Maitzen, Hans
Malaroda, Stella
Manteiga Outeiro, Minia
Sánchez, Almeida
Sanwal, Basant
Sareyan, Jean-Pierre
Sarre, Peter
Sasso, Clementina
Satoshi, Honda
Sbordone, Luca
Schild, Rudolph
Schroeder, Klaus
Schuh, Sonja
Schuler, Simon
Seggewiss, Wilhelm
Selam, Selim
Shetrone, Matthew
Shi, Huoming
Shi, Jianrong
Sholukhova, Olga
Shore, Steven
Simon, Theodore
Singh, Mahendra
Sinnerstad, Ulf
Smalley, Barry
Smith, Myron
Smith, Graeme
Smith, Verne
Snow, Theodore
Soderblom, David
Sonneborn, George
Sonti, Sreedhar
Spite, François
Spite, Monique
St-Louis, Nicole
Stalio, Roberto
Stateva, Ivanka
Stathakis, Raylee
Stawikowski, Antoni
Stecher, Theodore
Steffen, Matthias
Stefl, Stanislav
Stencel, Robert
Suntzeff, Nicholas
Svolopoulos, Sotirios
Swings, Jean-Pierre
Szeifert, Thomas
Takashi, Hasegawa
Talavera, Antonio
Tantalo, Rosaria
Tautvaišiene, Gražina
Thevenin, Frederic
Tomasella, Lina
Tomov, Toma
José Miguel Torrejón, José Miguel
Tuominen, Ilkka
Ulyanov, Oleg
Usenko, Igor
Utrobin, Victor
Utsumi, Kazuhiko
Valenti, Jeff
Valtier, Jean-Claude

Dolidze, Madona
Doppmann, Gregory
Dragunova, Alina
Drake, Natalia
Duncan, Douglas
Dworetsky, Michael
Edwards, Suzan
Elmhamdi, Abouazza
Faraggiana, Rosanna
Feast, Michael
Felenbok, Paul
Fernandez-Figueroa, M.
Fitzpatrick, Edward
Floquet, Michèle
Foing, Bernard
Foy, Renaud
Franchini, Mariagrazia
François, Patrick
Frandsen, Søren
Freire Ferrero, Rubens
Friedjung, Michael
Friel, Eileen
Fujita, Yoshio
Fulbright, Jon
Fullerton, Alexander
Gamen, Roberto
García, López
Garmany, Katy
Garrison, Robert
Gautier, Daniel
Gehren, Thomas
Gerbaldi, Michèle
Gershberg, R.
Gesicki, Krzysztof
Ghosh, Kajal
Giampapa, Mark
Gilra, Daya
Giovannelli, Franco
Glagolevskij, Yurij
Glazunova, Ljudmila
Glushneva, Irina
Goebel, John
Marilli, Ettore
Marsden, Stephen
Martinez Fiorenzano, Aldo
Massey, Philip
Mathys, Gautier
Matsuura, Mikako
Mazzali, Paolo
McDavid, David
McGregor, Peter
McNamara, Delbert
McSaveney, Jennifer
McSwain, Mary
Megessier, Claude
Melendez, Jorge
Melo, Claudio
Merlo, David
Mickaelian, Areg
Mikulasek, Zdeněk
Moffat, Anthony
Molaro, Paolo
Monin, Dmitry
Montes, David
Moos, Henry
Morossi, Carlo
Morrison, Nancy
Napiwotzki, Ralf
Nazarenko, Victor
Neckel, Heinz
Neiner, Coralie
Nicholls, Ralph
Niedzielski, Andrzej
Nielsen, Krister
Niemczura, Ewa
Nilsson, Hampus
Nishimura, Shiro
Norris, John
North, Pierre
Nugis, Tiit
O'Neal, Douglas
O'Toole, Simon
Okazaki, Atsuo
Oudmaijer, René
van der Hucht, Karel
van Eck, Sophie
van Winckel, Hans
van't Veer-Menneret, Claude
Vasu-Mallik, Sushma
Vennes, Stéphane
Verdugo, Eva
Verheijen, Marc
Vilhu, Osmi
Viotti, Roberto
Vladilo, Giovanni
Vogt, Nikolaus
Vogt, Steven
Vreux, Jean
Wade, Gregg
Wallerstein, George
Wang, Feilu
Waterworth, Michael
Wegner, Gary
Wehinger, Peter
Whelan, Emma
Williams, Peredur
Wing, Robert
Wolf, Bernhard
Wolff, Sidney
Wood, H.
Wright, Nicholas
Wyckoff, Susan
Yamashita, Yasumasa
Yoshioka, Kazuo
Yüce, Kutluay
Yushkin, Maxim
Zaggia, Simone
Zapatero-Osorio, Maria Rosa
Zhang, Huawei
Zhang, Bo
Zhang, Haotong
Zhang, Yanxia
Zhu, Zhenxi
Zorec, Juan
Zverko, Juraj

Division IX Commission 30 Radial Velocities

President Torres, Guillermo
Vice-President Pourbaix, Dimitri

Organizing Committee

Marcy, Geoffrey
Mathieu, Robert
Mazeh, Tsevi
Minniti, Dante
Pepe, Francesco
Turon, Catherine
Zwitter, Tomaž

Members

Abt, Helmut
Al-Malki, Mohammed
Andersen, Johannes
Arnold, Richard
Balona, Luis
Batten, Alan
Beavers, Willet
Beers, Timothy
Bernstein, Hans-Heinrich
Beuzit, Jean-Luc
Breger, Michel
Burki, Gilbert
Butkovskaya, Varvara
Cardoso Santos, Nuno
Carney, Bruce
Chadid, Merieme
Chen, Yuqin
Cochran, William
Couto da Silva, Telma
Crampton, David
Crifo, Françoise
da Costa, Luiz
Davis, Robert
Davis, Marc
de Jonge, J.
de Medeiros, José
De Souza Pellegrini, Paulo
Dravins, Dainis
Dubath, Pierre
Fekel, Francis
Fletcher, J.
Florsch, Alphonse
Foltz, Craig
Forveille, Thierry
García, Beatriz
Georgelin, Yvon
Gilmore, Gerard
Giovanelli, Riccardo
Gnedin, Yurij
Gonzalez, Jorge
Gouguenheim, Lucienne
Gray, David
Griffin, Roger
Halbwachs, Jean-Louis
Hearnshaw, John
Hewett, Paul
Hilditch, Ronald
Hill, Graham
Holmberg, Johan
Howard, Andrew
Hrivnak, Bruce
Hu, Zhong wen
Hube, Douglas
Hubrig, Swetlana
Huchra, John
Imbert, Maurice
Irwin, Alan
Jorissen, Alain
Kadouri, Talib
Karachentsev, Igor
Katz, David
Khalesseh, Bahram
Konacki, Maciej
Kraft, Robert
Latham, David
Levato, Orlando
Lewis, Brian
Lindgren, Harri
Lo Curto, Gaspare
Marschall, Laurence
Martinez Fiorenzano, Aldo
Maurice, Eric
Mayor, Michel
McMillan, Robert
Melnick, Gary
Meylan, Georges
Mink, Douglas
Missana, Marco
Mkrtichian, David
Morbey, Christopher
Morrell, Nidia
Naef, Dominique
Napolitano, Nicola
Oetken, L.
Pedoussaut, André
Pepe, Francesco
Perrier-Bellet, Christian
Peterson, Ruth
Philip, A.G.
Preston, George
Quintana, Hernan
Rastorguev, Aleksej
Ratnatunga, Kavan
Romanov, Yuri
Royer, Frédéric
Rubenstein, Eric
Rubin, Vera
Sachkov, Mikhail
Samus, Nikolaj
Scarfe, Colin
Schröder, Anja
Sivan, Jean-Pierre
Smith, Myron
Solivella, Gladys
Stefanik, Robert
Steinmetz, Matthias
Stickland, David
Strauss, Michael
Suntzeff, Nicholas
Szabados, Laśzló
Szécsényi-Nagy, Gábor
Tokovinin, Andrei
Tomasella, Lina
Tonry, John
van Dessel, Edwin
Verschueren, Werner
József Vinkó, Jozsef
Walker, Gordon
Wegner, Gary
Willstrop, Roderick
Yang, Stephenson
Yoss, Kenneth
Zaggia, Simone
Zhang, Haotong

Division I Commission 31 Time

President Manchester, Richard
Vice-President Hosokawa, Mizuhiko

Organizing Committee

Arias, Elisa
Gang, Zhang
Tuckey, Philip
Zharov, Vladimir

Members

Abele, Maris
Ahn, Youngsook
Allan, David
Alley, Carrol
Aoki, Shinko
Arakida, Hideyoshi
Archinal, Brent
Breakiron, Lee
Brentjens, Michiel
Bruyninx, Carine
Carter, William
Chou, Yi
Dehant, Véronique
Dick, Wolfgang
Dickey, Jean
Douglas, R.
Du, Lan
Fallon, Frederick
Fliegel, Henry
Foschini, Luigi
Fujimoto, Masa-Katsu
Fukushima, Toshio
Gambis, Daniel
Gang, Zhang
Gao, Yuping
Granveaud, Michel
Guinot, Bernard
Han, Tianqi
Hers, Jan
Hobbs, George
Hu, Yonghui
Hua, Yu
Iijima, Shigetaka
Ilyasov, Yuri
Ivanov, Dmitrij
Jin, WenJing
Kakuta, Chuichi
Klepczynski, William
Kolaczek, Barbara
Koshelyaevsky, Nikolay
Kovalevsky, Jean
Kwok, Sun
Li, Xiaohui
Lieske, Jay
Lu, BenKui
Lu, Xiaochun
Luck, John
Luo, Dingchang
Ma, Lihua
McCarthy, Dennis
Meinig, Manfred
Melbourne, William
Mendes, Virgilio
Morgan, Peter
Mueller, Ivan
Nelson, Robert
Newhall, X.
Noël, Fernando
Paquet, Paul
Pilkington, John
Pineau des Forêts, Guillaume
Pugliano, Antonio
Pushkin, Sergej
Ray, James
Ray, Paul
Robertson, Douglas
Rodin, Alexander
Sheikh, Suneel
Smylie, Douglas
Song, Jinan
Stanila, George
Sun, Fuping
Thomas, Claudine
van Leeuwen, Joeri
Vernotte, François
Vicente, Raimundo
Vilinga, Jaime
Wilkins, George
Wu, Guichen
Wu, Shouxian
Wu, Haitao
Wu, Dong
Yang, Xuhai
Yatskiv, Yaroslav
Ye, Shuhua
Zhang, Weiqun
Zhang, Haotong
Zheng, Yong

Division VII Commission 33 Structure & Dynamics of the Galactic System

President Wyse, Rosemary
Vice-President Nordström, Birgitta

Organizing Committee

Bland-Hawthorn, Jonathan
Feltzing, Sofia
Fuchs, Burkhard
Minniti, Dante

Members

Aarseth, Sverre
Acosta Pulido, José
Adamson, Andrew
Afanassiev, Viktor
Aguilar, Luis A.
Alcobé, Santiago
Alejandra, Recio-Blanco
Allende Prieto, Carlos
Altenhoff, Wilhelm
Ambastha, Ashok
Andersen, Johannes
Antonov, Vadim
Aoki, Shinko
Ardeberg, Arne
Ardi, Eliani
Arnold, Richard
Asteriadis, Georgios
Athanassoula, Evangelia
Babusiaux, Carine
Baek, Chang Hyun
Baier, Frank
Balázs, Lajos
Balbus, Steven
Balcells, Marc
Baldwin, John
Banhatti, Dilip
Baranov, Alexander
Barberis, Bruno
Bartašiute, Stanislava
Bash, Frank
Basu, Baidyanath
Baud, Boudewijn
Bellazzini, Michele
Bensby, Thomas
Berkhuijsen, Elly
Bienayme, Olivier
Binney, James
Blaauw, Adriaan
Blanco, Victor
Blitz, Leo
Bloemen, Hans
Blommaert, Joris
Bobylev, Vadim
Brand, Jan
Bronfman, Leonardo
Brown, Warren
Burke, Bernard
Green, Anne
Green, James
Grenon, Michel
Gupta, Sunil
Habe, Asao
Habing, Harm
Hakkila, Jon
Hamajima, Kiyotoshi
Hanami, Hitoshi
Hanson, Margaret
Hartkopf, William
Hayli, Abraham
Haywood, Misha
Heiles, Carl
Helmi, Amina
Herbst, William
Hernández-Pajares, Manuel
Hetem Jr., Annibal
Hobbs, Robert
Holmberg, Johan
Honma, Mareki
Hori, Genichiro
Hozumi, Shunsuke
Hron, Josef
Hu, Hongbo
Hulsbosch, A.
Humphreys, Roberta
Humphreys, Elizabeth
Hunter, Christopher
Iguchi, Osamu
Ikeuchi, Satoru
Ilya, Mandel
Inagaki, Shogo
Innanen, Kimmo
Israel, Frank
Ivezic, Zeljko
Iwaniszewska, Cecylia
Iye, Masanori
Jablonka, Pascale
Jackson, Peter
Jahreiss, Hartmut
Jalali, Mir Abbas
Jasniewicz, Gerard
Jiang, Dongrong
Jiang, Ing-Guey
Jog, Chanda
Johansson, Peter
Ostorero, Luisa
Ostriker, Jeremiah
Ostriker, Eve
Palmer, Patrick
Palouš, Jan
Pandey, A.
Pandey, Birendra
Papayannopoulos, Theodoros
Park, Byeong-Gon
Parmentier, Geneviève
Patsis, Panos
Pauls, Thomas
Peimbert, Manuel
Perek, Lubos
Perryman, Michael
Pesch, Peter
Philip, A.G.
Pier, Jeffrey
Pirzkal, Norbert
Polyachenko, Evgeny
Portinari, Laura
Price, R.
Rabolli, Monica
Raharto, Moedji
Ratnatunga, Kavan
Reid, Iain
Reif, Klaus
Reylé, Céline
Rich, Robert
Richter, Philipp
Riegel, Kurt
Roberts, Morton
Roberts Jr., William
Robin, Annie
Rocha-Pinto, Hélio
Rodrigues de Oliveira Filho, Irapuan
Rohlfs, Kristen
Rong, Jianxiang
Rubin, Vera
Ruelas-Mayorga, R.
Ruiz, Maria
Ružička, Adam
Rybicki, George
Saar, Enn
Sakano, Masaaki
Sala, Ferran
Sánchez Doreste, Néstor

Burton, W.
Butler, Raymond
Caldwell, John
Cane, Hilary
Cao, Zhen
Caretta, Cesar
Carollo, Daniela
Carpintero, Daniel
Carrasco, Luis
Caswell, James
Cesarsky, Diego
Cesarsky, Catherine
Cha, Seung-Hoon
Chakrabarty, Dalia
Chapman, Jessica
Chen, Li
Chen, Yuqin
Christodoulou, Dimitris
Churchwell, Edward
Cincotta, Pablo
Cioni, Maria-Rosa
Ciurla, Tadeusz
Clemens, Dan
Clube, S.
Cohen, Richard
Comins, Neil
Contopoulos, George
Corradi, Romano
Costa, Edgardo
Courtes, Georges
Crampton, David
Crawford, David
Crézé, Michel
Cropper, Mark
Croton, Darren
Cubarsi, Rafael
Cudworth, Kyle
Cuisinier, François
Cuperman, Sami
Dalla Bontaà, Elena
Dambis, Andrei
Dauphole, Bertrand
Davies, Rodney
Dawson, Peter
de Jong, Teije
Dejonghe, Herwig
Dekel, Avishai
Diaferio, Antonaldo
Diaz, Ruben
Dickel, Helene
Dickel, John
Dickman, Robert
Dieter Conklin, Nannielou
Djorgovski, Stanislav
do Nascimento Jr., José
Downes, Dennis
Drilling, John
Drimmel, Ronald
Ducati, Jorge

Johnson, Hugh
Jones, Derek
Kalandadze, N.
Kalnajs, Agris
Kang, Yong
Kasumov, Fikret
Kato, Shoji
Khovritchev, Maxim
Kim, Sungsoo
King, Ivan
Kinman, Thomas
Klare, Gerhard
Knapp, Gillian
Korchagin, Vladimir
Kormendy, John
Krajnovic, Davor
Kulsrud, Russell
Kutuzov, Sergej
Lafon, Jean-Pierre
Laloum, Maurice
Larson, Richard
Latham, David
Lecar, Myron
Lee, Myung Gyoon
Lee, Sang
Lee, Hyung
Lee, Kang Hwan
Lepine, Sebastien
Li, Jinzeng
Liebert, James
Lin, Chia
Lin, Qing
Lindblad, Per
Lockman, Felix
Loden, Kerstin
Luo, Ali
MacConnell, Darrell
Majumdar, Subhabrata
Manchester, Richard
Marochnik, L.
Martin, Christopher
Martinet, Louis
Martínez-Delgado, David
Martos, Marco
Mathewson, Donald
Matteucci, Francesca
Mavridis, Lyssimachos
Mayor, Michel
McClure-Griffiths, Naomi
McGregor, Peter
Méndez Bussard, René
Merrifield, Michael
Mezger, Peter
Mikkola, Seppo
Miller, Richard
Mirabel, Igor
Mishurov, Yuri
Miyamoto, Masanori
Moffat, Anthony

Sánchez-Saavedra, M.
Sandqvist, Aage
Santiago, Basilio
Santillan, Alfredo
Sanz, Jaume
Sargent, Annelia
Schechter, Paul
Schmidt, Maarten
Schmidt-Kaler, Theodor
Schoedel, Rainer
Seggewiss, Wilhelm
Seimenis, John
Sellwood, Jerry
Serabyn, Eugene
Shan, Hongguang
Shane, William
Shi, Huoming
Shu, Frank
Sigalotti, Leonardo
Simonson, S.
Sobouti, Yousef
Song, Liming
Song, Qian
Sotnikova, Natalia
Soubiran, Caroline
Sparke, Linda
Spergel, David
Spiegel, E.
Stecker, Floyd
Steinlin, Uli
Stibbs, Douglas
Stoehr, Felix
Strobel, Andrzej
Su, Cheng-yue
Subramaniam, Annapurni
Surdin, Vladimir
Svolopoulos, Sotirios
Sygnet, Jean-François
Takashi, Hasegawa
Tammann, Gustav
Terzides, Charalambos
Thé, Pik-Sin
Thielheim, Klaus
Thomas, Claudine
Tian, Wenwu
Tinney, Christopher
Tobin, William
Tomisaka, Kohji
Toomre, Alar
Toomre, Juri
Torra, Jordi
Tosa, Makoto
Tsujimoto, Takuji
Turon, Catherine
Upgren, Arthur
Urquhart, James
Valtonen, Mauri
van der Kruit, Pieter
van Woerden, Hugo

Ducourant, Christine
Dzigvashvili, R.
Egret, Daniel
Einasto, Jaan
Elmegreen, Debra
Esamdin, Ali
Esimbek, Jarken
Evangelidis, E.
Faber, Sandra
Fathi, Kambiz
Feast, Michael
Feitzinger, Johannes
Ferguson, Annette
Fernández, David
Figueras, Francesca
Foster, Tyler
Freeman, Kenneth
Fridman, Aleksej
Fuchs, Burkhard
Fujimoto, Masa-Katsu
Fujiwara, Takao
Galletto, Dionigi
Ganguly, Rajib
Garzón, Francisco
Gemmo, Alessandra
Genkin, Igor
Genzel, Reinhard
Georgelin, Yvon
Gilmore, Gerard
Goldreich, Peter
Gomez, Ana
Gordon, Mark
Gottesman, Stephen
Grayzeck, Edwin
Mohammed, Ali
Moitinho, André
Monet, David
Monnet, Guy
Morales Rueda, Luisa
Moreno Lupiañez, Manuel
Morris, Mark
Morris, Rhys
Muench, Guido
Nakasato, Naohito
Namboodiri, P.
Napolitano, Nicola
Neckel, Th.
Nelemans, Gijs
Nelson, Alistair
Newberg, Heidi
Nikiforov, Igor
Ninkovic, Slobodan
Nishida, Minoru
Nishida, Mitsugu
Norman, Colin
Nuritdinov, Salakhutdin
Oblak, Edouard
Ocvirk, Pierre
Oey, Sally
Oh, Kap-Soo
Oja, Tarmo
Ojha, Devendra
Okuda, Haruyuki
Olano, Carlos
Ollongren, A.
Orlov, Victor
Ortiz, Roberto
Vandervoort, Peter
Varela Perez, Antonia
Vega, E.
Venugopal, V.
Vergne, María
Verschuur, Gerrit
Villas da Rocha, Jaime
Vivas, Anna
Volkov, Evgeni
Voroshilov, Volodymyr
Wachlin, Felipe
Wagner, Alexander
Weaver, Harold
Weistrop, Donna
Westerhout, Gart
Whiteoak, John
Whittet, Douglas
Wielebinski, Richard
Wielen, Roland
Woltjer, Lodewijk
Woodward, Paul
Wouterloot, Jan
Wramdemark, Stig
Wright, Nicholas
Wünsch, Richard
Yamagata, Tomohiko
Yim, Hong-Suh
Yoshii, Yuzuru
Younis, Saad
Zachilas, Loukas
Zaggia, Simone
Zhang, Haotong
Zhou, Jianjun

Division VI Commission 34 Interstellar Matter

President Millar, Thomas
Vice-President Chu, You-Hua

Organizing Committee

Breitschwerdt, Dieter
Burton, Michael
Cabrit, Sylvie
Caselli, Paola
de Gouveia Dal Pino, Elisabete
Dyson, John
Ferland, Gary
Juvela, Mika
Koo, Bon-Chul
Kwok, Sun
Lizano, Susana
Rozyczka, Michal
Tóth, Laśzló
Tsuboi, Masato
Yang, Ji

Members

Aannestad, Per
Abgrall, Herve
Acker, Agnes
Adams, Fred
Aiad, A.
Aikawa, Yuri
Aitken, David
Akabane, Kenji
Akio, Inoue
Al-Mostafa, Zaki
Alcolea, Javier
Altenhoff, Wilhelm
Alves, João
Andersen, Anja
Andersen, Morten
Andersson, Bengt
Andrillat, Yvette
Andronov, Ivan
Anglada, Guillem
Ardila, David
Arkhipova, Vera
Arny, Thomas
Arthur, Jane
Audard, Marc
Avery, Lorne
Axford, W.
Azcarate, Diana
Baars, Jacob
Babkovskaia, Natalia
Bachiller, Rafael
Baek, Chang Hyun
Baker, Andrew
Baldwin, John
Ballesteros-Paredes, Javier
Balser, Dana
Baluteau, Jean-Paul
Bania, Thomas
Barlow, Michael
Barnes, Aaron
Baryshev, Andrey
Bash, Frank
Basu, Shantanu
Baudry, Alain
Bautista, Manuel
Bayet, Estelle
Guelin, Michel
Gürtler, Joachim
Güsten, Rolf
Guilloteau, Stéphane
Gull, Theodore
Günthardt, Guillermo
Guo, Jianheng
Guseinov, O.
Habing, Harm
Hackwell, John
Haisch Jr., Karl
Hanami, Hitoshi
Hardebeck, Ellen
Harrington, J.
Harris, Alan
Harris-Law, Stella
Harten, Ronald
Hartl, Herbert
Hartquist, Thomas
Harvey, Paul
Hatchell, Jennifer
Haverkorn, Marijke
Hayashi, Saeko
Haynes, Raymond
He, Jinhua
Hebrard, Guillaume
Hecht, James
Heikkilä, Arto
Heiles, Carl
Hein, Righini
Helfer, H.
Helmich, Frank
Helou, George
Henkel, Christian
Henney, William
Henning, Thomas
Herbstmeier, Uwe
Hernández, Jesùs
Herpin, Fabrice
Heydari-Malayeri, Mohammad
Heyer, Mark
Hidayat, Bambang
Hideko, Nomura
Higgs, Lloyd
Hildebrand, Roger
Penzias, Arno
Pequignot, Daniel
Perault, Michel
Persi, Paolo
Persson, Carina
Peters, William
Petrosian, Vahe
Petuchowski, Samuel
Philipp, Sabine
Philippe, Salome
Phillips, Thomas
Phillips, John
Pihlström, Ylva
Pihlström, Ylva
Pineau des Forêts, Guillaume
Pittard, Julian
Plume, René
Podio, Linda
Pöppel, Wolfgang
Pongracic, Helen
Pontoppidan, Klaus
Porceddu, Ignazio
Pottasch, Stuart
Pound, Marc
Pouquet, Annick
Prasad, Sheo
Preite Martinez, Andrea
Price, R.
Prochaska, Jason
Pronik, Iraida
Prusti, Timo
Puget, Jean-Loup
Qin, Zhihai
Radhakrishnan, V.
Raimond, Ernst
Ramírez, José
Ranalli, Piero
Rastogi, Shantanu
Ratag, Mezak
Rawlings, Jonathan
Rawlings, Mark
Raymond, John
Recchi, Simone
Redman, Matthew
Reipurth, Bo

Becklin, Eric
Beckman, John
Beckwith, Steven
Bedogni, Roberto
Benaydoun, Jean-Jacques
Bergeron, Jacqueline
Bergin, Edwin
Bergman, Per
Bergström, Lars
Berkhuijsen, Elly
Bernat, Andrew
Bertout, Claude
Bhat, Ramesh
Bhatt, H.
Bianchi, Luciana
Bieging, John
Bignall, Hayley
Bignell, R.
Binette, Luc
Birkle, Kurt
Black, John
Blades, John
Blair, Guy
Blair, William
Bless, Robert
Blitz, Leo
Bloemen, Hans
Bobrowsky, Matthew
Bocchino, Fabrizio
Bochkarev, Nikolai
Bode, Michael
Bodenheimer, Peter
Boeshaar, Gregory
Boggess, Albert
Bohlin, Ralph
Boisse, Patrick
Boland, Wilfried
Bontemps, Sylvain
Bordbar, Gholam
Borgman, Jan
Borkowski, Kazimierz
Boulanger, François
Boumis, Panayotis
Bourke, Tyler
Bouvier, Jerôme
Bowen, David
Brand, Jan
Brand, Peter
Briceño, Cesar
Brinkmann, Wolfgang
Bromage, Gordon
Brooks, Kate
Brouillet, Nathalie
Bruhweiler, Frederick
Bujarrabal, Valentin
Burke, Bernard
Burton, W.
Bychkov, Konstantin

Hillenbrand, Lynne
Hippelein, Hans
Hirano, Naomi
Hiriart, David
Hiroki, Chihara
Hiroko, Nagahara
Hiromoto, Norihisa
Hjalmarson, Ake
Hobbs, Lewis
Hoeglund, Bertil
Hollenbach, David
Hollis, Jan
Hong, Seung
Hora, Joseph
Horáček, Jiří
Houde, Martin
Houziaux, Leo
Hovhannessian, Rafik
Hua, Chon Trung
Hudson, Reggie
Huggins, Patrick
Hulsbosch, A.
Hutchings, John
Hutsemekers, Damien
Hyung, Siek
Il'in, Vladimir
Inutsuka, Shu-ichiro
Irvine, William
Israel, Frank
Issa, Issa
Itoh, Hiroshi
Jabir, Niama
Jackson, James
Jacoby, George
Jacq, Thierry
Jaffe, Daniel
Jahnke, Knud
Jenkins, Edward
Jiménez-Vicente, Jorge
Jin, Zhenyu
Johnson, Hugh
Johnson, Fred
Johnston, Kenneth
Johnstone, Douglas
Jones, Frank
Jones, Christine
Jourdain de Muizon, Marie
Jura, Michael
Just, Andreas
Justtanont-Liseau, Kay
Jørgensen, Jes
Kafatos, Menas
Kaftan, May
Kaifu, Norio
Kalenskii, Sergei
Kaler, James
Kamaya, Hideyuki
Kamijo, Fumio

Rengarajan, Thinniam
Rengel, Miriam
Reshetnyk, Volodymyr
Reyes, Rafael
Reynolds, Ronald
Reynolds, Cormac
Reynoso, Estela
Richter, Philipp
Rickard, Lee
Roberge, Wayne
Roberts, Douglas
Roberts Jr., William
Robinson, Garry
Rodrigues, Claudia
Rodríguez, Luis
Rodríguez, Monica
Rödiger, Elke
Roelfsema, Peter
Röser, Hans-Peter
Roger, Robert
Rogers, Alan
Rohlfs, Kristen
Rosa, Michael
Rosado, Margarita
Rose, William
Rouan, Daniel
Roxburgh, Ian
Rozhkovskij, Dimitrij
Rubin, Robert
Ryabov, Michael
Sabano, Yutaka
Sabbadin, Franco
Sahu, Kailash
Saigo, Kazuya
Sakano, Masaaki
Salama, Farid
Salinari, Piero
Salter, Christopher
Sami, Dib
Samodurov, Vladimir
Sánchez Doreste, Néstor
Sánchez-Saavedra, M.
Sancisi, Renzo
Sandell, Goran
Sandqvist, Aage
Sarazin, Craig
Sargent, Annelia
Sarma, N.
Sarre, Peter
Sato, Fumio
Sato, Shuji
Savage, Blair
Savedoff, Malcolm
Scalo, John
Scappini, Flavio
Schatzman, Evry
Scherb, Frank
Schilke, Peter

Bykov, Andrei
Bzowski, Maciej
Cai, Kai
Cambresy, Laurent
Cami, Jan
Canto, Jorge
Caplan, James
Cappa de Nicolau, Cristina
Capriotti, Eugene
Capuzzo Dolcetta, Roberto
Carretti, Ettore
Carruthers, George
Casasola, Viviana
Castañeda, Héctor
Castelletti, Gabriela
Caswell, James
Cattaneo, Andrea
Cecchi-Pellini, Cesare
Centurión Martin, Miriam
Cernicharo, José
Cerruti Sola, Monica
Cersosimo, Juan
Cesarsky, Diego
Cesarsky, Catherine
Cha, Seung-Hoon
Chandra, Suresh
Chen, Yang
Chen, Yafeng
Chen, Huei-Ru
Chen, Xuefei
Cheng, Kwang
Cherchneff, Isabelle
Chevalier, Roger
Chini, Rolf
Chopinet, Marguerite
Christopoulou, Panagiota
Churchwell, Edward
Ciardullo, Robin
Cichowolski, Silvina
Ciroi, Stefano
Clark, Frank
Clarke, David
Clegg, Robin
Codella, Claudio
Coffey, Deirdre
Cohen, Marshall
Colangeli, Luigi
Collin, Suzy
Colomb, Fernando
Combes, Françoise
Constantini, Elisa
Corbelli, Edvige
Corradi, Wagner
Corradi, Romano
Costero, Rafael
Courtes, Georges
Cowie, Lennox
Cox, Donald
Cox, Pierre

Kamp, Inga
Kanekar, Nissim
Kantharia, Nimisha
Kassim, Namir
Keene, Jocelyn
Kegel, Wilhelm
Keheyan, Yeghis
Kennicutt, Robert
Khesali, Ali
Kim, Jongsoo
Kimura, Toshiya
Kirkpatrick, Ronald
Kirshner, Robert
Klessen, Ralf
Knacke, Roger
Knapp, Gillian
Knezek, Patricia
Knude, Jens
Ko, Chung-Ming
Kobayashi, Naoto
Kohoutek, Lubos
Koike, Chiyoe
Kondo, Yoji
Kong, Xu
Koornneef, Jan
Korpi, Maarit
Kostyakova, Elena
Kozasa, Takashi
Krajnovic, Davor
Kramer, Busaba
Krautter, Joachim
Kravchuk, Sergei
Kreysa, Ernst
Krishna, Swamy
Krumholz, Mark
Kuan, Yi-Jehng
Kudoh, Takahiro
Kuiper, Thomas
Kulhánek, Petr
Kumar, C.
Kundu, Mukul
Kunth, Daniel
Kutner, Marc
Kwitter, Karen
Kylafis, Nikolaos
Lada, Charles
Lafon, Jean-Pierre
Lai, Shih-Ping
Laloum, Maurice
Langer, William
Latter, William
Laureijs, René
Laurent, Claudine
Lauroesch, James
Lazarian, Alexandre
Lazio, Joseph
Le Squeren, Anne-Marie
Leão, João Rodrigo
Lebron, Mayra

Schlemmer, Stephan
Schmid-Burgk, J.
Schmidt-Kaler, Theodor
Schröder, Anja
Schulz, R.
Schwartz, Richard
Schwartz, Philip
Schwarz, Ulrich
Scott, Eugene
Scoville, Nicholas
Seki, Munezo
Sellgren, Kristen
Sembach, Kenneth
Sen, Asoke
Seon, Kang-Il
Shadmehri, Mohsen
Shane, William
Shao, Cheng-yuan
Shapiro, Stuart
Sharpless, Stewart
Shaver, Peter
Shawl, Stephen
Shchekinov, Yuri
Shematovich, Valerij
Sherwood, William
Shields, Gregory
Shipman, Russell
Shmeld, Ivar
Shu, Frank
Shull, John
Shull, Peter
Shustov, Boris
Siebenmorgen, Ralf
Sigalotti, Leonardo
Silich, Sergey
Silk, Joseph
Silva, Laura
Silvestro, Giovanni
Sitko, Michael
Sivan, Jean-Pierre
Skilling, John
Skulskyj, Mychajlo
Slane, Patrick
Sloan, Gregory
Smith, Barham
Smith, Peter
Smith, Craig
Smith, Michael
Smith, Tracy
Smith, Randall
Smith, Robert
Snell, Ronald
Snow, Theodore
Sobolev, Andrej
Sofia, Ulysses
Sofia, Sabatino
Sofue, Yoshiaki
Šolc, Martin
Somerville, William

Coyne, George
Crane, Philippe
Crawford, Ian
Crovisier, Jacques
Cruvellier, Paul
Cuesta Crespo, Luis
Cunningham, Maria
Czyzak, Stanley
d'Hendecourt, Louis
d'Odorico, Sandro
Dahn, Conard
Dale, James
Dalgarno, Alexander
Danks, Anthony
Danly, Laura
Davies, Rodney
Davis, Christopher
de Almeida, Amaury
De Avillez, Miguel
De Bernardis, Paolo
de Boer, Klaas
De Buizer, James
de Gregorio-Monsalvo, Itziar
de Jong, Teije
de la Noë, Jérôme
de Marco, Orsola
Decourchelle, Anne
Deguchi, Shuji
Deharveng, Lise
Deiss, Bruno
Dennefeld, Michel
Dent, William
Dewdney, Peter
Di Fazio, Alberto
Dias da Costa, Roberto
Diaz, Ruben
Dickel, Helene
Dickel, John
Dickey, John
Dieleman, Pieter
Dinerstein, Harriet
Dinh, Van
Disney, Michael
Djamaluddin, Thomas
Docenko, Dmitrijs
Dokuchaev, Vyacheslav
Dokuchaeva, Olga
Dominik, Carsten
Donn, Bertram
Dopita, Michael
Dorschner, Johann
Dottori, Horacio
Downes, Dennis
Draine, Bruce
Dreher, John
Dubner, Gloria
Dubout, Renée
Dudorov, Alexander
Dufour, Reginald
Lee, Myung Gyoon
Lee, Hee-Won
Lee, Dae-Hee
Léger, Alain
Lehtinen, Kimmo
Leisawitz, David
Lépine, Jacques
Lequeux, James
Leto, Giuseppe
Leung, Chun
Li, Jinzeng
Liang, Yanchun
Ligori, Sebastiano
Likkel, Lauren
Liller, William
Limongi, Marco
Lin, Chia
Lin, Weipeng
Linke, Richard
Linnartz, Harold
Lis, Dariusz
Liseau, René
Liszt, Harvey
Liu, Sheng-Yuan
Lloyd, Myfanwy
Lo, Kwok-Yung
Lockman, Felix
Lodders, Katharina
Loinard, Laurent
López Garcia, José
Loren, Robert
Louise, Raymond
Lovas, Francis
Lozinskaya, Tatjana
Lucas, Robert
Luo, Shaoguang
Lynds, Beverly
Lyon, Ian
Ma, Jun
Mac Low, Mordecai-Mark
Maciel, Walter
MacLeod, John
Madsen, Gregory
Maihara, Toshinori
Makiuti, Sin'itirou
Malbet, Fabien
Mampaso, Antonio
Manchado, Arturo
Manchester, Richard
Manfroid, Jean
Maret, Sébastien
Marston, Anthony
Martin, Peter
Martin, Robert
Martin, Christopher
Martin-Pintado, Jesùs
Masson, Colin
Mather, John
Mathews, William
Song, In-Ok
Spaans, Marco
Stahler, Steven
Stanga, Ruggero
Stanghellini, Letizia
Stanimirovic, Snezana
Stapelfeldt, Karl
Stark, Ronald
Stasinska, Grazyna
Stecher, Theodore
Stecklum, Bringfried
Stenholm, Björn
Stone, James
Strom, Richard
Suh, Kyung-Won
Sun, Jin
Sutherland, Ralph
Suzuki, Tomoharu
Swade, Daryl
Sylvester, Roger
Szczerba, Ryszard
Tachihara, Kengo
Tafalla, Mario
Takahashi, Junko
Takakubo, Keiya
Takano, Toshiaki
Tamura, Motohide
Tamura, Shin'ichi
Tanaka, Masuo
Tantalo, Rosaria
Taylor, Kenneth
Teixeira, Paula
Tenorio-Tagle, Guillermo
Terzian, Yervant
Testi, Leonardo
Thaddeus, Patrick
Thé, Pik-Sin
Thompson, A.
Thonnard, Norbert
Thronson Jr., Harley
Tilanus, Remo
Tokarev, Yurij
Torrelles, José
Torres-Peimbert, Silvia
Tosi, Monica
Tothill, Nicholas
Townes, Charles
Trammell, Susan
Treffers, Richard
Trinidad, Miguel
Turner, Kenneth
Tyulbashev, Sergei
Ulrich, Marie-Helene
Urošević, Dejan
Urquhart, James
van de Steene, Griet
van den Ancker, Mario
van der Hulst, Jan
van der Laan, Harry

Duley, Walter
Dupree, Andrea
Dutrey, Anne
Duvert, Gilles
Dwarkadas, Vikram
Dwek, Eli
Edwards, Suzan
Egan, Michael
Ehlerová, Soňá
Eisloeffel, Jochen
Elia, Davide
Elitzur, Moshe
Elliott, Kenneth
Elmegreen, Bruce
Elmegreen, Debra
Elvius, Aina
Emerson, James
Encrenaz, Pierre
Esamdin, Ali
Escalante, Vladimir
Esimbek, Jarken
Esipov, Valentin
Esteban, César
Evans, Neal
Evans, Aneurin
Falceta-Gonçalves, Diego
Falgarone, Edith
Falk Jr., Sydney
Falle, Samuel
Federman, Steven
Feitzinger, Johannes
Felli, Marcello
Felten, James
Fendt, Christian
Ferlet, Roger
Fernandes, Amadeu
Ferriere, Katia
Ferrini, Federico
Fesen, Robert
Fiebig, Dirk
Field, George
Field, David
Fierro, Julieta
Fischer, Jacqueline
Flannery, Brian
Fleck, Robert
Florido, Estrella
Flower, David
Folini, Doris
Ford, Holland
Forster, James
Franco, José
Franco, Gabriel Armando
Fraser, Helen
Freimanis, Juris
Fridlund, Malcolm
Frisch, Priscilla
Fuente, Asuncion
Fukuda, Naoya
Mathewson, Donald
Mathis, John
Matsuhara, Hideo
Matsumoto, Tomoaki
Matsumura, Masafumi
Mattila, Kalevi
Mauersberger, Rainer
McCall, Marshall
McClure-Griffiths, Naomi
Mccombie, June
McCray, Richard
McGee, Richard
McGregor, Peter
McKee, Christopher
McNally, Derek
Mebold, Ulrich
Meier, Robert
Meixner, Margaret
Mellema, Garrelt
Melnick, Gary
Mennella, Vito
Menon, T.
Menzies, John
Meszaros, Peter
Meyer, Martin
Mezger, Peter
Millar, Thomas
Miller, Joseph
Milne, Douglas
Minier, Vincent
Minn, Young
Minter, Anthony
Mitchell, George
Mitsunobu, Kawada
Miyama, Syoken
Mizuno, Shun
Mo, Jinger
Monin, Jean-Louis
Montmerle, Thierry
Moore, Marla
Moreno-Corral, Marco
Moriarty-Schieven, Gerald
Morimoto, Masaki
Morris, Mark
Morton, Donald
Mouschovias, Telemachos
Muench, Guido
Mufson, Stuart
Mulas, Giacomo
Muller, Erik
Muller, Sebastien
Murthy, Jayant
Myers, Philip
Nagata, Tetsuya
Nakada, Yoshikazu
Nakagawa, Takao
Nakamoto, Taishi
Nakamura, Fumitaka
Nakano, Makoto
van der Tak, Floris
van Dishoeck, Ewine
van Gorkom, Jacqueline
van Loon, Jacco
van Woerden, Hugo
VandenBout, Paul
Varshalovich, Dmitrij
Vázquez, Roberto
Velázquez , Pablo
Verdoes Kleijn, Gijsbert
Verheijen, Marc
Verner, Ekaterina
Verschuur, Gerrit
Viala, Yves
Viallefond, François
Vidal, Jean-Louis
Vidal-Madjar, Alfred
Viegas, Sueli
Vijh, Uma
Vilchez, José
Villaver, Eva
Vink, Jacco
Viti, Serena
Volk, Kevin
Vorobyov, Eduard
Voronkov, Maxim
Voshchinnikov, Nikolai
Vrba, Frederick
Wakker, Bastiaan
Walker, Gordon
Walmsley, C.
Walsh, Wilfred
Walsh, Andrew
Walton, Nicholas
Wang, Hongchi
Wang, Q. Daniel
Wang, Jun-Jie
Wang, Hong-Guang
Wannier, Peter
Ward-Thompson, Derek
Wardle, Mark
Watt, Graeme
Weaver, Harold
Wei, Liu
Weiler, Kurt
Weinberger, Ronald
Weisheit, Jon
Wesselius, Paul
Weymann, Ray
Whelan, Emma
White, Glenn
White, Richard
Whitelock, Patricia
Whiteoak, John
Whittet, Douglas
Whitworth, Anthony
Wickramasinghe, N.
Wiebe, Dmitri
Wild, Wolfgang

Fukui, Yasuo
Fuller, Gary
Furniss, Ian
Furuya, Ray
Gaensler, Bryan
Galli, Daniele
Gao, Yu
Garcia, Paulo
Garcia-Lario, Pedro
Garnett, Donald
Gathier, Roel
Gaume, Ralph
Gay, Jean
Geballe, Thomas
Gehrels, Tom
Genzel, Reinhard
Georgelin, Yvon
Gérard, Eric
Gerin, Maryvonne
Gerola, Humberto
Gezari, Daniel
Ghanbari, Jamshid
Giacani, Elsa
Gibson, Steven
Gilra, Daya
Giovanelli, Riccardo
Glover, Simon
Goddi, Ciriaco
Godfrey, Peter
Goebel, John
Goldes, Guillermo
Goldreich, Peter
Goldsmith, Donald
Golovatyj, Volodymyr
Gomez, Gonzalez
Gonçalves, Denise
Gonzales-Alfonso, Eduardo
Goodman, Alyssa
Gordon, Courtney
Gordon, Mark
Gosachinskij, Igor
Goss, W. Miller
Graham, David
Granato, Gian Luigi
Grasdalen, Gary
Gredel, Roland
Green, James
Gregorio-Hetem, Jane
Greisen, Eric
Grewing, Michael
Nakano, Takenori
Natta, Antonella
Neugebauer, Gerry
Nguyen-Quang, Rieu
Nikoli'c, Silvana
Nishi, Ryoichi
Nordh, Lennart
Norman, Colin
Nulsen, Paul
Nürnberger, Dieter
Nussbaumer, Harry
Nuth, Joseph
O'Dell, Charles
O'Dell, Stephen
Oey, Sally
Ohtani, Hiroshi
Okuda, Haruyuki
Okumura, Shin-ichiro
Olofsson, Hans
Omont, Alain
Omukai, Kazuyuki
Onaka, Takashi
Onello, Joseph
Opendak, Michael
Orlando, Salvatore
Osborne, John
Ostriker, Eve
Ott, Juergen
Oudmaijer, René
Pagani, Laurent
Pagano, Isabella
Pak, Soojong
Palla, Francesco
Palmer, Patrick
Palumbo, Maria Elisabetta
Panagia, Nino
Pandey, Birendra
Pankonin, Vernon
Park, Yong-Sun
Parker, Eugene
Paron, Sergio
Parthasarathy, Mudumba
Pauls, Thomas
Pecker, Jean-Claude
Peeters, Els
Peimbert, Manuel
Pellegrini, Silvia
Pena, Miriam
Pendleton, Yvonne
Peng, Qingyu
Wilkin, Francis
Williams, David
Williams, Robert
Williams, Robin
Willis, Allan
Willner, Steven
Wilson, Robert
Wilson, Thomas
Wilson, Christine
Winnberg, Anders
Winnewisser, Gisbert
Witt, Adolf
Wolff, Michael
Wolfire, Mark
Wolstencroft, Ramon
Wolszczan, Alexander
Woltjer, Lodewijk
Wood, Douglas
Woodward, Paul
Woolf, Neville
Wootten, Henry
Wouterloot, Jan
Wright, Edward
Wu, Chi
Wünsch, Richard
Wynn-Williams, C.
Yabushita, Shin
Yamada, Masako
Yamamoto, Satoshi
Yamamura, Issei
Yamashita, Takuya
Yan, Jun
York, Donald
Yorke, Harold
Yoshida, Shigeomi
Younis, Saad
Yui, Yukari
Yun, Joao
Zavagno, Annie
Zealey, William
Zeilik, Michael
Zeng, Qin
Zhang, Cheng-Yue
Zhang, JiangShui
Zhang, Jingyi
Zhou, Jianjun
Zhu, Wenbai
Zimmermann, Helmut
Zinchenko, Igor
Zuckerman, Ben

Division IV Commission 35 Stellar Constitution

President Charbonnel, Corinne
Vice-President Limongi, Marco

Organizing Committee

Fontaine, Gilles
Leitherer, Claus
Weiss, Achim
Isern, Jorge
van Loon, Jacco
Yungel'son, Lev
Lattanzio, John

Members

Adams, Mark
Aiad, A.
Aizenman, Morris
Anand, S.
Angelov, Trajko
Antia, H.
Appenzeller, Immo
Arai, Kenzo
Arentoft, Torben
Argast, Dominik
Arimoto, Nobuo
Arnett, W.
Arnould, Marcel
Audouze, Jean
Baglin, Annie
Barnes, Sydney
Basu, Sarbani
Baym, Gordon
Bazot, Michael
Beaudet, Gilles
Becker, Stephen
Belmonte Avilés, Juan Antonio
Benz, Willy
Bergeron, Pierre
Bertelli, Gianpaolo
Berthomieu, Gabrielle
Bisnovatyi-Kogan, Gennadij
Blaga, Christina
Bocchia, Romeo
Bodenheimer, Peter
Bombaci, Ignazio
Bono, Giuseppe
Boss, Alan
Brassard, Pierre
Bravo, Eduardo
Bressan, Alessandro
Browning, Matthew
Brownlee, Robert
Bruenn, Stephen
Brun, Allan
Buchler, J.
Burbidge, Geoffrey
Busso, Maurizio
Callebaut, Dirk
Caloi, Vittoria
Canal, Ramon
Caputo, Filippina
Hachisu, Izumi
Hammond, Gordon
Han, Zhanwen
Hashimoto, Masa-aki
Hayashi, Chushiro
Heger, Alexander
Henry, Richard
Hernanz, Margarita
Hillier, Desmond
Hirschi, Raphael
Hollowell, David
Homeier, Derek
Huang, Runqian
Huggins, Patrick
Humphreys, Roberta
Iben Jr., Icko
Iliev, Ilian
Imbroane, Alexandru
Imshennik, Vladimir
Isern, Jorge
Ishizuka, Toshihisa
Itoh, Naoki
James, Richard
José, Jordi
Jørgensen, Jes
Kaehler, Helmuth
Käpylä, Petri
Karakas, Amanda
Kato, Mariko
Kiguchi, Masayoshi
King, David
Kippenhahn, Rudolf
Kiziloglu, Nilgun
Knoelker, Michael
Kochhar, Rajesh
Koester, Detlev
Konar, Sushan
Kosovichev, Alexander
Kovetz, Attay
Kozlowski, Maciej
Kumar, Shiv
Kwok, Sun
Labay, Javier
Lamb, Susan
Lamb Jr., Donald
Lamzin, Sergei
Langer, Norbert
Pongracic, Helen
Pontoppidan, Klaus
Porfir'ev, Vladimir
Poveda, Arcadio
Prentice, Andrew
Prialnik-Kovetz, Dina
Proffitt, Charles
Provost, Janine
Pulone, Luigi
Qu, Qinyue
Raedler, K.
Ramadurai, Souriraja
Rauscher, Thomas
Ray, Alak
Rayet, Marc
Reeves, Hubert
Renzini, Alvio
Reyniers, Maarten
Ritter, Hans
Roca Cortés, Teodoro
Rood, Robert
Rouse, Carl
Roxburgh, Ian
Ruiz-Lapuente, María
Sackmann, Ingrid
Saio, Hideyuki
Sakashita, Shiro
Santos, Filipe
Sarna, Marek
Sato, Katsuhiko
Savedoff, Malcolm
Savonije, Gerrit
Scalo, John
Schatten, Kenneth
Schatzman, Evry
Schónberner, Detlef
Schuler, Simon
Schutz, Bernard
Scuflaire, Richard
Seidov, Zakir
Sengbusch, Kurt
Shaviv, Giora
Shibahashi, Hiromoto
Shibata, Yukio
Shustov, Boris
Sienkiewicz, Ryszard
Siess, Lionel

Carson, T.
Castor, John
Chaboyer, Brian
Chabrier, Gilles
Chamel, Nicolas
Chan, Kwing
Chan, Roberto
Charpinet, Stéphane
Chechetkin, Valerij
Chiosi, Cesare
Chitre, Shashikumar
Chkhikvadze, Iakob
Christensen-Dalsgaard, Jørgen
Christy, Robert
Cohen, Judith
Cohen, Jeffrey
Connolly, Leo
Córsico, Alejandro
Cowan, John
Das, Mrinal
Daszynska-Daszkiewicz, Jadwiga
Davis Jr., Cecil
De Grève, Jean-Pierre
de Jager, Cornelis
de Loore, Camiel
de Medeiros, José
Dearborn, David
Deinzer, W.
Deliyannis, Constantine
Demarque, Pierre
Denisenkov, Pavel
Despain, Keith
Deupree, Robert
Di Mauro, Maria Pia
Dluzhnevskaya, Olga
Dominguez, Inma
Dupuis, Jean
Durisen, Richard
Dziembowski, Wojciech
Edwards, Alan
Edwards, Suzan
Eggenberger, Patrick
Eggleton, Peter
Elmhamdi, Abouazza
Eminzade, T.
Endal, Andrew
Eriguchi, Yoshiharu
Fadeyev, Yurij
Faulkner, John
Flannery, Brian
Forbes, J.
Fossat, Eric
Foukal, Peter
Fujimoto, Masayuki
Gabriel, Maurice
Gallino, Roberto
Garcia, Domingo
Gautschy, Alfred
Geroyannis, Vassilis
Laskarides, Paul
Lasota, Jean-Pierre
Lebovitz, Norman
Lebreton, Yveline
Lee, Thyphoon
Lee, William
Leitherer, Claus
Lépine, Jacques
Li, Li
Liebendörfer, Matthias
Lignieres, François
Limongi, Marco
Linnell, Albert
Littleton, John
Livio, Mario
Lucatello, Sara
Lugaro, Maria
Maeda, Keiichi
Maeder, André
Maheswaran, Murugesapillai
Masani, A.
Mathis, Stéphane
Matteucci, Francesca
Mazurek, Thaddeus
Mazzitelli, Italo
McDavid, David
Mendes, Luiz
Mestel, Leon
Meyer-Hofmeister, Eva
Meynet, Georges
Michaud, Georges
Mitalas, Romas
Miyaji, Shigeki
Möllenhoff, Claus
Mohan, Chander
Moiseenko, Sergey
Monaghan, Joseph
Monier, Richard
Monteiro, Mario João
Moore, Daniel
Morgan, John
Moskalik, Pawel
Moss, David
Mowlavi, Nami
Nadyozhin, Dmitrij
Nakamura, Takashi
Nakano, Takenori
Nakazawa, Kiyoshi
Narasimha, Delampady
Narita, Shinji
Nelemans, Gijs
Newman, Michael
Nishida, Minoru
Nobuyuki , Iwamoto
Noels, Arlette
Nomoto, Ken'ichi
O'Toole, Simon
Odell, Andrew
Ohyama, Noboru
Sigalotti, Leonardo
Signore, Monique
Sills, Alison
Silvestro, Giovanni
Sion, Edward
Smeyers, Paul
Smith, Robert
Sobouti, Yousef
Sofia, Sabatino
Sparks, Warren
Spiegel, E.
Sreenivasan, S.
Starrfield, Sumner
Stellingwerf, Robert
Stibbs, Douglas
Stringfellow, Guy
Strittmatter, Peter
Suda, Kazuo
Suda, Takuma
Sugimoto, Daiichiro
Sweigart, Allen
Taam, Ronald
Takahara, Mariko
Thielemann, Friedrich-Karl
Thomas, Hans
Tjin-a-Djie, Herman
Tohline, Joel
Toomre, Juri
Tornambe, Amedeo
Townsend, Richard
Trimble, Virginia
Truran Jr., James
Tscharnuter, Werner
Tuominen, Ilkka
Turck-Chieze, Sylvaine
Tutukov, Aleksandr
Uchida, Juichi
Ulrich, Roger
Unno, Wasaburo
Utrobin, Victor
van den Heuvel, Edward
van der Borght, René
van Horn, Hugh
van Loon, Jacco
van Riper, Kenneth
VandenBerg, Don
Vardya, M.
Vauclair, Gérard
Ventura, Paolo
Vila, Samuel
Vilhu, Osmi
Vilkoviskij, Emmanuil
Vink, Jorick
Ward, Richard
Weaver, Thomas
Webbink, Ronald
Weiss, Nigel
Wheeler, J.
Willson, Lee Anne

Giannone, Pietro
Gimenez, Alvaro
Girardi, Leo
Giridhar, Sunetra
Glatzmaier, Gary
Goedhart, Sharmila
Gonçalves, Denise
Goriely, Stéphane
Gough, Douglas
Goupil, Marie-Jo
Graham, Eric
Greggio, Laura
Guenther, David
Guzik, Joyce
Okamoto, Isao
Oliveira, Joana
Osaki, Yoji
Ostriker, Jeremiah
Oswalt, Terry
Oudmaijer, René
Pamyatnykh, Aleksej
Pande, Girish
Panei, Jorge
Papaloizou, John
Pearce, Gillian
Phillips, Mark
Pines, David
Pinotsis, Antonis
Wilson, Robert
Winkler, Karl-Heinz
Wood, Peter
Wood, Matthew
Woosley, Stanford
Xiong, Da Run
Yamaoka, Hitoshi
Yi, Sukyoung
Yorke, Harold
Yoshida, Shin'ichirou
Yoshida, Takashi
Yushkin, Maxim
Zahn, Jean-Paul
Ziolkowski, Janusz

Division IV Commission 36 Theory of Stellar Atmospheres

President Asplund, Martin
Vice-President Puls, Joachim

Organizing Committee

Allende Prieto, Carlos
Ayres, Thomas
Berdyugina, Svetlana
Gustafsson, Bengt
Hubeny, Ivan
Ludwig, Hans
Mashonkina, Lyudmila
Randich, Sofia

Members

Abbott, David
Afram, Nadine
Allard, France
Altrock, Richard
Andretta, Vincenzo
Ardila, David
Arpigny, Claude
Atanacković-Vukmanović, Olga
Athay, R.
Auer, Lawrence
Auman, Jason
Avrett, Eugene
Baade, Dietrich
Baird, Scott
Baliunas, Sallie
Balona, Luis
Barber, Robert
Barbuy, Beatriz
Baschek, Bodo
Basri, Gibor
Bell, Roger
Bennett, Philip
Bernat, Andrew
Bertone, Emanuele
Bertout, Claude
Biazzo, Katia
Bingham, Robert
Blanco, Carlo
Bless, Robert
Blomme, Ronny
Bodo, Gianluigi
Boesgaard, Ann
Bonifacio, Piercarlo
Bopp, Bernard
Bowen, George
Brown, Alexander
Brown, Douglas
Browning, Matthew
Buchlin, Eric
Bues, Irmela
Busà, Innocenza
Butkovskaya, Varvara
Cameron, Andrew
Carbon, Duane
Carlsson, Mats
Gussmann, Ernst
Hack, Margherita
Haisch, Bernard
Hall, Douglas
Hamann, Wolf-Rainer
Harper, Graham
Hartmann, Lee
Harutyunian, Haik
Heasley, James
Heber, Ulrich
Heiter, Ulrike
Hempel, Marc
Hill, Vanessa
Hillier, Desmond
Hoare, Melvin
Hoeflich, Peter
Höfner, Susanne
Holzer, Thomas
Homeier, Derek
House, Lewis
Huang, He
Hui bon Hoa, Alain
Hutchings, John
Ignace, Richard
Ignjatović, Ljubinko
Ito, Yutaka
Ivanov, Vsevolod
Jahn, Krzysztof
Jankov, Slobodan
Jatenco-Pereira, Vera
Jevremovic, Darko
Johnson, Hollis
Jones, Carol
Jordan, Stefan
Judge, Philip
Kadouri, Talib
Kalkofen, Wolfgang
Kamp, Lucas
Kamp, Inga
Kandel, Robert
Käpylä, Petri
Karp, Alan
Kašparová, Jana
Katsova, Maria
Kiselman, Dan
Pinsonneault, Marc
Pintado, Olga
Pinto, Philip
Piskunov, Nikolai
Plez, Bertrand
Pogodin, Mikhail
Pottasch, Stuart
Przybilla, Norbert
Pustõnski, Vladislav-Veniamin
Querci, François
Querci, Monique
Rachkovsky, D.
Ramsey, Lawrence
Rangarajan, K.
Rauch, Thomas
Reale, Fabio
Reimers, Dieter
Reiners, Ansgar
Rostas, François
Rovira, Marta
Rucinski, Slavek
Rutten, Robert
Ryabchikova, Tatiana
Rybicki, George
Sachkov, Mikhail
Saffe, Carlos
Saio, Hideyuki
Saito, Kuniji
Sakhibullin, Nail
Sapar, Arved
Sapar, Lili
Sarre, Peter
Sasselov, Dimitar
Sasso, Clementina
Sauty, Christophe
Sbordone, Luca
Schaerer, Daniel
Scharmer, Goeran
Schmalberger, Donald
Schmid-Burgk, J.
Schmutz, Werner
Schónberner, Detlef
Scholz, M.
Schrijver, Karel
Sedlmayer, Erwin

Carson, T.
Cassinelli, Joseph
Castelli, Fiorella
Castor, John
Catala, Claude
Catalano, Franco
Cayrel, Roger
Cayrel de Strobel, Giusa
Chan, Kwing
Christlieb, Norbert
Chugai, Nikolaj
Cidale, Lydia
Conti, Peter
Cowley, Charles
Cram, Lawrence
Cruzado, Alicia
Cugier, Henryk
Cui, Wen-Yuan
Cuntz, Manfred
Cuny, Yvette
Curé, Michel
Dall, Thomas
Daszyńska-Daszkiewicz, Jadwiga
Davis Jr., Cecil
de Koter, Alex
Decin, Leen
Deliyannis, Constantine
Dimitrijevic, Milan
Donati, Jean-François
Doyle, John
Drake, Stephen
Dravins, Dainis
Dreizler, Stefan
Duari, Debiprosad
Dufton, Philip
Dupree, Andrea
Edvardsson, Bengt
Elmhamdi, Abouazza
Elste, Gunther
Eriksson, Kjell
Evangelidis, E.
Faraggiana, Rosanna
Faurobert-Scholl, Marianne
Feigelson, Eric
Ferreira, João
Fitzpatrick, Edward
Fluri, Dominique
Fontaine, Gilles
Fontenla, Juan
Forveille, Thierry
Foy, Renaud
Freire Ferrero, Rubens
Fremat, Yves
Freytag, Bernd
Frisch, Helene
Frisch, Uriel
Froeschlé, Christiane
Gail, Hans-Peter
Gallino, Roberto
Klein, Richard
Kochukhov, Oleg
Kodaira, Keiichi
Koester, Detlev
Kolesov, Aleksandr
Kondo, Yoji
Kontizas, Evangelos
Korčáková, Daniela
Korn, Andreas
Kraus, Michaela
Krikorian, Ralph
Krishna, Swamy
Krtička, Jiří
Kubát, Jiří
Kudritzki, Rolf-Peter
Kuhi, Leonard
Kumar, Shiv
Kupka, Friedrich
Kurucz, Robert
Lambert, David
Lamers, Henny
Lanz, Thierry
Lee, Jae-Woo
Leibacher, John
Leitherer, Claus
Li, Ji
Liebert, James
Linnell, Albert
Linsky, Jeffrey
Liu, Caipin
Loskutov, Viktor
Lucatello, Sara
Luck, R.
Luo, Qinghuan
Luttermoser, Donald
Lyubimkov, Leonid
Machado, Maria
Madej, Jerzy
Magazzù, Antonio
Magnan, Christian
Marlborough, J.
Marley, Mark
Martins, Fabrice
Massaglia, Silvano
Mathys, Gautier
Mauas, Pablo
Medupe, Rodney
Merlo, David
Michaud, Georges
Mihajlov, Anatolij
Mihalas, Dimitri
Molaro, Paolo
Muench, Guido
Musielak, Zdzislaw
Mutschlecner, J.
Nagirner, Dmitrij
Najarro de la Parra, Francisco
Narasimha, Delampady
Nariai, Kyoji
Sengupta, Sujan
Shi, Jianrong
Shine, Richard
Shipman, Harry
Short, Christopher
Sigut, Aaron
Simon, Theodore
Simon, Klaus
Simonneau, Eduardo
Skumanich, Andrew
Sneden, Chris
Snezhko, Leonid
Socas-Navarro, Hector
Soderblom, David
Spiegel, E.
Spite, François
Spruit, Henk
Stalio, Roberto
Stauffer, John
Stee, Philippe
Steffen, Matthias
Stein, Robert
Stępień, Kazimierz
Stern, Robert
Stibbs, Douglas
Strom, Stephen
Stuik, Remko
Szécsényi-Nagy, Gábor
Takeda, Yoichi
Thejll, Peter
Thompson, Rodger
Toomre, Juri
Townsend, Richard
Tsuji, Takashi
Tuominen, Ilkka
Ueno, Sueo
Uesugi, Akira
Ulmschneider, Peter
Unno, Wasaburo
Utrobin, Victor
Vakili, Farrokh
van't Veer-Menneret, Claude
van't Veer, Frans
Vardavas, Ilias
Vardya, M.
Vasu-Mallik, Sushma
Vaughan, Arthur
Velusamy, T.
Vennes, Stéphane
Vieytes, Mariela
Viik, Tõnu
Vilhu, Osmi
Vink, Jorick
Walter, Frederick
Watanabe, Tetsuya
Waters, Laurens
Weber, Stephen
Wehrse, Rainer
Weidemann, Volker

García, López
Gebbie, Katharine
Gesicki, Krzysztof
Giampapa, Mark
Gigas, Detlef
Gonzalez, Jean-François
Gordon, Charlotte
Gough, Douglas
Gratton, Raffaele
Gray, David
Grevesse, Nicolas
Grinin, Vladimir
Güdel, Manuel
Niemczura, Ewa
Nikoghossian, Arthur
Nishimura, Masayoshi
Nordlund, Aake
Owocki, Stanley
Pacharin-Tanakun, P.
Pakhomov, Yury
Pandey, Birendra
Panek, Robert
Pavlenko, Yakov
Pecker, Jean-Claude
Peraiah, Annamaneni
Peters, Geraldine
Werner, Klaus
White, Richard
Wickramasinghe, N.
Willson, Lee Anne
Wilson, Peter
Wöhl, Hubertus
Wolff, Sidney
Yanovitskij, Edgard
Yengibarian, Norair
Yorke, Harold
Začs, Laimons
Zahn, Jean-Paul
Zhang, Bo

Division VII Commission 37 Star Clusters & Associations

President Elmegreen, Bruce
Vice-President Carraro, Giovanni

Organizing Committee

Da Costa, Gary
de Grijs, Richard
Deng, LiCai
Lada, Charles
Lee, Young
Minniti, Dante
Sarajedini, Ata
Tosi, Monica

Members

Aarseth, Sverre
Abou'el-ella, Mohamed
Ahumada, Javier
Ahumada, Andrea
Aiad, A.
Akeson, Rachel
Alcaino, Gonzalo
Alfaro, Emilio
Alksnis, Andrejs
Allen, Christine
Allen, Lori
Andreuzzi, Gloria
Aparicio, Antonio
Armandroff, Taft
Aurière, Michel
Bailyn, Charles
Balázs, Béla
Barmby, Pauline
Barrado Navascués, David
Bartašiute, Stanislava
Baume, Gustavo
Baumgardt, Holger
Bell, Roger
Bellazzini, Michele
Biazzo, Katia
Bijaoui, Albert
Blum, Robert
Boily, Christian
Bonatto, Charles
Bosch, Guillermo
Bragaglia, Angela
Brown, Anthony
Buonanno, Roberto
Burderi, Luciano
Burkhead, Martin
Butler, Raymond
Butler, Dennis
Buzzoni, Alberto
Byrd, Gene
Calamida, Annalisa
Callebaut, Dirk
Caloi, Vittoria
Cantiello, Michele
Caputo, Filippina
Capuzzo Dolcetta, Roberto
Carney, Bruce
Harvel, Christopher
Hawarden, Timothy
Heggie, Douglas
Henon, Michel
Herbst, William
Hesser, James
Heudier, Jean-Louis
Hilker, Michael
Hillenbrand, Lynne
Hills, Jack
Hodapp, Klaus-Werner
Hünsch, Matthias
Hut, Piet
Iben Jr., Icko
Illingworth, Garth
Ilya, Mandel
Ishida, Keiichi
Janes, Kenneth
Jeon, Young-Beom
Joshi, Umesh
Kalirai, Jason
Kamp, Lucas
Karakas, Amanda
Kilambi, G.
Kim, Sungsoo
Kim, Seung-Lee
King, Ivan
Kitsionas, Spyridon
Ko, Chung-Ming
Kontizas, Evangelos
Kontizas, Mary
Kraft, Robert
Kroupa, Pavel
Krumholz, Mark
Kun, Maria
Kundu, Arunav
Kurtev, Radostin
Landolt, Arlo
Lapasset, Emilio
Larsson-Leander, Gunnar
Laval, Annie
Lee, Jae-Woo
Lee, Kang Hwan
Leisawitz, David
Leonard, Peter
Li, Jinzeng
Petrovskaya, Margarita
Peykov, Zvezdelin
Phelps, Randy
Piatti, Andrés
Pilachowski, Catherine
Piskunov, Anatolij
Platais, Imants
Porras, Bertha
Portegies Zwart, Simon
Poveda, Arcadio
Pritchet, Christopher
Pulone, Luigi
Raimondo, Gabriella
Ravindranath, Swara
Rebull, Luisa
Renzini, Alvio
Rey, Soo-Chang
Richer, Harvey
Richtler, Tom
Rodrigues de Oliveira Filho, Irapuan
Rothberg, Barry
Rountree, Janet
Royer, Pierre
Russeva, Tatjana
Sagar, Ram
Salukvadze, G.
Samus, Nikolaj
Sánchez Béjar, Victor
Sanders, Walt
Santiago, Basilio
Santos Jr., João
Schoedel, Rainer
Schuler, Simon
Schweizer, François
Seitzer, Patrick
Semkov, Evgeni
Shawl, Stephen
Sher, David
Shi, Huoming
Shobbrook, Robert
Shu, Chenggang
Simoda, Mahiro
Skinner, Stephen
Smith, Graeme
Smith, J.
Song, Inseok

Chaboyer, Brian
Chavarria-K, Carlos
Chen, Huei-Ru
Cheng, Kwang
Chiosi, Cesare
Christian, Carol
Chryssovergis, Michael
Chun, Mun-Suk
Claria, Juan
Clementini, Gisella
Colin, Jacques
Covino, Elvira
Cropper, Mark
D'Amico, Nicolò
D'Antona, Francesca
Dale, James
Danford, Stephen
Dapergolas, Anastasios
Daube-Kurzemniece, Ilga
Davies, Melvyn
De Marchi, Guido
DeGioia-Eastwood, Kathleen
Dehghani, Mohammad
Demarque, Pierre
Demers, Serge
Di Fazio, Alberto
Djupvik, Anlaug
Dluzhnevskaya, Olga
Drissen, Laurent
Durrell, Patrick
El Basuny, Ahmed
Fall, S.
Feinstein, Alejandro
Forbes, Douglas
Forte, Juan
Friel, Eileen
Fukushige, Toshiyuki
Fusi-Pecci, Flavio
García, Beatriz
Gascoigne, S.
Geffert, Michael
Geisler, Douglas
Giersz, Miroslav
Giorgi, Edgard
Glushkova, Elena
Golay, Marcel
Goodwin, Simon
Gouliermis, Dimitrios
Gratton, Raffaele
Green, Elizabeth
Griffiths, William
Grundahl, Frank
Guetter, Harry
Haisch Jr., Karl
Hanes, David
Hanson, Margaret
Harris, Gretchen
Harris, Hugh
Liu, Michael
Lloyd Evans, Thomas
Lodieu, Nicolas
Loktin, Alexhander
Lu, Phillip
Lucatello, Sara
Lynden-Bell, Donald
Maccarone, Thomas
Maeder, André
Makalkin, Andrei
Makino, Junichiro
Mamajek, Eric
Marco, Amparo
Mardling, Rosemary
Markkanen, Tapio
Markov, Haralambi
Marraco, Hugo
Marsden, Stephen
Marshall, Kevin
Martinez Roger, Carlos
Martins, Donald
Menon, T.
Menzies, John
Meylan, Georges
Milone, Eugene
Moehler, Sabine
Mohan, Vijay
Moitinho, André
Mould, Jeremy
Muminov, Muydinjon
Murray, Andrew
Muzzio, Juan
Navone, Hugo
Naylor, Tim
Nemec, James
Nesci, Roberto
Neuhäuser, Ralph
Newell, Edward
Ninkov, Zoran
Nürnberger, Dieter
Oey, Sally
Ogura, Katsuo
Oliveira, Joana
Origlia, Livia
Ortolani, Sergio
Osman, Anas
Oudmaijer, René
Pandey, A.
Park, Byeong-Gon
Parmentier, Geneviève
Parsamyan, Elma
Patten, Brian
Paunzen, Ernst
Pavani, Daniela
Pedreros, Mario
Penny, Alan
Pérez, Mario
Peterson, Charles
Spurzem, Rainer
Stauffer, John
Stetson, Peter
Stringfellow, Guy
Stuik, Remko
Subramaniam, Annapurni
Sugimoto, Daiichiro
Sung, Hwankyung
Suntzeff, Nicholas
Szécsényi-Nagy, Gábor
Tadross, Ashraf
Takahashi, Koji
Takashi, Hasegawa
Taş, Günay
Taylor, John
Terranegra, Luciano
Terzan, Agop
Thoul, Anne
Tikhonov, Nikolai
Tornambe, Amedeo
Tripicco, Michael
Trullols, I.
Tsvetkov, Milcho
Tsvetkova, Katja
Turner, David
Twarog, Bruce
Upgren, Arthur
van Altena, William
van den Bergh, Sidney
VandenBerg, Don
Vazquez, Rub'en
Veltchev, Todor
Ventura, Paolo
Verschueren, Werner
Vesperini, Enrico
von Hippel, Theodore
Walker, Merle
Walker, Gordon
Warren Jr., Wayne
Weaver, Harold
Wehlau, Amelia
Wielen, Roland
Wofford, Aida
Wramdemark, Stig
Wright, Nicholas
Wu, Hsin-Heng
Xiradaki, Evangelia
Yakut, Kadri
Yi, Sukyoung
Yim, Hong-Suh
Yumiko, Oasa
Zaggia, Simone
Zakharova, Polina
Zapatero-Osorio, Maria Rosa
Zhang, Fenghui
Zhao, Jun Liang
Zinn, Robert

Division X Commission 40 Radio Astronomy

President Nan, Ren-Dong
Vice-President Taylor, Russell

Organizing Committee

Carilli, Christopher
Chapman, Jessica
Dubner, Gloria
Garrett, Michael
Goss, W. Miller
Hills, Richard
Hirabayashi, Hisashi
Rodríguez, Luis
Shastri, Prajval
Torrelles, José

Members

Abdulla, Shaker
Abraham, Peter
Ade, Peter
Akabane, Kenji
Akujor, Chidi
Alberdi, Antonio
Alexander, Joseph
Alexander, Paul
Allen, Ronald
Aller, Margo
Aller, Hugh
Altenhoff, Wilhelm
Altunin, Valery
Ambrosini, Roberto
Andernach, Heinz
Anglada, Guillem
Antonova, Antoaneta
Aparici, Juan
Argo, Megan
Arnal, Edmundo
Asareh, Habibolah
Aschwanden, Markus
Assousa, George
Aubier, Monique
Augusto, Pedro
Aurass, Henry
Avery, Lorne
Axon, David
Baan, Willem
Baars, Jacob
Bååth, Lars
Babkovskaia, Natalia
Bachiller, Rafael
Backer, Donald
Bagri, Durgadas
Bailes, Matthew
Bajaja, Esteban
Bajkova, Anisa
Baker, Joanne
Baker, Andrew
Balasubramanian, V.
Balasubramanyam, Ramesh
Baldwin, John
Balklavs-Grinhofs, Arturs
Ball, Lewis
Bally, John
Hibbard, John
Higgs, Lloyd
Hirota, Tomoya
Hjalmarson, Ake
Ho, Paul
Hoang, Binh
Hobbs, Robert
Hobbs, George
Hoeglund, Bertil
Hofner, Peter
Högbom, Jan
Hogg, David
Hojaev, Alisher
Hollis, Jan
Hong, Xiaoyu
Hopkins, Andrew
Horiuchi, Shinji
Hotan, Aidan
Howard, William
Huchtmeier, Walter
Hughes, Philip
Hughes, David
Hulsbosch, A.
Humphreys, Elizabeth
Hunstead, Richard
Huynh, Minh
Hwang, Chorng-Yuan
Ibrahim, Zainol Abidin
Iguchi, Satoru
Ikhsanov, Robert
Imai, Hiroshi
Inatani, Junji
Inoue, Makoto
Ipatov, Aleksandr
Irvine, William
Ishiguro, Masato
Israel, Frank
Ivanov, Dmitrij
Iwata, Takahiro
Jackson, Carole
Jackson, Neal
Jacq, Thierry
Jaffe, Walter
Janssen, Michael
Jauncey, David
Jenkins, Charles
Pogrebenko, Sergei
Polatidis, Antonios
Pompei, Emanuela
Ponsonby, John
Pooley, Guy
Porcas, Richard
Porras, Bertha
Prandoni, Isabella
Preston, Robert
Preuss, Eugen
Price, R.
Puschell, Jeffery
Radford, Simon
Radhakrishnan, V.
Rahimov, Ismail
Raimond, Ernst
Ransom, Scott
Rao, A.
Raoult, Antoinette
Ray, Thomas
Ray, Paul
Razin, Vladimir
Readhead, Anthony
Redman, Matthew
Reich, Wolfgang
Reid, Mark
Reif, Klaus
Reyes, Francisco
Reynolds, John
Reynolds, Cormac
Rhee, Myung-Hyun
Ribó, Marc
Richer, John
Rickard, Lee
Ridgway, Susan
Riihimaa, Jorma
Riley, Julia
Rioja, Maria
José, José
Roberts, Morton
Roberts, David
Robertson, James
Robertson, Douglas
Robinson Jr., Richard
Roeder, Robert
Roelfsema, Peter

Balonek, Thomas
Banhatti, Dilip
Barrow, Colin
Bartel, Norbert
Barthel, Peter
Bartkiewicz, Anna
Barvainis, Richard
Baryshev, Andrey
Bash, Frank
Basu, Dipak
Baudry, Alain
Baum, Stefi
Bayet, Estelle
Beasley, Anthony
Beck, Rainer
Benaglia, Paula
Benn, Chris
Bennett, Charles
Benz, Arnold
Berge, Glenn
Berkhuijsen, Elly
Bhandari, Rajendra
Bhat, Ramesh
Bieging, John
Biermann, Peter
Biggs, James
Bignall, Hayley
Bignell, R.
Biraud, François
Biretta, John
Birkinshaw, Mark
Blair, David
Blandford, Roger
Bloemhof, Eric
Blundell, Katherine
Boboltz, David
Bock, Douglas
Bockelee-Morvan, Dominique
Bolatto, Alberto
Bondi, Marco
Boonstra, Albert
Bos, Albert
Bottinelli, Lucette
Bower, Geoffrey
Bowers, Phillip
Branchesi, Marca
Breahna, Iulian
Bregman, Jacob
Brentjens, Michiel
Bridle, Alan
Brinks, Elias
Britzen, Silke
Broderick, John
Bronfman, Leonardo
Brooks, Kate
Broten, Norman
Brouw, Willem
Brown, Jo-Anne
Browne, Ian
Jewell, Philip
Jin, Zhenyu
Johnson, Donald
Johnston, Kenneth
Johnston, Helen
Joly, François
Jones, Paul
Jones, Dayton
Josselin, Eric
Jung, Jae
Kaftan, May
Kahlmann, Hans
Kaidanovski, Mikhail
Kaifu, Norio
Kakinuma, Takakiyo
Kalberla, Peter
Kaltman, Tatyana
Kameya, Osamu
Kandalyan, Rafik
Kanekar, Nissim
Kang, Gon-Ik
Kardashev, Nicolay
Kassim, Namir
Kasuga, Takashi
Kaufmann, Pierre
Kawabata, Kinaki
Kawabe, Ryohei
Kawaguchi, Kentarou
Kawamura, Akiko
Kedziora-Chudozer, Lucyna
Kellermann, Kenneth
Kesteven, Michael
Khaikin, Vladimir
Kijak, Jaroslaw
Kilborn, Virginia
Killeen, Neil
Kim, Hyun-Goo
Kim, Tu-Whan
Kim, Kwang-tae
Kim, Bong
Kim, Sang Joon
Kislyakov, Albert
Kitaeff, Vyacheslav
Kiyoaki, Wajima
Klein, Ulrich
Klein, Karl
Knudsen, Kirsten
Ko, Hsien
Kobayashi, Hideyuki
Kocharovsky, Vitalij
Koda, Jin
Kohno, Kotaro
Kojima, Masayoshi
Kolomiyets, Svitlana
Kondratiev, Vladislav
Konovalenko, Alexander
Kopylova, Yulia
Korzhavin, Anatolij
Kovalev, Yuri
Roennaeng, Bernt
Röser, Hans-Peter
Roger, Robert
Rogers, Alan
Rogstad, David
Rohlfs, Kristen
Romanov, Andrey
Romero, Gustavo
Romney, Jonathan
Rosa González, Daniel
Rowson, Barrie
Rubin, Robert
Rubio, Monica
Rudnick, Lawrence
Rudnitskij, Georgij
Russell, Jane
Rydbeck, Gustaf
Rys, Stanislaw
Sadler, Elaine
Saikia, Dhruba
Sakamoto, Seiichi
Salter, Christopher
Samodurov, Vladimir
Sanamian, V.
Sandell, Goran
Sanders, David
Sargent, Annelia
Sarma, N.
Sarma, Anuj
Sastry, Ch.
Sato, Fumio
Savage, Ann
Savolainen, Tuomas
Sawada, Tsuyoshi
Sawada-Satoh, Satoko
Sawant, Hanumant
Scalise Jr., Eugenio
Schilizzi, Richard
Schilke, Peter
Schlickeiser, Reinhard
Schmidt, Maarten
Schröder, Anja
Schuch, Nelson
Schulz, R.
Schwartz, Philip
Schwarz, Ulrich
Scott, Paul
Scott, John
Seaquist, Ernest
Seielstad, George
Sekido, Mamoru
Sekimoto, Yutaro
Setti, Giancarlo
Shaffer, David
Shaposhnikov, Vladimir
Shaver, Peter
Shen, Zhiqiang
Shepherd, Debra
Sheridan, K.

Brunetti, Gianfranco
Bryant, Julia
Bujarrabal, Valentin
Burbidge, Geoffrey
Burderi, Luciano
Burke, Bernard
Campbell, Robert
Campbell-Wilson, Duncan
Caproni, Anderson
Carlqvist, Per
Caroubalos, Constantinos
Carr, Thomas
Carretti, Ettore
Carvalho, Joel
Casasola, Viviana
Casoli, Fabienne
Castelletti, Gabriela
Castets, Alain
Caswell, James
Cecconi, Baptiste
Celotti, Anna Lisa
Cernicharo, José
Chan, Kwing
Chandler, Claire
Charlot, Patrick
Chen, Yongjun
Chen, Huei-Ru
Chen, Xuefei
Chen, Zhiyuan
Chengalur, Jayaram
Chikada, Yoshihiro
Chin, Yi-nan
Chini, Rolf
Cho, Se-Hyung
Choudhury, Tirthankar
Christiansen, Wayne
Chung, Hyun
Chyzy, Krzysztof Tadeusz
Cichowolski, Silvina
Ciliegi, Paolo
Clark, Barry
Clark, David
Clark, Frank
Clegg, Andrew
Clemens, Dan
Cohen, Marshall
Cohen, Richard
Coleman, Paul
Colomb, Fernando
Colomer, Francisco
Combes, Françoise
Combi, Jorge
Condon, James
Conklin, Edward
Contreras, Maria
Conway, Robin
Conway, John
Cordes, James
Costa, Marco
Corbel, Stéphane

Kovalev, Yuri
Kramer, Michael
Kramer, Busaba
Kreysa, Ernst
Krichbaum, Thomas
Krishna, Gopal
Krishnan, Thiruvenkata
Kronberg, Philipp
Krügel, Endrik
Krygier, Bernard
Kuan, Yi-Jehng
Kuijpers, H.
Kuiper, Thomas
Kulkarni, Prabhakar
Kulkarni, Vasant
Kulkarni, Shrinivas
Kumkova, Irina
Kundt, Wolfgang
Kundu, Mukul
Kunert-Bajraszewska, Magdalena
Kuril'chik, Vladimir
Kus, Andrzej
Kutner, Marc
Kwok, Sun
La Franca, Fabio
Lada, Charles
Lai, Shih-Ping
Lal, Dharam
Landecker, Thomas
Lang, Kenneth
Langer, William
Langston, Glen
LaRosa, Theodore
Lasenby, Anthony
Lawrence, Charles
Le Squeren, Anne-Marie
Leahy, J.
Lebron, Mayra
Lee, Youngung
Lee, Chang-Won
Lee, Yong Bok
Legg, Thomas
Lehnert, Matthew
Lépine, Jacques
Lequeux, James
Lesch, Harald
Lestrade, Jean-François
Leung, Chun
Levreault, Russell
Li, Hong-Wei
Li, Zhi
Liang, Shiguang
Likkel, Lauren
Lilley, Edward
Lim, Jeremy
Lindqvist, Michael
Linke, Richard
Lis, Dariusz
Liseau, René

Shevgaonkar, R.
Shibata, Katsunori
Shinnaga, Hiroko
Shmeld, Ivar
Shone, David
Shulga, Valery
Sieber, Wolfgang
Singal, Ashok
Sinha, Rameshwar
Skillman, Evan
Slade, Martin
Slee, O.
Smith, Dean
Smith, Niall
Smolentsev, Sergej
Smolkov, Gennadij
Snellen, Ignas
Sobolev, Yakov
Soboleva, Natalja
Sodin, Leonid
Sofue, Yoshiaki
Sokolov, Konstantin
Somanah, Radhakhrishna
Song, Qian
Sorochenko, Roman
Spencer, Ralph
Spencer, John
Spoelstra, T.
Sramek, Richard
Sridharan, Tirupati
Stahr-Carpenter, M.
Stairs, Ingrid
Stanghellini, Carlo
Stanley, G.
Steffen, Matthias
Steinberg, Jean-Louis

Stewart, Ronald
Stewart, Paul
Stil, Jeroen
Stone, R.
Storey, Michelle
Strom, Richard
Strukov, Igor
Subrahmanya, C.
Subrahmanyan, Ravi
Sugitani, Koji
Sukumar, Sundarajan
Sullivan, III, Woodruff
Sunada, Kazuyoshi
Swarup, Govind
Swenson Jr., George
Szymczak, Marian
Takaba, Hiroshi
Takakubo, Keiya
Takano, Toshiaki
Takano, Shuro
Tanaka, Riichiro
Tapping, Kenneth

Cotton Jr., William
Crane, Patrick
Crawford, Fronefield
Crovisier, Jacques
Crutcher, Richard
Cunningham, Maria
D'Amico, Nicolò
Dagkesamansky, Rustam
Daishido, Tsuneaki
Daisuke, Iono
Dallacasa, Daniele
Davies, Rodney
Davis, Michael
Davis, Robert
de Bergh, Catherine
De Bernardis, Paolo
de Gregorio-Monsalvo, Itziar
de Jager, Cornelis
de la Noë, Jérôme
de Lange, Gert
de Ruiter, Hans
de Vicente, Pablo
de Young, David
Degaonkar, S.
Delannoy, Jean
Denisse, Jean-François
Dent, William
Deshpande, Avinash
Despois, Didier
Dewdney, Peter
Dhawan, Vivek
Diamond, Philip
Dickel, Helene
Dickel, John
Dickey, John
Dickman, Robert
Dieter Conklin, Nannielou
Dixon, Robert
Dobashi, Kazuhito
Dodson, Richard
Doubinskij, Boris
Dougherty, Sean
Downes, Dennis
Downs, George
Drake, Frank
Drake, Stephen
Dreher, John
Duffett-Smith, Peter
Dulk, George
Dutrey, Anne
Dwarakanath, K.
Dyson, Freeman
Eales, Stephen
Edelson, Rick
Ehle, Matthias
Ekers, Ronald
Elia, Davide
Ellingsen, Simon
Ellis, Graeme

Lister, Matthew
Litvinenko, Leonid
Liu, Xiang
Liu, Sheng-Yuan
Lo, Kwok-Yung
Locke, Jack
Lockman, Felix
Loiseau, Nora
Longair, Malcolm
Loren, Robert
Loukitcheva, Maria
Lovell, Sir Bernard
Lovell, James
Lozinskaya, Tatjana
Lubowich, Donald
Luks, Thomas
Luo, Xianhan
Lyne, Andrew
Macchetto, Ferdinando
MacDonald, Geoffrey
MacDonald, James
Machalski, Jerzy
Mack, Karl-Heinz
MacLeod, John
Maehara, Hideo
Malofeev, Valery
Malumian, Vigen
Manchester, Richard
Mandolesi, Nazzareno
Mantovani, Franco
Mao, Rui-Qing
Maran, Stephen
Marcaide, Juan-Maria
Mardyshkin, Vyacheslav
Marecki, Andrzej
Markoff, Sera
Marques, Dos
Marscher, Alan
Marti, Josep
Martin, Robert
Martin, Christopher
Martin-Pintado, Jesùs
Marvel, Kevin
Masheder, Michael
Maslowski, Jozef
Mason, Paul
Massardi, Marcella
Masson, Colin
Masumichi , Seta
Matsakis, Demetrios
Matsuo, Hiroshi
Matsushita, Satoki
Matthews, Henry
Matthews, Brenda
Mattila, Kalevi
Matveenko, Leonid
Mauersberger, Rainer
Maxwell, Alan
May, J.

Tarter, Jill
Tatematsu, Ken'ichi
te Lintel Hekkert, Peter
Tello Bohorquez, Camilo
Terasranta, Harri
Terzian, Yervant
Theureau, Gilles
Thomasson, Peter
Thompson, A.
Thum, Clemens
Tian, Wenwu
Tingay, Steven
Tiplady, Adrian
Tlamicha, Antonin
Tofani, Gianni
Tolbert, Charles
Tornikoski, Merja
Tosaki, Tomoka
Tovmassian, Hrant
Townes, Charles
Trigilio, Corrado
Trinidad, Miguel
Tritton, Keith
Troland, Thomas
Truong, Bach
Trushkin, Sergej
Tsuboi, Masato
Tsutsumi, Takahiro
Tuccari, Gino
Turlo, Zygmunt
Turner, Kenneth
Turner, Jean
Turtle, A.
Tyulbashev, Sergei
Tzioumis, Anastasios
Udaltsov, Vyacheslav
Udaya, Shankar
Ukita, Nobuharu
Ulrich, Marie-Helene
Ulvestad, James
Ulyanov, Oleg
Umana, Grazia
Umemoto, Tomofumi
Unger, Stephen
Unwin, Stephen
Urama, Johnson
Urošević, Dejan
Urquhart, James
Uson, Juan
Val'tts, Irina
Vallee, Jacques
Valtaoja, Esko
Valtonen, Mauri
van der Hulst, Jan
van der Kruit, Pieter
van der Laan, Harry
van der Tak, Floris
van Driel, Wim
van Gorkom, Jacqueline

Emerson, Darrel
Enome, Shinzo
Epstein, Eugene
Erickson, William
Esamdin, Ali
Eshleman, Von
Esimbek, Jarken
Ewing, Martin
Ezawa, Hajime
Facondi, Silvia
Falcke, Heino
Fanaroff, Bernard
Fanti, Roberto
Feain, Ilana
Fedotov, Leonid
Feigelson, Eric
Feldman, Paul
Felli, Marcello
Felten, James
Feretti, Luigina
Fernandes, Francisco
Ferrari, Attilio
Fey, Alan
Fharo, Toshihiro
Field, George
Filipovic, Miroslav
Finkelstein, Andrej
Fleischer, Robert
Florkowski, David
Foley, Anthony
Fomalont, Edward
Fort, David
Forveille, Thierry
Fouqué, Pascal
Frail, Dale
Frater, Robert
Frey, Sandor
Friberg, Per
Fuerst, Ernst
Fukui, Yasuo
Gabuzda, Denise

Gaensler, Bryan
Gallego, Juan Daniel
Gallimore, Jack
Galt, John
Gao, Yu
Garay, Guido
Gasiprong, Nipon
Gaume, Ralph
Gaylard, Michael
Geldzahler, Bernard
Gelfreikh, Georgij
Geng, Lihong
Genzel, Reinhard
Gérard, Eric
Gergely, Tomas
Gervasi, Massimo
Ghigo, Francis

McAdam, Bruce
McConnell, David
McCulloch, Peter
McKenna, Lawlor
McLean, Donald
McMullin, Joseph
Mebold, Ulrich
Meeks, M.
Meier, David
Menon, T.
Menten, Karl
Meyer, Martin
Mezger, Peter
Michalec, Adam
Mikhailov, Andrey
Miley, George
Miller, Neal
Mills, Bernard
Milne, Douglas
Milogradov-Turin, Jelena
Minakov, Anatoliy
Mirabel, Igor
Miroshnichenko, Alla
Mitchell, Kenneth
Miyawaki, Ryosuke
Miyazaki, Atsushi
Miyoshi, Makoto
Mizuno, Akira
Mizuno, Norikazu
Moellenbrock III, George
Moffett, David
Momjian, Emmanuel
Momose, Muntake
Montmerle, Thierry
Morabito, David
Moran, James
Morgan, Lawrence
Morganti, Raffaella
Morimoto, Masaki
Morison, Ian
Morita, Koh-ichiro
Morita, Kazuhiko
Moriyama, Fumio
Morras, Ricardo
Morris, David
Morris, Mark
Moscadelli, Luca
Muller, Erik
Muller, Sebastien
Mundy, Lee
Murdoch, Hugh
Murphy, Tara
Mutel, Robert
Muxlow, Thomas
Myers, Philip
Nadeau, Daniel
Nagnibeda, Valerij
Nakajima, Junichi
Nakano, Takenori

van Langevelde, Huib
van Leeuwen, Joeri
van Woerden, Hugo
VandenBout, Paul
Vats, Hari
Vaughan, Alan
Velusamy, T.
Venturi, Tiziana
Venugopal, V.
Verheijen, Marc
Verkhodanov, Oleg
Vermeulen, René
Véron, Philippe
Verschuur, Gerrit
Verter, Frances
Vestergaard, Marianne
Vilas, Faith
Vilas-Boas, José
Vivekanand, M.
Vogel, Stuart
Volvach, Alexander
Voronkov, Maxim
Walker, Robert
Wall, Jasper
Wall, William
Walmsley, C.
Walsh, Wilfred
Walsh, Andrew
Wang, Shouguan
Wang, Na
Wang, Hong-Guang
Wang, Shujuan
Wang, Jingyu
Wannier, Peter
Ward-Thompson, Derek
Wardle, John
Warmels, Rein
Warner, Peter
Warwick, James
Watson, Robert
Wehrle, Ann
Wei, Mingzhi
Weigelt, Gerd
Weiler, Kurt
Weiler, Edward
Welch, William
Wellington, Kelvin
Wen, Linqing
Wenlei, Shan
Westerhout, Gart
Westmeier, Tobias
Whiteoak, John
Whiting, Matthew
Wickramasinghe, N.
Wielebinski, Richard
Wiik, Kaj
Wiklind, Tommy
Wild, Wolfgang
Wilkinson, Peter

Ghosh, Tapashi
Gil, Janusz
Gimenez, Alvaro
Gioia, Isabella
Giovannini, Gabriele
Giroletti, Marcello
Gitti, Myriam
Goddi, Ciriaco
Goedhart, Sharmila
Goldwire Jr., Henry
Gomez, Gonzalez
Gómez Fernández, José
Gopalswamy, Natchimuthuk
Gordon, Mark
Gorgolewski, Stanislaw
Gorschkov, Aleksandr
Gosachinskij, Igor
Gottesman, Stephen
Gower, Ann
Graham, David
Green, David
Green, Anne
Green, James
Gregorini, Loretta
Gregorio-Hetem, Jane
Gregory, Philip
Grewing, Michael
Gu, Xuedong
Gubchenko, Vladimir
Guelin, Michel
Güsten, Rolf
Guidice, Donald
Guilloteau, Stéphane
Gulkis, Samuel
Gull, Stephen
Guo, Jianheng
Gupta, Yashwant
Gurvits, Leonid
Gwinn, Carl
Hall, Peter
Han, Jin-Lin
Hanasz, Jan
Handa, Toshihiro
Hanisch, Robert
Hankins, Timothy
Hardee, Philip
Harnett, Julienne
Harris, Daniel
Harten, Ronald
Haschick, Aubrey
Hasegawa, Tetsuo
Haverkorn, Marijke
Hayashi, Masahiko
Haynes, Raymond
Haynes, Martha
He, Jinhua
Heald, George
Heeralall-Issur, Nalini
Nakashima, Jun-ichi
Neeser, Mark
Nguyen-Quang, Rieu
Nicastro, Luciano
Nice, David
Nicolson, George
Nikolić, Silvana
Nishio, Masanori
Norris, Raymond
Nürnberger, Dieter
O'Dea, Christopher
O'Sullivan, John
Ogawa, Hideo
Ohashi, Nagayoshi
Ohishi, Masatoshi
Ojha, Roopesh
Okoye, Samuel
Okumura, Sachiko
Olberg, Michael
Onishi, Toshikazu
Onuora, Lesley
Orchiston, Wayne
Otmianowska-Mazur, Katarzyna
Ott, Juergen
Owen, Frazer
özel, Mehmet
Pacholczyk, Andrzej
Padman, Rachael
Palmer, Patrick
Panessa, Francesca
Pankonin, Vernon
Paragi, Zsolt
Paredes Poy, Josep
Parijskij, Yurij
Park, Yong-Sun
Parma, Paola
Paron, Sergio
Parrish, Allan
Pasachoff, Jay
Pashchenko, Mikhail
Patel, Nimesh
Pauls, Thomas
Payne, David
Pearson, Timothy
Peck, Alison
Pedersen, Holger
Pedlar, Alan
Peng, Bo
Peng, Qingyu
Penzias, Arno
Perez, Fournon
Perez-Torres, Miguel
Perley, Richard
Persson, Carina
Peters, William
Petrova, Svetlana
Pettengill, Gordon
Philipp, Sabine
Willis, Anthony
Wills, Beverley
Wills, Derek
Willson, Robert
Wilner, David
Wilson, Robert
Wilson, Thomas
Wilson, William
Windhorst, Rogier
Winnberg, Anders
Winnewisser, Gisbert
Wise, Michael
Witzel, Arno
Wolszczan, Alexander
Woltjer, Lodewijk
Wood, Douglas
Woodsworth, Andrew
Wootten, Henry
Wright, Alan
Wrobel, Joan
Wu, Yuefang
Wu, Xinji
Wu, Nailong
Wucknitz, Olaf
Yang, Ji
Yang, Zhigen
Yao, Qijun
Yasuhiro, Koyama
Yasuhiro, Murata
Ye, Shuhua
Yin, Qi-Feng
Yonekura, Yoshinori
Younis, Saad
Yu, Zhiyao
Yusef-Zadeh, Farhad
Zabolotny, Vladimir
Zainal Abidin, Zamri
Zaitsev, Valerij
Zanichelli, Alessandra
Zannoni, Mario
Zarka, Philippe
Zavala, Jr., Robert
Zensus, J-Anton
Zhang, Jian
Zhang, Xizhen
Zhang, Hongbo
Zhang, Qizhou
Zhang, JiangShui
Zhang, Jingyi
Zhang, Haiyan
Zhao, Jun
Zheleznyak, Alexander
Zheleznyakov, Vladimir
Zheng, Xinwu
Zhou, Jianfeng
Zhou, Jianjun
Zhu, LiChun
Zhu, Wenbai

Heeschen, David
Heiles, Carl
Helou, George
Henkel, Christian
Herpin, Fabrice
Heske, Astrid
Hewish, Antony
Philippe, Salome
Phillips, Thomas
Phillips, Christopher
Pick, Monique
Ping, Jinsong
Pisano, Daniel
Planesas, Pere
Zieba, Stanislaw
Zinchenko, Igor
Zlobec, Paolo
Zlotnik, Elena
Zuckerman, Ben
Zwaan, Martin
Zylka, Robert

Division XII Commission 41 History of Astronomy

President Ruggles, Clive
Vice-President Kochhar, Rajesh
Secretary Belmonte Avilés, Juan Antonio

Organizing Committee

Corbin, Brenda
de Jong, Teije
Norris, Raymond
Pigatto, Luisa
Sôma, Mitsuru
Sterken, Christiaan
Sun, Xiaochun

Members

Abalakin, Viktor
Abt, Helmut
Acharya, Bannanje
Ahn, Youngsook
Andrews, David
Ansari, S.M.
Aoki, Shinko
Ashok, N.
Babu, G.
Babul, Arif
Badolati, Ennio
Bailey, Mark
Ball, Lewis
Ballabh, Goswami
Bandyopadhyay, A.
Barlai, Katalin
Batten, Alan
Bennett, Jim
Berendzen, Richard
Bertola, Francesco
Bessell, Michael
Bhatia, Vishnu
Bhatt, H.
Bhattacharjee, Pijush
Bien, Reinhold
Bishop, Roy
Blaauw, Adriaan
Boccaletti, Dino
Boerngen, Freimut
Bònoli, Fabrizio
Botez, Elvira
Bougeret, Jean-Louis
Bowen, David
Braccesi, Alessandro
Brecher, Kenneth
Brooks, Randall
Brosche, Peter
Brouw, Willem
Brunet, Jean-Pierre
Burman, Ronald
Cai, Kai
Cannon, Russell
Caplan, James
Carlson, John
Chang, Heon-Young
Han, Wonyong
Hasan, S. Sirajul
Hasegawa, Ichiro
Haubold, Hans
Haupt, Hermann
Hayli, Abraham
Haynes, Raymond
Haynes, Roslynn
Hearnshaw, John
Heck, Andre
Hemenway, Mary
Herrmann, Dieter
Hers, Jan
Hidayat, Bambang
Hingley, Peter
Hirai, Masanori
Hockey, Thomas
Høg, Erik
Hollow, Robert
Holmberg, Gustav
Hopkins, Andrew
Hu, Tiezhu
Huan, Nguyen
Hughes, David
Hunstead, Richard
Hurukawa, Kiitiro
Hwang, Chorng-Yuan
Hysom, Edmund
Hyung, Siek
Ibrahim, Alaa
Idlis, Grigorij
Jafelice, Luiz
Jahreiss, Hartmut
Jarrell, Richard
Jauncey, David
Jeong, Jang
Jiang, Xiaoyuan
Jiménez-Vicente, Jorge
Jones, Paul
Kapoor, Ramesh
Kawabata, Kinaki
Keay, Colin
Keller, Hans-Ulrich
Kellermann, Kenneth
Kerschbaum, Franz
Kilambi, G.
Mickelson, Michael
Mikhail, Joseph
Milne, Douglas
Milone, Eugene
Min, Wang
Moesgaard, Kristian
Molnar, Michael
Morimoto, Masaki
Mumford, George
Nadal, Robert
Narlikar, Jayant
Nefed'ev, Yury
Nguyen, Mau
Nguyen, Mau
Nicolaidis, Efthymios
Norris, Raymond
Nussbaumer, Harry
Oh, Kyu
Ôhashi, Yukio
Ohashi, Nagayoshi
Olivier, Enrico
Olowin, Ronald
Oproiu, Tiberiu
Osório, José
Page, Arthur
Pang, Kevin
Papathanasoglou, Dimitrios
Pasachoff, Jay
Pati, Ashok
Peterson, Charles
Pettersen, Bjørn
Pfleiderer, Jorg
Pinigin, Gennadiy
Polyakhova, Elena
Poulle, Emmanuel
Pozhalova, Zhanna
Prokakis, Theodore
Proverbio, Edoardo
Rafferty, Theodore
Ray, Thomas
Reboul, Henri
Ruggles, Clive
Satterthwaite, Gilbert
Schaefer, Bradley
Schmadel, Lutz
Segonds, Alain-Philippe

Chapman, Jessica
Chen, Meidong
Chin, Yi-nan
Chinnici, Ileana
Choudhary, Debi Prasad
Cornejo, Alejandro
Cui, Zhenhua
Cui, Shizhu
Dadic, Zarko
Danezis, Emmanuel
Das, P.
Davies, Rodney
Davis, A. E. L.
Davoust, Emmanuel
Débarbat, Suzanne
Deeming, Terence
Dekker, E.
Denisse, Jean-François
DeVorkin, David
Dewhirst, David
Dick, Wolfgang
Dick, Steven
Dorokhova, Tetyana
Dorschner, Johann
Duerbeck, Hilmar
Duffard, René
Dumont, Simone
Dworetsky, Michael
Eddy, John
Ehgamberdiev, Shuhrat
Esteban, César
Evans, Robert
Fernie, J.
Field, J
Fierro, Julieta
Firneis, Maria
Flin, Piotr
Florides, Petros
Fluke, Christopher
Freeman, Kenneth
Freitas Mourão, Ronaldo
Gangui, Alejandro
Geffert, Michael
Geyer, Edward
Gillingham, Peter
Gingerich, Owen
Glass, Ian
Goss, W. Miller
Graham, John
Green, David
Green, Anne
Green, Daniel
Gussmann, Ernst
Hadrava, Petr
Kim, Yong-Cheol
Kim, Chun
Kim, Yonggi
Kim, Young-Soo
King, David
Kippenhahn, Rudolf
Koch, Robert
Kollerstrom, Nicholas
Kolomiyets, Svitlana
Komonjinda, Siramas
Koribalski, Bärbel
Kosovichev, Alexander
Krajnovic, Davor
Kreiner, Jerzy
Krisciunas, Kevin
Krishnan, Thiruvenkata
Krupp, Edwin
Lanfranchi, Gustavo
Lang, Kenneth
Launay, Françoise
Le Guet Tully, Françoise
Lee, Eun-Hee
Lee, Woo-baik
Lee, Yong-Sam
Lee, Yong Bok
Lerner, Michel-Pierre
Leung, Kam
Levy, Eugene
Li, Yong
Liller, William
Liritzis, Ioannis
Liu, Ciyuan
Locher, Kurt
Lomb, Nicholas
Longo, Giuseppe
Lopes-Gautier, Rosaly
Lopez, Carlos
Luminet, Jean-Pierre
Malin, David
Mallamaci, Claudio
Malville, J.
Manchester, Richard
Marco, Olivier
Marsden, Brian
Mathewson, Donald
McAdam, Bruce
McConnell, David
McKenna, Lawlor
McLean, Donald
Meadows, A.
Meech, Karen
Mendillo, Michael
Menon, T.
Shank, Michael
Shankland, Paul
Shingareva, Kira
Shore, Steven
Shukla, K.
Signore, Monique
Sima, Zdislav
Simpson, Allen
Sobouti, Yousef
Šolc, Martin
Soonthornthum, Boonrucksar
Stathopoulou, Maria
Steele, John
Steinle, Helmut
Stephenson, F.
Stoev, Alexey
Storey, Michelle
Sullivan, III, Woodruff
Sun, Xiaochun
Sundman, Anita
Svolopoulos, Sotirios
Swerdlow, Noel
Theodossiou, Efstratios
Tignalli, Horacio
Tobin, William
Trimble, Virginia
Usher, Peter
Valdés Parra, José
van Gent, Robert
Vargha, Magda
Vass, Gheorghe
Vavilova, Iryna
Verdet, Jean-Pierre
Verdun, Andreas
Volyanskaya, Margarita
Wang, Rongbin
Whitaker, Ewen
White, Graeme
Whiteoak, John
Wilkins, George
Williams, Thomas
Wilson, Curtis
Wolfschmidt, Gudrun
Woudt, Patrick
Xi, Zezong
Xiong, Jianning
Yamaoka, Hitoshi
Yang, Hong-Jin
Yau, Kevin
Yeomans, Donald
Zanini, Valeria
Zhang, Shouzhong
Zhou, Yonghong

Division V Commission 42 Close Binary Stars

President Ribas, Ignasi
Vice-President Richards, Mercedes

Organizing Committee

Bradstreet, David
Harmanec, Petr
Kaluzny, Janusz
Mikolajewska, Joanna
Munari, Ulisse
Niarchos, Panayiotis
Olah, Katalin
Pribulla, Theodor
Scarfe, Colin
Torres, Guillermo

Members

Al-Naimiy, Hamid
Andersen, Johannes
Antipova, Lyudmila
Antokhin, Igor
Antonopoulou, Evgenia
Anupama, G.
Aquilano, Roberto
Arefiev, Vadim
Aungwerojwit, Amornrat
Awadalla, Nabil
Baba, Hajime
Babkovskaia, Natalia
Bailyn, Charles
Baptista, Raymundo
Baran, Andrzej
Barkin, Yuri
Barone, Fabrizio
Bartolini, Corrado
Batten, Alan
Bell, Steven
Bianchi, Luciana
Blair, William
Blundell, Katherine
Boffin, Henri
Bolton, Charles
Bonazzola, Silvano
Bookmyer, Beverly
Bopp, Bernard
Borisov, Nikolay
Boyd, David
Boyle, Stephen
Bozic, Hrvoje
Bradstreet, David
Brandi, Elisande
Broglia, Pietro
Brownlee, Robert
Bruch, Albert
Bruhweiler, Frederick
Budaj, Ján
Budding, Edwin
Bunner, Alan
Burderi, Luciano
Busà, Innocenza
Busso, Maurizio
Callanan, Paul
Canalle, João
Herczeg, Tibor
Hill, Graham
Hills, Jack
Hillwig, Todd
Hoard, Donald
Holmgren, David
Holt, Stephen
Honeycutt, R.
Horiuchi, Ritoku
Hric, Ladislav
Hrivnak, Bruce
Huang, Runqian
Hube, Douglas
Hutchings, John
Ibanoglu, Cafer
Ikhsanov, Nazar
Ilya, Mandel
Imamura, James
Imbert, Maurice
Jabbar, Sabeh
Jasniewicz, Gerard
Jeffers, Sandra
Jeong, Jang
Jin, Zhenyu
Jonker, Peter
Joss, Paul
Kadouri, Talib
Kaitchuck, Ronald
Kalomeni, Belinda
Kang, Young-Woon
Karetnikov, Valentin
Kato, Taichi
Kawabata, Shusaku
Kenji, Nakamura
Kenny, Harold
Kenyon, Scott
Khalesseh, Bahram
Kilpio, Elena
Kim, Chun
Kim, Ho-il
Kim, Young-Soo
King, Andrew
Kitamura, Masatoshi
Kjurkchieva, Diana
Kley, Wilhelm
Koch, Robert
Park, Hong
Parthasarathy, Mudumba
Patkos, Laśzló
Pavlenko, Elena
Pavlovski, Kresimir
Pearson, Kevin
Peters, Geraldine
Piccioni, Adalberto
Piirola, Vilppu
Pojmanski, Grzegorz
Polidan, Ronald
Pollacco, Don
Popov, Sergey
Postnov, Konstantin
Potter, Stephen
Pribulla, Theodor
Pringle, James
Prokhorov, Mikhail
Pustõnski, Vladislav-Veniamin
Qiao, Guojun
Rafert, James
Rahunen, Timo
Rakos, Karl
Ramsey, Lawrence
Ransom, Scott
Rao, Pasagada
Rasio, Frederic
Reglero Velasco, Victor
Rey, Soo-Chang
Ringwald, Frederick
Ritter, Hans
Robb, Russell
Robertson, John
Robinson, Edward
Rodrigues, Claudia
Rovithis, Peter
Rovithis-Livaniou, Helen
Roxburgh, Ian
Ruffert, Maximilian
Russo, Guido
Sadik, Aziz
Sahade, Jorge
Saijo, Keiichi
Samec, Ronald
Sanyal, Ashit
Sarty, Gordon

Cester, Bruno
Chambliss, Carlson
Chapman, Robert
Chaty, Sylvain
Chaubey, Uma
Chen, An-Le
Chen, Xuefei
Cherepashchuk, Anatolij
Chochol, Drahomir
Choi, Kyu-Hong
Choi, Chul-Sung
Chou, Yi
Ciardi, David
Claria, Juan
Clausen, Jens
Collins, George
Cornelisse, Remon
Corradi, Romano
Cowley, Anne
Cropper, Mark
Cui, Wen-Yuan
Cutispoto, Giuseppe
D'Amico, Nicolò
D'Antona, Francesca
Dall, Thomas
De Grève, Jean-Pierre
de Loore, Camiel
Del Santo, Melania
Delgado, Antonio
Demircan, Osman
Diaz, Marcos
Dobrzycka, Danuta
Dorfi, Ernst
Dougherty, Sean
Drechsel, Horst
Dubus, Guillaume
Dümmler, Rudolf
Duerbeck, Hilmar
Dupree, Andrea
Durisen, Richard
Duschl, Wolfgang
Eaton, Joel
Edalati Sharbaf, Mohammad
Eggleton, Peter
Elias II, Nicholas
Etzel, Paul
Eyres, Stewart
Fabrika, Sergei
Faulkner, John
Fekel, Francis
Ferluga, Steno
Fernández Lajs, Eduardo
Ferrario, Lilia
Ferrer, Osvaldo
Flannery, Brian
Fors, Octavi
Frank, Juhan
Fredrick, Laurence
Kolb, Ulrich
Kolesnikov, Sergey
Komonjinda, Siramas
Konacki, Maciej
Kondo, Yoji
Koubsky, Pavel
Kraft, Robert
Kraicheva, Zdravka
Krautter, Joachim
Kreiner, Jerzy
Kreykenbohm, Ingo
Kruchinenko, Vitaliy
Kruszewski, Andrzej
Krzeminski, Wojciech
Kudashkina, Larisa
Kumsiashvily, Mzia
Kurpinska-Winiarska, M.
Kwee, K.
Lacy, Claud
Lamb Jr., Donald
Landolt, Arlo
Lapasset, Emilio
Larsson, Stefan
Larsson-Leander, Gunnar
Lavrov, Mikhail
Lee, Woo-baik
Lee, Yong-Sam
Lee, William
Leedjärv, Laurits
Leung, Kam
Li, Zhongyuan
Li, Ji
Li, Zhi
Lim, Jeremy
Lin, Yi-qing
Linnell, Albert
Linsky, Jeffrey
Liu, Qingzhong
Livio, Mario
Lucy, Leon
MacDonald, James
Maceroni, Carla
Malasan, Hakim
Malkov, Oleg
Manimanis, Vassilios
Mardirossian, Fabio
Maria de Garcia, J.M.
Marilli, Ettore
Markoff, Sera
Markworth, Norman
Marsh, Thomas
Mason, Paul
Mathieu, Robert
Mayer, Pavel
Mazeh, Tsevi
McCluskey Jr., George
Meintjes, Petrus
Melia, Fulvio
Savonije, Gerrit
Schartel, Norbert
Schiller, Stephen
Schmid, Hans
Schmidtke, Paul
Schmidtobreick, Linda
Schober, Hans
Seggewiss, Wilhelm
Selam, Selim
Semeniuk, Irena
Shafter, Allen
Shahbaz, Tariq
Shakura, Nikolaj
Shaviv, Giora
Shu, Frank
Sima, Zdislav
Sistero, Roberto
Skopal, Augustin
Slovak, Mark
Smak, Józef
Smith, Robert
Sobieski, Stanley
Söderhjelm, Staffan
Solheim, Jan
Song, Liming
Sonti, Sreedhar
Sowell, James
Sparks, Warren
Srivastava, J.
Srivastava, Ram
Stagg, Christopher
Stanishev, Vallery
Starrfield, Sumner
Steiman-Cameron, Thomas
Steiner, João
Stencel, Robert
Sterken, Christiaan
Stringfellow, Guy
Sugimoto, Daiichiro
Sundman, Anita
Svechnikov, Marij
Szkody, Paula
Taam, Ronald
Tan, Huisong
Taş, Günay
Tauris, Thomas
Taylor, John
Teays, Terry
Terrell, Dirk
Todoran, Ioan
Tout, Christopher
Tremko, Jozef
Trimble, Virginia
Turolla, Roberto
Tutukov, Aleksandr
Ureche, Vasile
Vaccaro, Todd
van den Heuvel, Edward

Friedjung, Michael
Gänsicke, Boris
Gallagher III, John
Gamen, Roberto
Garcia, Lia
Garcia-Lorenzo, Maria
Garmany, Katy
Gasiprong, Nipon
Geldzahler, Bernard
Geyer, Edward
Giannone, Pietro
Gies, Douglas
Giovannelli, Franco
Goldman, Itzhak
Gomboc, Andreja
Gonzalez Martinez Pais, Ignacio
Gosset, Eric
Graffagnino, Vito
Groot, Paul
Grygar, Jiří
Gu, Wei-Min
Guinan, Edward
Gulliver, Austin
Gün, Gulnur
Gunn, Alastair
Guo, Jianheng
Guseinov, O.
Hadrava, Petr
Hakala, Pasi
Hall, Douglas
Hammerschlag-Hensberge, Godelieve
Hanawa, Tomoyuki
Hantzios, Panayiotis
Hassall, Barbara
Haswell, Carole
Hayasaki, Kimitake
Hazlehurst, John
He, Jinhua
Hegedues, Tibor
Hellier, Coel
Helt, Bodil
Hempelmann, Alexander
Hensler, Gerhard
Meliani, Mara
Mennickent, Ronald
Mereghetti, Sandro
Meyer-Hofmeister, Eva
Mezzetti, Marino
Mikolajewska, Joanna
Mikulasek, Zdeněk
Milano, Leopoldo
Milone, Eugene
Mineshige, Shin
Miyaji, Shigeki
Mochnacki, Stephan
Morales Rueda, Luisa
Morgan, Thomas
Morrell, Nidia
Mouchet, Martine
Mumford, George
Munari, Ulisse
Murray, James
Mutel, Robert
Nakamura, Yasuhisa
Nakao, Yasushi
Nariai, Kyoji
Nather, R.
Naylor, Tim
Neff, James
Nelemans, Gijs
Nelson, Burt
Newsom, Gerald
Nha, Il-Seong
Norton, Andrew
Ogłoza, Waldemar
Oh, Kyu
Okazaki, Akira
Oliveira, Alexandre
Oliver, John
Olson, Edward
Osaki, Yoji
Ozkan, Mustafa
Padalia, T.
Pandey, Uma
Panei, Jorge
Parimucha, Štefan
van Hamme, Walter
van Kerkwijk, Marten
van't Veer, Frans
Vaz, Luiz Paulo
Vennes, Stéphane
Vilhu, Osmi
Voloshina, Irina
Wachter, Stefanie
Wade, Richard
Walder, Rolf
Walker, William
Wang, Xunhao
Ward, Martin
Warner, Brian
Webbink, Ronald
Weiler, Edward
Wheatley, Peter
Wheeler, J.
White, James
Williamon, Richard
Williams, Robert
Williams, Glen
Wilson, Robert
Xue, Li
Yamaoka, Hitoshi
Yamasaki, Atsuma
Yoon, Tae
Yüce, Kutluay
Zakirov, Mamnum
Zamanov, Radoslav
Zavala, Jr., Robert
Zeilik, Michael
Zhang, Er-Ho
Zhang, Bo
Zharikov, Sergey
Zhou, Daoqi
Zhou, Hongnan
Zhu, Liying
Ziolkowski, Janusz
Zola, Stanislaw
Zwitter, Tomaž

Division XI Commission 44 Space & High Energy Astrophysics

President Hasinger, Günther
Vice-President Jones, Christine

Organizing Committee

Braga, João
Brosch, Noah
de Graauw, Thijs
Gurvits, Leonid
Helou, George
Howarth, Ian
Kunieda, Hideyo
Montmerle, Thierry
Okuda, Haruyuki
Salvati, Marco
Singh, Kulinder

Members

Abramowicz, Marek
Acharya, Bannanje
Acton, Loren
Agrawal, P.
Ahluwalia, Harjit
Ahmad, Imad
Alexander, Joseph
Allington-Smith, Jeremy
Almleaky, Yasseen
Amati, Lorenzo
Andersen, Bo Nyborg
Antonelli, Lucio Angelo
Apparao, K.
Arafune, Jiro
Arefiev, Vadim
Ramírez Arnaud, Monique
Arnould, Marcel
Arons, Jonathan
Aschenbach, Bernd
Asseo, Estelle
Asvarov, Abdul
Audard, Marc
Audouze, Jean
Awaki, Hisamitsu
Axford, W.
Aya, Bamba
Ayres, Thomas
Baan, Willem
Bailyn, Charles
Balikhin, Michael
Baliunas, Sallie
Baring, Matthew
Barret, Didier
Barstow, Martin
Baskill, Darren
Basu, Dipak
Baym, Gordon
Bazzano, Angela
Becker, Robert
Becker, Werner
Beckmann, Volker
Begelman, Mitchell
Beiersdorfer, Peter
Belloni, Tomaso
Hauser, Michael
Hawkes, Robert
Hayama, Kazuhiro
Haymes, Robert
Heckathorn, Harry
Heger, Alexander
Hein, Righini
Heise, John
Helfand, David
Helmken, Henry
Hempel, Marc
Henoux, Jean-Claude
Henriksen, Richard
Henry, Richard
Hensberge, Herman
Heske, Astrid
Hicks, Amalia
Hidenori, Kokubun
Hill, Adam
Hoffman, Jeffrey
Holberg, Jay
Holloway, Nigel
Holt, Stephen
Holz, Daniel
Hora, Joseph
Hörandel, Jörg
Hornschemeier, Ann
Hornstrup, Allan
Houziaux, Leo
Hoyng, Peter
Hsu, Rue-Ron
Hu, Wenrui
Huang, YongFeng
Huang, Jiasheng
Huber, Martin
Hulth, Per
Hurley, Kevin
Hutchings, John
Hwang, Chorng-Yuan
Ibrahim, Alaa
Ichimaru, Setsuo
Ikhsanov, Nazar
Illarionov, Andrei
Imamura, James
Pethick, Christopher
Petkaki, Panagiota
Petro, Larry
Petrosian, Vahe
Phillips, Kenneth
Pian, Elena
Pinkau, K.
Pinto, Philip
Pipher, Judith
Piran, Tsvi
Piro, Luigi
Polidan, Ronald
Polletta, Maria del Carmen
Popov, Sergey
Porquet, Delphine
Pottschmidt, Katja
Pounds, Kenneth
Poutanen, Juri
Pozanenko, Alexei
Prasanna, A.
Preuss, Eugen
Price, Stephan
Produit, Nicolas
Protheroe, Raymond
Prouza, Michael
Prusti, Timo
Qiu, Yulei
Qu, Qinyue
Quintana, Hernan
Quirós, Israel
Radhakrishnan, V.
Raiteri, Claudia
Ramadurai, Souriraja
Ramírez, José
Ranalli, Piero
Rao, Arikkala
Rao, Ramachandra
Rasmussen, Ib
Rasmussen, Jesper
Raubenheimer, Barend
Ray, Paul
Raychaudhury, Somak
Reale, Fabio
Rees, Martin

Bender, Peter
Benedict, George
Benford, Gregory
Bennett, Charles
Bennett, Kevin
Benvenuto, Omar
Bergeron, Jacqueline
Bernacca, Pierluigi
Berta, Stefano
Beskin, Gregory
Beskin, Vasily
Bhattacharjee, Pijush
Bhattacharya, Dipankar
Bhattacharyya, Sudip
Bianchi, Luciana
Bicknell, Geoffrey
Biermann, Peter
Bignami, Giovanni
Bingham, Robert
Biswas, Sukumar
Blamont, Jacques-Emile
Blandford, Roger
Bleeker, Johan
Bless, Robert
Blinnikov, Sergey
Bloemen, Hans
Blondin, John
Bocchino, Fabrizio
Boer, Michel
Boggess, Albert
Boggess, Nancy
Bohlin, Ralph
Boksenberg, Alec
Bonazzola, Silvano
Bonnet, Roger
Bonnet-Bidaud, Jean-Marc
Bonometto, Silvio
Borozdin, Konstantin
Bougeret, Jean-Louis
Bowyer, C.
Bradley, Arthur
Branchesi, Marca
Bumba, VáclavSøren
Brandt, John
Brandt, William
Brecher, Kenneth
Breslin, Ann
Brinkman, Bert
Brown, Alexander
Bruhweiler, Frederick
Bruner, Marilyn
Brunetti, Gianfranco
Bumba, Václav
Bunner, Alan
Buote, David
Burbidge, Geoffrey
Burderi, Luciano
Burenin, Rodion

Imhoff, Catherine
in't Zand, Johannes
Inoue, Hajime
Inoue, Makoto
Ioka, Kunihito
Ipser, James
Ishida, Manabu
Israel, Werner
Ito, Kensai
Itoh, Masayuki
Jackson, John
Jaffe, Walter
Jakobsson, Pall
Jamar, Claude
Janka, Hans
Jaranowski, Piotr
Jenkins, Edward
Jokipii, J.
Jones, Frank
Jones, Thomas
Jonker, Peter
Jordan, Carole
Jordan, Stuart
Joss, Paul
Kafatos, Menas
Kahabka, Peter
Kalemci, Emrah
Kaneda, Hidehiro
Kaper, Lex
Kapoor, Ramesh
Karakas, Amanda
Karpov, Sergey
Kaspi, Victoria
Kasturirangan, K.
Katarzyński, Krzysztof
Kato, Tsunehiko
Kato, Yoshiaki
Katsova, Maria
Katz, Jonathan
Kawai, Nobuyuki
Kellermann, Kenneth
Kellogg, Edwin
Kelly, Brandon
Kembhavi, Ajit
Kenji, Nakamura
Kessler, Martin
Khumlemlert, Thiranee
Killeen, Neil
Kilpio, Elena
Kim, Yonggi
Kimble, Randy
Kinugasa, Kenzo
Kirk, John
Kiyoshi, Hayashida
Klinkhamer, Frans
Klose, Sylvio
Knapp, Johannes
Ko, Chung-Ming

Reeves, Hubert
Reichert, Gail
Reiprich, Thomas
Rengarajan, Thinniam
Rense, William
Revnivtsev, Mikhail
Rhoads, James
Riegler, Günter
Robba, Natale
Roberts, Timothy
Roman, Nancy
Romano, Patrizia
Roming, Peter
Rosendhal, Jeffrey
Rosner, Robert
Rovero, Adrián
Rubino-Martin, José Alberto
Ruder, Hanns
Ruffini, Remo
Ruffolo, David
Ruszkowski, Mateusz
Rutledge, Robert
Sabau-Graziati, Lola
Safi-Harb, Samar
Sagdeev, Roald
Sahade, Jorge
Sahlén, Martin
Sáiz, Alejandro
Sakano, Masaaki
Sakelliou, Irini
Samimi, Jalal
Sanchez, Norma
Sanders, Wilton
Santos, Nilton
Santos-Lleó, Maria
Sartori, Leo
Saslaw, William
Sato, Katsuhiko
Savage, Blair
Savedoff, Malcolm
Sazonov, Sergey
Sbarufatti, Boris
Scargle, Jeffrey
Schaefer, Gerhard
Schartel, Norbert
Schatten, Kenneth
Schatzman, Evry
Schilizzi, Richard
Schmitt, Juergen
Schnopper, Herbert
Schreier, Ethan
Schulz, Norbert
Schwartz, Daniel
Schwartz, Steven
Schwehm, Gerhard
Sciortino, Salvatore
Scott, John
Seielstad, George

Burger, Marijke
Burke, Bernard
Burrows, David
Burrows, Adam
Bursa, Michal
Burton, William
Butler, Christopher
Butterworth, Paul
Caccianiga, Alessandro
Cai, Michael
Camenzind, Max
Campbell, Murray
Cao, Li
Cappi, Massimo
Caraveo, Patrizia
Cárdenas, Rolando
Cardini, Daniela
Carlson, Per
Carpenter, Kenneth
Carroll, P.
Casandjian, Jean-Marc
Cash Jr., Webster
Casse, Michel
Castro-Tirado, Alberto
Catura, Richard
Cavaliere, Alfonso
Celotti, Anna Lisa
Cesarsky, Catherine
Chakrabarti, Sandip
Chakraborty, Deo
Chang, Hsiang-Kuang
Chang, Heon-Young
Channok, Chanruangrit
Chapman, Robert
Chapman, Sandra
Charles, Philip
Chartas, George
Chechetkin, Valerij
Chen, Lin-wen
Chenevez, Jerome
Cheng, Kwongsang
Cheung, Cynthia
Chian, Abraham
Chiappetti, Lucio
Chikawa, Michiyuki
Chitre, Shashikumar
Chochol, Drahomir
Choe, Gwangson
Choi, Chul-Sung
Chou, Yi
Chubb, Talbot
Chupp, Edward
Churazov, Evgenij
Ciotti, Luca
Clark, George
Clark, Thomas
Clay, Roger
Cohen, Jeffrey
Collin, Suzy

Kobayashi, Shiho
Koch-Miramond, Lydie
Kocharov, Grant
Koide, Shinji
Koji, Mori
Kojima, Yasufumi
Kolb, Edward
Kondo, Yoji
Kondo, Masaaki
Kong, Albert
Koshiba, Masatoshi
Koupelis, Theodoros
Koyama, Katsuji
Kozlowski, Maciej
Kozma, Cecilia
Kraushaar, William
Kretschmar, Peter
Kreykenbohm, Ingo
Kristiansson, Krister
Kryvdyk, Volodymyr
Kuiper, Lucien
Kulsrud, Russell
Kumagai, Shiomi
Kuncic, Zdenka
Kundt, Wolfgang
Kunz, Martin
Kurt, Vladimir
Kusunose, Masaaki
La Franca, Fabio
Lagache,,, Guilaine
Lal, Dharam
Lamb, Frederick
Lamb, Susan
Lamb Jr., Donald
Lamers, Henny
Lampton, Michael
Lapington, Jonathan
Lasher, Gordon
Lattimer, James
Lea, Susan
Leckrone, David
Lee, Wo-Lung
Lee, William
Leighly, Karen
Lemaire, Philippe
Levin, Yuri
Levine, Robyn
Lewin, Walter
Li, Zhongyuan
Li, Tipei
Li, Yuanjie
Li, Xiangdong
Li, Li-Xin
Liang, Edison
Lin, Xuan-bin
Linsky, Jeffrey
Loaring, Nicola
Lochner, James
Long, Knox

Selvelli, Pierluigi
Semerák, Oldrich
Seon, Kwang-Il
Sequeiros, Juan
Setti, Giancarlo
Severgnini, Paola
Seward, Frederick
Shaham, Jacob
Shahbaz, Tariq
Shakhov, Boris
Shakura, Nikolaj
Shapiro, Maurice
Shaver, Peter
Shaviv, Giora
Shen, Zhiqiang
Shibai, Hiroshi
Shibanov, Yuri
Shibazaki, Noriaki
Shields, Gregory
Shigeyama, Toshikazu
Shimura, Toshiya
Shin, Watanabe
Shivanandan, Kandiah
Shukre, C.
Shustov, Boris
Signore, Monique
Sikora, Marek
Silvestro, Giovanni
Simon, Vojtech
Simon, Paul
Sims, Mark
Skilling, John
Skinner, Stephen
Skjraasen, Olaf
Smale, Alan
Smith, Bradford
Smith, Barham
Smith, Peter
Smith, Linda
Smith, Nigel
Snow, Theodore
Sofia, Sabatino
Sokolov, Vladimir
Somasundaram, Seetha
Song, Qian
Sonneborn, George
Sood, Ravi
Spallicci di Filottrano, Alessandro
Sreekumar, Parameswaran
Srinivasan, Ganesan
Srivastava, Dhruwa
Stachnik, Robert
Staubert, Rüdiger
Stecher, Theodore
Stecker, Floyd
Steigman, Gary
Steinberg, Jean-Louis
Steiner, João
Stencel, Robert

Comastri, Andrea
Condon, James
Constantini, Elisa
Contopoulos, Ioannis
Corbet, Robin
Corbett, Ian
Corcoran, Michael
Cordova, France
Cornelisse, Remon
Courtes, Georges
Courvoisier, Thierry
Cowie, Lennox
Cowsik, Ramanath
Crannell, Carol
Crocker, Roland
Cropper, Mark
Croton, Darren
Cruise, Adrian
Culhane, John
Cunniffe, John
Curir, Anna
Cusumano, Giancarlo
D'Amico, Flavio
da Costa, José
da Costa, Antonio
da Silveira, Enio
Dadhich, Naresh
Dai, Zigao
Dalla Bontaà, Elena
Damian, Audley
Darriulat, Pierre
Davidson, William
Davis, Michael
Davis, Robert
Dawson, Bruce
de Aguiar, Odylio
de Felice, Fernando
de Jager, Cornelis
de Martino, Domitilla
de Young, David
Del Santo, Melania
Del Zanna, Luca
Della Ceca, Roberto
Dempsey, Robert
den Herder, Jan-Willem
Dennerl, Konrad
Dennis, Brian
Dermer, Charles
Di Cocco, Guido
Diaz Trigo, Maria
Digel, Seth
Disney, Michael
Dokuchaev, Vyacheslav
Dolan, Joseph
Domingo, Vicente
Dominis Prester, Dijana
Donea, Alina
Dong, Xiao-Bo
Dotani, Tadayasu
Longair, Malcolm
Lovelace, Richard
Lovell, Sir Bernard
Lu, Tan
Lu, Jufu
Lu, Fangjun
Lu, Ye
Lüst, Reimar
Luminet, Jean-Pierre
Luo, Qinghuan
Lutovinov, Alexander
Lynden-Bell, Donald
Lyubarsky, Yury
Ma, YuQian
Maccacaro, Tommaso
Maccarone, Thomas
Macchetto, Ferdinando
Machida, Mami
Maggio, Antonio
Mainieri, Vincenzo
Majumdar, Subhabrata
Makarov, Valeri
Malesani, Daniele
Malitson, Harriet
Malkan, Matthew
Manara, Alessandro
Mandolesi, Nazzareno
Mangano, Vanessa
Maran, Stephen
Marar, T.
Maričić, Darije
Markoff, Sera
Marov, Mikhail
Marranghello, Guilherme
Martin, Inácio
Martínez Bravo, Oscar
Martinis, Mladen
Masahiro, Tsujimoto
Masai, Kuniaki
Masnou, Jean-Louis
Mason, Keith
Mason, Glenn
Mather, John
Matsumoto, Ryoji
Matsumoto, Hironori
Matsuoka, Masaru
Matt, Giorgio
Matz, Steven
Mazurek, Thaddeus
McBreen, Brian
McCluskey Jr., George
McCray, Richard
Mead, Jaylee
Medina, José
Medina Tanco, Gustavo
Meier, David
Meiksin, Avery
Melatos, Andrew
Melia, Fulvio
Stéphane, Corbel
Stephens, S.
Stern, Robert
Stevens, Ian
Stier, Mark
Still, Martin
Stockman Jr., Hervey
Stoehr, Felix
Stone, R.
Straumann, Norbert
Stringfellow, Guy
Strohmayer, Tod
Strong, Ian
Struminsky, Alexei
Füzfa, André, Zdeněk
Sturrock, Peter
Su, Cheng-yue
Šubr, Ladislav
Suleimanov, Valery
Sun, Wei-Hsin
Sunyaev, Rashid
Suzuki, Hideyuki
Swank, Jean
Tagliaferri, Gianpiero
Takahara, Fumio
Takahashi, Masaaki
Takahashi, Tadayuki
Takayoshi, Kohmura
Takei, Yoh
Tanaka, Yasuo
Tashiro, Makoto
Tatehiro, Mihara
Tavecchio, Fabrizio
Terashima, Yuichi
Terrell, James
Thomas, Roger
Thorne, Kip
Thronson Jr., Harley
Tian, Wenwu
Tomimatsu, Akira
Torres, Diego
Torres, Carlos Alberto
Totsuka, Yoji
Tovmassian, Hrant
Traub, Wesley
Trimble, Virginia
Trümper, Joachim
Truran Jr., James
Trussoni, Edoardo
Tsuguya , Naito
Tsunemi, Hiroshi
Tsuru, Takeshi
Tsuruta, Sachiko
Tsygan, Anatolij
Tuerler, Marc
Tylka, Allan
Ueda, Yoshihiro
Ulyanov, Oleg
Underwood, James

Dovciak, Michal
Downes, Turlough
Drake, Frank
Drury, Luke
Dubus, Guillaume
Duorah, Hira
Dupree, Andrea
Durouchoux, Philippe
Duthie, Joseph
Edelson, Rick
Edwards, Paul
Ehle, Matthias
Eichler, David
Eilek, Jean
El Raey, Mohamed
Elvis, Martin
Elyiv, Andrii
Emanuele, Alessandro
Enlin, Torsten
Esamdin, Ali
Esimbek, Jarken
Ettori, Stefano
Eungwanichayapant, Anant
Evans, W.
Evans, Daniel
Fabian, Andrew
Fabricant, Daniel
Fang, Li-Zhi
Fang, Liu
Faraggiana, Rosanna
Fatkhullin, Timur
Fazio, Giovanni
Feldman, Paul
Felten, James
Fendt, Christian
Ferrari, Attilio
Fichtel, Carl
Field, George
Fisher, Philip
Fishman, Gerald
Florido, Estrella
Foing, Bernard
Fomin, Valery
Fonseca Gonzalez, Maria
Forman, William
Foschini, Luigi
Franceschini, Alberto
Frandsen, Søren
Frank, Juhan
Fransson, Claes
Fredga, Kerstin
Fujimoto, Shin-ichiro
Fujita, Mitsutaka
Furniss, Ian
Fyfe, Duncan
Gabriel, Alan
Gaensler, Bryan
Gaisser, Thomas
Galeotti, Piero
Galloway, Duncan
Melnick, Gary
Melnyk, Olga
Melrose, Donald
Méndez, Mariano
Mereghetti, Sandro
Merlo, David
Mestel, Leon
Meszaros, Peter
Meyer, Friedrich
Meyer, Jean-Paul
Micela, Giuseppina
Michel, F.
Miller, Michael
Miller, Guy
Miller, John
Min, Yuan
Minakov, Anatoliy
Mineo, Teresa
Miroshnichenko, Alla
Miyaji, Shigeki
Miyaji, Takamitsu
Miyata, Emi
Mizumoto, Yoshihiko
Mizutani, Kohei
Moderski, Rafal
Modisette, Jerry
Mollá, Mercedes
Monet, David
Moon, Shin
Moos, Henry
Morgan, Thomas
Mori, Masaki
Morton, Donald
Mota, David
Motch, Christian
Motizuki, Yuko
Mourão, Ana Maria
Mucciarelli, Paola
Mulchaey, John
Murakami, Hiroshi
Murakami, Toshio
Murdock, Thomas
Murtagh, Fionn
Murthy, Jayant
Nagataki, Shigehiro
Nakayama, Kunji
Neff, Susan
Ness, Norman
Neuhäuser, Ralph
Neupert, Werner
Nichols, Joy
Nicollier, Claude
Nielsen, Krister
Nikołajuk, Marek
Nishimura, Osamu
Nitta, Shin-ya
Nityananda, Rajaram
Nomoto, Ken'ichi
Norci, Laura
Nordh, Lennart
Upson, Walter
Uslenghi, Michela
Usov, Vladimir
Vahia, Mayank
Valtonen, Mauri
van den Heuvel, Edward
van der Hucht, Karel
van der Walt, Diederick
van Duinen, R.
van Putten, Maurice
van Riper, Kenneth
Vaughan, Simon
Vercellone, Stefano
Vestergaard, Marianne
Vial, Jean-Claude
Vidal, Nissim
Vidal-Madjar, Alfred
Vignali, Cristian
Vikhlinin, Alexey
Vilhu, Osmi
Villata, Massimo
Vink, Jacco
Viollier, Raoul
Viotti, Roberto
Völk, Heinrich
Vrtilek, Saeqa
Wagner, Alexander
Walker, Helen
Wanas, Mamdouh
Wang, Shouguan
Wang, Shui
Wang, Zhenru
Wang, Yi-ming
Wang, Jiancheng
Wang, Ding-Xiong
Wang, Hong-Guang
Wang, Feilu
Wang, Shujuan
Warner, John
Watanabe, Ken
Watarai, Kenya
Watts, Anna
Waxman, Eli
Weaver, Kimberly
Weaver, Thomas
Webster, Adrian
Wehrle, Ann
Wei, Daming
Weiler, Kurt
Weiler, Edward
Weinberg, Jerry
Weisheit, Jon
Weisskopf, Martin
Wells, Donald
Wen, Linqing
Wentzel, Donat
Wesselius, Paul
Wheatley, Peter
Wheeler, J.
Whitcomb, Stanley

Gao, Yu
Garmire, Gordon
Gaskell, C.
Gathier, Roel
Gehrels, Neil
Gendre, Bruce
Georgantopoulos, Ioannis
George, Kosugi
Geppert, Ulrich
Gezari, Daniel
Ghia, Piera Luisa
Ghirlanda, Giancarlo
Ghisellini, Gabriele
Giacconi, Riccardo
Gilra, Daya
Gioia, Isabella
Giroletti, Marcello
Gitti, Myriam
Glaser, Harold
Goldsmith, Donald
Goldwurm, Andrea
Gomboc, Andreja
Gómez de Castro, Ana
Gonzales, Walter
Gotthelf, Eric
Götz, Diego
Graffagnino, Vito
Grebenev, Sergej
Greenhill, John
Gregorio, Anna
Grenier, Isabelle
Grewing, Michael
Greyber, Howard
Griffiths, Richard
Grindlay, Jonathan
Grosso, Nicolas
Guessoum, Nidhal
Gull, Theodore
Gumjudpai, Burin
Gün, Gulnur
Gunn, James
Guseinov, O.
Gutiérrez, Carlos
Guziy , Sergiy
Hack, Margherita
Hakkila, Jon
Halevin, Alexander
Hallam, Kenneth
Hameury, Jean-Marie
Hannikainen, Diana
Hardcastle, Martin
Harms, Richard
Harris, Daniel
Harvey, Christopher
Harvey, Paul
Harwit, Martin
Hasan, Hashima
Hatsukade, Isamu
Haubold, Hans
Haugbølle, Troels

Norman, Colin
Noyes, Robert
Nulsen, Paul
O'Brien, Paul
O'Connell, Robert
O'Mongain, Eon
O'Sullivan, Denis
Ogawara, Yoshiaki
Ogelman, Hakki
Okeke, Pius
Okoye, Samuel
Okuda, Toru
Olthof, Henk
Onken, Christopher
Oohara, Ken-ichi
Orellana, Mariana
Orford, Keith
Orio, Marina
Orlandini, Mauro
Orlando, Salvatore
Osborne, Julian
Osten, Rachel
Ostriker, Jeremiah
Ostrowski, Michal
Ott, Juergen
Owen, Tobias
Ozaki, Masanobu
özel, Mehmet
Pacholczyk, Andrzej
Paciesas, William
Pacini, Franco
Page, Clive
Page, Mathew
Pak, Soojong
Paltani, Stéphane
Palumbo, Giorgio
Pandey, Uma
Panessa, Francesca
Papadakis, Iossif
Paragi, Zsolt
Park, Myeong-gu
Parker, Eugene
Parkinson, John
Parkinson, William
Patten, Brian
Paul, Biswajit
Pavlov, George
Peacock, Anthony
Pearce, Mark
Pearson, Kevin
Pellegrini, Silvia
Pellizza, Leonardo
Peng, Qiuhe
Peng, Qingyu
Pérez, Mario
Perola, Giuseppe
Perry, Peter
Peters, Geraldine
Peterson, Bruce
Peterson, Laurence

White, Nicholas
Wijers, Ralph
Wijnands, Rudy
Will, Clifford
Willis, Allan
Willner, Steven
Wilms, Jörn
Winkler, Christoph
Wise, Michael
Wolfendale, Arnold
Wolstencroft, Ramon
Wolter, Anna
Woltjer, Lodewijk
Worrall, Diana
Wray, James
Wu, Chi
Wu, Xuejun
Wu, Shaoping
Wu, Jiun-Huei
Wunner, Günter
Xu, Renxin
Xu, Dawei
Yadav, Jagdish
Yakut, Kadri
Yamada, Shoichi
Yamamoto, Yoshiaki
Yamasaki, Tatsuya
Yamasaki, Noriko
Yamashita, Kojun
Yamauchi, Makoto
Yamauchi, Shigeo
Yock, Philip
Yoichiro, Suzuki
Yoshida, Atsumasa
You, Junhan
Yu, Wang
Yu, Wenfei
Yuan, Ye-fei
Yuan, Feng
Zacchei, Andrea
Zamorani, Giovanni
Zane, Silvia
Zannoni, Mario
Zarnecki, John
Zdziarski, Andrzej
Zezas, Andreas
Zhang, William
Zhang, Jialu
Zhang, Shuang Nan
Zhang, Li
Zhang, JiangShui
Zhang, Jingyi
Zhang, You-Hong
Zhang, Yanxia
Zheng, Wei
Zheng, Xiaoping
Zhou, Jianfeng
Zombeck, Martin

Division IV Commission 45 Stellar Classification

President Gray, Richard
Vice-President Nordström, Birgitta

Organizing Committee

Burgasser, Adam
Eyer, Laurent
Gupta, Ranjan
Hanson, Margaret
Irwin, Michael
Soubiran, Caroline

Members

Albers, Henry
Allende Prieto, Carlos
Ardeberg, Arne
Arellano Ferro, Armando
Babu, G.
Baglin, Annie
Bartaya, R.
Bartkevicius, Antanas
Bell, Roger
Bidelman, William
Blanco, Victor
Buser, Roland
Celis, Leopoldo
Cester, Bruno
Cherepashchuk, Anatolij
Christy, James
Claria, Juan
Cowley, Anne
Crawford, David
Dal Ri Barbosa, Cassio
Divan, Lucienne
Drilling, John
Eglitis, Ilgmars
Egret, Daniel
Faraggiana, Rosanna
Feast, Michael
Feltzing, Sofia
Fitzpatrick, Edward
Fukuda, Ichiro
Gamen, Roberto
Garmany, Katy
Garrison, Robert
Gerbaldi, Michèle
Geyer, Edward
Giorgi, Edgard
Gizis, John
Glagolevskij, Yurij
Golay, Marcel
Grenon, Michel
Grosso, Monica
Guetter, Harry
Gupta, Ranjan
Hack, Margherita
Hallam, Kenneth
Hanson, Margaret
Hauck, Bernard
Hayes, Donald
Holmberg, Johan
Houk, Nancy
Humphreys, Roberta
Kato, Ken-ichi
Kurtanidze, Omar
Kurtz, Michael
Kurtz, Donald
Lasala Jr., Gerald
Lattanzio, John
Lee, Sang
Leggett, Sandy
Lepine, Sebastien
Levato, Orlando
Li, Jinzeng
Lloyd Evans, Thomas
Loden, Kerstin
Lu, Phillip
Luo, Ali
Luri, Xavier
Lutz, Julie
MacConnell, Darrell
Maehara, Hideo
Malagnini, Maria
Malaroda, Stella
McNamara, Delbert
Mead, Jaylee
Mendoza, V.
Morossi, Carlo
Morrell, Nidia
Nicolet, Bernard
North, Pierre
Notni, Peter
Oja, Tarmo
Olsen, Erik
Osborn, Wayne
Oswalt, Terry
Pakhomov, Yury
Parsons, Sidney
Philip, A.G.
Pizzichini, Graziella
Preston, George
Pulone, Luigi
Rautela, B.
Roman, Nancy
Rountree, Janet
Schild, Rudolph
Schmidt-Kaler, Theodor
Sharpless, Stewart
Shore, Steven
Shvelidze, Teimuraz
Sinnerstad, Ulf
Sion, Edward
Smith, J.
Sonti, Sreedhar
Soubiran, Caroline
Steinlin, Uli
Straizys, Vytautas
Strobel, Andrzej
Upgren, Arthur
von Hippel, Theodore
Walborn, Nolan
Walker, Gordon
Warren Jr., Wayne
Weaver, William
Weiss, Werner
Williams, John
Wing, Robert
Wright, Nicholas
Wu, Hsin-Heng
Wyckoff, Susan
Yamashita, Yasumasa
Yoss, Kenneth
Yumiko, Oasa
Zdanavicius, Kazimeras
Zhang, Yanxia

Division XII Commission 46 Astronomy Education & Development

President Ros, Rosa
Vice-President Hearnshaw, John

Organizing Committee

Balkowski-Mauger, Chantal	Haubold, Hans	Marschall, Laurence
De Grève, Jean-Pierre	Jones, Barrie	Miley, George
Deustua, Susana	Kochhar, Rajesh	Pasachoff, Jay
Guinan, Edward	Malasan, Hakim	Torres-Peimbert, Silvia

Members

Acker, Agnes	Gingerich, Owen	Onuora, Lesley
Aguilar, Maria	Gouguenheim, Lucienne	Oowaki, Naoaki
Aiad, A.	Govender, Kevindran	Orchiston, Wayne
Al-Naimiy, Hamid	Gray, Richard	Osborn, Wayne
Albanese, Lara	Gregorio, Anna	Osório, José
Alexandrov, Yuri	Gregorio-Hetem, Jane	Oswalt, Terry
Alsabti, Abdul Athem	Hafizi, Mimoza	Pandey, Uma
Alvarez, Rodrigo	Haque Copilah, Shirin	Pantoja, Carmen
Alvarez Pomares, Oscar	Havlen, Robert	Parisot, Jean-Paul
Anandaram, Mandayam	Haywood, J.	Penston, Margaret
Andersen, Johannes	Hemenway, Mary	Perez, Maria
Andrews, Frank	Heudier, Jean-Louis	Perez-Torres, Miguel
Ansari, S.M.	Hicks, Amalia	Perrin, Marshall
Arcidiacono, Carmelo	Hidayat, Bambang	Picazzio, Enos
Arellano Ferro, Armando	Hobbs, George	Pompea, Stephen
Arion, Douglas	Hockey, Thomas	Ponce, Gustavo
Aslan, Zeki	Hoff, Darrel	Popov, Sergey
Aubier, Monique	Hollow, Robert	Porras, Bertha
Babu, G.	Hotan, Aidan	Proverbio, Edoardo
Badescu, Octavian	Houziaux, Leo	Quamar, Jawaid
Baek, Chang Hyun	Hsu, Rue-Ron	Querci, François
Bailey Mathae, Katherine	Huan, Nguyen	Quirós, Israel
Barclay, Charles	Huang, Tianyi	Radeva, Veselka
Baret, Bruny	Hüttemeister, Susanne	Ramadurai, Souriraja
Barlow, Nadine	Hughes, Stephen	Ravindranath, Swara
Barthel, Peter	Ibrahim, Alaa	Reboul, Henri
Baskill, Darren	Ilyas, Mohammad	Rijsdijk, Case
Batten, Alan	Impey, Christopher	Roberts, Morton
Bernabeu, Guillermo	Inglis, Michael	Robinson, Leif
Birlan, Mirel	Ishizaka, Chiharu	Roca Cortés, Teodoro
Bittar, Jamal	Iwaniszewska, Cecylia	Rosa González, Daniel
Black, Adam	Jafelice, Luiz	Rosenzweig-Levy, Patrica
Bobrowsky, Matthew	Jørgensen, Henning	Roslund, Curt
Bochonko, D.	Kablak, Nataliya	Routly, Paul
Bojurova, Eva	Kalemci, Emrah	Roy, Archie
Booth, Roy	Karetnikov, Valentin	Russo, Pedro
Borchkhadze, Tengiz	Karttunen, Hannu	Sabra, Bassem
Botez, Elvira	Kay, Laura	Sadat, Rachida
Bottinelli, Lucette	Keller, Hans-Ulrich	Safko, John
Braes, Lucien	Khan, J.	Sahade, Jorge
Brieva, Eduardo	Kiasatpour, Ahmad	Samodurov, Vladimir
Brosch, Noah	Kim, Yoo Jea	Sampson, Russell
Budding, Edwin	Klinglesmith III, Daniel	Sandqvist, Aage
Cai, Michael	Koechlin, Laurent	Sandrelli, Stefano
Calvet, Nuria	Kolenberg, Katrien	Saraiva, Maria de Fatima

Cannon, Wayne
Capaccioli, Massimo
Caretta, Cesar
Carter, Brian
Celebre, Cynthia
Chamcham, Khalil
Chen, An-Le
Chen, Alfred
Chen, Lin-wen
Chitre, Dattakumar
Christensen, Lars
Christlieb, Norbert
Ciroi, Stefano
Clarke, David
Coffey, Deirdre
Colafrancesco, Sergio
Corbally, Christopher
Cottrell, Peter
Couper, Heather
Couto da Silva, Telma
Covone, Giovanni
Crawford, David
Cui, Zhenhua
Cunningham, Maria
Dall'Ora, Massimo
Daniel, Jean-Yves
Danner, Rolf
Darhmaoui, Hassane
Darriulat, Pierre
De Grève, Jean-Pierre
DeGioia-Eastwood, Kathleen
Del Santo, Melania
Delsanti, Audrey
Demircan, Osman
Devaney, Martin
Diego, Francisco
Donahue, Megan
Doran, Rosa
Ducati, Jorge
Dukes Jr., Robert
Dupuy, David
Duval, Marie-France
Dworetsky, Michael
El Eid, Mounib
Eze, Romanus
Falceta-Gonçalves, Diego
Fernández, Julio
Fernandez-Figueroa, M.
Fienberg, Richard
Fierro, Julieta
Fleck, Robert
Florsch, Alphonse
Forbes, Douglas
Fu, Hsieh-Hai
Gabriel, Carlos
Gallino, Roberto
Gangui, Alejandro
Ganguly, Rajib
Kolka, Indrek
Kolomiyets, Svitlana
Komonjinda, Siramas
Kong, Xu
Kononovich, Edvard
Kozai, Yoshihide
Kramer, Busaba
Kreiner, Jerzy
Krishna, Gopal
Krupp, Edwin
Kuan, Yi-Jehng
Lago, Maria
Lai, Sebastiana
Lanciano, Nicoletta
Lanfranchi, Gustavo
Lee, Kang Hwan
Lee, Yong Bok
Leung, Kam
Leung, Chun
Lin, Weipeng
Linden-Vørnle, Michael
Little-Marenin, Irene
Lomb, Nicholas
Luck, John
Ma, Xingyuan
Maciel, Walter
Maddison, Ronald
Madsen, Claus
Mahoney, Terence
Mallamaci, Claudio
Mamadazimov, Mamadmuso
Marco, Olivier
Martinet, Louis
Martinez, Peter
Martínez Bravo, Oscar
Massey, Robert
Mavridis, Lyssimachos
Maza, José
McKinnon, David
McNally, Derek
Meidav, Meir
Merlo, David
Milogradov-Turin, Jelena
Mizuno, Takao
Moreels, Guy
Morimoto, Masaki
Murphy, John
Najid, Nour-Eddine
Narlikar, Jayant
Navone, Hugo
Nayar, S.R.Prabhakaran
Nguyen-Quang, Rieu
Nha, Il-Seong
Nicolson, Iain
Nikolov, Nikola
Ninkovic, Slobodan
Noels, Arlette
Norton, Andrew
Sattarov, Isroil
Satterthwaite, Gilbert
Saxena, P.
Schleicher, David
Schlosser, Wolfhard
Schmitter, Edward
Schroeder, Daniel
Seeds, Michael
Shen, Chun-Shan
Shipman, Harry
Siher, El Arbi
Slater, Timothy
Smail, Ian
Solheim, Jan
Soriano, Bernardo
Štefl, Vladimir
Stenholm, Björn
Stoev, Alexey
Straizys, Vytautas
Sukartadiredja, Darsa
Švestka, Jiří
Swarup, Govind
Szécsényi-Nagy, Gábor
Szostak, Roland
Tignalli, Horacio
Torres, Jesùs Rodrigo
Touma, Jihad
Trinidad, Miguel
Tugay, Anatoliy
Ugolnikov, Oleg
Urama, Johnson
Valdés Parra, José
van den Heuvel, Edward
van Santvoort, Jacques
Vauclair, Sylvie
Vilinga, Jaime
Vilks, Ilgonis
Viñuales Gavín, Ederlinda
Voelzke, Marcos
Vujnovic, Vladis
Walker, Constance
Walsh, Wilfred
Wang, Shouguan
Ward, Richard
Wentzel, Donat
West, Richard
Whelan, Emma
Whitelock, Patricia
Williamon, Richard
Willmore, A.
Wolfschmidt, Gudrun
Xie, Xianchun
Xiong, Jianning
Ye, Shuhua
Yim, Hong-Suh
Yumiko, Oasa
Zakirov, Mamnum
Zealey, William

Gasiprong, Nipon
Germany, Lisa
Ghobros, Roshdy
Gill, Peter
Gimenez, Alvaro
Ödman, Carolina
Oja, Heikki
Okeke, Pius
Okoye, Samuel
Olsen, Hans
Zeilik, Michael
Zhang, You-Hong
Zhao, Jun Liang
Zimmermann, Helmut

Division VIII Commission 47 Cosmology

President Padmanabhan, Thanu
Vice-President Schmidt, Brian

Organizing Committee

Bunker, Andrew
Ciardi, Benedetta
Jing, Yipeng
Koekemoer, Anton
Lahav, Ofer
Lefevre, Olivier
Scott, Douglas

Members

Abbas, Ummi
Abu Kassim, Hasan
Adami, Christophe
Adams, Jenni
Alard, Christophe
Alcaniz, Jailson
Alimi, Jean-Michel
Allan, Peter
Allington-Smith, Jeremy
Almaini, Omar
Amendola, Luca
Andersen, Michael
Andreani, Paola
Aretxaga, Itziar
Argüeso, Francisco
Atrio Barandela, Fernando
Audouze, Jean
Aussel, Herve
Avelino, Pedro
Azuma, Takahiro
Babul, Arif
Baddiley, Christopher
Bagla, Jasjeet
Bahcall, Neta
Bajtlik, Stanislaw
Baker, Andrew
Balbi, Amedeo
Baldwin, John
Balland, Christophe
Bamford, Steven
Banday, Anthony
Banerji, Sriranjan
Banhatti, Dilip
Barberis, Bruno
Barbuy, Beatriz
Bardeen, James
Bardelli, Sandro
Barger, Amy
Barkana, Rennan
Barrow, John
Bartelmann, Matthias
Barthel, Peter
Barton, Elizabeth
Baryshev, Andrey
Basa, Stéphane
Bassett, Bruce
Basu, Dipak
Gregory, Stephen
Greve, Thomas
Greyber, Howard
Griest, Kim
Grishchuk, Leonid
Gudmundsson, Einar
Gumjudpai, Burin
Gunn, James
Gutiérrez, Carlos
Guzzo, Luigi
Haehnelt, Martin
Hagen, Hans-Juergen
Hall, Patrick
Hamilton, Andrew
Hannestad, Steen
Hansen, Frode
Hardy, Eduardo
Harms, Richard
Hau, George
Haugbølle, Troels
Hayashi, Chushiro
He, XiangTao
Heavens, Alan
Heinamaki, Pekka
Hellaby, Charles
Heller, Michael
Hendry, Martin
Henriksen, Mark
Hernández, Xavier
Hewett, Paul
Hewitt, Adelaide
Heyrovský, David
Hicks, Amalia
Hideki, Asada
Hiroshi, Yoshida
Hnatyk, Bohdan
Hoekstra, Hendrik
Holz, Daniel
Hu, Esther
Hu, Hongbo
Huang, Jiasheng
Huchra, John
Hudson, Michael
Hughes, David
Hütsi, Gert
Huynh, Minh
Hwang, Jai-chan
Petitjean, Patrick
Petrosian, Vahe
Pimbblet, Kevin
Plionis, Manolis
Podolský, Jiří
Polletta, Maria del Carmen
Pompei, Emanuela
Popescu, Nedelia
Portinari, Laura
Power, Chris
Prandoni, Isabella
Premadi, Premana
Press, William
Puetzfeld, Dirk
Puget, Jean-Loup
Puy, Denis
Qin, Bo
Qu, Qinyue
Quirós, Israel
Rahvar, Sohrab
Ramella, Massimo
Ranalli, Piero
Ravindranath, Swara
Rebolo, Rafael
Reboul, Henri
Rees, Martin
Reeves, Hubert
Reiprich, Thomas
Revnivtsev, Mikhail
Rey, Soo-Chang
Riazi, Nematollah
Riazuelo, Alain
Ribeiro, Marcelo
Ribeiro, Andre Luis
Richter, Philipp
Ricotti, Massimo
Ridgway, Susan
Rindler, Wolfgang
Rivolo, Arthur
Roberts, David
Robinson, I.
Rocca-Volmerange, Brigitte
Roeder, Robert
Roettgering, Huub
Romano-Diaz, Emilio
Romeo, Alessandro
Romer, Anita

Bechtold, Jill
Beckman, John
Beesham, Aroonkumar
Belinski, Vladimir
Bennett, Charles
Bennett, David
Bergeron, Jacqueline
Bergvall, Nils
Berman, Marcelo
Berta, Stefano
Bertola, Francesco
Bertschinger, Edmund
Betancor Rijo, Juan
Bharadwaj, Somnath
Bhavsar, Suketu
Bianchi, Simone
Bicknell, Geoffrey
Bignami, Giovanni
Binetruy, Pierre
Birkinshaw, Mark
Biviano, Andrea
Blakeslee, John
Blanchard, Alain
Bleyer, Ulrich
Blundell, Katherine
Böhringer, Hans
Boksenberg, Alec
Bolzonella, Micol
Bond, John
Bongiovanni, Angel
Bonnor, W.
Borgani, Stefano
Boschin, Walter
Bouchet, François
Bowen, David
Boyle, Brian
Branchesi, Marca
Brecher, Kenneth
Bridle, Sarah
Brinchmann, Jarle
Brown, Michael
Bryant, Julia
Buote, David
Burbidge, Geoffrey
Burns Jr., Jack
Calvani, Massimo
Cao, Li
Cappi, Alberto
Cárdenas, Rolando
Carr, Bernard
Carretti, Ettore
Castagnino, Mario
Cattaneo, Andrea
Cavaliere, Alfonso
Cesarsky, Diego
Chae, Kyu-Hyun
Chang, Kyongae
Chang, Heon-Young
Chen, DaMing
Hwang, Chorng-Yuan
Icke, Vincent
Ikeuchi, Satoru
Im, Myungshin
Impey, Christopher
Inada, Naohisa
Iovino, Angela
Ishihara, Hideki
Iwata, Ikuru
Iyer, B.
Jahnke, Knud
Jakobsson, Pall
Jannuzi, Buell
Jaroszynski, Michal
Jauncey, David
Jaunsen, Andreas
Jedamzik, Karsten
Jensen, Brian
Jetzer, Philippe
Jones, Heath
Jones, Christine
Jordán, Andrés
Jovanovic, Predrag
Junkkarinen, Vesa
Juszkiewicz, Roman
Kajino, Toshitaka
Kanekar, Nissim
Kang, Hyesung
Kapoor, Ramesh
Karachentsev, Igor
Kato, Shoji
Kauffmann, Guinevere
Kaul, Chaman
Kawabata, Kinaki
Kawabata, Kiyoshi
Kawasaki, Masahiro
Kellermann, Kenneth
Kembhavi, Ajit
Khare, Pushpa
Khmil, Sergiy
Kim, Jik
King, Lindsay
Kirilova, Daniela
Kiyotomo, Ichiki
Kneib, Jean-Paul
Kodama, Hideo
Kokkotas, Konstantinos
Kolb, Edward
Kompaneets, Dmitrij
Koopmans, Leon
Kormendy, John
Kovalev, Yuri
Kovetz, Attay
Kozai, Yoshihide
Kozlovsky, Ben
Krasinski, Andrzej
Kriss, Gerard
Kudrya, Yury
Kunth, Daniel
Rosa González, Daniel
Rosquist, Kjell
Rothberg, Barry
Rowan-Robinson, Michael
Roxburgh, Ian
Rubin, Vera
Rubino-Martin, José Alberto
Rudnick, Lawrence
Rudnicki, Konrad
Ruffini, Remo
Ruszkowski, Mateusz
Ružička, Adam
Saar, Enn
Sadat, Rachida
Sahlén, Martin
Sahni, Varun
Sáiz, Alejandro
Salvador-Sole, Eduardo
Salzer, John
Santos-Lleó, Maria
Sanz, José
Sapar, Arved
Saracco, Paolo
Sargent, Wallace
Sasaki, Misao
Sasaki, Shin
Sato, Shinji
Sato, Katsuhiko
Sato, Humitaka
Sato, Jun'ichi
Savage, Ann
Saviane, Ivo
Sazhin, Mikhail
Scaramella, Roberto
Schartel, Norbert
Schatzman, Evry
Schaye, Joop
Schechter, Paul
Schindler, Sabine
Schmidt, Maarten
Schneider, Peter
Schneider, Donald
Schneider, Jean
Schneider, Raffaella
Schramm, Thomas
Schuch, Nelson
Schucking, Engelbert
Schumacher, Gerard
Scodeggio, Marco
Seielstad, George
Semerák, Oldrich
Serjeant, Stephen
Setti, Giancarlo
Severgnini, Paola
Shandarin, Sergei
Shanks, Thomas
Shao, Zhengyi
Sharp, Nigel
Shaver, Peter

Chen, Jiansheng
Chen, Hsiao-Wen
Chen, Xuelei
Chen, Lin-wen
Chen, Pisin
Cheng, Fuzhen
Chiba, Takeshi
Chincarini, Guido
Chitre, Dattakumar
Chodorowski, Michal
Choudhury, Tirthankar
Christensen, Lise
Chu, Yaoquan
Ciliegi, Paolo
Claeskens, Jean-François
Claria, Juan
Clarke, Tracy
Clowe, Douglas
Clowes, Roger
Cocke, William
Cohen, Jeffrey
Cohen, Ross
Colafrancesco, Sergio
Cole, Shaun
Coles, Peter
Colless, Matthew
Colombi, Stéphane
Condon, James
Cooray, Asantha
Cora, Sofia
Corsini, Enrico
Courbin, Frederic
Courteau, Stéphane
Covone, Giovanni
Crane, Patrick
Crane, Philippe
Crawford, Steven
Cristiani, Stefano
Croom, Scott
Croton, Darren
Curran, Stephen
Cypriano, Eduardo
D' Odorico, Valentina
Da Costa, Gary
Da Rocha, Cristiano
Dadhich, Naresh
Dahle, Haakon
Daigne, Frederic
Daisuke, Iono
Dalla Bontaà, Elena
Danese, Luigi
Das, P.
Davidson, William
Davies, Paul
Davies, Roger
Davis, Michael
Davis, Marc
Davis, Tamara
De Bernardis, Paolo
Kunz, Martin
La Barbera, Francesco
La Franca, Fabio
Lacey, Cedric
Lachieze-Rey, Marc
Lagache, Guilaine
Lake, Kayll
Lanfranchi, Gustavo
Larionov, Mikhail
Lasota, Jean-Pierre
Layzer, David
Leão, João Rodrigo
Lee, Wo-Lung
Lehnert, Matthew
Lequeux, James
Leubner, Manfred
Levin, Yuri
Levine, Robyn
Lewis, Geraint
Li, Li-Xin
Lian, Luo
Liang, Yanchun
Liddle, Andrew
Liebscher, Dierck-E
Lilje, Per
Lilly, Simon
Lin, Weipeng
Liou, Guo Chin
Liu, Yongzhen
Lombardi, Marco
Longair, Malcolm
Longo, Giuseppe
Lonsdale, Carol
Lo'pez, Sebastian
Lopez Corredoira, Martin
Loveday, Jon
Lu, Tan
Lubin, Lori
Lukash, Vladimir
Luminet, Jean-Pierre
Lynden-Bell, Donald
Ma, Jun
Maartens, Roy
Maccagni, Dario
MacCallum, Malcolm
Maddox, Stephen
Maeda, Kei-ichi
Maharaj, Sunil
Maia, Marcio
Mainieri, Vincenzo
Majumdar, Subhabrata
Malesani, Daniele
Mamon, Gary
Mandolesi, Nazzareno
Mangalam, Arun
Mann, Robert
Manrique, Alberto
Mansouri, Reza
Mao, Shude
Shaviv, Giora
Shaya, Edward
Shibata, Masaru
Shivanandan, Kandiah
Siebenmorgen, Ralf
Signore, Monique
Silk, Joseph
Silva, Laura
Sironi, Giorgio
Sistero, Roberto
Slosar, Anze
Smail, Ian
Smette, Alain
Smith, Rodney
Smith, Nigel
Smoot III, George
Sokolowski, Lech
Sollerman, Jesper
Song, Doo
Souradeep, Tarun
Souriau, Jean-Marie
Spinoglio, Luigi
Spyrou, Nicolaos
Squires, Gordon
Srianand, Roghunathan
Stavrev, Konstantin
Stecker, Floyd
Steigman, Gary
Stewart, John
Stoehr, Felix
Stolyarov, Vladislav
Storrie-Lombardi, Lisa
Straumann, Norbert
Strauss, Michael
Stritzinger, Maximilian
Struble, Mitchell
Strukov, Igor
Stuchlík, Zdeněk
Stuik, Remko
Subrahmanya, C.
Suginohara, Tatsushi
Sugiyama, Naoshi
Suhhonenko, Ivan
Sunyaev, Rashid
Surdej, Jean
Szalay, Alex
Szydlowski, Marek
Tagoshi, Hideyuki
Takahara, Fumio
Tammann, Gustav
Tanabe, Kenji
Tarter, Jill
Taruya, Atsushi
Tatekawa, Takayuki
Temporin, Sonia
Thuan, Trinh
Tifft, William
Tipler, Frank
Toffolatti, Luigi

de Lapparent-Gurriet, Valérie
de Lima, José
de Petris, Marco
de Ruiter, Hans
de Silva, Lindamulage
de Zotti, Gianfranco
Dekel, Avishai
Dell'Antonio, Ian
Demianski, Marek
Désert, François-Xavier
Deustua, Susana
Dhurandhar, Sanjeev
Diaferio, Antonaldo
Diaz, Ruben
Dietrich , Joerg
Dionysiou, Demetrios
Djorgovski, Stanislav
Dobbs, Matt
Dobrzycki, Adam
Domínguez, Mariano
Dong, Xiao-Bo
Dressler, Alan
Drinkwater, Michael
Dultzin-Hacyan, Deborah
Dunsby, Peter
Dyer, Charles
Eales, Stephen
Edsjö, Joakim
Efstathiou, George
Einasto, Jaan
Elgaroy, Oystein
Elizalde, Emilio
Ellis, George
Ellis, Richard
Elvis, Martin
Elyiv, Andrii
Enginol, Turan
Ettori, Stefano
Eungwanichayapant, Anant
Faber, Sandra
Falk Jr., Sydney
Fall, S.
Fan, Zuhui
Fang, Li-Zhi
Fassnacht, Christopher
Fatkhullin, Timur
Fedeli, Cosimo
Fedorova, Elena
Felten, James
Feng, Long Long
Field, George
Filippenko, Alexei
Florides, Petros
Focardi, Paola
Fomin, Piotr
Fong, Richard
Ford, Holland
Forman, William
Fouqué, Pascal
Maoz, Dan
Marano, Bruno
Mardirossian, Fabio
Marek, John
Maris, Michèle
Marranghello, Guilherme
Martínez-González, Enrique
Martinis, Mladen
Martins, Carlos
Massardi, Marcella
Mather, John
Mathez, Guy
Matsumoto, Toshio
Matzner, Richard
Mavrides, Stamatia
Mellier, Yannick
Melnyk, Olga
Melott, Adrian
Meneghetti, Massimo
Menéndez-Delmestre, Karin
Merighi, Roberto
Meszaros, Peter
Meszaros, Attila
Meyer, David
Meyer, Martin
Meylan, Georges
Meza, Andres
Mezzetti, Marino
Minakov, Anatoliy
Miralda-Escudé, Jordi
Miralles, Joan-Marc
Miranda, Oswaldo
Misner, Charles
Miyazaki, Satoshi
Miyoshi, Shigeru
Mo, Houjun
Mohr, Joseph
Mollá, Mercedes
Monaco, Pierluigi
Moore, Ben
Moreau, Olivier
Moscardini, Lauro
Mota, David
Motta, Veronica
Mourão, Ana Maria
Muanwong, Orrarujee
Mücket, Jan
Müller, Andreas
Muller, Richard
Murakami, Izumi
Murphy, John
Nakamichi, Akika
Nambu, Yasusada
Narasimha, Delampady
Narlikar, Jayant
Naselsky, Pavel
Nasr-Esfahani, Bahram
Neves de Araujo, José
Nicoll, Jeffrey
Toft, Sune
Tomimatsu, Akira
Tomita, Kenji
Tonry, John
Tormen, Giuseppe
Totani, Tomonori
Tozzi, Paolo
Tremaine, Scott
Treu, Tommaso
Trevese, Dario
Trimble, Virginia
Trotta, Roberto
Trujillo Cabrera, Ignacio
Tsamparlis, Michael
Tugay, Anatoliy
Tully, Richard
Turner, Edwin
Turner, Michael
Turnshek, David
Tyson, John
Tytler, David
Tyulbashev, Sergei
Ugolnikov, Oleg
Umemura, Masayuki
Uson, Juan
Vagnetti, Fausto
Vaidya, P.
Väisänen, Petri
Valls-Gabaud, David
Valtchanov, Ivan
van der Laan, Harry
van Haarlem, Michiel
Vedel, Henrik
Verdoes Kleijn, Gijsbert
Vestergaard, Marianne
Vettolani, Giampaolo
Viana, Pedro
Viel, Matteo
Vishniac, Ethan
Vishveshwara, C.
von Borzeszkowski, H.
Waddington, Ian
Wagoner, Robert
Wainwright, John
Wambsganss, Joachim
Wanas, Mamdouh
Wang, Huiyuan
Watson, Darach
Webb, Tracy
Webster, Adrian
Weilbacher, Peter
Weinberg, Steven
Wesson, Paul
West, Michael
White, Simon
Whiting, Alan
Widrow, Larry
Will, Clifford
Wilson, Albert

Fox, Andrew
Franceschini, Alberto
Frenk, Carlos
Friaca, Amancio
Fukugita, Masataka
Fukui, Takao
Furlanetto, Steven
Füzfa, Andre
Fynbo, Johan
Galletto, Dionigi
Gangui, Alejandro
Ganguly, Rajib
Garilli, Bianca
Garrison, Robert
Gavignaud, Isabelle
Geller, Margaret
Geng, Lihong
Germany, Lisa
Ghirlanda, Giancarlo
Giallongo, Emanuele
Gioia, Isabella
Gitti, Myriam
Glazebrook, Karl
Glover, Simon
Goldsmith, Donald
Gonzalez, Alejando
Gonzalez-Solares, Eduardo
Goobar, Ariel
Goret, Philippe
Gosset, Eric
Gottlöber, Stefan
Gouda, Naoteru
Govinder, Keshlan
Goyal, Ashok
Grainge, Keith
Granato, Gian Luigi
Gray, Richard
Gray, Meghan
Green, Anne
Gregorio, Anna
Nishida, Minoru
Noerdlinger, Peter
Noh, Hyerim
Noonan, Thomas
Norman, Colin
Norman, Dara
Nottale, Laurent
Novello, Mario
Novikov, Igor
Novosyadlyj, Bohdan
Novotný, Jan
Nozari, Kourosh
O'Connell, Robert
Ocvirk, Pierre
Oemler Jr., Augustus
Okoye, Samuel
Oliver, Sebastian
Olowin, Ronald
Omnes, Roland
Onuora, Lesley
Oscoz, Alejandro
Ostorero, Luisa
Ott, Heinz-Albert
Ozsvath, I.
Page, Don
Paragi, Zsolt
Parnovsky, Sergei
Partridge, Robert
Pecker, Jean-Claude
Pedersen, Kristian
Pedrosa, Susana
Peebles, P.
Pello, Roser
Pen, Ue-Li
Penzias, Arno
Péroux, Céline
Perryman, Michael
Persides, Sotirios
Persson, Carina
Peterson, Bruce
Windhorst, Rogier
Wise, Michael
Wold, Margrethe
Wolfe, Arthur
Woltjer, Lodewijk
Woszczyna, Andrzej
Wright, Edward
Wu, Xiangping
Wu, Jiun-Huei
Wu, Wentao
Wu, Jianghua
Wucknitz, Olaf
Wyithe, Stuart
Xiang, Shouping
Xu, Chongming
Xu, Dawei
Yasuda, Naoki
Yeşilyurt, Ibrahim
Yi, Sukyoung
Yoichiro, Suzuki
Yokoyama, Jun-ichi
Yoshii, Yuzuru
Yoshikawa, Kohji
Yoshioka, Satoshi
Yushchenko, Alexander
Zacchei, Andrea
Zamorani, Giovanni
Zanichelli, Alessandra
Zannoni, Mario
Zaroubi, Saleem
Zhang, Tong-Jie
Zhang, Jialu
Zhao, Donghai
Zhou, Youyuan
Zhou, Hongyan
Zhu, Xingfeng
Zieba, Stanislaw
Zou, Zhenlong
Zucca, Elena

Division II Commission 49 Interplanetary Plasma & Heliosphere

President Gopalswamy, Natchimuthuk
Vice-President Mann, Ingrid

Organizing Committee

Briand, Carine
Lallement, Rosine
Lario, David
Manoharan, P.
Shibata, Kazunari
Webb, David

Members

Ahluwalia, Harjit
Anderson, Kinsey
Andretta, Vincenzo
Balikhin, Michael
Barnes, Aaron
Barrow, Colin
Bárta, Miroslav
Benz, Arnold
Bertaux, Jean-Loup
Blackwell, Donald
Blandford, Roger
Bochsler, Peter
Bonnet, Roger
Brandt, John
Briand, Carine
Browning, Philippa
Bruno, Roberto
Burlaga, Leonard
Buti, Bimla
Cairns, Iver
Cecconi, Baptiste
Channok, Chanruangrit
Chapman, Sandra
Chashei, Igor
Chassefiere, Eric
Chitre, Shashikumar
Choe, Gwangson
Chou, Chih-Kang
Couturier, Pierre
Cramer, Neil
Cuperman, Sami
Daglis, Ioannis
Dalla, Silvia
Dasso, Sergio
de Jager, Cornelis
de Keyser, Johan
de Toma, Giuliana
Del Zanna, Luca
Dinulescu, Simona
Dorotovič, Ivan
Dryer, Murray
Duldig, Marc
Durney, Bernard
Eshleman, Von
Eviatar, Aharon
Grzedzielski, Stanislaw
Habbal, Shadia
Harvey, Christopher
Heras, Ana
Heynderickx, Daniel
Heyvaerts, Jean
Hollweg, Joseph
Holzer, Thomas
Huber, Martin
Humble, John
Inagaki, Shogo
Ivanov, Evgenij
Jokipii, J.
Joselyn, Jo
Kakinuma, Takakiyo
Keller, Horst
Khan, J
Ko, Chung-Ming
Kojima, Masayoshi
Kozak, Lyudmyla
Lafon, Jean-Pierre
Lai, Sebastiana
Landi, Simone
Lapenta, Giovanni
Lario, David
Levy, Eugene
Li, Bo
Lotova, Natalja
Lüst, Reimar
Lundstedt, Henrik
MacQueen, Robert
Malandraki, Olga
Malara, Francesco
Mangeney, André
Manoharan, P.
Marsch, Eckart
Mason, Glenn
Matsuura, Oscar
Mavromichalaki, Helen
Meister, Claudia
Melrose, Donald
Mestel, Leon
Michel, F.
Moncuquet, Michel
Morabito, David
Quemerais, Eric
Raadu, Michael
Readhead, Anthony
Reinard, Alysha
Reshetnyk, Volodymyr
Riddle, Anthony
Riley, Pete
Ripken, Hartmut
Rosa, Reinaldo
Rosner, Robert
Roth, Michel
Roxburgh, Ian
Ruffolo, David
Russell, Christopher
Sagdeev, Roald
Sáiz, Alejandro
Sarris, Emmanuel
Sastri, Hanumath
Satoshi, Nozawa
Sawyer, Constance
Schatzman, Evry
Scherb, Frank
Schindler, Karl
Schmidt, H.
Schreiber, Roman
Schwartz, Steven
Schwenn, Rainer
Setti, Giancarlo
Shea, Margaret
Smith, Dean
Sonett, Charles
Stone, R.
Struminsky, Alexei
Sturrock, Peter
Suess, Steven
Tritakis, Basil
Tyulbashev, Sergei
Usoskin, Ilya
Vainshtein, Leonid
Verheest, Frank
Vinod, S.
Voitsekhovska, Anna
Vucetich, Héctor
Wang, Yi-ming
Watanabe, Takashi

Fahr, Hans
Fernandes, Francisco
Feynman, Joan
Fichtner, Horst
Field, George
Fraenz, Markus
Galvin, Antoinette
Gedalin, Michael
Gleisner, Hans
Goldman, Martin
Gosling, John
Moussas, Xenophon
Munetoshi, Tokumaru
Nickeler, Dieter
Pandey, Birendra
Paresce, Francesco
Parhi, Shyamsundar
Parker, Eugene
Perkins, Francis
Pflug, Klaus
Pneuman, Gerald
Podesta, John
Watari, Shinichi
Weller, Charles
Wu, Shi
Yang, Jing
Yang, Lei
Yeh, Tyan
Yeşilyurt, Ibrahim
Yi, Yu
Zharkova, Valentina

Division XII Commission 50 Protection of Existing & Potential Observatory Sites

President	van Driel, Wim
Vice-President	Green, Richard

Organizing Committee

Alvarez del Castillo, Elizabeth
Blanco, Carlo
Crawford, David
Metaxa, Margarita
Ohishi, Masatoshi
Sullivan, III, Woodruff
Tzioumis, Anastasios

Members

Alvarez del Castillo, Elizabeth
Ardeberg, Arne
Arsenijevic, Jelisaveta
Baan, Willem
Baddiley, Christopher
Baskill, Darren
Bazzano, Angela
Benkhaldoun, Zouhair
Bensammar, Slimane
Bhattacharyya, J.
Blanco, Victor
Boonstra, Albert
Brown, Robert
Burstein, David
Carramiñana, Alberto
Carrasco, Bertha
Cayrel, Roger
Cinzano, Pierantonio
Clegg, Andrew
Colas, François
Costero, Rafael
Coyne, George
Davis, Donald
Davis, John
de Greiff, J.
Dommanget, Jean
Dukes Jr., Robert
Edwards, Paul
Galán, Maximino
García, Beatriz
Garcia-Lorenzo, Maria
Gergely, Tomas
Gibson, David
Goebel, Ernst
Green, Richard
Hänel, Andreus
Heck, Andre
Helmer, Leif
Hempel, Marc
Hidayat, Bambang
Ilyasov, Sabit
Kadiri, Samir
Kahlmann, Hans
Kołomański, Sylwester
Kontizas, Evangelos
Kontizas, Mary
Kovalevsky, Jean
Kozai, Yoshihide
Kramer, Busaba
Leibowitz, Elia
Lewis, Brian
Lomb, Nicholas
Malin, David
Markkanen, Tapio
Mattig, W.
McNally, Derek
Mendoza-Torres, José-Eduardo
Menzies, John
Mitton, Jacqueline
Murdin, Paul
Nelson, Burt
Osório, José
Owen, Frazer
Özel, Mehmet
Pankonin, Vernon
Percy, John
Pound, Marc
Sánchez, Francisco
Schilizzi, Richard
Shetrone, Matthew
Siebenmorgen, Ralf
Smith, Malcolm
Smith, Robert
Spoelstra, T.
Stencel, Robert
Storey, Michelle
Suntzeff, Nicholas
Torres, Carlos
Tremko, Jozef
Tzioumis, Anastasios
Upgren, Arthur
van den Bergh, Sidney
van Driel, Wim
Vernin, Jean
Walker, Merle
Walker, Constance
Wang, Xunhao
Whiteoak, John
Woolf, Neville
Woszczyk, Andrzej
Yano, Hajime
Zhang, Haiyan
Žižňovský, Jozef

Division III Commission 51 Bio-Astronomy

President Irvine, William
Vice-President Ehrenfreund, Pascale

Organizing Committee

Cosmovici, Cristiano
Kwok, Sun
Levasseur-Regourd, Anny-Chantal
Morrison, David
Udry, Stéphane

Members

Al-Naimiy, Hamid
Allard, France
Almar, Ivan
Alsabti, Abdul Athem
Ando, Hiroyasu
Apai, Daniel
Balázs, Béla
Balbi, Amedeo
Ball, John
Bania, Thomas
Barbieri, Cesare
Basu, Dipak
Basu, Baidyanath
Baum, William
Beaudet, Gilles
Beckman, John
Beckwith, Steven
Beebe, Reta
Benest, Daniel
Bennett, David
Berendzen, Richard
Bernacca, Pierluigi
Billingham, John
Biraud, François
Bless, Robert
Bond, Ian
Bouchet, Patrice
Bowyer, C.
Boyce, Peter
Broderick, John
Buccino, Andrea
Burke, Bernard
Cai, Kai
Calvin, William
Campusano, Luis
Cárdenas, Rolando
Cardoso Santos, Nuno
Carlson, John
Carr, Thomas
Chaisson, Eric
Chou, Kyong
Ćirković, Milan
Clark, Thomas
Colomb, Fernando
Connes, Pierre
Corbet, Robin
Coudé du Foresto, Vincent
Couper, Heather
Coustenis, Athena
Gregory, Philip
Gulkis, Samuel
Gunn, James
Haghighipour, Nader
Haisch, Bernard
Hale, Alan
Hart, Michael
Heck, Andre
Heeschen, David
Herczeg, Tibor
Hershey, John
Heudier, Jean-Louis
Hinners, Noel
Hirabayashi, Hisashi
Hoang, Binh
Högbom, Jan
Hollis, Jan
Holm, Nils
Horowitz, Paul
Howard, Andrew
Hunten, Donald
Hunter, James
Hysom, Edmund
Idlis, Grigorij
Israel, Frank
Jastrow, Robert
Jayawardhana, Ray
Jones, Eric
Jugaku, Jun
Kafatos, Menas
Kaifu, Norio
Kane, Stephen
Kapisinsky, Igor
Kardashev, Nicolay
Kaufmann, Pierre
Keay, Colin
Keheyan, Yeghis
Keller, Hans-Ulrich
Kellermann, Kenneth
Kilston, Steven
Klahr, Hubert
Knowles, Stephen
Kocer, Dursun
Koch, Robert
Koeberl, Christian
Ksanfomality, Leonid
Kuan, Yi-Jehng
Kuiper, Thomas
Kwok, Sun
Ostriker, Jeremiah
Owen, Tobias
Pacini, Franco
Pallé Bagó, Enric
Parijskij, Yurij
Pascucci, Ilaria
Perek, Lubos
Pollacco, Don
Ponsonby, John
Qiu, Yaohui
Quintana, Hernan
Quintana, José
Quirrenbach, Andreas
Rajamohan, R.
Rawlings, Mark
Rees, Martin
Riihimaa, Jorma
Robinson, Leif
Rodríguez, Luis
Rood, Robert
Rowan-Robinson, Michael
Rubin, Robert
Russell, Jane
Sakurai, Kunitomo
Sánchez Béjar, Victor
Sancisi, Renzo
Sarre, Peter
Scargle, Jeffrey
Schatzman, Evry
Schild, Rudolph
Schneider, Jean
Schober, Hans
Schuch, Nelson
Seielstad, George
Shapiro, Maurice
Shen, Chun-Shan
Shostak, G.
Sims, Mark
Singh, Harinder
Sivaram, C.
Snyder, Lewis
Sofue, Yoshiaki
Song, In-Ok
Sozzetti, Alessandro
Stalio, Roberto
Stein, John
Sterzik, Michael
Straizys, Vytautas
Sturrock, Peter

Cuesta Crespo, Luis
Cunningham, Maria
Cuntz, Manfred
da Silveira, Enio
Daigne, Gerard
Davis, Michael
de Jager, Cornelis
de Jonge, J.
de Loore, Camiel
de Vincenzi, Donald
Deeg, Hans
Delsemme, Armand
Dent, William
Despois, Didier
Dick, Steven
Dixon, Robert
Dorschner, Johann
Doubinskij, Boris
Downs, George
Drake, Frank
Dutil, Yvan
Dyson, Freeman
Ellis, George
Epstein, Eugene
Evans, Neal
Fazio, Giovanni
Fejes, Istvan
Feldman, Paul
Field, George
Firneis, Maria
Firneis, Friedrich
Fisher, Philip
Fraser, Helen
Fredrick, Laurence
Freire Ferrero, Rubens
Fujimoto, Masa-Katsu
Gargaud, Muriel
Gatewood, George
Gehrels, Tom
Ghigo, Francis
Giovannelli, Franco
Godoli, Giovanni
Golden, Aaron
Goldsmith, Donald
Gott, J.
Goudis, Christos
Lafon, Jean-Pierre
Lamontagne, Robert
Laques, Pierre
Lazio, Joseph
Lee, Sang
Léger, Alain
Levasseur-Regourd, Anny-Chantal
Lilley, Edward
Lineweaver, Charles
Lippincott Zimmerman, Sarah
Liu, Sheng-Yuan
Lodieu, Nicolas
Lovell, Sir Bernard
Lyon, Ian
Margrave Jr., Thomas
Marov, Mikhail
Martin, Maria
Martinez-Frias, Jesùs
Marzo, Giuseppe
Matsakis, Demetrios
Matsuda, Takuya
Matthews, Clifford
Mavridis, Lyssimachos
Mayor, Michel
McAlister, Harold
McDonough, Thomas
Melott, Adrian
Mendoza, V.
Merín Martín, Bruno
Minh, Young
Minn, Young
Minniti, Dante
Mirabel, Igor
Mitsunobu, Kawada
Mokhele, Khotso
Moore, Marla
Morimoto, Masaki
Morris, Mark
Muller, Richard
Naef, Dominique
Nelson, Robert
Neuhäuser, Ralph
Niarchos, Panayiotis
Norris, Raymond
Ohishi, Masatoshi
Ollongren, A.
Sullivan, III, Woodruff
Tähtinen, Leena
Takaba, Hiroshi
Takada-Hidai, Masahide
Tarter, Jill
Tavakol, Reza
Tedesco, Edward
Tejfel, Viktor
Terzian, Yervant
Thaddeus, Patrick
Tolbert, Charles
Toro, Tibor
Tovmassian, Hrant
Townes, Charles
Trimble, Virginia
Turner, Kenneth
Turner, Edwin
Vallee, Jacques
Valtaoja, Esko
Varshalovich, Dmitrij
Vauclair, Gérard
Vázquez, Manuel
Vázquez, Roberto
Venugopal, V.
Verschuur, Gerrit
Vogt, Nikolaus
von Braun, Kaspar
von Hippel, Theodore
Wallis, Max
Walsh, Wilfred
Walsh, Andrew
Watson, Frederick
Welch, William
Wellington, Kelvin
Wesson, Paul
Wielebinski, Richard
Williams, Iwan
Willson, Robert
Wilson, Thomas
Wolstencroft, Ramon
Wright, Ian
Wright, Alan
Xu, Weibiao
Ye, Shuhua
Zapatero-Osorio, Maria Rosa
Zuckerman, Ben

Division I Commission 52 Relativity in Fundamental Astronomy

President Petit, Gérard
Vice-President Soffel, Michael

Organizing Committee

Brumberg, Victor
Capitaine, Nicole
Fienga, Agnès
Fukushima, Toshio
Guinot, Bernard
Huang, Cheng
Mignard, François
Seidelmann, P.
Wallace, Patrick

Members

Abbas, Ummi
Antonelli, Lucio Angelo
Arakida, Hideyoshi
Ashby, Neil
Bazzano, Angela
Boucher, Claude
Bursa, Michal
Calabretta, Mark
Chae, Kyu-Hyun
de Felice, Fernando
Domínguez, Mariano
Efroimsky, Michael
Evans, Daniel
Folkner, William
Foschini, Luigi
Giorgini, Jon
Gray, Norman
Gumjudpai, Burin
Hackman, Christine
Hestroffer, Daniel
Hilton, James
Hobbs, David
Hobbs, George
Holz, Daniel
Hsu, Rue-Ron
Hu, Hongbo
Huang, Tianyi
Ilya, Mandel
Kaplan, George
Khumlemlert, Thiranee
Kopeikin, Sergei
Luzum, Brian
Manchester, Richard
Marranghello, Guilherme
McCarthy, Dennis
Melnyk, Olga
Morabito, David
Mota, David
Müller, Andreas
Nelson, Robert
Orellana, Mariana
Osório, José
Panessa, Francesca
Pireaux, Sophie
Pitjeva, Elena
Podolský, Jiří
Ray, James
Schartel, Norbert
Standish, E.
van Leeuwen, Joeri
Vityazev, Veniamin
Watts, Anna
Wen, Linqing
Wu, Jiun-Huei
Wucknitz, Olaf
Yakut, Kadri

Division III Commission 53 Extrasolar Planets (WGESP)

President Boss, Alan
Vice-President Lecavelier des Etangs, Alain

Organizing Committee

Bodenheimer, Peter
Cameron, Andrew
Kokubo, Eiichiro
Mardling, Rosemary
Minniti, Dante
Queloz, Didier

Members

Allard, France
Apai, Daniel
Ardila, David
Bacham, Eswar
Baran, Andrzej
Baryshev, Andrey
Bennett, David
Biazzo, Katia
Bodenheimer, Peter
Borde, Pascal
Buccino, Andrea
Cameron, Andrew
Cottini, Valeria
Coustenis, Athena
Cuesta Crespo, Luis
Dall, Thomas
Delplancke, Françoise
Dominis Prester, Dijana
Emelyanenko, Vacheslav
Esenoğlu, Hasan
Fernández Lajús, Eduardo
Giani, Elisabetta
Homeier, Derek
Howard, Andrew
Hu, Zhong wen
Ireland, Michael
Iro, Nicolas
Jeffers, Sandra
Kane, Stephen
Kim, Yoo Jea
Kitiashvili, Irina
Klahr, Hubert
Konacki, Maciej
Kozak, Lyudmyla
Lacour, Sylvestre
Lagage, Pierre-Olivier
Liu, Michael
Lodieu, Nicolas
Marchi, Simone
Martinez Fiorenzano, Aldo
Mendillo, Michael
Merín Martín, Bruno
Millan Gabet, Rafael
Misato, Fukagawa
Moerchen, Margaret
Pallé Bagó, Enric
Park, Byeong-Gon
Patten, Brian
Pérez, Mario
Perrin, Marshall
Persson, Carina
Plavchan, Jr., Peter
Reffert, Sabine
Reiners, Ansgar
Rojo, Patricio
Saffe, Carlos
Sánchez Béjar, Victor
Selam, Selim
Shankland, Paul
Sozzetti, Alessandro
Stam, Daphne
Sterzik, Michael
Villaver, Eva
von Braun, Kaspar
Wallace, James
Wang, Xiao-bin
Whelan, Emma
Yeşilyurt, Ibrahim
Yüce, Kutluay
Yumiko, Oasa
Yutaka, Shiratori
Zapatero-Osorio, Maria Rosa
Zarka, Philippe
Zhang, You-Hong

Division IX Commission 54 Optical & Infrared Interferometry

President	Ridgway, Stephen
Vice-President	van Belle, Gerard

Organizing Committee

Duvert, Gilles	Hummel, Christian	Tuthill, Peter
Genzel, Reinhard	Lawson, Peter	Vakili, Farrokh
Haniff, Christopher	Monnier, John	

Members

Acke, Bram	Eisner, Josh	Paumard, Thibaut
Arcidiacono, Carmelo	Florido, Estrella	Perez-Torres, Miguel
Babkovskaia, Natalia	Giani, Elisabetta	Pott, Jorg-Uwe
Baddiley, Christopher	Gong, Xuefei	Rajagopal, Jayadev
Bakker, Eric	Gu, Xuedong	Rivinius, Thomas
Bakker, Eric	Hora, Joseph	Rousset, Gérard
Bamford, Steven	Hu, Zhong wen	Schinckel, Antony
Barrientos, Luis	Ireland, Michael	Schoedel, Rainer
Bendjoya, Philippe	Tallon-Bosc, Isabelle	Schuller, Peter
Benson, James	Iwata, Ikuru	Tallon, Michel
Boonstra, Albert	Jordán, Andrés	Tycner, Christopher
Borde, Pascal	Jørgensen, Anders	Urquhart, James
Bryant, Julia	Jørgensen, Anders	Vaňko, Martin
Buscher, David	Kalomeni, Belinda	Verhoelst, Tijl
Chen, Zhiyuan	Kim, Young-Soo	von Braun, Kaspar
Ciliegi, Paolo	Köhler, Rainer	Wallace, James
Coffey, Deirdre	Konacki, Maciej	Wang, Guo-min
Crawford, Steven	Lacour, Sylvestre	Wang, Shen
Cuby, Jean-Gabriel	Lebohec, Stephan	Wang, Jingyu
Cuillandre, Jean-Charles	Lehnert, Matthew	Woillez, Julien
Danchi, William	Leisawitz, David	Yang, Dehua
de Lange, Gert	Malbet, Fabien	Yuan, Xiangyan
Delplancke, Françoise	Mason, Brian	Zavala, Jr., Robert
Dieleman, Pieter	Millan Gabet, Rafael	Zhang, You-Hong
Dutrey, Anne	Mitsunobu, Kawada	Zhou, Jianfeng

Division XII Commission 55 Communicating Astronomy with the Public

President Crabtree, Dennis
Vice-President Christensen, Lars

Organizing Committee

Alvarez Pomares, Oscar
Damineli Neto, Augusto
Fienberg, Richard
Green, Anne
Kembhavi, Ajit
Russo, Pedro
Sekiguchi, Kazuhiro
Whitelock, Patricia
Zhu, Jin

Members

Accomazzi, Alberto
Apai, Daniel
Argo, Megan
Arion, Douglas
Babul, Arif
Baek, Chang Hyun
Bailey Mathae, Katherine
Balbi, Amedeo
Bamford, Steven
Barlow, Nadine
Beckmann, Volker
Briand, Carine
Cattaneo, Andrea
Coffey, Deirdre
Couto da Silva, Telma
Cuesta Crespo, Luis
Daisuke, Iono
Darhmaoui, Hassane
de Grijs, Richard
de Lange, Gert
Delsanti, Audrey
Domínguez, Mariano
Donahue, Megan
Doran, Rosa
Ehle, Matthias
Feain, Ilana
Gangui, Alejandro
Garcia-Lorenzo, Maria
George, Martin
Gills, Martins
Govender, Kevindran
Gumjudpai, Burin
Günthardt, Guillermo
Hau, George
Haverkorn, Marijke
Hempel, Marc
Hollow, Robert
Ibrahim, Alaa
Jafelice, Luiz
Jahnke, Knud
Jeon, Young-Beom
Jiménez-Vicente, Jorge
Johnston, Helen
Kalemci, Emrah
Kołomański, Sylwester
Komonjinda, Siramas
Loaring, Nicola
Longo, Giuseppe
López-Sánchez, Angel
Madsen, Claus
Majumdar, Subhabrata
Marchi, Simone
Maričić, Darije
Martínez-Delgado, David
Merín Martín, Bruno
Minier, Vincent
Mourão, Ana Maria
Müller, Andreas
Mújica, Raul
Niemczura, Ewa
Ocvirk, Pierre
Ödman, Carolina
Olivier, Enrico
Pantoja, Carmen
Pavani, Daniela
Perrin, Marshall
Politi, Romolo
Reynolds, Cormac
Roberts, Douglas
Sampson, Russell
Sandrelli, Stefano
Santos-Lleó, Maria
Sharp, Nigel
Stam, Daphne
Tignalli, Horacio
Tomasella, Lina
Tugay, Anatoliy
Verdoes Kleijn, Gijsbert
Walker, Constance
Wu, Jiun-Huei

Printed in the United States
By Bookmasters